A Complete Guide to the Environment, Climate Change, and Disaster Management

A Complete Guide to the Environment, Climate Change, and Disaster Management

By

Rupak Dey and Seema B. Sharma

Cambridge Scholars Publishing

A Complete Guide to the Environment, Climate Change, and Disaster Management

By Rupak Dey and Seema B. Sharma

This book first published 2023

Cambridge Scholars Publishing

Lady Stephenson Library, Newcastle upon Tyne, NE6 2PA, UK

British Library Cataloguing in Publication Data
A catalogue record for this book is available from the British Library

ISBN (10): 1-5275-2862-6
ISBN (13): 978-1-5275-2862-8

TABLE OF CONTENTS

Part 2: Ecology

Part 4: Climate Change and Current Environmental Issues

PART 1:

PRINCIPLES AND CONCEPTS

CHAPTER 1

ENVIRONMENT

Environmental science is a broad field of study that includes geological features, hydrogeology, atmospheric science, chemistry, physics, medical advances, bioscience, agriculture, public health, hygienic engineering, and much more. It is the study of things that happen in the physical world. It looks at the origins, interactions, movements, effects, and fortunes of physical and biological species in the atmosphere, water, and soil, as well as how human actions affect them (Fig. 1.1).

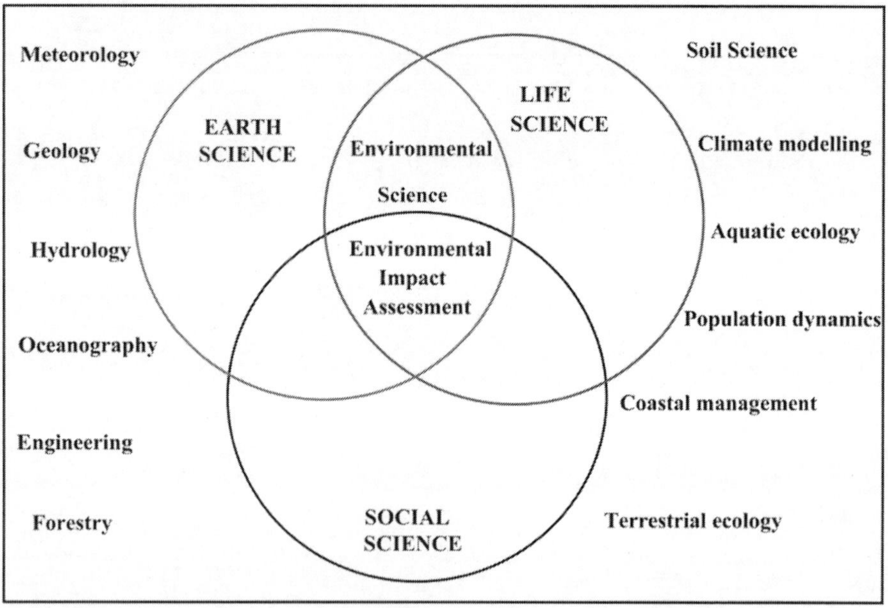

Fig. 1.1: Interaction of various disciplines of Environmental Science

The Definition of Environment

Environment is the sum of all the things that are going on around us at a certain time and place. It is made up of systems of physical, biological, and cultural factors that interact with each other and with each other as a whole. The environment is all of the things that an organism needs to live or keep its life process going. It has an effect on how organisms grow and change.

The Scope of Environment

The environment is comprised of four basic parts as explained below:

1. The atmosphere: this is a layer or several layers of gases that circle a planet and are held in place by the gravity of the planet. A planet keeps its atmosphere when the gravity is strong and the temperature of the atmosphere is low. About 78 percent of the air around the Earth is nitrogen, 21 percent is oxygen, 0.9 percent is argon, 0.3 percent is carbon, and 0.1 percent is made up of other gases. These other gases include methane, water vapour, Helium, Krypton, Hydrogen, and Neon.
   a) It keeps life on earth alive.
   b) It protects earth from the potentially dangerous environment of space.
   c) It absorbs a significant portion of the electromagnetic energy from the sun as well as the majority of cosmic rays from outer space.
   d) It protects us from the Sun's harmful UV rays as well as against minor meteorite strikes from space.

2. The Hydrosphere: this is the term used to describe the total amount of water that exists on a planet. The hydrosphere is made up of the water found on the surface of the planet, as well as the water found below the surface and in the atmosphere. A planet's hydrosphere may be made up of liquid, vapour, or ice, depending on the temperature and density of the planet. Seas, lakes, and rivers are the three primary types of bodies of liquid water found on the surface of the Earth.
   a) The oceans are home to 97 percent of the water that may be found on Earth.

b) The polar icecaps and glaciers hold 2 percent of the world's water supply.
c) Only about 1 percent of fresh surface water—rivers, lakes, streams, and ground water—is available for human consumption and other purposes.
d) Water is in a liquid state when the temperature rises above 32 degrees Fahrenheit and falls below 212 degrees Fahrenheit, and it governs climate.
e) The hydrosphere originated in the primordial atmosphere as a consequence of the condensation and solidification of water vapour induced by the high temperatures that caused the water to be in the form of vapour. The hydrosphere on Earth is thought to be 4 billion years old.

3. The Lithosphere: the solid crust of the Earth, or the lithosphere, extends from the surface to a depth of around 80–100 kilometres. The crust and the brittle uppermost layer of the mantle make up the lithosphere. The atmosphere and asthenosphere surround it on its top and lower sides, respectively. It features a varied topography, including mountains, plateaus, deserts, plains, valleys, and other landforms. The crust, mantle, and core are the three main layers that make up the Earth's interior.
   a) The Moho discontinuity is the boundary that divides the crust from the mantle. "Discontinuity" refers to the point on the surface where seismic waves change velocity and accelerate.
   b) There exists a conrad discontinuity between the upper and lower crusts.
   c) The upper and lower mantles are separated by a repiti discontinuity.
   d) Between the mantle and the outer core, there is a Guttenberg discontinuity.
   e) Between the outer and inner core, there is a Lehmann discontinuity.

4. The Biosphere: all of the Earth's environments that are home to life together make up the biosphere. The biosphere includes a wide range of ecosystems, including deep tree roots, the deep ocean, beautiful tropical rainforests, high mountainous regions, and transitional regions like estuaries where ocean and terrestrial organisms coexist. For the survival of species and their mutual

interactions, the biosphere is essential. It is a crucial part of climate regulation. Climate changes as a result of changes in the biosphere.

a) The geologist Eduard Suess coined the term biosphere in 1875. The biosphere, according to him, is the portion of the earth's surface where life is present.
b) The biosphere is thought to have developed via a process called biopoiesis or biogenesis at least 3.5 billion years ago. While biogenesis is the process by which life originates from living matter, biopoiesis is the process by which life develops naturally from non-living components.
c) All over the world, there are many biosphere reserves where people work together to protect the environment. In fact, there are 669 biosphere reserves spread over 120 different countries.
d) It is difficult to pinpoint the exact outer edge of the biosphere. This is because the biosphere's depth is uncertain and certain species, such the Ruppell's vulture, may fly as high as 11,300 metres in the air. In the Puerto Rico Trench, fish have been found to live as deep as 8,300 metres.
e) According to statistics from 2022, India has 18 biosphere reserves, 12 of which are also a member of the UNESCO MAB Program.

Elements of Environment

The environment is made up of interdependent systems of physical, biological, and cultural elements that are linked to one another both individually and collectively in a wide range of ways. The following is an explanation of these elements:

1) Physical elements: space, land forms, bodies of water, climatic soils, rocks, and minerals are all examples of physical components. They explain how different the human environment is and what it can and can't do.
2) Biological elements: the biosphere is made up of biological elements such as plants, animals, microorganisms, and humans.
3) Cultural elements: economic, social, and political dimensions are ultimately man-made traits that comprise the cultural world.

Aside from that, the environment may be divided into two categories:(a) microenvironment and (b) macroenvironment. It may also be separated into two further categories:(i) physical environment and (ii) biotic environment. The term "micro environment" refers to the organism's immediate local surroundings, while "macro environment" refers to all of the physical and biotic variables that surround the organism outside.

Humans and the Environment

Humans cannot be seen in isolation from their environment. People still have a wide range of needs across the world. In the past, the majority of environmental issues that were taken into account by international organisations were those that were acknowledged by developed nations, such as the need to reduce pollution, the importance of preserving of biological and ecological diversity, and the value of protecting the world's natural beauty. Since the United Nations Conference on the Human Environment, however, it has become more and more clear that environmental issues also pose a serious threat to developing countries, and that in a large portion of the world, the environmental issues still revolve around poverty—poor housing, terrible public health, malnutrition, and inadequate employment. The development and understanding of environmental challenges are highly dependent on how society is formed, as well as on its ideals and goals. Changes in the structure and goals of society have a significant impact on how man interacts with his physical surroundings. Man's goal must be to create a society that is inherently compatible with its environment if he is to avoid the situation where a lot of time and money are spent fixing component flaws.

With the growth of theological beliefs, numerous faiths began to see the established natural order as a means of human enslavement. It is said that man was created in God's likeness, given dominion over nature, and the ability to accomplish almost whatever he wanted. Religious moral standards backed activities undertaken by people who destroyed the environment on a large scale without taking into account the need to preserve the biotic and abiotic components of the ecosystem. Among the important ones are:

a) Excessive use of natural resources.
b) Disruption of bio-geochemical cycles.
c) Pollution of water bodies, especially the rivers.
d) Other forms of pollution of the environment.

The Sixth Mass Extinction

A mass extinction occurs when the Earth loses the majority of its species in a short period of time. There have been five significant mass extinctions throughout the Earth's history. The Holocene extinction is sometimes known as the "sixth great extinction" or the "Anthropocene extinction." It is an ongoing extinction event of species caused by human actions during the modern Holocene epoch. The current rate of extinction is predicted to be 100 to 1,000 times higher than natural backdrop extinction rates. At least five human actions may contribute to extinction, and they are as follows:

a) Continuous greenhouse gas emission.
b) Hazardous chemicals of agriculture, mines, factories and industry.
c) Manmade nuclear arms.
d) Biotechnological misuse.
e) Manmade radioactive waste.

Global Environmental Problems

The environment refers to anything that surrounds or effects an organism throughout its existence. It is made up of both living (biotic) and non-living (abiotic) components. Human civilization and globalisation are the primary sources of continual change in the current world environment. Pollution, global warming, ozone depletion, acid rain, depletion of natural resources, overcrowding, waste disposal, deforestation, and biodiversity loss are all factors that may be considered to be contributing to global environmental issues. Almost all of these processes are the result of unsustainable use of natural resources. Our ecosystem is facing a variety of challenges, many of which seem to be worsening with time, placing us on the verge of a true environmental disaster. As a result, it is becoming more important to raise awareness of the prevalence of these challenges, as well as what can be done to prevent their negative impact. Some of the major issues are:

a) Pollution: Pollution of the air, water, and soil caused by pollutants such as plastics, heavy metals, and nitrates generated by companies, combustion of fossil fuels, acid rain, oil spills, and industrial waste.
b) Global warming: Human activity promotes global warming by emitting greenhouse gases, which causes an increase in

temperature, which ultimately leads to rising sea levels, melting polar ice caps, flash floods, and desertification.

c) Overpopulation: We are running out of resources such as food, water, and fuel to feed the world's growing population, particularly in developing countries. Intensive agriculture, in an attempt to mitigate the problem, causes further damage via the use of chemical fertilisers, herbicides, and insecticides.
d) Waste disposal: A massive amount of garbage is produced and dumped into the oceans. Trash from nuclear power plants, as well as waste from polymers and technology, is among the most dangerous waste types.
e) Ocean acidification: The rise in acidity of the seas, which is caused by a rise in the emission of carbon dioxide by humans, has a significant impact on marine life and has severe repercussions.
f) Loss of biodiversity: Human actions are responsible for the extinction of several species and habitats. As a result, natural processes such as pollination become unbalanced, endangering ecosystems and causing harm to coral reefs in particular.
g) Deforestation: Trees are cut down to make place for residential, industrial, or commercial buildings, which reduces oxygen production and alters temperature and rainfall.
h) Ozone layer depletion: The emission of chlorofluorocarbons (CFCs) into the atmosphere causes pollution, which in turn causes a hole to form in the ozone layer, which protects the earth from potentially harmful ultraviolet radiation.
i) Acid rain: Acid rain is caused by pollutants in the atmosphere, such as sulphur dioxide and nitrogen oxides, and can have negative effects on humans, animals, and aquatic species. Acid rain can also cause damage to buildings.
j) Public health problems: One of the most serious environmental issues today is a lack of pure water. Air pollution may lead to consequences such as respiratory sickness and cardiovascular disease.

The Importance of Environmental Science

Environmental science encourages the investigation and comprehension of environmental processes, including those that occur naturally and those that are induced by humans. Students develop their talents in critical thinking, problem solving, and creative thinking through studying and investigating challenging environmental situations. This helps students strengthen their existing skills. The study of the environment teaches us about the need to preserve and conserve the environment, especially in light of the indiscriminate discharge of toxins into the environment by humans. There are now a large number of ecological concerns that, over time, have grown in scale and complexity, putting the continued survival of humanity on earth in jeopardy. Along with researching solutions to these problems, our focus in Environment Studies is also on these concerns. The following are some of the reasons why environmental studies are becoming more important:

a) Environmental issues must be of global significance: It is clear that environmental problems like global warming, ozone depletion, acid rain, marine pollution, and the loss of biodiversity are not just national problems, but also global problems that must be solved by people from all over the world working together.
b) As a result of development, new problems have emerged: In its wake, development spawned urbanisation, industrial growth, transportation systems, agriculture, and housing, among others. However, it has been steadily eliminated in modern nations. To sanitise their own environment, the Global North has succeeded in shifting the 'dirty' industries to the Global South. When the West developed, it likely did so without considering the environmental ramifications of its actions. Even if the rising world follows such a route, it is manifestly unrealistic and unacceptable.
c) Exponential Rise in Pollution: One in seven people on this planet live in India, according to a world census. With 16 percent of the world's population and only 2.4 percent of its land area, it is clear that there is a lot of pressure on natural resources, especially land. Agricultural experts have noticed problems with the health of the soil, such as the loss of micronutrients and organic matter, soil salinity, and the breakdown of its structure.
d) Call for a Better Approach: Especially for developing countries, the establishment of alternative routes leading to alternative goals is of the greatest priority. We need an aim which encompasses the following:
    1) A target, which in the end becomes the primary purpose of the development being carried out in order to ensure long-term sustainability.
    2) A goal that should unite all of the inhabitants on our planet.
e) Need To Save Earth from Extinction: We have a moral obligation to save humanity from extinction. In the name of progress, we have constrained the environment and depleted the biosphere.

f) Need For Smart Management and Development: We depend on the environment for both our life and for nourishment. In any strategy for development, the extraction of resources, the processing, and the use of the product all need to be coordinated with the natural cycles. Our actions should be organised ecologically for the purpose of maintaining the environment while simultaneously advancing development. With this in mind, Misra (1991) identified the four fundamental principles of ecology: holism, ecosystem, succession, and conversation. He has emphasised the following four crucial aspects of environmental management, which are as follows:
    1) Impact of human activities on the environment
    2) Value system
    3) Plan and design for sustainable development
    4) Environmental education

Need for Awareness

There is no doubt that people are to blame for a lot of the damage to the environment. Since the beginning of the industrial revolution, the amount of greenhouse gases in the air has gone up by more than a third. So, it's up to us to learn more about the environment and change the way we act. The greenhouse effect is gaining strength because more greenhouse gases are being released into the air. This has caused temperatures around the world to rise. The air, land, and water are also harmed by these gases. Pollution lowers the quality of food, which could cause people to eat compounds that are bad for them. Poor air quality also poses serious health risks to people.

Destruction of forests is another way that people cause damage to the environment. The things we need most, like food and medicine, come from forests. Trees are also important "carbon sinks," which means that they help keep the amount of carbon dioxide in the air in balance by taking it in when they photosynthesise. By cutting down more trees, we are making the greenhouse effect worse, which means that temperatures around the world are going up. This causes big changes in the weather, like storms, droughts, and floods. Climate change and other problems are also putting many species at risk of going extinct because they can't adapt to the changing climate. All ecosystems are connected, so the loss of a species, no matter how small it may seem, can have big effects on humans. For example, many people don't know that the loss of honeybees can cause a lot of serious problems. Some people might even be happy about the idea of bees going extinct because they are known for having painful stings. People don't care much about bees, except when they discover that honey in oats makes the best meal ever. 35 percent of the 100 different crop species that provide us with the majority of our food are pollinated by them. So, getting rid of honeybees has a direct effect on how long people will be able to keep eating. This shows how something we might think of as annoying is actually important to our survival as a species. Even if there isn't a clear link between what we do and the serious environmental problems, that doesn't mean that the results won't affect us. This is why it's so important to take care of the environment whenever and wherever we can.

CHAPTER 2

COMPONENTS OF ENVIRONMENT

Introduction

The physical limits imposed by the environment are unique. It consists of four major components: the atmosphere, the hydrosphere, the lithosphere, and the biosphere. It has been shown that interactions between abiotic-abiotic, abiotic-biotic, and biotic-biotic components are continuous. These interactions involve the movement or transformation of elements, compounds, and other forms of energy (Fig. 2.1).

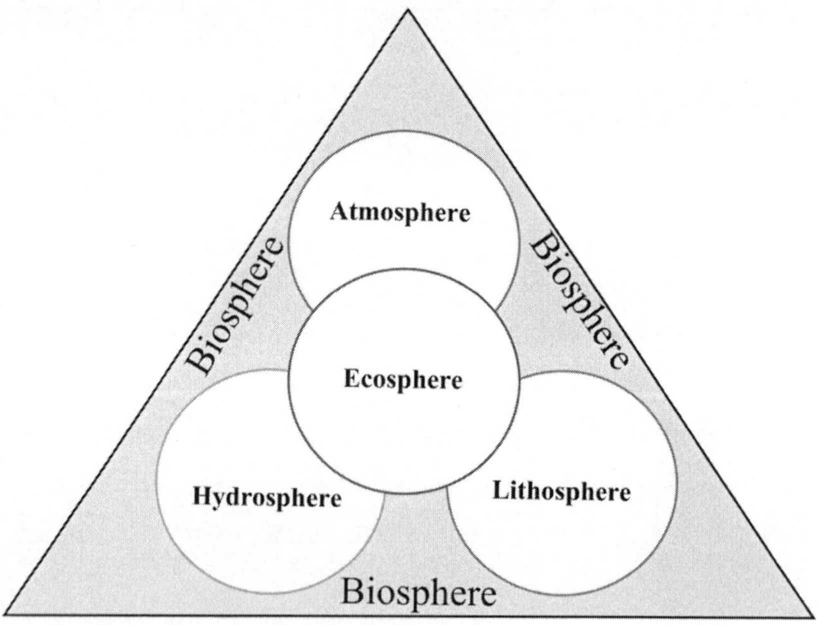

Fig. 2.1: Environmental components interaction

Atmosphere

When Earth first formed, there was no atmosphere, which is a fluid system of gases and particles in the air. Today's atmosphere was made by chemical and biological interactions on the planet itself. The fluid system makes a gaseous envelope around the globe, but it is hard to say where its limits are. They can be called anything you want, but in this book, we call them the Earth's atmosphere and space.

The air we breathe is a stable mix of 78 percent nitrogen, 21 percent oxygen, about 1 percent argon, and small amounts of carbon dioxide and water vapour (Table 2.1). But our planet's atmosphere was very different 4.6 billion years ago when it was first formed. Most likely, early on in the Earth's history, its atmosphere was made up of hydrogen, helium, methane, ammonia, carbon dioxide, and water vapour. These were the most common gases in the early solar system. Hydrogen and helium, which were the lightest of these gases, couldn't stay on Earth because it wasn't strong enough to hold them. The strong solar winds of a young, active Sun probably sent most of the remaining gases into space. The first permanent atmosphere on Earth was made by a process called "outgassing" in which gases trapped inside the planet are released into space. On a global scale, the gases that come out of hundreds of active volcanoes are still an important part of how the planet works. Early on, the planet's interior must have been very hot and moving like a fluid, which must have caused a lot of gas to escape. Based on what we know about volcanic eruptions today, Earth's original atmosphere was probably made up of water vapour, carbon dioxide, and sulphur dioxide, along with small amounts of other gases but no nitrogen. The most important problem was that there was no oxygen to breathe.

Gases	Formula	Percent by volume
Nitrogen	N_2	78.084
Oxygen	O_2	20.946
Argon	Ar	0.934
Carbon dioxide	CO_2	0.036
Neon	Ne	0.00182
Helium	He	0.000524
Krypton	Kr	0.000114
Hydrogen	H_2	0.00005

Table 2.1: Permanent gases of the Atmosphere

The earth's surface is covered by a layer of atmosphere, which is made up of gases and different solid and liquid particles. The gaseous materials go up several hundred kilometres, but it's not clear what the highest point of the atmosphere is. The first 80 kilometres of the Earth's atmosphere hold more than 99 percent of the mass of the whole atmosphere. Based on how the temperature changes up and down, the atmosphere is divided into five layers: the troposphere, stratosphere, mesosphere, thermosphere, and ionosphere (Fig. 2.2). In addition to the layers that have different temperatures at different heights, there are also other layers or zones in the atmosphere. The atmosphere is usually divided into two layers, the homosphere and the heterosphere, based on their chemical makeup. From the surface of the Earth to a height of about 80 kilometres, the amounts of the gases that make up the air are the same. This layer is called the homosphere, which means "zone of uniform composition." In contrast, the atmosphere above 80 kilometres is not uniform even though it is not very thick. Because it is made up of different things, it is called a heterosphere. Here, the gases are put into four almost-spherical shells, each of which has a different mix of gases. The lowest layer is mostly made up of molecules of nitrogen. The next layer is mostly made up of atoms of oxygen. The next layer is mostly made up of helium atoms, and the last layer is mostly made up of hydrogen atoms. The gases that make up the heterosphere are arranged in layers that depend on how heavy they are. Since molecular nitrogen is the heaviest, it is also the one at the bottom. Hydrogen is the most common and outer gas. It is also the lightest.

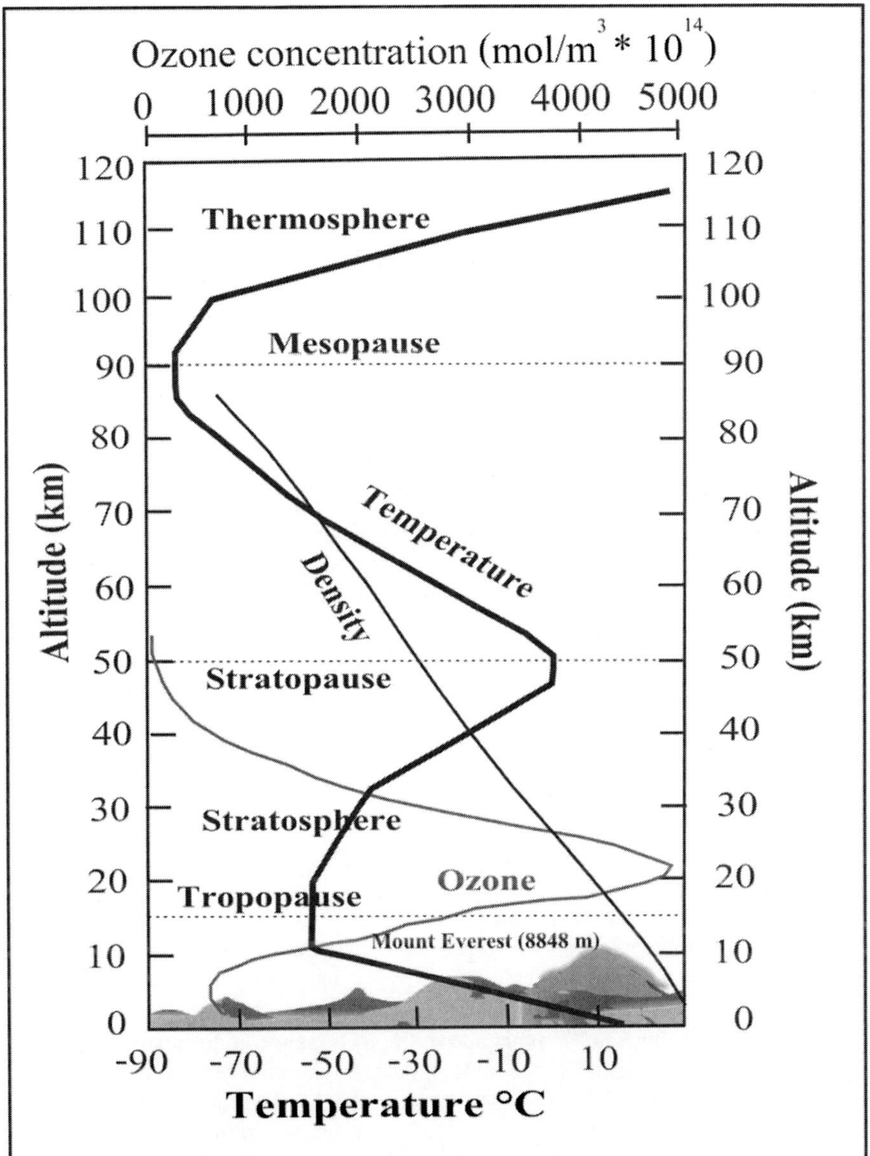

Fig. 2.2: Vertical distribution of atmospheric layers

1. Troposphere:

The troposphere is the lowest layer of the atmosphere in which humans live, with temperatures decreasing as height increases. The term was coined in 1908 by Teisserrenc de Bort, and it literally means "the place where air turns over," referring to the strong vertical mingling of air in this lower zone.

The environmental lapse rate is the rate at which temperature decreases in the troposphere. It has a typical lapse rate of 6.5 degrees Celsius per kilometre as its average value. It should be emphasised, however, that the environmental lapse rate is not constant and may be quite variable, necessitating regular monitoring. Radiosondes are used to calculate the real environmental lapse rate and gather data on vertical fluctuations in air pressure, wind, and humidity. The environmental lapse rate may fluctuate during the day owing to weather changes, as well as seasonally and from location to location. Shallow layers with temperatures that actually increase with height are sometimes detected in the troposphere. When such a reversal occurs, a temperature inversion is said to exist. The temperature fall continues up to around 12 kilometres in altitude. The troposphere's thickness, however, varies around the world. It may reach heights of more than 16 kilometres in the tropics, while it is more docile in the polar regions, reaching just 9 kilometres or less. The troposphere is meteorologists' primary concern since it is in this layer that almost all significant weather occurrences occur. All clouds, almost certainly all precipitation, and all of our severe storms develop in this stratum of the atmosphere. As a result, the troposphere is frequently referred to as the "weather sphere."

2. Stratosphere:

The stratosphere comes after the troposphere. The line where the troposphere and the stratosphere meet is called the tropopause. Large-scale turbulence and mixing can easily pass on characteristics of the atmosphere below the tropopause, but they can't do this above the tropopause, in the stratosphere. At first, the temperature in the stratosphere stays almost the same until a height of about 20 kilometres. After that, the temperature starts to rise quickly and stays that way until the stratopause, which is about 50 kilometres above the Earth's surface. Because ozone is concentrated in the stratosphere, this layer of the atmosphere experiences higher temperatures. UV light from the Sun is absorbed by ozone. The Sun heats the stratosphere because of this. Even though the highest concentration of ozone is between 15 and 30 kilometres, the lower levels of ozone above this height absorb enough UV.

3. Mesosphere:

The mesosphere is situated beneath the thermosphere and directly above the stratosphere. It covers a distance of from 50 to 85 kilometres. The mesosphere's temperature drops with height. Near the top of this layer, temperatures as cold as -90° C are recorded in the Earth's atmosphere. The mesopause is the transition zone between the mesosphere and thermosphere. The stratopause, which marks the transition between the mesosphere and the stratosphere, is located near the bottom of the mesosphere. The mesosphere still contains some meteoric debris, which causes a relatively high concentration of iron and other metal atoms in this layer. Near the poles, rather rare high-altitude clouds known as "noctilucent clouds" or "polar mesospheric clouds" may be found. These distinct clouds arise at a considerably greater altitude than other types of clouds. The mesosphere is one of the parts of the atmosphere that has received the least attention due to access issues. The reason for this is because it is inaccessible to both the lowest orbiting satellites and the research balloons with the highest altitudes. Only recently have technology advancements started to close this knowledge gap.

4. Thermosphere:

The fourth layer extends from the mesopause and has no clear upper limit. It is the thermosphere, a layer that accounts for just a tiny fraction of the mass of the atmosphere. Temperatures increase again in this outermost layer's very rarified air owing to the absorption of very shortwave, high-energy solar radiation by oxygen and nitrogen atoms. Temperatures in the thermosphere may reach very high levels of more than 1000 degrees Celsius. However, such temperatures are not comparable to those found at the Earth's surface. The average speed at which molecules travel is used to describe temperature. The temperature is particularly high because the gases of the thermosphere are moving at incredibly rapid speeds. However, the gases are so rare that they can only store a little quantity of heat collectively. As a result, the temperature of a satellite orbiting Earth in the thermosphere is mostly determined by the amount of solar energy it absorbs rather than the high temperature of the almost nonexistent surrounding air. The air in this stratum would not feel warm if an astronaut within exposed his or her hand to it.

5. Ionosphere:

The ionosphere, a layer that is electrically charged and lies between 80 to 400 kilometres in height, overlaps with the lowest reaches of the thermosphere and heterosphere. Here, when exposed to high-energy shortwave sunlight, oxygen and nitrogen atoms and molecules are swiftly ionised. Every molecule or atom that is influenced by this process loses one or more electrons and changes into a positively charged ion, allowing the remaining electrons to be released and flow as electric currents. Positively charged ions and negative electrons are most concentrated in the range of 80 to 400 kilometres, while ionisation may occur at heights of up to 1000 kilometres and as low as possibly 50 kilometres. Because the majority of the short-wavelength energy needed for ionisation has already been used up, the ion concentration below this zone is not significant. The ionosphere has an irregular electrical structure. It has three layers with varying ion densities. These layers are referred to as the D, E, and F layers, in that order. Because the formation of ions needs direct solar energy, the concentration of charged particles varies from day to night, particularly in the D and E zones. In other words, these layers weaken and disappear at night, then reappear throughout the day. On the other hand, the top layer, or F layer, is present throughout the day and night. Since the atmosphere's density is fairly low in this layer, positive ions and electrons do not collide and recombine as quickly as they do at lower altitudes, where density is greater. In light of this, the F layer's ion and electron concentrations do not change quickly, and the layer, while being thin, lasts throughout the night.

Hydrosphere

The surface water and its surroundings are included in the hydrosphere. The survival of life depends on it. On the surface of the world, water is the most common substance. More than 70 percent of the Earth's surface is covered by oceans, and other regions are covered in glaciers and ice. Water is also present in streams, lakes, soils, and aquifers, the atmosphere, and the bodies of all living things. A commonly recognised theory is that Earth began to form some 4.5 billion years ago via the accretion of small, cold pieces of rock and metal that were encircling the Sun. Due to gravitational compression, heat from collisional impact, and radioactive decay of materials like uranium, the Earth heated up as it formed. As the Earth's temperature increased, the accretions that formed it melted and the iron "fall" to the planet's centre to form its core. During this time, there was violent volcanic activity that released enormous volumes of water vapour and other gases from the Earth's core and may have even covered the surface in a thick, sea of lava. Earth began to cool as its internal reorganisation and growth halted and as the amount of radioactive material decreased due to decay. After a while, the surface of the Earth solidified into rock and became cold enough for the billowing clouds of volcanic water vapour to condense into liquid water, which then fell as rain. Thus, around 4 billion years ago, the present oceans were formed. Some of the oldest rocks found on Earth provide evidence for this, such as pillow basalts, which reveal underwater eruptions, and rocks that were originally sedimentary and formed in a shallow sea. As the Earth continued to be degassed by volcanic activity, the seas increased in size and saltiness as a result of the water absorbing chlorine from other volcanic gases and sodium (as well as calcium and magnesium) from the chemical weathering of minerals on the surface of the planet.

The mechanism that defines how water is distributed and moves between the planet and its atmosphere is known as the hydrological cycle. The idea includes how water is constantly moving between the oceans, the atmosphere, plants, and land. The evaporation of water from the ocean's surface initiates the hydrologic cycle. Clouds are produced when moist air is lifted, cools, and water vapour condenses. The world is covered with moisture, which eventually returns to the surface as precipitation. One of two processes may take place once the water touches the ground: either the water would collect on the surface and turn into groundwater, or some of it would evaporate back into the atmosphere. Either groundwater leaks into oceans, rivers, and streams or it is released back into the atmosphere via transpiration. Runoff, which flows into lakes, rivers, and streams and is carried back to the oceans where the cycle begins, accounts for the remaining water on the earth's surface. A great illustration of the hydrological mechanisms at action is lake effect snowfall. This cycle is being disrupted by the effects of climate change and human activity, and as a result, demand for global water supplies is rising daily while the amount of clean fresh water available—which is just 2.5 percent of the total—is sharply declining. As a result, there is a need for efficient use of the available clean freshwater as well as for adequate storage and conservation. Fig. 2.3 depicts the hydrologic cycle in a concise manner.

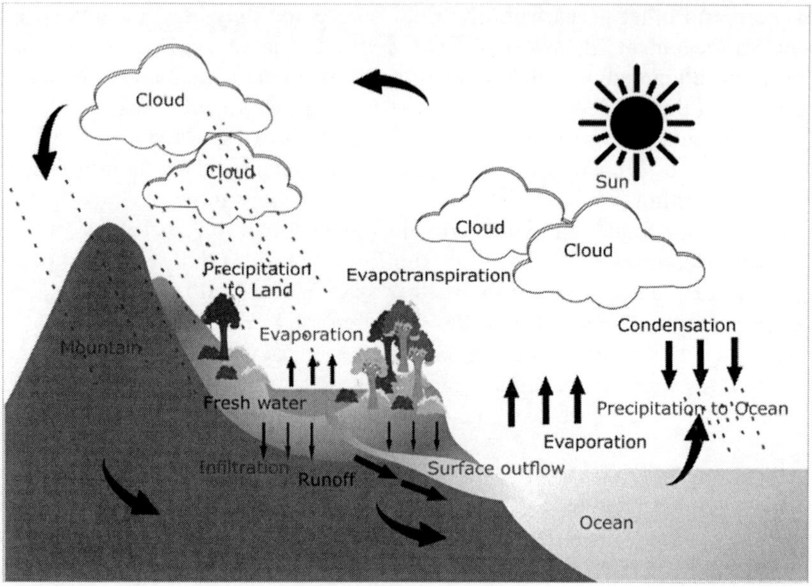

Fig.2.3: Hydrological cycle

Lithosphere

The Earth's solid outer layer is known as the lithosphere. The crust, mantle, and core are its three primary layers. The topmost layers of the Earth's structure, the crust and the brittle top of the mantle, are together referred to as the lithosphere. The asthenosphere, another layer of the upper mantle that lubricates plate tectonics and where convection cells develop, borders it from below and above. The lithosphere's rocks are not viscous, despite the fact that they are still thought of as elastic. The asthenosphere is viscous, and geologists and rheologists—scientists who study the flow of matter—mark the difference in ductility between the two layers of the upper mantle at the lithosphere-asthenosphere boundary (LAB). The ability of a solid substance to stretch or deform under pressure is measured by ductility. Compared to the asthenosphere, the lithosphere is substantially less ductile.

There are two types of lithospheres: continental and oceanic. Continental lithosphere has a granitic composition and oceanic lithosphere is made mostly of basalt with a mean density of 3.0. Oceanic lithosphere is slightly denser than continental lithosphere and is linked to oceanic crust. The density of the Earth as a whole is 5.5 gm/cc. The lithosphere extends to a depth of 100 km and is mostly made up of the earth's cold, hard, rocky crust. The continental lithosphere's rocks have an average density of 2.7 g/cm3 and are almost entirely composed of 11 elements, which together make up around 99.5 percent of its mass (Fig. 2.4).

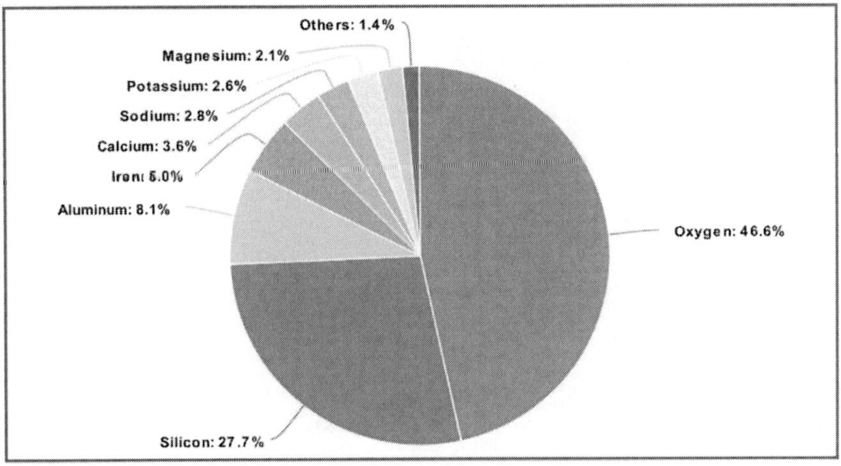

Fig.2.4: Elements found in Earth's crust

Earth's Internal Structure

Scientists were able to identify the three main zones of the Earth's interior via the indirect study of seismic refraction and seismic reflection as well as data from other sources including meteorites, mining, and volcanic eruptions. The crust is the top layer of rock that covers the Earth's surface like a thin skin. The mantle, a substantial rock layer below the crust, is what separates it from the core. The Earth's centre is its core. It is most likely metallic and is where the Earth's magnetic field originates (Fig. 2.5).

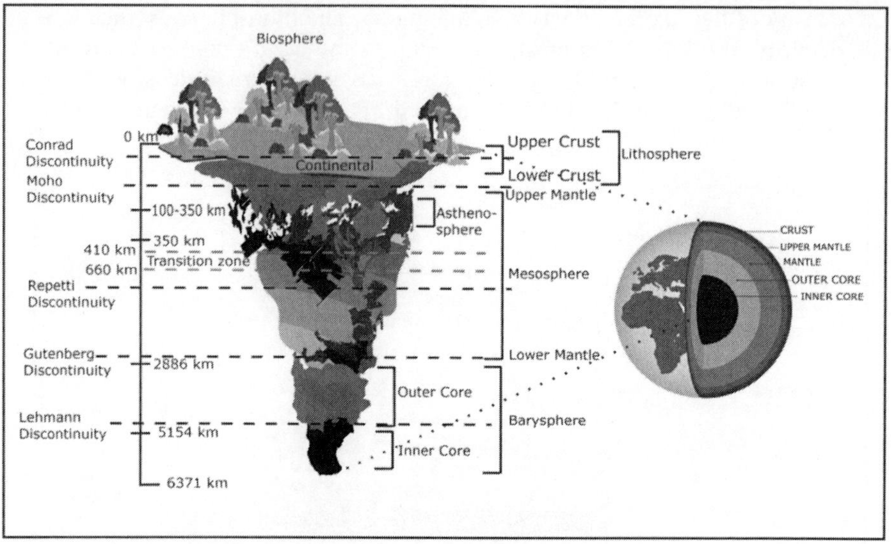

Fig. 2.5: Internal structure of the EarthCrust

The crust, which is the topmost layer of the Earth, is solid and hard. There are two different forms of crust: continental crust and oceanic crust. The continental crust is more substantial, predominantly felsic in composition, and contains silica-rich minerals. Due to the fact that continental crust is thinner than oceanic crust, the difference in composition is substantial. The majority of the mafic minerals make up the thinner oceanic crust. Mafic rocks include minerals with less silica and more iron and magnesium. Mafic rocks are denser than felsic rocks in the continental crust, and so make up the oceanic crust. The thickness of the continental crust is substantially greater than that of the oceanic crust; it ranges from 20 to 70 kilometres, on average. Seismic P waves move at a slower rate across continental crust—around 6 kilometres per second, the same as they do through granite and gneiss. Although the word "granitic" is often used to describe continental crust, it should be used with caution since granite does not make up the majority of the exposed rocks on land. Crust floats on the mantle. Continental crust floats higher in the mantle than oceanic crust because it is less dense. Due to the significant difference in densities, tectonic plates will force the plate with oceanic crust into the mantle under the plate with continental crust in the event of a collision. Following is an overview of the properties of the continental and oceanic crusts:

	Oceanic Crust	Continental Crust
Average thickness	7 km	20 to 70 km (thickest under mountains like the Himalayas)
Seismic P wave Velocity	7 km / sec	6 km / sec (higher in lower crust)
Density	3.0 gm / cm^3	2.7 gm / cm^3
Probable composition	Gabbro overlapped by basalt	Granite and other plutonic rocks with sedimentary cover at the top

Mantle

It is believed that rocks rich in olivine make up the majority of the Earth's mantle. It has variable temperatures depending on depth. As you move deeper, the temperature increases from its lowest point close under the crust. At the highest temperatures, the mantle material comes into contact with the heat-generating core. The progressive increase in temperature with depth is known as the geothermal gradient. The geothermal gradient produces various rock behaviours, and it uses these rock behaviours to divide the mantle into two zones. While the lower mantle rocks are warm and mushy, the higher mantle rocks are cold and brittle (but not molten). Rocks in the upper mantle are brittle enough to break under pressure, resulting in earthquakes. On the other hand, the soft rocks in the lower mantle flow when under stress rather than breaking. The lowest point of brittle behaviour is where the upper and lower mantles meet. The mantle is around 2900 kilometres thick.

Around 84 percent of the earth's volume and 67 percent of its mass are made up of the mantle. With a density of between 3.3 and 5.4 grammes per cubic centimetre, the layer is denser than the crust. The whole of the Earth's crust plus the uppermost solid part of the mantle make up the lithosphere. The upper mantle's asthenosphere, which lies between 80 and 200 kilometres below the lithosphere, is a very viscous, weakly elastic, and ductile region that is capable of deforming. The asthenosphere is where most magma is generated and where continental and lithospheric plates move through (plate tectonics). The division between the top and bottom mantles is known as a Repetti Discontinuity. The mesosphere is the area of the mantle slightly below the lithosphere and asthenosphere but above the core.

Core

Many scientists believe that the bulk of the Earth's core is composed of iron and nickel alloys. Calculations of density and the fact that many meteorites—which are thought to be fragments of a planetary body's core—are iron-nickel alloys are the foundations for this composition. The earth's core is a source of internal heat because it contains radioactive substances that release heat as they transform into more stable molecules. In the middle, there are two different zones. The outer core is a liquid because the temperatures are high enough to melt the iron-nickel alloy. Even though the inner core's temperature is higher than the outer core's, it is still a solid. The atoms are forced firmly together by the immense pressure produced by the weight of the surrounding rocks, which prevents the liquid state from developing. It is the layer that envelops the earth's core that is the thickest. The core and mantle are separated by Guttenberg's Discontinuity. Since iron (Fe) and nickel make up the majority of its composition, it is also known as NIFE (Ni).

About 15 percent of the planet's volume and 32.5 percent of its mass are made up by the core. The densest layer is the earth's core, which ranges in density from 9.5 to 14.5 g/cm3. The two sub-layers that make up the Core are the inner core and the outer core. While the inner core is solid, the outside core is in a liquid state (or semi-liquid). The Lehmann Discontinuity is a separation between the upper core and the lower core. Both the earth's core and its whole interior are referred to as the "barysphere" in different contexts.

Biosphere

The biosphere is made up of all the living things on Earth, like humans, animals, plants, and bacteria, as well as the organic matter they make and all the non-living abiotic materials needed for life. All of these things are connected in some way, from the nutrients carried by deep ocean currents to the animals that move around the world to the tiny interactions between organisms living in a cave far away.

Eduard Suess coined the term "biosphere" in 1875, but Vladimir Vernadsky refined it in the 1920s to signify its modern scientific meaning. The biosphere's organizational structure is divided into five tiers:

Organisms → Population → Community → Ecosystem → Biome

Biomes: These are different places that make up different parts of the biosphere. Biomes are the largest of the five levels that make up the structure of the biosphere. Scientists divide biomes into five categories: aquatic, desert, forest, grassland, and tundra. The main reason for categorizing the biosphere into biomes is to show how physical geography and environmental conditions affect the populations of living things.

Ecosystem: These are the second organisational category when considering the five levels of the biosphere. All the interactions between biotic (animals and plants) and abiotic (oxygen, nitrogen, and carbon) factors make up an ecosystem. Ecosystems are categorised based on how they communicate and transfer energy. Within each ecosystem, energy is used and stuff is cycled between different groups of organisms and their environment in the form of chemicals and nutrients. Simple examples of how primary producers, like plants, use the sun's energy include photosynthesis. Consumers, such as herbivores and omnivores, eat the plants to get energy. Decomposers, like fungi, transform the organic matter of deceased animals into substances that improve the soil and encourage the growth of plants.

Species Community: A community is the third level of structure in the biosphere. The populations of several species make form a community. A certain habitat or ecosystem is shared by communities. Abiotic factors in the area, such as temperature, pH, and nutrients in the air and soil, limit the kind of organisms that may survive there. Communities of species may be constrained by biotic factors such as predators and readily available food sources.

Population: A population is the fourth level of biosphere organisation, and it consists of all members of a particular species that are present in a certain habitat. A population might consist of thousands of individuals or only one. A population's growth or decline may have significant effects on the environment. While the presence of keystone species may have massive effects on the ecosystem as a whole, the existence of indicator species may have significant effects on the ecosystem as a whole.

Organisms: Living entities that reproduce via DNA are considered to be organisms, the highest level of biosphere structure. Depending on their energy source or cellular makeup, organisms are categorised in one of two ways. Cellular structure distinguishes between prokaryotes, which have free-floating DNA in cells without a nucleus, and eukaryotes, whose DNA is located in the cell's nucleus. Autotrophic species, like plants, get their energy from self-feeding, while heterotrophic organisms, like mammals, get their energy from eating other organisms.

CHAPTER 3

CLIMATE, TOPOGRAPHY AND SOIL SCIENCES

The term "environment" has been derived from the French "environia," which means "surrounding." Environment is a mix of variables that surrounds an organism (light, temperature, soil, water, etc.). Environmental factors are any external force, substance, or condition that affect an organism's existence. Environmental, ecological, andother factors are used to characterise these biotic and abiotic aspects. An organism's environment has living and non-living components. The habitat of an organism is a set of environmental factors called the environmental complex. The ecological components of this habitat complex have been characterised and examined in two ways:

1. Direct factors: A few examples of these factors include light, temperature, humidity, soil water, and the dynamics of soil nutrients.
2. Indirect factors: These factors include height, wind, slope, wind direction, soil organisms, and soil structure.

Generally, Scientists prefer to four categories of ecological factors which are given below:

1. Climatic factors:
   a) Light
   b) Temperature
   c) Precipitation
   d) Humidity
   e) Atmosphere
2. Topographic factors, including the factors which are related to geography and geology simultaneously:
   a) Altitude
   b) Direction of mountains and valleys
   c) Steepness and exposure of slopes etc.

3. Edaphic factors or soil science: this is related to the formation of the soil as well as its origin as well as the physical and chemical characteristics of the soil.
4. Biotic factors: these cover the interaction between diverse living forms, i.e., plants, animals, bacteria, fungus etc.

Climate

Climate may be defined as the long-term pattern of temperature and precipitation averages as well as extremes in a particular location. Meteorological elements, such as air relative humidity, temperature, wind pressures, and evaporation, are the primary contributors to the climate of any given region. In general, climate-related factors are what determine the identity of a location. There are a number of factors, including but not limited to light, air temperature, air humidity, rainfall, and the gaseous composition of the atmosphere, which includes wind, which all have an impact on the habitat of life in the air.

Light

It is well known that light can have an effect on the most fundamental physiological processes in plants, such as photosynthesis, transpiration, seed germination, and flowering. As a consequence of this, light is of great interest in ecology, where it plays a key role in species composition and the growth of vegetation. There is a wide range of variation in the amount of light in the form of EMR that reaches the surface of the planet. The amount of light that reaches the surface of the planet displays large geographic changes, which are influenced by a variety of factors including the following:

1. Gases in the atmosphere, notably nitrogen and oxygen, have the ability to absorb and scatter very small parts of light with shorter wavelengths.
2. Particles in suspension, such as dust and smoke, or solid particles distributed in water, such as clay, silt, plankton, and so on, have a screening effect. This may be seen in a variety of environments.
3. Water strata and layers
4. Vegetation found mostly in terrestrial environments, such as forests and grasslands, that is arranged in layers.
5. Variations in intensity and duration of intensity throughout the day are significantly affected by topographic features such as the slope and direction of the land surface.

When the radiant energy from the sun reaches the earth, it is transformed into many different forms of energy, such as thermal energy, mechanical energy, electrical energy, and so on. These transformations, in turn, cause alterations in a wide variety of physical processes that occur both in the atmosphere and on the ground. There is a substantial correlation between the temperature and the relative humidity of the air and the amount of light that is present. As a consequence of this, the effects of light on organisms may be either direct or indirect. A significant quantity of light from the sun is taken in by the atmosphere. As a direct consequence of this, the amount of light that is accessible at higher altitudes is more than what is available at sea level. This is because the air layers that make up an area are thinner at higher altitudes. It is possible for the light intensity to be reduced to 4 percent of what it would normally be on an overcast day. In ecology, the quality of light is given less weight despite the fact that the red and blue parts of the spectrum are essential for photosynthesis. This is due to the fact that plant dispersion is a more crucial factor.

Light intensities are typically lower in the morning and evening, as well as during the winter months. During the summer and winter months on the equator, the amount of time spent in daylight is approximately 12 hours out of every 24. The length of a summer day might extend far above 12 hours as one approaches the pole.

Light in relevance to plants

Light affects plants in the following ways directly or indirectly

1. Chlorophyll production: Only a selected number of plant species, such as mosses, algae, seedlings of conifers, and early fronds of ferns, are able to produce chlorophyll in the absence of light. However, the majority of plant species need light in order to produce chlorophyll. Plants take in between 1 and 3 percent of the total amount of solar energy that strikes the earth each day.
2. Heating effect: The effect that light has on a plant's metabolism is that it causes that plant's temperature to increase when it is exposed to it.
3. Transpiration rate: The increase in temperature that is brought about by light has an indirect impact on the rates of transpiration. As a direct consequence of this, arid conditions characterised by high rates of transpiration are always associated with high levels of light intensity. During the course of a plant's growing season, a single leaf will lose a quantity of water that is many times more

than its own weight. On a daily basis, a single acre of maize will expel between 11,400 to 15,100 litres of water, whereas a single massive oak tree has the potential to expel 151,000 litres of water.

4. Stomatal movement: Stomatal movement is connected to transpiration and absorption and here again light controls the opening and closure of stomata.
5. Plant distribution: The light conditions at the poles differ from those in other areas of the globe. As a consequence, the total amount of radiation reaching the earth's surface changes with latitude. This could be one of the factors for the variation in vegetation between the poles and the rest of the world.
6. Plant components overall vegetative development: Plants are sometimes categorised ecologically depending on their relative light demands. The categories are:
   i. Heliophytes: are plants that thrive in bright, direct sunshine. Some heliophytes thrive in direct sunshine, but others, known as facultative sciophytes, are able to flourish in shady conditions as well.
   ii. Sciophytes: These plants thrive in dim light although there are a few species that are able to thrive in dim light as well as full sunshine; these plants are referred to as facultative heliophtes.
7. Photoperiodism: The response of a plant to the day-and-night cycles is referred to as photoperiodism. Some plants, known as long day plants (Brassica rapa, Sorghum vulgare), need to be exposed to light for a period of time that is longer than a predetermined critical duration before they begin blooming. Other plants, known as short day plants, need to be exposed to light for a period of time that is shorter than this critical length before they begin blooming (Salvia splendens, cannabis sativa). In day-neutral plants, such as Cucumis sativa and Nicotiana tabacum, there is no correlation between the amount of time a plant is exposed to light and whether or not it begins to produce flowers in response to that light. The brightness or darkness of an area may be gauged by looking at the leaves. It is only when the plants are subjected to the necessary inductive photoperiod that a hormone-like substance is able to move from the leaves to the tips of the shoots, which in turn causes flowering.
8. Succession: The process of photosynthesis will be hampered if there is insufficient light. This will have an effect not only on the distribution of plants but also of the animals that consume plants. Some plants are able to adjust to low light conditions by developing

bigger leaves, which increases the amount of photosynthesis they can do. As a consequence of this, the development of new plant tissue is inhibited. The process of plant succession may be broken down into nine distinct stages: (1) Nudation, (2) Migration, (3) Germination, (4) Ecesis, (5) Colonisation and Aggregation, (6) Competition and Co-action, (7) Invasion, (8) Reaction, and (9) Stabilisation.

Light in relation to animals

Light may have a variety of effects on many aspects of an animal's life and existence. Insects, birds, fish, reptiles, and mammals are all subject to the effects of light on their growth, pigmentation, migration, and reproduction, as well as their diapause. There are a lot of animals that like being in the dark, like badgers, but there are also plenty of species that cannot survive in the dark, like hydroids. The following is a brief summary of some of them:

1. Metabolism: The amount of light an organism receives has a significant influence on the rate at which it burns calories. An increase in light intensity causes an increase in the protoplasmic solubility of salts and minerals, as well as an increase in the activity of enzymes and the overall metabolic rate. On the other hand, as the amount of light that is present grows stronger, the solubility of gases will decrease. Animals that live in caves have been shown to have sluggish behaviours and a slower rate of metabolism than other animals.
2. Reproduction: Light is necessary for the stimulation of gonads and the early stages of yearly ovulation in many animals. This is because gonads are responsible for fertility (e.g., birds). It has been demonstrated that during the summer, when there is more light, the gonads of birds become more active, whereas this activity decreases during the winter months when there is less sunlight.
3. Development process: In some cases, such as with Salmon larvae and Mytilus larvae, light has the effect of accelerating the development process, while in others, it has the opposite effect and slows it down.
4. Locomotion: The pace of movement in many lower animals is determined by their exposure to light. The scientific term for this phenomenon is called photokinesis. This phenomenon, in which animals move in response to light, is known as phototaxis, and movements of this kind are referred to as phototactic. Light plays a

role in the direction that certain animals move when they are on the move. Earthworms, slugs, and certain zooplanktons like copepods, for example, are examples of organisms that are negatively phototactic. On the other hand, animals such as Euglena and Rantara, and others like them, are positively phototactic and migrate toward light sources. In response to light, certain animals, such as numerous tubiculous worms and polyps of many coelenterates, exhibit a behaviour known as phototropism, which involves the movement of only a portion of their bodies.

5. Photoperiodism: It has been discovered that the behaviour and metabolism of many different kinds of organisms are affected by the 24-hour cycles of light (day) and darkness that occur on a daily basis (night). These climatic cycles of light and darkness are driven, in large part, by the movements of the earth in regard to the sun and the moon. Alternating periods of night and day are brought about by the rotation of the earth on its axis. Both the axial tilt of the earth and its annual rotation around the sun are responsible for the changing of the seasons. The study of how different types of organisms respond to the alternating periods of light and dark in their surroundings is referred to as photoperiodism. The photoperiod is a period of the daily cycle that begins with a phase of light and then changes into a period of darkness.
6. Protoplasm: Despite the fact that the majority of animal bodies are covered by some kind of body covering, which shields animal tissues from the potentially damaging effects of ultraviolet radiation from the sun, protoplasm is still present. On the other hand, solar rays can occasionally pass through such coverings, which can result in the excitation, activation, ionisation, and heating of protoplasm in a variety of cells throughout the body. It has been shown that exposure to UV radiation may cause mutations to occur in the DNA of a number of different species.
7. Pigmentation: The amount of melanin an animal has can be affected by the amount of light it receives. Cave mammals are characterised by their lack of dermal pigmentation. They restore the skin's pigmentation if they are protected from the sun for an extended period of time. It can be deduced from the fact that people who live in the tropics have darker skin pigmentation that exposure to sunlight plays a role in the formation of skin pigmentation. The sun's rays are necessary for the production of pigment in human skin. Light is responsible for several animals' distinctive pigment patterns, which play a role in sexual dimorphism and in the

development of protective colouration. Despite the fact that the animals that dwell in the deepest sections of the ocean are pigmented, the animals' colouring does not exhibit any patterns since the environment in which they live is uniform.

Temperature

It has been found that temperature has a number of effects on living beings, including a large influence on cells, morphology, physiology, behaviour, growth, ontogenetic development, and distribution of animals and plants. These effects may be broken down into subcategories as follows: Several of them are discussed in more detail below:

1. Metabolism: Temperature has an effect on the activity of many enzymes, which in turn has an effect on the majority of the metabolic processes that take place in microbes, plants, and animals. As a direct result, an increase in temperature, up to a certain extent, induces an increase in the activity level of enzymes, which in turn causes an increase in the rate at which metabolism occurs. As an example, it has been discovered that the activity of the liver arginase enzyme on arginine amino acid gradually and steadily increases from 17 degrees Celsius to 48 degrees Celsius as the temperature rises. Nevertheless, if the temperature is raised over 48 degrees Celsius, this will have a negative impact on the metabolism of this enzymatic activity, as in, it will quickly slow down.
2. Reproduction: There is a particular temperature at which gametogenesis, the growth of gonads, and gamete libration all take place. This temperature varies from species to species. Some species, for instance, are capable of reproducing at any time of the year, whilst others can do so only during the summer or winter months, and yet others have two distinct mating seasons: one in the spring and one in the autumn. The mating seasons of the vast majority of organisms are consequently controlled by temperature. Temperature has an influence on the reproductive capacity of animals as well. The reproductive capacity of an animal, also known as its fecundity, may be thought of as the total number of offspring that the animal will produce over the course of its lifespan. For example, females of the acridid insect Chrotogonustrachyplerus attained sexual maturity more quickly at temperatures of 30 and 35 degrees Celsius than they did at 25 degrees Celsius, and they laid the most eggs per female when the

temperature was 30 degrees Celsius. The number of eggs dropped from 243 to 190 when the temperature was raised to between 30 and 35 degrees Celsius.

3. Growth and development: The rate of development of many different kinds of plants and animals may be affected by temperature. For example, after they reach adulthood, trout consume very little and do not grow until the water temperature exceeds 10 degrees Celsius. The length of the body of the oyster Ostraea virginica grows from 1.4 millimetres to 10.3 millimetres when the temperature is increased from 10 degrees Celsius to 20 degrees Celsius. Both the sea snail (Urosalpinx cinerea) and the sea urchin (Echinus esculcntus) attain their full growth potential in waters that are warmer. Corals are most successful in waters with temperatures that are cooler than 21 degrees Celsius. Both very low temperatures and excessively high temperatures might hinder the growth of plants. A few instances of cold injuries brought on by low temperatures include desiccation, chilling injury, and freezing injury. The ability of some perennials to endure conditions of very low temperature is what's meant by the term "cold resistance." The detrimental effects of very high temperatures on a range of physiological processes, including as respiration, transpiration, protein metabolism, and so on, cause plants to become stunted and eventually die. This happens when the processes are disrupted. This kind of damage is known as heat injury.
4. Crossing over: In animals like fruit flies and other species of Drospohilia, scientists have shown that changes in temperature may affect both the crossing over of genes and the somatic manifestations of inherited traits. It is more difficult for wings, eyes, and other body components to grow in larvae or pupae if the temperature in which they are kept is either too warm or too cold. When it's hot outside, the wings become longer, but when it's cold, they are stunted.
5. Sex ratio: In certain kinds of animals, the temperature of their natural habitat has a role in determining the proportion of males to females within a species. For instance, it seems that the sex ratio of the copepod Maerocyclopsalbidu changes depending on the temperature. When the temperature rises, there is a proportionately larger increase in the number of males. The plague flea, Xenopsyllacheopis, had a greater number of males than females on days when the mean temperature stayed between 21 and 25 degrees Celsius. On days when it is cooler, though, the scenario is reversed.

6. Colouration: The size of animals as well as the colouring of their fur may be affected by temperature. In warm and humid environments, the skin of many animals, including insects, birds, and mammals, develops a darker pigmentation than other varieties of the same species that live in cold and dry conditions. This phenomenon is referred to by its name, the Gloger rule.
7. Morphology: According to Bergman's rule, temperature has an effect not only on the absolute size of an animal but also on the relative characteristics of its various body portions. Birds and mammals, for example, grow larger in colder temperatures than they do in warmer regions, and colder conditions promote the existence of species that are larger. On the other hand, poikilotherms shrink in size as they move to colder locations. Desert environments are best suited for the survival of smaller creatures for a number of reasons. According to Allen's rule, the extremities of animals, such as the legs, nose, and tail, are shorter in colder areas than they are in warmer ones. Other examples of this include ears and nose. According to Rensch's rule, the bird species that inhabit regions with cooler temperatures have wings that are more acuminate and smaller than those that inhabit regions with higher average temperatures. According to Jordan's rule, temperature has been shown to have an influence on the morphology of some fishes and has been connected to the number of vertebrae.

In some cases, the flora of the world is divided into distinct classifications based on temperature circumstances. One example of this is the Megatherms region, which is characterised by consistently high temperatures throughout the year and a primary vegetation type consisting of tropical rain forest. Mesotherms are characterised by high temperatures that are interspersed with periods of lower temperatures, and the predominant kind of vegetation is tropical deciduous forest. Microtherms are regions that have temperatures that are lower than average and have mixed coniferous forest vegetation. Heikistotherms are defined by their extreme sensitivity to temperature and their reliance on alpine plants as their primary food source.

The range of temperatures has an impact on the circumstances in which terrestrial plants grow. Alterations in temperature, likewise, have an effect on the habitats of aquatic organisms. The most prevalent places to see thermal stratification are in aquatic settings. The change in temperature profile that occurs with increasing depth in a body of water is referred to as thermal stratification. There are three distinct classifications that may be used to thermal stratifications. 1. The warmer water that is found in the top layer, known as the epilimnion. 2. The metalimnion is the intermediate layer, and it has a zone with a decreasing temperature. 3. The Hypolimnion is the layer of colder water that may be found at the ocean floor.

Effect of temperature on the distribution of animals

Temperature variations, as previously established, have a considerable influence on plant distribution in a number of ways. Animals are also impacted. All organisms are categorised into two categories:

1. Homoiothermic or endothermic (warm-blooded) creatures: These animals, such as birds and mammals, are able to keep their internal temperatures stable independent of the temperature of their surrounding environment.
2. Poikilothermic or ectothermic (cold-blooded) creatures: Animals such as reptiles, fishes, and amphibians, whose internal temperatures change in reaction to changes in the temperature of their surrounding environment, fall into this category.

Precipitation

Rainfall is the primary source of water for the earth's soil. The water that plants and animals get from the soil comes from precipitation in the form of rain. The hydrologic cycle, also known as the water cycle, is the process by which water moves from the surface of the earth to the atmosphere. The two most important processes that take place throughout this cycle are evaporation and precipitation, respectively.

In fact, rain provides organisms with very little or even no benefit from a water source. It is not put to use until after it has been incorporated into the soil. Precipitation is produced when the gravitational force of the atmosphere acts upon the vapour in the atmosphere. There is a wide variety of forms that precipitation can take. Examples of precipitation include drizzle, rain, snow, dew and frost, sleet, and hail. A drizzle is a light form of liquid precipitation that is made up of droplets of liquid water that are much smaller than raindrops, typically having a diameter of less than 0.5 millimetres. The predominant cloud types responsible for drizzle are low stratiform clouds and stratocumulus clouds. Rain is liquid water that has condensed into droplets and become heavy enough to fall to the ground after initially existing as water vapour in the atmosphere. In contrast to dew and frost, which are formed when moisture condenses directly on the surfaces of objects, plants, animals, and soil, individual ice crystals form while they are suspended in the air—typically within clouds—and then eventually fall to the ground, where they go through additional transformations. Ice crystals are formed in the atmosphere, whereas dew and frost are formed on the ground. When liquid water droplets, such as raindrops, freeze, they produce microscopic ice particles that become sleet. Sleet is composed of these microscopic ice particles. Sleet is also sometimes referred to as ice pellets. Hail is a kind of frozen precipitation that may grow to large dimensions owing to the accumulation of water that freezes onto the surface of the hailstone. Hail can be distinguished from snow by its lack of crystalline structure.

The most important form of precipitation is rain, followed by snow and then hail. It has an effect on the relative humidity of the air and is a source of water for the soil. The South-west monsoons are the most probable causes for India's rainy weather. In the course of annual precipitation, around 45 percent of the water that is accessible is carried by rivers, 20 percent is absorbed by the ground, and the remaining 35 percent is lost to evaporation. The amount of precipitation that an area receives on an annual basis is a major factor in determining the kind of plants that may grow there. For example, evergreen forests are the most common kind of vegetation in tropical regions that see significant levels of rainfall throughout the year. In regions that get a lot of precipitation in the winter but very little throughout the summer, you'll find sclerophyllous forests. The plants resemble shrubs and have leaves that are evergreen, leathery, and thick, but their growth is restricted. Grasslands are located in regions that get a lot of rain in the summer but very little rain in the winter. This allows for optimal conditions for plant growth. In regions where precipitation is limited, xerophytic flora and deserts may sometimes be found.

Humidity of Air

It is one of the many different kinds of water that can be found in the natural world. The term for moisture that exists in the atmosphere in the form of an invisible vapour is called humidity. The principles that are assigned for relative humidity are used to measure how humid the air is. It is the percentage of the total amount of moisture in the air that the air can hold at saturation at the temperature that is currently being experienced. The intensity of the sun's radiation, the temperature, the height, the wind, the exposure, the cover, and the condition of the soil water all have a significant impact on humidity. When temperatures rise, the ability of the air to retain moisture also increases; consequently, relative humidity drops as a consequence of this. The ability of the air to retain moisture is reduced when temperatures are low, which results in an increase in the relative humidity. To determine the relative humidity of an environment, one can use a psychrometer, a paper strip hygrometer, or a thermo hydrograph. The daily variation in relative humidity levels is determined by the kind of habitat conditions that are present. There is a possibility that it will change throughout the day in plains and deserts, but on oceanic islands, where the climate is constant throughout the year, there is very little variation. As a consequence of this, humidity has a wide range of effects on the life of plants. Plants react in a manner that is analogous to how they react to a decrease in the intensity of the light that they receive when exposed to moist air. Many types of plants, including orchids, lichens, and mosses, absorb moisture from the air directly. It plays an essential role in the early stages of the life cycle of fungus and other microorganisms, beginning with the germination of spores and continuing through subsequent phases.

Atmosphere: Gases and Wind

The term "atmosphere" refers to the dense gaseous layer that surrounds the earth and encompasses all of space. As we'll see in a little while, the ground has its very own microclimate. Between the soil particles and the soil aggregates, there are spaces called pores, and it is these spaces that give the soil its atmosphere. In the ongoing conversation about the atmosphere's role as an ecological component, it is important to take into account the effects that main gaseous elements that are present in the atmosphere and wind have on living organisms.

Atmospheric gases

Up to a height of about 300 kilometres above the surface of the earth, there lies a thick gaseous layer known as the atmosphere. As altitude is increased above sea level, there is a proportional decrease in the air pressure. Up to a height of around 20 kilometres above the surface of the earth, approximately 95 percent of the total air is present in the atmosphere, with the remaining 5 percent reaching a height of approximately 280 kilometres above the earth's surface. It's possible that the gaseous layer contains a variety of distinct gases, all present in varying concentrations. Nitrogen, oxygen, and carbon dioxide are the primary components of these gases. The following table 3.1 presents a comparison of the relative amounts of numerous gases when measured in volume:

Gases	Percent (By volume)	Gases	Percent (By volume)
Nitrogen	78.08	**Crypton**	0.00011
Oxygen	20.9	**Xenon**	0.00009
Argon	0.93	**Hydrogen**	0.00006
Carbon dioxide	0.03	**Methane**	0.0002
Neon	0.00182	**Nitrous oxide**	0.00005
Helium	0.00052	**Ozone**	0.000004

Table 3.1: Atmospheric gases by volume

Wind

Wind may be defined as the movement of air. Because of the effect it has on plant life, it is an important component of the atmosphere's biological makeup. This is especially true in areas with relatively flat terrain, near coastlines, and at higher altitudes in mountain ranges. Wind is an active participant in transpiration, causing a variety of different types of mechanical damage as well as disseminating pollen, seeds, and fruit. Additionally, the water dynamics and illumination conditions in a particular area are altered as a result of this alteration. Wind speed may be affected by a number of factors, including geographical location, topography, and plant densities, as well as position in relation to coastlines. At low elevations by the coast and at high elevations atop mountains, plants are susceptible to substantial damage by the wind. One region with high pressure gives way to another region with lower pressure. The differential heating of the atmosphere is the primary contributor to the pressure differences that are seen. The equatorial regions get more heat than the northern or southern regions, which causes low pressure to form at lower latitudes. The air then travels all the way from the poles to the equator.

The following is a list of the many physical, anatomical, and physiological changes that are brought about in plants as a result of wind:

1. Breaking and uprooting: Branches can be snapped off live trees, and the trees themselves can be uprooted, if there is a strong enough wind. On the other hand, such effects are very rare in forests, since the canopies of individual trees at a variety of heights reduce the speed of the wind by around 80 percent. As a consequence of this, the flora in these types of forests functions as

natural windbreaks. Single trees or groups of trees in the forest that have been uprooted by the wind are referred to as wind tosses or windfalls.

2. Deformation: Exposure to strong winds that blow in the same direction for an extended period of time may result in irreversible shifts in the shape and placement of the shoots. Deformation is common in trees that are found growing on slopes and in close proximity to beaches.
3. Lodging: Lodging is a kind of wind damage that happens when powerful gusts of wind force grasses such as maize, oats, sugarcane, and other herbaceous plants against the ground and cause them to get pressed against the surface.
4. Abrasion: Particles of dirt or ice that are carried by the wind may act as a tremendous abrasive force, eroding the buds and other parts of the plants as they pass through the plant. The risk of this form of damage is increased significantly for crops that are grown on sandy soils.
5. Erosion and deposition: When strong winds blow, dirt is blown away, and this exposes the roots of plants whose bark or other protective coverings are weak. In addition, the areas where these additional soils are deposited become unsuitable for plant growth as a result of this process.
6. Salt spray: The salts of the water are carried along the seashore by the strong winds that blow in the region where the ocean is located. These salts may be harmful to the health of certain of the plants that are able to flourish there.

Anatomical and physiological effects

1. Compression wood: This robust and ruddy kind of xylem, known as compression wood, develops in the boles of some trees as a result of wind distortion.
2. Desiccation: The rates of evaporation and transpiration increase when there are high winds present. As a direct consequence of these circumstances, desiccation will take place in any plant that is incapable of preserving its own internal water balance. It's possible that some plants with tightly packed branches won't be damaged by the wind. Cushion plants are well named since they thrive in breezy environments and are often seen there.
3. Shrinkage: Plants that are growing in an environment with drying winds often have difficulties with dehydration and consequent loss of turgidity, which may lead to dwarfing. As a consequence of

these disorders, their organs become abnormally small. This may be seen in a significant number of trees along the coast, in the arctic, and at the timberline in the mountains.

Topographic factors

It has been hypothesised that the dynamic interactions between solar radiation, air temperature, relative humidity, and annual precipitation are what determine the climate of any particular location. In addition to the geographic latitude and altitude of the location, the terrain of the area also has a considerable bearing on each of these climatic aspects. The topographic characteristics of the area are used to describe the earth's physical topography at this particular location. The surface of the Earth may exhibit a wide range of irregularities and does not have a consistent appearance anywhere on the planet. These types of topographic characteristics have an effect on the vegetation of an area because they cause shifts in the weather patterns that prevail there. Changes in climate that are brought on by these factors may sometimes bring up diverse local climates or even microclimates, in addition to changes in the conditions of the soil. Therefore, topographic factors have an affect, but it is mostly indirect because of the effects that they have on climatic parameters. The following are the primary topographic factors:

1. The height of the mountain: Plate tectonic activity are responsible for the formation of mountains, hills, hillocks, valleys, and other landforms. It is possible that the influence of differing heights will be more readily seen on mountains. As one ascends above sea level, one will experience changes in temperature, pressure, wind speed, relative humidity, and the amount of solar radiation. Because of these shifts, the vegetation at different altitudes looks quite different from one another. The best example of this may be observed in the Himalayas, which trend from east to west, since the height of the Himalayan terrain is always changing, which causes the flora to shift as well.
2. The orientation of the mountains and valleys: Mountains may deflect or direct winds in a variety of ways, and certain slopes on mountains can collect precipitation brought on by the wind. As a direct consequence of this, water vapour may preferentially condense in particular favoured orientations, resulting in the formation of clouds from which rain will eventually fall. It is possible that this is the reason why there are thick forests on certain

sides and at certain heights, while there is just a scant covering of flora on other sides and at lower altitudes. Plate tectonics are responsible for the world's greatest rainfall, which occurs in Meghalaya's Cherrapunji and Mawsynram regions. This is because the region's topography mountains act as a barrier, preventing precipitation from the Bay of Bengal from escaping into the atmosphere.

3. The steepness of the slope: The slope is the most defining aspect of a mountain. The steepness of a slope has an effect on the rate of water flow, which in turn has repercussions for the amount of solar energy that is absorbed during the day as well as the characteristics of the soil. At higher heights in the northern hemisphere, the steepness of the slope causes the surface to be exposed to the sun for a greater amount of time on the southern slopes, while the temperature remains lower on the northern slopes. This is due to the fact that the rays of the midday sun fall almost exactly at right angles on the steep southern slopes, whereas the northern slopes receive only oblique rays during the morning and evening hours, and sometimes none at all except for a short period of time during the summer. The reason for this phenomenon can be explained by the fact that the steep southern slopes face south. Because of the different amounts of light exposure and temperatures on the two slopes, the types of plants that grow there are also distinct from one another (Fig. 3.1). The steepness of the slope is also connected to the amount of soil that is eroded.
4. The exposure of the slope: The kinds of plants that are able to thrive on a slope are largely determined by the amount of light and wind that the slope receives. In general, the types of plants that are supported by vegetation that is more or less exposed to the sun and wind, or that is not exposed to any of these factors at all, will be entirely different from one another. Because of this, in areas with a temperate climate, greenhouses and hot beds are often oriented such that they face the sun or are located on southern slopes, both of which get a significant amount of heat from the sun.

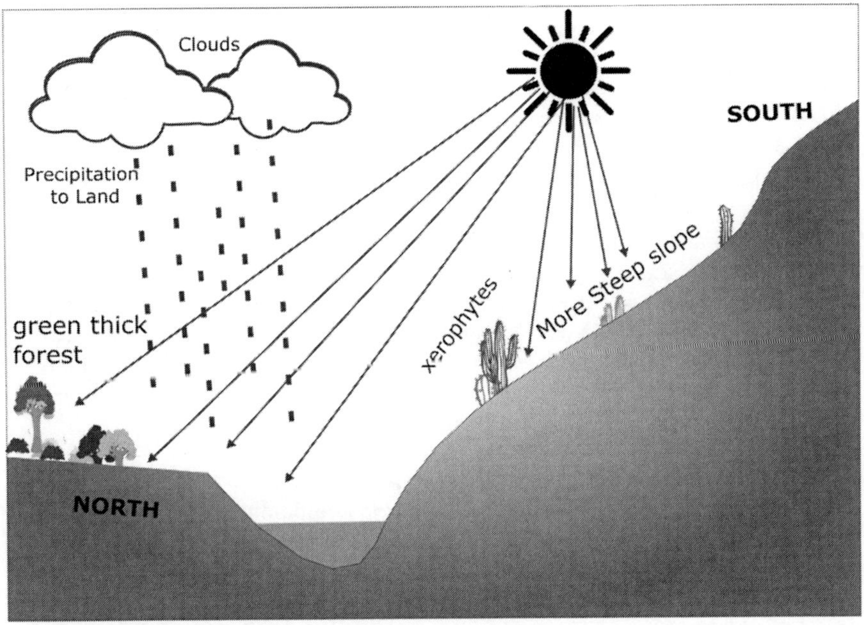

Fig. 3.1: Diagram showing the effect of steepness of slope on climate and relation of the angle of incoming solar rays.

Soil Science

Soil science examines not just the structure and composition of soil, but also its physical and chemical characteristics. Instead of being a single entity, soil is a complicated system that is made up of a number of different components. It is the loose surface layer that covers the majority of land. The study of soil involves doing research on the properties of the many elements that comprise the soil complex.

Importance of Soil

Our life support system is topsoil. The roots of plants attach themselves to the soil, while the soil acts as a reservoir for water and nutrients. In addition to the plethora of microorganisms that fix nitrogen and break down organic matter in soil, legions of small critters such as earthworms and termites are also responsible for these processes. The health of the planet's ecosystem depends heavily on the quality of its soil. Without soil, the life of humans would be impossible. Soil provides plants with a footing for their roots and contains the required nutrients for plants to grow; it filters rainfall and controls the release of surplus precipitation, avoiding floods; it is capable of storing vast quantities of organic carbon; it buffers pollutants, so preserving groundwater quality; it gives humans vital building and manufacturing materials, i.e., we build our homes with clay bricks, and we sip coffee from a cup made of soil; it protects groundwater quality; it provides humans with some key construction and manufacturing. Warming (1909) emphasised the significance of soil characteristics and proposed five ecological groups of plants. These groups are as follows: (a) oxylophytes, which are typically found on acid soils; (b) halophyhtes, which are typically found on saline soils; (c) psammophytes, which are typically found on sand; (d) lithophytes, which are typically found on rock surfaces; and (e) chasmophytes, which are usually found in rock crevices.

The capacities of the soil are referred to as its functions, and they play an important role in a wide number of fields, including agriculture, the environment, nature conservation, landscape architecture, and urban planning. The following is a list of the six most significant functions of soil:

1. The generation of food and other forms of biomass.
2. Interaction with the surrounding environment, including storing, filtering, and transformation.
3. The genetic makeup of the population as well as the biological environment.
4. Acquiring raw materials.
5. Heritage, in both the cultural and physical senses.
6. Other man-made projects, such as roads, buildings, and bridges, may be constructed on top of this platform.

Composition of Soil

Therefore, soil is not just a mixture of different types of mineral particles and other rock fragments. In addition to this, it has a biological system that is comprised of live creatures, in addition to other components. As a result, it is suggested that we refer to it as a soil complex, which includes the five categories of components that are listed below.

1. Mineral matter: Mineral matter is a matrix of mineral particles that is formed as a result of varying degrees of rock's disintegration into its parent material.

2. Soil Orgainic matter: Soil organic matter, also known as humus, is an organic component that is created when material from organisms that live above and below the earth, such as plants, animals, and microorganisms, is added to the soil over a long or short period of time.
3. Soil water: Soil water, also known as soil solution, refers to all of the water that is present inside soil, together with any solids, liquids, or gases that have dissolved in it. Both capillary and absorptive forces are at work between and on the surface of the soil particles to keep water in the soil where it belongs.
4. Soil atmosphere: It is found in the pore space between the soil particles, which is not currently filled with water at any one time. Its make-up differs from that of the atmosphere above ground in the sense that, on average, it contains less oxygen and a higher concentration of carbon dioxide than the atmosphere above ground.
5. Biological system: In addition to the aforementioned, there is also the possibility of adding the biological system, due to the fact that every soil has its own unique flora and fauna, which may include bacteria, fungi, algae, protozoa, rotifers, molluscs, and arthropods.

Formation of soil

A well-known Russian pedologist named Dokuchayev (1889) said that "the soil is the outcome of the actions and reciprocal affects of parent rocks, climate, topography, plants, animals, and the chronology of the land." (Fig. 3.2). The following mathematical formula may be used to represent it:

$S= (g.e.b.)\Delta t$

Where, s-soil, g- geology, e- environment, b- biological influences and t-time

The entire process of soil formation can be broken down into two stages: the first stage is called weathering, and it refers to the process by which larger rocks are broken down into finer and smaller mineral particles. The second stage is called pedogenesis, and it refers to the process by which mineral matter is modified as a result of interactions between biological, topographic, and climate factors, leading to the formation of any one of a broad variety of potential soil types.

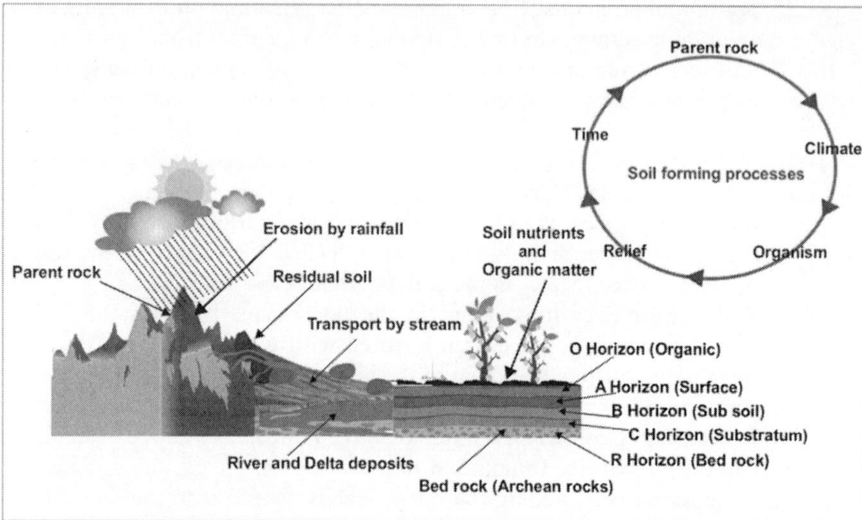

Fig. 3.2: Diagram showing soil forming processes

Weathering process

Weathering is the process by which rocks are broken down to produce soil. This process might take millions of years. Rocks and other types of sedimentary geological material are the primary components of soil (the materials from which soils have formed). The earth is home to a wide variety of rocks, some of which have an alkaline composition, some of which have an acidic composition, some of which have a coarse texture like sand, and others which have a fine texture and a clayey composition. The rocks and sediments that compose a soil are responsible for imparting its distinctive texture. When you see rocks scattered throughout the landscape, it is much simpler to conceptualise the amount of time necessary to break them down into soil. In fact, the formation of only one millimetre of soil from some of the most resistant rocks may take upwards of five hundred years. To our good fortune, during the Ice Age, which occurred over 10,000 years ago, enormous quantities of rocks were broken down and transformed into clays, sands, or gravels, making it easier to develop soils out of these materials.

There are three primary types of weathering: physical, chemical, and biological. Physical weathering is the most common. The term "physical weathering" refers to the process by which rocks and other materials are broken down into smaller and finer particles as a consequence of a variety of processes, including freezing and thawing, wetting and drying, shrinking and swelling. The term "chemical weathering" refers to the process by which rocks are worn away by a number of chemical processes including acidification, dissolution, and oxidation. Some minerals, which are stable inside bedrock, become less stable when exposed to the environment. As a result, changes begin to occur in rocks that are close to the surface, which causes the rocks to become less stable. Biological weathering is the term given to the process by which living organisms contribute to the degradation of rock. This encompasses, for example, the activities of plant roots and the organisms that live in soil. The process of carbon dioxide respiration by plant roots may result in the production of carbonic acid, which may chemically attack rocks and sediments and assist in the transformation of these materials into soils. A number of different weathering processes are active close to the surface of the soil, and they are all cooperating with one another to break down rocks and minerals and form soil. These weathering processes are responsible for the formation of the majority of the earth's soils.

The weathering processes are physical as well as chemical

1. Physical: These may be:
   a) Wetting-drying: It's induced by the enlargement of layer lattice minerals while they're moist.
   b) Heating-cooling: This process results in the fracture of heterogeneous crystalline rocks because the inclusions within these rocks have different thermal expansion coefficients. The sun's heat causes large boulders to flake off near the surface, which is particularly common in dry regions.
   c) Freezing: When water freezes, its expansion causes porous, lamellar, or vesicular rocks to fracture, which is known as frost shatter.
   d) Glaciation: Larger snow and ice glaciers have the potential to cause physical erosion of rocks via a process known as grinding as they descend.
   e) Solution: The solution is that calcium chlorides, sulphates, and other mobile rock components may be readily dissolved away by solvents such as water.
   f) Sand blast: In arid, desert situations, the rocks are broken by wind, sand, and other forms of physical force.

2. Chemical: These processes include:
   a) Hydration: Hydration is a sort of chemical weathering that takes place when a mineral reacts with water and modifies its complicated construction. This process takes place when a mineral is exposed to water. Hydration occurs as a result of the interaction of the mineral anhydrite with groundwater. Anhydrite, when combined with water, may be transformed into gypsum, which is one of the minerals that is found in the greatest abundance on our planet.
   b) Hydrolysis: Rainwater seeps through rocks, causing the hydrogen ion (H+) in the water to react with metallic ions in the rocks, causing rock minerals to breakdown. For example, hydrolysis of orthoclase ($K_2Al_2Si_6O_{16}$) to kaolinite ($Al_2O_3.2SiO_2.2H_2O$).
   c) Oxidation-reduction: Because Fe^{2+} is more soluble than Fe^{3+}, certain oxidation-reduction chemical processes, such as the reversible transformation of Fe^{3+} to Fe^{2+}, induce rock disintegration.
   d) Carbonation: During the process of carbonation, carbon dioxide from the atmosphere and from living things dissolves in the water to generate carbonic acid. This results in the water found in rocks becoming acidic, which in turn causes more chemical reactions to occur with the minerals found in rocks.
   e) Chelation: It is possible to degrade minerals and rocks by eliminating metallic cations via the biological process of chelation, which is carried out by organisms. In order to promote weathering, organisms might alter the soil's moisture regime. Chelates stop nutrients from being lost by leaching. Chelation enhances the mobility of nutrients in the soil. This increased mobility facilitates plants' nutrient uptake. By limiting the amount of iron that pathogens may access, chelating compounds prevent the growth of plant diseases.

Pedogenesis

Throughout the process of weathering, the rocks are worn away until they are reduced to very little pieces. However, considering that this is not real soil, it will not support the growth of any plants. The material that has been exposed to the environment goes through a sequence of transformations that are collectively referred to as pedogenesis, which is another name for the process of soil formation. The process of pedogenesis is essentially a biological one, while the weathering of rocks is more of a physical and chemical occurrence. Living creatures such as lichens, bacteria, fungi, algae, micro-arthropods, mulluscs, and others affect geochemical, biochemical, and biophysical processes as a result of the release of organic acids, enzymes, creation of CO2, and the addition of organic matter after their death. As a consequence of all of these processes, weathered rock debris crusts are changed into genuine soils that have a complex mineral matrix, a variety of organic compounds, and a rich microbial population. These soils may support a wide spectrum of plant and animal life.

As a result, during the process of pedogenesis, a wide variety of organic things, including dead organic materials, living organisms, and other organic compounds, are added to the mineral mass. As a result of mineralization of dead organic matter, minerals are gradually added to different layers of developing soil. This process is called "soil stratification." When it has reached its full maturity, this soil will consist of a number of layers or horizons of soil; this structure is referred to as the soil profile (see Fig. 3.3).

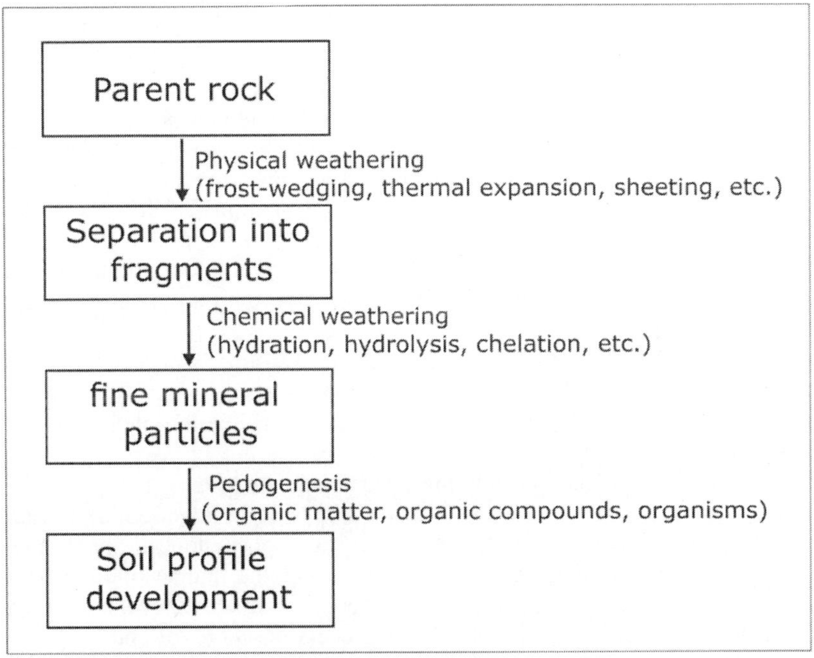

Fig. 3.3: Process of development of soil profile

Factors affecting Soil formation

The nature of the parent material and the ways in which it interacts with factors such as climate, terrain, organisms, and other environmental factors have a significant role in determining the character and type of soil that forms in a given location. The development of soil is determined by a number of distinct elements, which may be broken down into the following five categories:

a) Parent material: The term "parent material" refers to the oldest solid material that contributes to the formation of the soil. The term "parent material" may refer to either consolidated substances like rocks or unconsolidated particles like water deposits, volcanic ash, or biological components. Rocks are one example of consolidated substances. The content of the parent material has an effect on the formation of the soil since it determines the composition of the soil itself; for instance, rocks that contain iron generate soil that is rich in iron and has a higher pH and a darker colour. Wind, water, and volcanic activity are common sources of accumulating parent materials, which may lead to variations in the fundamental makeup of the rock. The effect of the parent material may also be observed in the surrounding soils, which often have different soil profiles since the parent materials came from different places. Alterations in the parent material may occur either abruptly or gradually throughout the course of the process of soil formation.

b) Topography: The process of soil formation is influenced by topography, which includes characteristics such as the geological structure of the elevation above sea level, configuration, and slope. During the process of pedogenesis, the placement of parent material or soil has an influence on the hydrologic cycle, transpiration, and other processes of a similar kind. It has been observed that the soil profiles on the tops of the concave slopes are often deeper, with a greater number of distinguishable sublayers. On the other hand, the soil profiles on the convex slopes are generally shallower. On the other hand, because of the runoff, the organic matter in the soil at lower slopes has a larger concentration than the organic matter in the soil at higher slopes. Changes in topography might take place throughout the course of time as a result of processes such as soil erosion and earthquakes, which would then have an effect on the process of soil formation.

c) Climate: The climate is a significant factor in determining pedogenesis, which is the process by which the soil evolves over long periods of time. Examples of these kinds of characteristics include rainfall, temperature, and the patterns of storms. The climate has a direct influence on the development of soil via the interaction of water and solar radiation. The amount of water present in the soil is affected by the amount of sunlight that hits it. Water plays an important role in the life cycles of a wide variety of soil organisms. Desert soils, which are prevalent in the equatorial region owing to high levels of sun and water energy, provide a vivid example of how the impact of climate on soil formation may be detected and observed clearly. On the other hand, the atmosphere in temperate regions is humid, which produces tropical soil that has an adequate amount of moisture.

d) Time: Despite the fact that topography and parent materials are site-related, climate and organisms are flux-related, which means that they get input from the environment around them. The passage of time is a factor that is unaffected by either the location or the flux, yet it may nonetheless result in surprisingly significant shifts. Time is considered to be an independent variable since it is an abstract variable, which indicates that the growth of soil may change even in the absence of any other external inputs. The effects that time has had on the soil profile may be discernible by the composition of the soil. As a result of downward translocation, clay and lime tend to collect in the sublayers of the soil. It's possible that the amount of humus present in different soil strata will change as time passes.

e) Organism: The soil formation process is significantly influenced by a variety of organisms, including human population, animal habitat, and plant life. In terms of the effects that vegetation has, research has revealed that the soils found under trees are substantially more acidic and contain a great deal less humus than those that are found beneath grass. These disparities are the result of differences in the types of litter that are produced by the two different plant species.

Soil Profile

If you look closely enough in a cut by the side of the road or in a soil pit, you may figure out different layers of the soil. Horizons are the name given to these layers of soil. The profile of a soil describes the order in which the different strata of the soil are found. Soil scientists, also known as pedologists, are responsible for doing research on and characterising soil profiles and soil layers in order to classify and interpret the soil for a variety of different uses. Horizons of soil may be distinguished from one another by a wide range of easily apparent soil characteristics, including colour, texture, structure, and thickness. Other features are less visible. Specific laboratory testing is required for qualities such the chemical and mineral composition, cohesiveness, and reactivity of the material. All of these characteristics are used when describing different types of soil horizons.

In the diagram shown in Fig. 3.4, the master horizons are denoted by the capital letters O, A, B, C, and R. The differences between these horizons are shown by the use of lowercase letters. The surface horizon (A), the subsoil (B), and the substratum horizon (C) are the three major horizons that may be found in most soils. An organic horizon, denoted by the letter O, may be present on the surface of some soils; however, this horizon may also be buried. The master horizon, denoted by the letter E, is used to subsurface horizons that have seen a significant amount of mineral eluviation (eluviation). The letter R is used to denote hard bedrock, which is distinct from soil.

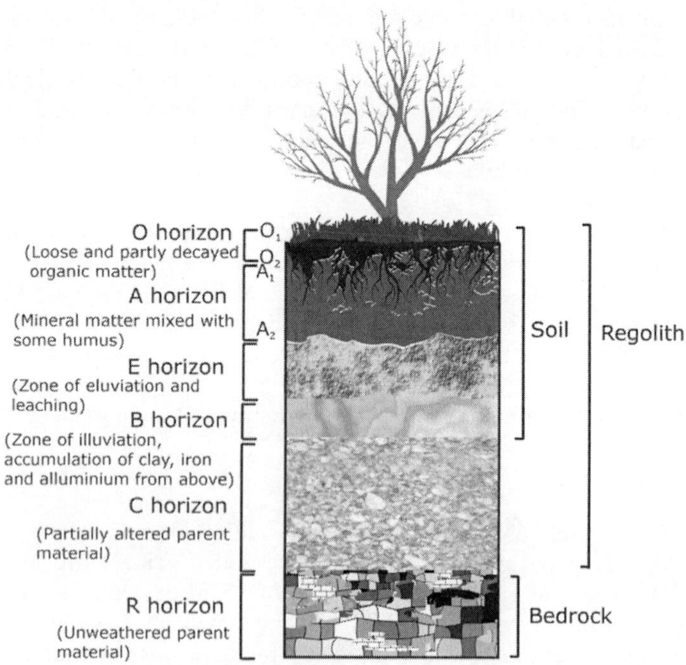

Fig. 3.4: Hypothetical diagram of soil profile

The horizons are explained below:

1. O (humus or organic): This layer is composed mostly of organic matter, such as leaves in the process of decomposition. Some soils have an O horizon that is sparse, while others have an O horizon that is thick, and yet others have no O horizon at all. It is further broken down into the O1 and O2 regions.
2. A (topsoil): This lay consists mostly of minerals derived from the source material and also contains organic material. A substance that ensures the continued existence of plants and other organisms. The categories that are further subdivided are A1 and A2.
3. E (eluviated): Soil that has been leached of clay, minerals, and organic matter, leaving a predominance of sand and silt particles of quartz or other resistant minerals. This is often present in older soils, as well as in forest soils, and is not found in certain soils.
4. B (subsoil): The B layer, which is the subsoil, is abundant in minerals that have leached down from the A or E levels and gathered here. In certain circles, it is also referred to as the zone of illuviation.
5. C (parent material): The deposit at the Earth's surface from which the soil developed.
6. R (bedrock): A mass of rock that comprises the parent material for different soils, such as granite, basalt, quartzite, limestone, or sandstone if the bedrock is close enough to the surface to be weathered. This material is not soil and may be found below the C horizon.

Characteristics of Soil to the climate type

The following are some of the processes that are closely related with the formation of soil as a result of the influence of climate and other factors. These are characteristics that are associated with a certain kind of climate. The following are some of those that stand out as being very prominent:

1. Lateralization: In tropical and subtropical soils, lateralization is a pedogenic phenomenon. High temperatures and copious amounts of precipitation accelerate the weathering of rocks and minerals. Eluviation and leaching are caused by large amounts of water permeating the soil. Nearly all of the results of weathering, which are nutritional ions or very simple small molecules, are translocated out of the soil profile by leaching if they are not used by plants for nourishment. The only compounds exempt from this procedure are

those made of iron and aluminium. The unique red colour of tropical soils is caused by iron oxides. These soils have an acidic pH as a consequence of significant leaching due to the net loss of base cations.

2. Melanization: Melanization is a common occurrence in low-humidity areas where humus formed from the mixing of water and organic matter in the soil's A horizon. The A horizon develops a dark tint as a result.
3. Podzolization: This phenomenon is linked to coniferous vegetation and damp, cold mid-latitude temperatures. Pine litter decomposition and heavy summer rains produce an extremely acidic soil solution. The removal of soluble base cations, aluminium, and iron compounds from the A horizon is caused by the eluviation and leaching processes being sped up by this acidic soil solution. In the A horizon, this process forms a silica sand sub-layer that ranges in colour from white to grey.
4. Gleization: A pedogenic condition called gleization results from poor drainage. The accumulation of organic compounds in the deeper layers of the soil is a part of this process. Lower strata's mineral layers have a blue colour due to the chemical reduction of iron. A common pedogenic regime of soil that occurs in a moist or chilly and cold climate is gleization.
5. Calcification: When evapotranspiration exceeds rainfall, dissolved alkaline salts in groundwater increase, leading to calcification. Likewise, the salts fall as a result of rainwater moving by. The translocated cations are therefore deposited in the B horizon. These deposits may solidify under certain conditions to form the caliche layer. The substance used in this method most often is calcium carbonate. The grasses of the grassland are prone to a build-up of calcium.
6. Salinization: This process functions in a manner similar to calcification. The formation of salt deposits on or near the soil surface sets it apart from calcification. Even in areas that are much drier, salinization does take place.

Soil classification

There are various ways of soil categorization. On the basis of the mechanism of formation, soils are categorized as:

Residual soils: The term "residual soil" refers to any soil in which the whole process of soil formation, including pedogenesis and weathering, takes place in the same site. Therefore, the development of soil takes place in these soils in the same location as the materials that served as their parent. For instance, bentonite is a kind of chemically weathered volcanic ash that may be found on the parent rock that it is formed from.

Transported soils: Soils are said to be transported after the process of soil formation via pedogenesis has reached its conclusion and the weathered components have been removed from the area by transporting forces such as rivers, glaciers, and other natural features. The following types of soils may be transmitted depending on the characteristics of the transporting agents:

a) Colluvial Soils: Colluvial soils are made up of locally transported detritus components of soil horizons and parent materials of sloping terrains. These materials are brought down the slopes from the top regions of the slopes by water erosion or landslides.
b) Alluvial: The river plains are covered with a significant amount of alluvial soil, which is an extremely important kind of soil. They come into existence as a result of the deposition of materials or sediments (also known as alluvium), which are transported in downriver by rivers and may include silt, sand, clay, and other materials. The alluvial soil that is found in river deltas is referred to as deltaic alluvium because of its location.
c) Glacial: The soil is made up of boulder clays, moraines, and other features that were most likely formed during the Pleistocene epoch as a result of the action of ice.
d) Eolian soil is a kind of soil that may range from sand dunes to loess deposits. The particles in eolian soil are mostly of the silt size, and it is formed as a result of wind activity.

It is not feasible, given by the competitive curriculum of this book, to describe the features of the geography of different groups of soils and the typical sort of flora that each of these groups can maintain. The United States Department of Agriculture (USDA) has provided the most recent classification, which identifies 12 orders. This classification has been made available to the public. The following is a list of the distinguishing qualities of each order's horizons, on the basis of which the order is recognised:

Soil order	Characteristics	Global area
Entisols	Very poorly developed soils	16.2 percent
Vertisols	Cracking clay soils	2.4 percent
Inceptisols	Less developed, dark forest soils	9.8 percent
Aridisols	Soils of desert region	12.0 percent
Mollisols	Mainly grassland soils with black, organic matter enriched surface horizons	6.9 percent
Spodosols	Acidic forest soils with B horizons high in iron, aluminum and humus.	2.6 percent
Alfisols	Weakly acidic primarily forest soils with clay-enriched B horizons	9.7 percent
Ultisols	Acidic, degraded forest soils of mid-latitudes	8.5 percent
Oxisols	Highly weathered soils of the tropics	7.5 percent
Histosols	Peatland soil	1.2 percent
Gelisols	Permafrost region	8.6 percent
Andisols	Soils formed by volcanic deposits	0.7 percent

Soil Erosion

Earthen materials (soil, rocks, and sediments) are eroded away and moved over time by natural forces like water or wind; sometimes, this process is sped up by poor land management or other human impacts. Erosion is a geological process. The topsoil, a fertile material vital to life, is the top layer of the ground that is lost as a result of soil erosion. Numerous factors, such as the kind of plant present, the amount of wind and rain, and the nature of the soil all affect how quickly soil erosion occurs. We have the ability (and the duty) to address one of the planet's most pressing environmental issues since our own activities may also affect how quickly soil erosion occurs. When the ground is left exposed to strong winds, savage rains, and flowing water, soil erosion often occurs. In certain cases, human activities like farming and clearing land for development leave soil vulnerable to erosion. For instance, farmers may expose the soil to the environment for weeks or months when they till (plough) the soil before or after farming a season of crops. Large areas of land may become devoid of the ground-covering plants that would normally maintain the soil in place due to the overgrazing of agricultural animals like cattle and sheep. Deforestation, particularly clearcutting, a common practise of the industrial logging industry, is another activity that has severe effects on soil health. When a tree is removed, the earth is left exposed to the elements without the safety of the tree's roots to keep the soil from being blown away. Erosion is largely influenced by the climate. Changes in rainfall and water levels may shift soil, large temperature changes may increase the susceptibility of topsoil to erosion, and protracted droughts may inhibit plant growth, exposing soil to increased exposure. Arable land (area that may be used to raise crops) and soil ecosystems are both impacted by soil erosion (land that can be used to grow crops).

Kinds of Soil Erosion

Various forms of soil erosion are identified. However, depending on the pace at which soils loss takes place, there are two basic categories of soil erosion:

1. Normal or geologic erosion: It takes place in typical, natural circumstances all by itself, undisturbed by any actions taken by humans. Because it is a natural process, it is an extremely slow one that may take hundreds of years.
2. Accelerated soil erosion: This particular kind of erosion is particularly quick, and it never manages to keep up with the development of soil. This erosion is often caused by the intrusion of agency such as man and other animals, and it is responsible for causing the most significant dangers and losses.

Agents of Soil Erosion

Rocks are broken apart by the process known as weathering, which paves the way for the rocks to be carried away through the process known as erosion. The following is an explanation of how water, wind, ice, and waves all play a role in the process of erosion, which causes the surface of the Earth to wear away over time:

1. Water erosion: The most significant agent of erosion is water, which does its work most often in the form of moving water in streams. Erosion may be caused by water in any of its forms. Raindrops induce splash erosion, which in turn moves little soil particles from one location to another (particularly in dry areas). When water collects on the surface of the soil and then travels

towards smaller rivulets and streams, a process known as sheet erosion takes place. In streams, the most important component contributing to erosion is water. The higher the volume of material that rivers and streams are able to carry and transport, the faster they move. The rate of erosion at which this occurs is known as the critical erosion velocity. There are three different processes that lead to the erosion of the banks of streams: 1) the water itself moves the sediments, 2) the water corrodes the sediments by removing ions and causing them to dissolve, and 3) particles in the water impact bedrock and cause it to erode. There are three different ways that stream water may erode the stream bed: headward erosion erodes the channel upslope, down cutting erodes the stream bed deeper, and lateral erosion erodes the sediment that is located on the sides of the stream channel. Depending on the nature of the lost soil, it might be any of the following:

a) Sheet erosion: The removal of soil in uniformly thin layers caused by the impact of raindrops and the shallow flow of surface water (sheets). If the material has not been deposited nearby or the damage has not already reached a severe level, it may be difficult to detect sheet erosion. During the erosion process, the micro particles of soil, which are responsible for holding the bulk of the essential nutrients and organic elements, are washed away.
b) Rill erosion: Rill erosion is a kind of erosion that results in the formation of thin, well-defined channels that are often less extensive than gully erosion channels. Erosion caused by gullies may disappear with time, or in more severe cases, it may be covered up by tilling and become less noticeable.
c) Gully erosion: the process by which topsoil is washed away over unprepared ground by means of deep gullies or channels The term "gully erosion" may refer to either the process of soil being carried away by man-made drainage lines or the process of soil running through grooves formed by heavy rains. Both of these processes can lead to the loss of soil.

2. Wind Erosion: Aeolian erosion, also known as eolian erosion, is the kind of wind erosion that gets its name from Aeolus, the Greek deity of winds. Aeolian erosion almost often takes place in desert environments. Sand dunes are partially caused by the wind-driven erosion of desert sand that occurs due to the presence of eolian activity. The force of the wind may erode rock and sand to various

degrees. This might refer to any of the following categories, depending on the overall mechanism of the soil removal process:

a) Saltation: Saltation is a kind of wind erosion that includes the movement of very fine soil particles from one site to another as a result of the action of the wind. The majority of the soil particles in a certain area are shifted as a result of this saltation process.

b) Suspension: The process of minuscule particles smaller than 0.1 millimetres in diameter being pushed into the air by saltation and then being transported aloft by turbulence to produce dust storms is referred to as suspension. Sand grains, particles of clay, and bits of biological material are some of the things that make up the particles. On the other hand, not all of the dust that is kicked up from the ground is perpetually dispersed through the atmosphere. Larger dust particles, measuring between 0.05 and 0.1 millimetres in size, may be deposited within a few kilometres of the erosion site. While particles with a diameter of 0.001 mm have the potential to travel thousands of kilometres, those with a diameter of 0.01 mm are only capable of travelling hundreds of kilometres.

c) Surface creep: Surface creep happens when large particles with sizes of between 0.5 mm and 2 mm roll along the surface of an area that has been affected by wind erosion. As a consequence of this, they collide with other particles, which causes those particles to get dislodged. As a result of surface creep and wind erosion, these enormous particles only travel a few metres at most.

d) Slip erosion: This kind of natural disaster is called as "slip erosion," and it occurs when the ground from a higher place slides down and crashes onto a lower surface. It is a mass movement or slope movement that occurs when soil, rock, and other materials flow downslope mostly because of the force of gravity. The natural phenomenon known as a landslide is sometimes referred to as slip erosion.

e) Stream bank erosion: Erosion of stream banks occurs when floodwaters from rivers slap up against the sides of the streambed and tear through the soil. The power of the water is so great, particularly at bends, that it causes the bank to collapse along the side. This kind of erosion is also known as riparian erosion in other contexts.

f) Deforestation and overgrazing: It has been shown beyond a reasonable doubt that both of these practises contribute to the erosion of soil. Deforestation and overgrazing are the root causes of soil degradation, which includes the disappearance of organic matter and the depletion of nutrients, as well as acidification, salinization, and chemical pollution, and the deterioration of the soil's physical qualities, such as its susceptibility to water and wind erosion. Because it kills the plants that contribute to keeping the topsoil in place, overgrazing is a major contributor to soil erosion.

3. Wave Erosion: Erosion of coastlines is caused by waves generated by oceans and other large bodies of water. The energy of sea waves is incredible, for example, large storm waves may sometimes create a pressure that is 2,000 pounds per square foot. The eroding of the rock along the beach is caused by the combination of the sheer power of the waves and the chemical make-up of the water. The waves have a much easier time carrying out the process of sand erosion, and there is often a yearly cycle that occurs in which sand is removed from a beach during one season, only to be replaced by waves during another season.
4. Ice Erosion: The erosive power of moving ice is really a little bit higher than the power of water, but since water is so much more abundant, it is responsible for a far greater amount of erosion on the surface of the world. Glaciers are capable of performing erosive activities; they pluck and abrade the landscape. The process of plucking is carried out when water seeps into fissures under the glacier, freezes, and then breaks off pieces of rock that are then transported by the glacier. Under the glacier, abrasion wears away at the granite, which, in turn, causes the surface of the rock to become more polished and smooth.

Soil Conservation

A collection of strategies aimed at preserving the soil are referred to as "soil conservation." Soil loss and a decline in soil fertility may be attributed to a number of issues, including overuse of the land, erosion, salinization, and chemical contamination. Some of the less developed nations in the world employ methods of land clearing and subsistence farming known as slash-and-burn agriculture, which may lead to major deforestation, the loss of soil nutrients, severe erosion, and even desertification. Erosion of the soil depletes the topsoil, which is vital for the growth and flourishing of plants, by removing organic matter, nutrients, and bacteria from the soil. A good example of a method that protects the soil against erosion is something called "soil conservation." The use of pesticides and fertilisers on agricultural land ultimately pollutes the surrounding waterways and aquatic ecosystems.

One of the most essential objectives of soil conservation is to maintain the diversity of the ecocommunities that exist on it and, in their own particular ways, contribute to the fertility of the soil. They do this by introducing organic matter into the water, breaking down dead organisms in order to liberate nutrients, and introducing organic matter into the water as well. Because microorganisms transform organic resources to meet the needs of plants, it is essential to provide suitable conditions for the living organisms that are found in the soil in order for vegetation to flourish there.

Methods of soil conservation: The various methods may be broadly arranged into the following types:

1. Biological methods: By the use of plants and animals they are divided into:
   a) Agronomic practices: Planting vegetation in such a manner as to prevent excessive loss of soil results in the provision of natural protection. The following techniques qualify as such:
      i. Contour farming: Contour farming involves ploughing, planting, and weeding in a manner that follows the contour of the land, as opposed to going up and down hills. The term "contour line" refers to a line that goes over a hill slope while keeping the same height during the journey and does not run either uphill or downhill.
      ii. Mulching: Mulching is a process or practise that involves covering the soil or ground in order to increase the growth of plants, the development of plants, and the efficiency with which crops are produced. The term "soil coating" may also be referred to as "mulch."
      iii. Crop rotation: Crop rotation, often known as crop rotation, is the practise of farming a variety of crops in the same site over the course of several growing seasons. Crop rotation is a strategy that can help you manage your soil and fertility while also lowering erosion rates, enhancing the health of your soil, and increasing the amount of nutrients that are accessible to your crops.

iv. Strip cropping: Strip cropping is a method of farming that is used when there is no other way to prevent soil erosion and the slope of the land is either too steep or too long.

b) Dry farming: This strategy is excellent for croplands established in places with little or moderate rainfall, which pose a danger to traditional farming.
c) Agrostological methods: Cynodondactylon and other types of grasses are examples of plants that can withstand erosion. They are often sown in narrow strips between rows of crops. They perform the role of stabilisers when they are allowed to grow in gullies and sodding. The following is a list of some of the most important agronomic practises that may be used to minimise soil erosion:
   i. Grassland cultivation (ley farming).
   ii. Land retirement.
   iii. Afforestation and reforestation
   iv. Monitoring of overgrazing
2. Mechanical methods: usage as additives to biological processes and are comprised of the following:
   a) Basin listing: In this method of soil and water conservation, the formation of basins is accomplished with the use of a particular device known as a basin-lister. The basins may be found all the way up the slope. When the basin is listed, there is a greater possibility that precipitation may soak into the ground.
   b) Contour terracing: This involves building a canal down the slopes to collect and reroute runoff water. Contour terracing is another name for slope stabilisation. It's possible that this area is a bench terrace, but it might also be a channel terrace or a ridge terrace with a large base.
3. Other methods include:
   a) Gully control: Gully control structures, such as spillways and check dams, are constructed in order to prevent the heads and fingers of gullies from advancing further and to improve the moisture regime in the command regions. This ultimately leads to an increase in the amount of yield that can be harvested from agricultural land.
   b) Stream bank protection: This includes the monitoring the cutting and caving of river banks, building drainage, constructing concrete or stone pitches, and other similar activities.

c) Reforestation: In arid regions, trees are often planted to act as windbreaks and therefore slow the speed of the wind. Windbreaks are often planted at an angle of ninety degrees to the wind that blows most frequently across the area. They keep a close check on the progression of desert conditions and the growth of sand dunes, as well as the erosion of the fertile top soil.

CHAPTER 4

ORIGIN AND EVOLUTION OF BIOSPHERE

Introduction

During the 3.8 billion years that life has been on Earth, species have made big changes to the chemical makeup of the biosphere. At the same time, the way the biosphere is made up chemically has determined what kinds of life can live there. The chemical make-up of the biosphere has changed over time because the rates at which nutrients are changed have not always been in balance. For instance, when life first started, there was a lot more carbon dioxide in the air than there is now, and there was almost no free oxygen. After photosynthesis started, there was a big drop in the amount of carbon dioxide in the air and a rise in the amount of oxygen. Much of the carbon that used to be in the air as CO2 is now in fossil fuel reserves and limestone rock. The rise in the amount of oxygen in the air had a big effect on how life evolved. Multicellular creatures couldn't form until oxygen levels were about the same as they are now (about 21 percent by volume). In order to keep up with how fast they breathe, these organisms need a lot of oxygen.

Since Darwin first put forward his theory of evolution in 1859, it has been improved and strengthened. Still, Darwin's basic idea of natural selection is still the most important part of modern evolutionary theory. Darwin pointed out that natural selection needs two forces that work against each other. The first is a way to make the huge amount of variety he saw among members of the same species. The second is a process in which some people are able to pass on their genes to the next generation while others are not. In this way, nature picks traits that work better in the environment, and these traits are more likely to be passed on. Changes in the environment and the slow build-up of new features over long periods of time are what evolution is all about. Darwin didn't know what made each member of a species different from the others. But as the study of genetics has grown in the twentieth century, many of his ideas have been proven and explained. We now know that mutations and rearrangements of genes, which are strings of nucleotides that code for proteins, cause differences between people. This difference is so big that, besides identical twins, it's unlikely that any two people have ever had the same genome.

Since the 1950s, molecular biology has helped bring the study of evolution down to the level of molecules. This discovery has added a level of complexity to the theory of evolution that was mostly unexpected. Gene and protein sequencing have shown that there is much more variety at the molecular level than scientists had thought. Scientists have found that genes can change their order and move from one creature to another in ways that were not known before. Molecular biology has shown that only a small part of a complex organism's genome codes for proteins. This raises the question of what role noncoding parts of the genome play in evolution. It has also shown how important DNA regulatory regions are for evolution, since the proteins made by different species differ not so much in what they do but in how much and when they do it. Traditional studies of evolution have also been affected by molecular biology in a big way. It has made important contributions to the field of systematics, which is about putting organisms into groups and explaining how they are related. When DNA or protein sequences are looked at, they may show small differences between species that are hard to see from how they look or act. Molecular biology has also helped investigations in the field. It has been used to study the structure and activities of populations by keeping track of how genes move between groups that breed. The development of new species, which is still a big part of evolutionary biology, can now be looked at genetically. With the help of molecular technology, the evolution of microbes, which used to be a very vague topic in biology, is now much clearer. At the molecular level, it is not possible to explain a lot of what is known about evolution. For example, natural selection is not a molecular process. Instead, it is the result of how many different creatures interact in a complex environment. In general, evolutionary biology theories cannot be summed up in a single level of organisation. They will have to use many different tools.

These levels of structure in living things go all the way down to the most basic level, which is the whole planet and all the living things on it. Organisms don't just change to fit their surroundings; they also change their surroundings, which changes the course of evolution. Humans are not the first organisms to be able to change the whole planet and its ecosystem. It has been done before, but by much weaker creatures than us.

Ancestral Life forms: Fossils

It is essential to have an understanding of the early lifeforms that existed on the surface of the earth, also known as the fossilised remains of earlier life forms, in order to trace and comprehend the evolutionary process that we believe led to the development of living systems on Earth. The fossil evidence, which is made up of the remains or impressions of species that were preserved in the earth's crust, gives us information on the kinds of organisms and paleo-environment that existed at multiple angles in the geologic record. The age of the rock layers in which fossils are found can be used to estimate the age of the fossils themselves. This is typically accomplished through the use of a technique known as radio isotope dating. The decay rate of a radioactive element is adequately predictable that it can be used to estimate the age of a material. The rubidium-strontium method has been used to date rock samples extensively throughout the field of geology. There are many different types of rocks that make up the crust of the Earth, and each one formed in its own unique way. It should come as no surprise that the vast majority of fossils are unearthed in sedimentary rocks, given that metamorphic and igneous rocks are completely fossil-free. The history of the earth can be broken down into four primary eras, each of which is determined by the age of the rocks: the Precambrian, the Paleozoic, the Mesozoic, and the Cenozoic. The oldest rock ever found was found in Precambrian rocks and dates back approximately 3.2 billion years. These fossils may contain traces of a diverse assortment of minute organisms as well as organic compounds.

The Geological Time Scale

EON	ERA	PERIOD	EPOCH	MILLIONS OF YEARS AGO	MAJOR EVENTS
Phanerozoic	Cenozoic	Quaternary	Holocene	2.4 to recent	Rise of Man; ice age glaciation; and extinction of many species of large mammals and birds
			Pleistocene		
		Tertiary	Pliocene	65.5 to 2.4	Rise of mammals such as whales, horses and monkeys; and spread of grassy ecosystems
			Miocene		
			Oligocene		
			Eocene		
			Paleocene		
	Mesozoic	Cretaceous		145 to 65.5	Modern seed-bearing plants
		Jurassic		200 to 145	Dinosaurs diversify; and first birds
		Triassic		251 to 200	Cycads; first dinosaurs; and the breaking up of pangaea
	Paleozoic	Permian		299 to 251	First reptiles
		Carboniferous	Pennsylvanian	359 to 299	Trees; seed ferns; first insects; coal forming swamps; and sharks abundant
			Mississippian		

		Devonian	416 to 359	Fish diversity; and first amphibians
		Silurian	423 to 416	First vascular plants; and earliest land animals
		Ordovician	488 to 416	Early bony fish; and diversification of animal life
		Cambrian	542 to 488	Invertebrate animals; brachiopods; andtrilobites
Proterozoic		Also known as Precambrian	2500 to 542	Beginning of life in sea; bacteria; algae; jellyfish; abundant carbonate rocks; first iron deposits; and oxygen began to accumulate in the atmosphere
Archean			4000 to 2500	Oldest rocks; and the Earth's crust forms
Hadean			4543 to 4000	Earth forms

Geologists, evolutionary biologists, and other Earth scientists utilise the geologic time scale (GTS), a method of chronological dating that connects geological strata (stratigraphy) to time, to explain the timing and linkages of events that have taken place throughout Earth's history.

Evolution of Life forms through ages

The Hadean Eon, named after the Greek deity Hades, goes back 4.5–4.0 billion years. During this epoch, Earth had a molten surface, volcanism, and asteroid strikes. As the Earth cooled, minerals crystallised and settled, forming the crust, mantle, and core. The source of ocean water is unknown. Volcanic outgassing, comets, and meteorites explain Earth's water. The Archean Eon began the rock record 4.0–2.5 billion years ago. Plate tectonics need continents to function. The age of continental material complicates this chicken-and-egg question. Craton formation began in the Archean Eon. As plate-tectonic processes pushed lighter mantle material to the surface, early protocontinents collected felsic minerals. Archean continents gave birth to today's Proterozoic continents. Hypotheses for life's origin include the early atmosphere and ocean, hydrothermal vents, and comets or other things. Proterozoic means "early life" and is 2.5 billion to 541 million years old. During this epoch, the continents' cores developed and plate tectonics began. Chemosynthesis might have used Earth's heat to sustain life. Comets or other things may have brought life to Earth. The largest difference between the Archean and Proterozoic is atmospheric oxygenation. Free oxygen and atmospheric methane produced carbon dioxide. Greenhouse gases trap heat in the Earth's atmosphere. The expansion of biological variety through evolution helped scientists grasp Earth's ecosystems. Eukaryotes gave birth to fungi, plants, and animals. Ediacaran animals likely resembled jellyfish or worms. Phanerozoic species possessed claws, scales, shells, and bones that fossilised readily. Precambrian rocks are scarce because the creatures had soft body parts. The Carboniferous Period contained some of Earth's largest forests. Their fossils fuelled the industrial revolution. Pangea was formed by tectonic processes involving continental collisions and ocean-basin closures. Most animal phyla appeared during the Cambrian Period 541–485 million years ago. By the end of the Paleozoic Era, continental collision and Pangea created a towering mountain range on the east coast of North America. The Paleozoic extinction is the greatest in earth's history. In the Carboniferous, oxygen levels rose and reptiles appeared. The Mesozoic began 252 million years after the Permian Mass Extinction. Reptiles, birds, and blooming plants evolved when Pangea broke apart. Siberian

Traps may have released poisonous gases, aerosols, and greenhouse gases. As the Mid-Atlantic Ridge grew, the Atlantic Ocean opened. Many contemporary animal groupings appeared in the Triassic. This comprises frogs, turtles, ichthyosaurs, plesiosaurs, mammals, and archosaurs. End-Triassic mass extinction was fourth in geologic history. The end-Triassic extinction drove the emergence of mammals, pterosaurs, and dinosaurs. By the Late Jurassic, dinosaurs ruled. Cretaceous dinosaurs and fauna diversified, specialised, and dominated. Pangea's split influenced current geography and biodiversity. This extinction catastrophe was likely triggered by a massive bolide that impacted with earth.

Cenozoic animals, including human ancestors, became dominant. Birds prospered in dinosaurs' empty niches. Volcanism triggered by tectonic movements in the west turned the subduction zone into a transform boundary. The Alps, Zagros, and Himalaya helped close the Tethys Sea. Five types of early mammals are known from fossil teeth, the hardest vertebrate bone. Our present ice age began 2.58 million years ago, marking the commencement of the Quaternary epoch. During this time, cold-adapted megafauna (including gigantic sloths, saber-tooth cats, and woolly mammoths) arose and became extinct when the Earth warmed from the last glacial maximum. This widespread migration and the hominid evolutionary tree are complicated and need their own path. Many experts believe the emergence of human civilisation ended the Holocene period and began the Anthropocene. Extinctions, increasing tritium from nuclear testing, and rising carbon dioxide are evidence. Some claim agriculture began 12,000 years ago, based on geologic formations. Others correlate the industrial revolution to carbon dioxide emissions.

PART 2:

ECOLOGY

CHAPTER 5

ECOLOGY

The Definition and Historical Background of Ecology

The field of biology known as ecology examines the complex relationships that exist between living things and the settings in which they live. In essence, ecology is the study of how natural systems are connected. It is the fundamental principle that underpins the whole field of environmental science. Biologists have understood the fundamentals of ecology since long before the phrases "environment" and "development" came into common use. In the middle of the nineteenth century, biologists demonstrated the existence of dynamic interactions between creatures and the settings in which they lived. At the same time, they popularised a wide range of words to characterise this kind of connection. Isodore Geoffroy St. Hilaaire, a French naturalist, is credited with coining the term "ethology" in 1859. This term refers to the study of the interactions that exist between living things within the context of their families, societies, and communities. Ecology is a term that was formed by combining two Greek words: oikos, which means "home" or "dwelling place," and logos, which means "the study of." This was done in order to convey the connection that exists between creatures and the environment in which they live. It is generally accepted that German biologist Ernst Haeckel was the first to explain the importance of the term, despite the fact that there is considerable disagreement as to who really came up with the term.

Historical background

References to ecological concepts can be traced back to prehistoric man, who used environmental knowledge for food, shelter, medicines, and other necessities in order to survive nature's challenges. Although modern ecology emerged mostly around the year 1900, allusions to ecological concepts can be traced back even further. There is a possibility that ecological concepts are mentioned in Indian classics such as the Vedic, Epic, and Pauranic, and Chakra brought attention to the relevance of Vayu (gases and air), Jala (water), Desha (topography), and Time in the control of plant life. In their works on natural history, Greek philosophers and scientists such as Hippocrates (known as the "father of medicine"), Aristotle, Theophrastus, and Reaumur chronicled the behaviours of animals and plants that thrived in diverse locations.

In their volumes titled "Natural History," authors Linnaeus (1707-1778) and Buffon (1756) made substantial contributions to the field of ecology. The nineteenth century saw an increase in the number of sites that were researched for their flora and fauna due mostly to the travels of B.A. von Humboldt, Richard Spruce, Edward Forbes, and J. Hooker, all of whom visited many continental regions. When it comes to plant life, Le cog Sendtner and Kemer were the forefathers of the community ecology notion. Karl Mobius developed this perspective further with regard to animal populations. Many people, including S.A. Forbes, Warming, Cowles, and Clements, made important contributions to the study of the communities of plants and animals. Synecology was first brought to widespread attention by Schroeter and Kirchner in 1896.

During the first half of the twentieth century, specialised areas of ecology such as paleo-ecology, zoogeography, oceanography, and limnology, amongst others, started to develop as separate academic disciplines. After then, the concept of a population was developed, and methods from statistical study and sampling were used to find solutions to issues affecting communities. In common parlance, this period of time is referred to as the "period of population and community ecologists." The first published use of the word "ecosystem" by Tansley, which occurred in 1935, is generally regarded as the beginning of the era known as the "ecosystem approach to ecology." It is possible that this may result in the development of ideas in ecology such as productivity, energy exchanges, and other similar concepts. This line of thinking is frequently referred to as the bioenergetic approach. Many people have made major contributions to this field, including Odum, Whittaker, Kendeigh, Pearsall, Ovington, Rodin, Kira, and many others.

The present state of ecology may be seen from the perspective that it plays an important role in the resolution of socioeconomic and political issues. This entails making efforts toward pollution control and field management, in addition to conserving the soil.

Basic concepts of Ecology

The four basic concepts of ecology are:

1. Holism: The hierarchical levels at which we discuss interacting units of ecology are, individual < population < community < ecosystem < biome < biosphere. According to Williams Ophuls (1974), the true foundation of ecology is holism. It focuses on the interrelationship system concept. Jan Christian Smuts, a South African leader, created the term holism (from the Greek word holon, meaning entity) in his book Holism and Evolution in 1926 to explain the process of evolution by the coming together of smaller wholes in order to build the universe. In ecology, this process is envisioned as a guiding and creative principle that operates from beginning to end. Smuts described holism as "nature's drive to generate wholes larger than the sum of their parts through creative evolution." Epistemologically, holism contradicts certain scientists' reductionism. As one of the fundamental principles of ecology, holism or holistic philosophy describes the features of units (integrated wholes) at progressively higher levels of organisation of organisms. The three ecological levels of organisation, population, community, and ecosystem, differ in some ways. A population is a collection of individual organisms of the same species living in the same region, as well as the manner in which the members of the population interact with their environment. A community is a collection of diverse species' populations in a certain location. It may comprise all of the populations in that region –including plants, animals, and microbes – or it may be defined more narrowly as a specific group, such as the fern community or the seed-eating bird community. The number of species and the number of individuals of each species in a community are used to calculate diversity.
2. Ecosystem: A region's whole biotic community as well as its abiotic surroundings make up an ecosystem. Due to this, it includes all species as well as the physical and chemical characteristics of sediments, water, and gases. Ecosystem ecology emphasises the exchange of nutrients and energy between biotic and abiotic ecosystem components. Intricate food webs may be created by connecting any biotic component. Food webs are the core ecological components of ecosystems. Therefore, populations come first in ecology, followed by ecosystems. The fact that energy and nutrients travel across food webs makes them basic units. Along with interactions with living things, there are also interactions with the abiotic environment. Basic energy and nutrient flow patterns take place in a generalised ecosystem. In their interactions with the

abiotic environment and the ecosystem boundaries, energy and nutrient flow patterns behave quite differently. Ecosystems transport energy, gaining it from the outside in the form of light energy from the sun and losing it directly in the form of heat spread by the breathing of every member of the community. In ecosystems, the cycle of nutrients is far more extensive.

3. Succession: All living things and their environments are connected and affect each other in different ways. In an ecosystem, animal populations, flora, and vegetation all affect and are affected by each other. Succession is an ecological process that happens when a previous environment is replaced by one with more biomass and different types of food chains. It is often a long-term process that takes years for a time-series procession of increasingly resistant systems of producer, consumer, and decomposer species to lead to a stable system along with biotically changed terrain or habitat in a certain climate.
4. Conservation: All living systems need ecological conservation, which means preserving the continuity of the dynamic ecosystem. The resilience or homeostasis of the system is ensured through the trophic management of populations. In fluxes that affect structure and function, it encourages variation and stability. While dominance keeps the system stable, species diversity increases the system's production efficiency.

The Significance of Ecology for Humans

Ecology benefits the environment and is crucial to the prosperity and well-being of people. It advances our knowledge of the interdependence of environment and humanity, which is important for maintaining clean water and air as well as biodiversity in a changing climate. In order to understand a given relationship, ecology takes into account factors like population growth, competition, symbiotic ecological relationships (mutualism), trophic relations (the transfer of energy from one level of the food chain to another), biodiversity, migration, and physical environment interactions. Because ecology includes both the physical and chemical habitats of all living things on Earth. It makes it possible for us to understand the effects of our actions on the environment. With this information, we can lead conservation efforts by first identifying the main causes of the problems we face in our ecosystem, and then after this identification process, we can decide where our efforts would have the most effect. Ecology also shows how much environmental damage people really do and provides projections of how bad the damage may become. These indications create a sense of urgency among the public, motivating people to actively engage in conservation efforts and ensure the survival of the planet. Additionally, ecology helps us understand how each organism fits into the ecosystem's web of interconnectedness. With this information, we can identify the resources that different species depend on to survive. Analysing human needs, which have the most effects on the ecology, is crucial. For instance, the carbon footprint of the environment has increased as a consequence of human dependency on fossil fuels. Ecology makes it possible for humans to detect these problems, which makes it necessary for them to make informed decisions about how to modify their resource requirements in order to prevent overloading the ecosystem with unsustainable demands.

All of the data and research gained from ecology is ultimately used to enhance eco-friendliness. Because of their understanding of the environment, they increase people's awareness of it and inspire them to live environmentally friendly lifestyles. Given that survival and quality of life are reliant on environmental sustainability, it follows that people eventually choose to behave less selfishly and exert more effort to protect the interests of all living things. As a consequence, it encourages peace of mind and guarantees the survival of all species.

Major divisions of Ecology

The concept of ecology makes it perfectly transparent that studying plants, animals, and their surroundings is necessary for gaining a more comprehensive understanding of nature. Reviewing ecology's evolutionary development reveals that ecological studies are often centred on three main factors: taxonomic groupings, habitat, and levels of organisation. As a consequence, these methods could lead to the development of the following significant ecological divisions:

1. Based on Taxonomic Groups: In the early stages of ecology, botanists and zoologists conducted separate studies on the ecology of plants and animals. As a consequence, subfields including plant ecology and animal ecology were developed.
2. Based on Habitat: Ecologists have thought about researching ecosystems and how they affect living things. Numerous habitats, such as freshwater, marine, estuaries, grasslands, woodlands, arid places, etc., have been selected. The possible relationship between these and the kinds of creatures present is then carefully studied. A

new branch of ecology called habitat ecology was formed as a result of this habitat-based methodology.

3. Based on levels of Organisation: They are often referred to as individual organisms or groups of organisms when using such an approach. The two categories of ecology depending on organisational levels are autecology and synecology which are explain below:

Autecology: This is also called "individual ecology," and it is the study of how different organisms interact with their surroundings. So, at any given time, the focus is on the needs and responses of a certain species, as well as how the environment affects that species. In an autecological approach, each species is the unit of study. These are studied to find out things like where they live, how they look, where they fit in the taxonomy, and how long they live, as well as how different ecological conditions may affect different parts of their life cycles.

Synecology: Under natural conditions, however, species (plants, animals, microorganisms, and so on) live together as a natural group and affect each other's lives in many ways. So, things get more complicated when the units of study are groups of organisms, called communities, instead of single species: A synecological method is one in which groups of organisms are used as the units of study. Depending on the level of organisation, synecology can be about a population, a community, a biome, or an ecosystem.

a) Population ecology: In this instance, pure stands of individuals from a single species population serve as the study units.
b) Community ecology: In contrast to population ecology, the unit of study in this scenario is a group of individuals from several species, including both plants and animals. The nature of interdependencies between individuals of various species is typically explored when looking at the living (biotic) components of the community.
c) Biome ecology: In nature, we often find a complex of many communities, some of which are in various phases of succession and others of which are at their climax, and which all thrive in an area with more or less similar climatic conditions. As a result, interactions between various populations in a region are examined using biomes as study units.
d) Ecosystem ecology: The most recent ecological development is this. The interaction between living (biotic) and non-living (abiotic) elements of nature is widely established. An integrated system-ecosystem, ecological complex, or ecological system is formed as a

result of these biotic and abiotic elements interacting with one another. This is sometimes referred to as the "bioenergetic method" in contemporary ecology.

CHAPTER 6

STRUCTURE AND FUNCTIONS OF AN ECOSYSTEM

An ecosystem is the whole integration of living things that interact with one another and their surroundings. It often involves an open system with continuous but variable energy and material loss and intake. It consists of both biotic and abiotic elements interacting with one another, both of which are necessary for the continuation of life on Earth. It is a basic, functional unit with no boundaries (Fig. 6.1). The highest degree of energy-based ecological integration is therefore represented by an ecosystem, which is a functional entity capable of energy transformation, accumulation, and circulation. Its main ecological function is to emphasise necessary connections, interdependence, and causal relationships.

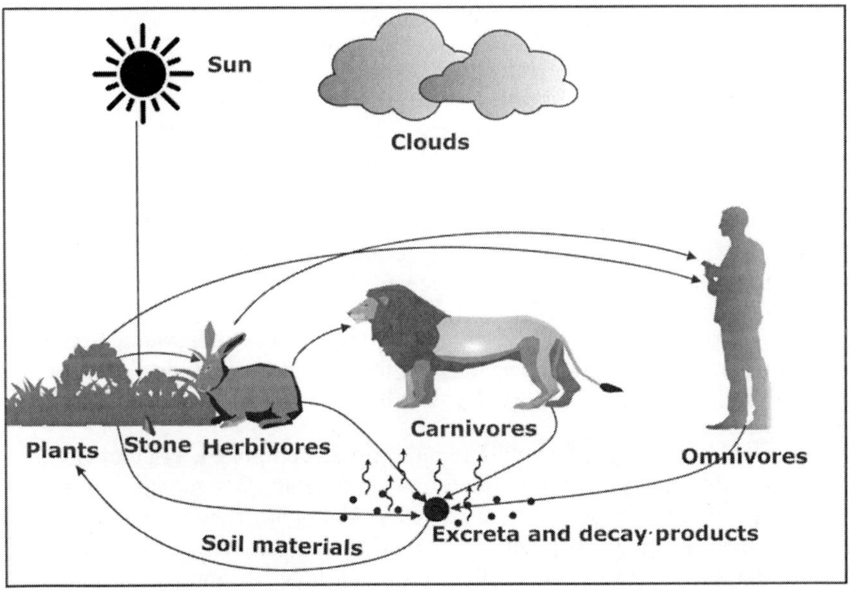

Fig. 6.1: Diagram showing a generalized interaction among different biotic components of an ecosystem

Ecosystem Types: Ecosystems can be broadly classified as a natural or artificial ecosystem. These are explained in further detail below:

Natural Ecosystem: These operate independently in their native environments with little or no human influence. These are further divided into aquatic (water-open), which may be further divided into freshwater or marine, and terrestrial (forest, grassland, desert, etc.) habitat types. Freshwater may be lentic (as in a lake) or marine, which encompasses both shallow and deepwater bodies, such as an estuary, a sea, and so on and lotic (like in a spring, stream, or river).

Artificial Ecosystem: The natural balance is continually upset by the input of energy and deliberate manipulations in these, which are artificially maintained by man. In croplands, such as maize, wheat, and rice fields, man tries to control both the biotic community and the physicochemical environment. These are instances of artificial ecosystems.

Ecosystem structure

The primary two components of an ecosystem are as follows.

Abiotic Component: Abiotic components include the amount of inorganic chemicals included in material cycles, such as P, S, C, N, and H. The amount of these inorganic substances present at any given time in an ecosystem is referred to as the standing state or standing quality, which is the amount and dispersion of inorganic chemicals, like chlorophylls, as well as organic materials, such as proteins, carbohydrates, lipids, and so on, existing either in the biomass or in the surroundings, i.e. biochemical framework that helps connect the biotic and abiotic components of the ecosystem, and the climate of the particular region.

Biotic Component: Every ecosystem has a trophic structure in which living creatures are distinguished depending on their nutritional connections. An ecosystem is made up of two trophic (nutritional) components:

a) Autotrophic component: The autotrophic component is dominated by solar energy fixation, the use of simple inorganic molecules, and the creation of complex compounds. The bulk of the component is made up of green plants and photosynthetic bacteria. To a lesser degree, chemosynthetic bacteria contribute to the build-up of organic matter. Producers are autotrophic component members.
b) Heterotrophic component: The usage, rearrangement, and breakdown of complex components dominate the heterotrophic component.

Consumers are the animals engaged since they consume the stuff generated by producers (autotrophs).

Additional categories for consumers include the following:

The term "macroconsumers" refers to the consumers, which may be categorised as herbivores, carnivores, or omnivores based on the order in which they occur in the food chain. Macroconsumers are at the top of the food chain (or omnivores). Herbivores, often known as primary consumers, are animals that primarily devour plants. Carnivorous or omnivorous diets are typical of secondary and tertiary consumers, if they ever exist. All of these shots come from animals that feed on other organic and organic particle components.

Microconsumers are also often referred to as decomposers in the scientific community. They are known as saprotrophs (osmotrophs) and are composed of fungus, bacteria, and actinomycetes for the most part. They decompose complexes of dead or living protoplasm, absorb some of the by-products of decomposition and breakdown, and then release inorganic nutrients into the environment, making them once again accessible to autotrophs.

Ecological pyramids

A graphical representation of the interactions that exist between different living creatures at different trophic levels is referred to as an ecological pyramid. The term was first given by G. Evylen Hutchinson and Raymond Lindeman. These pyramids have the shape of true pyramids; the base is the widest part of the structure, and it is covered by the trophic level that is lowest in the food web, which is producers. The main consumers represent the next trophic level, which is occupied by the secondary consumers, and so on. Because a sample space consisting of just a few numbers or a few species would result in a significant degree of error, every calculation that goes into the formation of these kinds of ecological pyramids has to take into account every creature that exists inside a certain trophic level. There are three distinct types of ecological pyramids, these are as follows: (a) pyramids of numbers, which depict the number of individual organisms present at each level; (b) pyramids of biomass, which depict the total dry weight as well as other suitable measures of total living matter; and (c) pyramids of energy, which depict the rate at which energy is transferred.

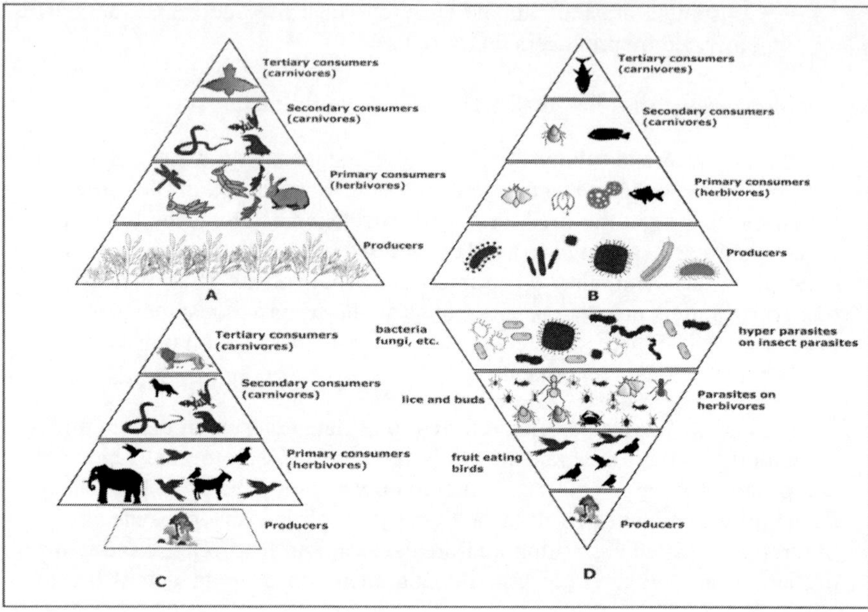

Fig. 6.2: Diagram showing pyramids of numbers in different kinds of ecosystem and food chains

a) Pyramids of Numbers: They illustrate the statistical relationship that exists between producers, herbivores, and carnivores at increasing trophic levels. The number pyramids may be seen in three different kinds of ecosystems, as shown in Fig. 6.2. (A-C). Always present in the greatest number in a grassland (Fig. 6.2 A) are the producers, which are comprised almost entirely of grasses. This number then falls towards the apex as a result of the fact that primary consumers (herbivores) such as rabbits, mice, and so on are less in number than grasses, and secondary consumers such as snakes and lizards are fewer in number than rabbits and mice. Last but not least, the top consumers, known as tertiary consumers, such as hawks and other birds, are even fewer in number. Because of this, the pyramid is now standing upright. Likewise, the pyramid is standing upright in the environment of a pond (Fig. 6.2 B). In fact, the numerical pyramids are not an accurate depiction of the food chain since they are not especially practical. They do not demonstrate the proportionate effect that the 'geometric,' 'food chain,' and'size' features of the creatures have on the ecosystem.

Nevertheless, the form of the number pyramid is one that can only be found in a forest ecosystem (see Fig. 6.2 C). Large trees, also known as producers, are fewer in number yet form the foundation of the pyramid because of their size. However, in a parasitic food chain (Fig. 6.2 D), the pyramids are always turned upside down or inverted. This is because a single plant is responsible for the growth of a large number of herbivores, and each herbivore, in turn, may provide nutrition to a number of parasites, which in turn may support a large number of hyperparasites.

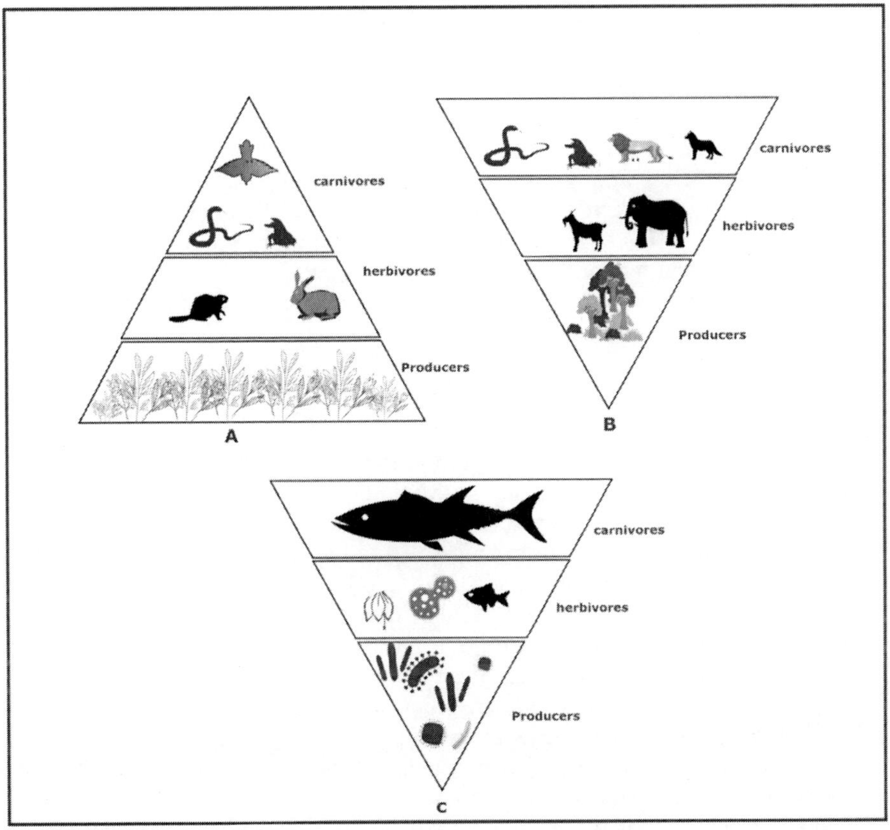

Fig. 6.3: Diagram showing pyramids of biomass in different kinds of ecosystem.

b) Pyramids of Biomass: They illustrate quantitative connections of standing rather than geometric elements, which makes them more fundamental when compared to the other models. The pyramids of biomass seen in a wide variety of habitats are shown in Fig. 9.4. (A-C). There is a general decrease in the biomass of organisms at successive tiers from producers to top predators in grassland and forest (Fig. 6.3. A, B). This drop may be seen in both grassland and forest. The end consequence is that the pyramids are standing upright. However, since the producers in a pond (shown in Fig. 6.3 C) are little organisms, their biomass is the lowest. This number gradually rises as one moves towards the top of the pyramid, which causes the pyramid to be inverted.

c) Pyramid of Energy: Out of the three different types of ecological pyramids, the energy pyramids provide the most precise illustration of the overall characteristics of an ecosystem. In this scenario, the number of organisms on each level, as well as their weight, are governed not by the amount of fixed energy that is present in the level below at any given moment, but rather by the pace at which food is produced. Due to the one-way nature of energy transmission within a food chain, the energy pyramid is the only kind of ecological pyramid that can never be found in an inverted position. When the number of trophic levels increases, there is also a loss of energy in the surrounding environment.

Functions of the Ecosystem

Studying functions is another important step in acquiring a more holistic understanding of the natural world. Minerals (such as carbon, hydrogen, oxygen, nitrogen, phosphorus, lithium, calcium, magnesium, zinc, and iron) that green plants get from their soil and the air around them are used by the plants to fix radiant energy and to produce complex organic matter. Minerals continue to travel in a circular fashion whereas energy moves in a non-cyclical (unidirectional) route from the sun to decomposers through producers and macro consumers (herbivores and carnivores). Energy goes from the sun to decomposers. In order to accomplish mineral cycling, many distinct biogeochemical cycles are superimposed on the unidirectional flow of energy that occurs via the biotic component of the ecosystem. Later on, it would become clear that energy does not merely flow in a unidirectional direction but rather that it is also removed from the system in a number of different ways, and that minerals also demonstrate a net loss in a variety of different ways.

Productivity of Ecosystem

An ecosystem's productivity may be defined as the rate of production or the quantity of organic matter that has accumulated in a certain period of time. The many types of productivity include the following:

Primary Productivity

It has a connection to autotrophic producers, the most majority of which are photosynthetic, and chemosynthetic bacteria, although to a much smaller extent. This group includes green plants, upper and lower macrophytes, phytoplanktons, and certain photosynthetic microbes. Primary productivity may be defined as the rate at which radiant energy is stored by the actions of producers such as photosynthetic and chemosynthetic processes. Primary productivity may be broken down into the following categories:

Gross primary productivity (GPP): It refers to the rate of photosynthesis as a whole over the duration of the measurement, taking into account the organic matter that is consumed through respiration. The same concept may also be referred to using the phrases total (gross) photosynthesis and total assimilation. The quantity of chlorophyll that is present is the factor that decides this.

Net primary productivity (NPP): During the measurement period, the rate of organic matter accumulation in plant tissues was higher than the rate of plant respiration. This is the rate at which biomass increases and is sometimes referred to as apparent photosynthesis or net assimilation. It is the balance between overall respiration and photosynthesis, as well as other plant losses like growth and reproduction. NPP = GPP - R, where R represents the amount of energy used during respiration.

Secondary productivity: It refers to consumers, who are also called heterotrophs in certain circles. These are the rates of energy storage that are applicable at the consumer level. Gross and net quantities are not broken out separately for secondary productivity. The rate at which different types of consumers generate new organic matter is referred to as secondary productivity.

Net productivity: This is the rate at which organic matter is stored that is not consumed by heterotrophs. This rate is equal to the difference between the amount of organic matter that is consumed by heterotrophs and the amount of organic matter that is produced by primary producers during a unit period, such as a year or season. Therefore, the rate of growth in primary producers' biomass is what consumers have neglected to account for in their calculations. The production of C g/m2/day is a common measure of net productivity.

Ecosystem's Food chain

A food chain is the transfer of food energy from producers to consumers via a variety of species (herbivores to carnivores to decomposers) with recurring eating and being eaten by one another. This takes place over time. During the process of photosynthesis, producers make use of the radiant energy from the sun, which is then changed into ATP, a chemical form. Because they reside at the first trophic (nutritional) level, also known as the producer level, green plants are considered to be the main producers in any given food chain. The energy that is stored in the food item that is created by green plants is afterwards consumed by herbivores, also known as primary consumers, who make up the second trophic level, which is the level of the major consumers, and are referred to as the main consumers level (herbivores). Herbivores are referred to as secondary consumers because carnivores, who are members of the third trophic level and the secondary consumers level, devour them for food. After that, other carnivores at the tertiary consumer level, also known as the tertiary consumers, might end up eating them (carnivores). A fundamental chain of eating and being eaten is what the food chain represents. In a grassland ecosystem, the food chain begins with the grasses and shrubs and progresses through grasshoppers, frogs, snakes, and hawks in a logical order based on food habits. On the other hand, in a pond ecosystem, the food chain begins with phytoplanktons and progresses through water fleas, smaller fish, larger fish, birds, and larger animals, and so on. The two kinds of food chains that often appear in natural settings are as follows:

a) Grazing Food chain: This particular kind of food chain starts with living, green plants, then goes on to grazing herbivores (who are predators that feed on living plant components), and finally ends with carnivores (animal eaters). Ecosystems that include this kind of food chain are totally dependent on the energy that comes from the sun. Because of this, the completion of this kind of cycle is dependent on the acquisition of energy through autotrophy and the subsequent transmission of that energy to herbivores. The majority of the world's natural ecosystems have a food chain that looks like this. The transfer of energy across these networks is of the utmost importance. The phytoplanktons, zooplanktons, and fish sequence and the grasses, rabbits, and foxes sequence are also examples of grazing food chains.

b) Detritus Food chain: This sort of food chain begins with dead organic matter, then passes on to bacteria, detritivores, and predators of detritivores. As a consequence, such ecosystems are less dependent on direct solar energy. These are largely dependant on the influx of organic materials created in another system. In a temperate woods, for example, such a food chain may be observed digesting accumulated garbage.

Food web

On the other hand, food chains in nature never function as separate sequences; rather, they are coupled to create a food web, which is a sort of pattern that consists of interlocking patterns. The organisation of food chains in a linear fashion is uncommon under natural settings; yet, these chains are connected in some way by the presence of a wide variety of species that reside at various trophic levels. Within the context of a grassland's grazing food chain, mice have the potential to devour grass in the event that rabbits are not there. It's possible that the hawk would then consume the mouse on its own, but it's also possible that the snake may consume the mouse first and then the hawk. As a consequence of this, there have been discoveries made in nature of alternatives that, when combined, constitute a system of interrelated components known as the food web (Fig6.4).

Biogeochemical cycles in Ecosystems

The circulation of energy and nutrients throughout an ecosystem is essential to the continued existence of life on Earth. Both of these factors have an effect on the total number of species that are found there as well as the metabolic rate of those species and the complexity of the ecosystem. The transfer of energy throughout ecosystems makes it possible for species to carry out a wide variety of tasks; nevertheless, this energy is ultimately wasted in the form of heat, which lowers the overall value of the system. On the other hand, the nutrients included in food stuff can never be exhausted. They may be used again and over and over again without end. For instance, every time we take a breath, we could take in several million atoms of elements that our ancestors or other animals might have taken in. As elements and compounds, carbon, hydrogen, oxygen, nitrogen, and phosphorus make up 97 percent of the mass of human bodies and more than 95 percent of the mass of all living organisms. In addition to these, life and good health need around 15 to 25 more elements in some form from both plants and animals. These elements or mineral nutrients are continuously cycling through an ecosystem, moving from non-living to living components, and then back to non-living components in a pattern that is more or less cyclical. This pattern is referred to as a biogeochemical cycle (the word "bio" stands for life, while "geo" refers to the atmosphere). The process through which an element or compound, such as water, travels between its many living and non-living forms and locations within the biosphere is referred to as a biogeochemical cycle.

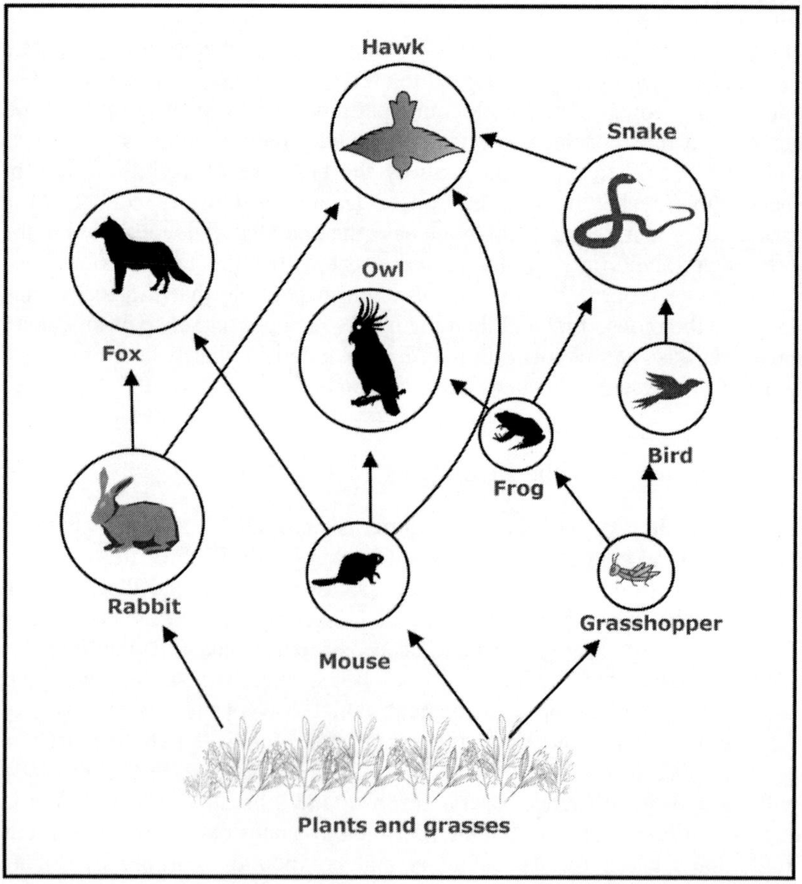

Fig. 6.4: Diagrammatic sketch showing a food web in a grassland ecosystem.

Depending on the make-up of the reservoir, there are two different types of cycles: gaseous and sedimentary cycles. The reservoir for the Gaseous Cycle is either the atmosphere or the hydrosphere, while the reservoir for the Sedimentary Cycle is the crust of the planet.

Atmospheric cycles

Water cycle

Due to the fact that it follows the route of a compound, water, this cannot be considered an elemental cycle in the strictest sense. Despite this, the circulation of water both within and across ecosystems is essential to the comprehension of the nutrient cycle for a number of reasons:

a) It is a source of hydrogen that plants can use in the process of photosynthesis.
b) A significant amount of water is required for both the upkeep of plants' hydrostatic skeletons and the transport of chemicals throughout their bodies.
c) Plants absorb elements in aqueous phase from soil.

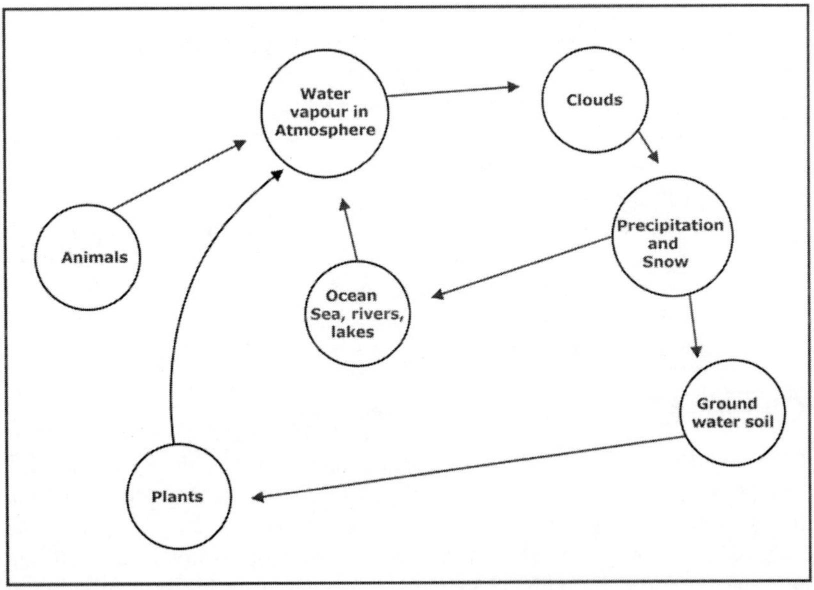

Fig. 6.5: Global water cycle

The components of the global hydrological cycle as well as the flows between them are depicted in Fig. 6.5. The majority of the storage is contained in the oceans, and the processes that cause the most significant fluxes are evaporation from the oceans and precipitation that falls onto the oceans. However, there is a net flow of water vapour from the oceans to land, where it eventually condenses into precipitation in the form of rain, hail, or snow. The stability of the cycle is ensured by the flow of water from the land, either in the form of surface runoff or as the movement of groundwater into rivers and then back into the ocean. In spite of the fact that volcanic eruptions cause only a negligible amount of water to be released into the atmosphere, the crust of the earth contains a significant amount of water.

Impact of Human activities

The water cycle is altered in the following ways:

a) We get a significant amount of our potable water from underground sources, as well as lakes, rivers, and streams (wells).
b) We strip land of its vegetation in order to make way for agriculture, mining, the construction of roads and buildings, and other uses. On occasion, we cover the ground with buildings, concrete, or asphalt. This leads to an increase in runoff and a decrease in infiltration, all of which serve to raise the risk of flooding, hasten the process of soil erosion and landslides, and recharge groundwater sources. Wetlands, which act like sponges by soaking up and holding excess water, are being destroyed at an alarming rate, which also contributes to flooding.
c) We pollute the water by adding various contaminants and nutrients, such as phosphates, nitrates, and other substances that come from fertilisers. This excess of nutrients and toxins has the potential to change or disrupt the natural ecological balances that are responsible for the purification of water.

Carbon cycle

The biosphere's carbon cycle is primarily concerned with the gas carbon dioxide found in the atmosphere, its incorporation into organic matter via the process of photosynthesis, and its subsequent release through the metabolic processes of all living things (Fig. 6.6). Additionally, carbon may be found in the rock of the earth, most often in the form of calcium and magnesium carbonates. The majority of these compounds have an organic composition, and they are the mineralized residues of the skeletons of marine animals. Through the process of weathering, carbonate rocks on land, which formed as a consequence of later geological action, contribute to the nutrients of soil and the nutrition of plants. In compared to the exchanges that occur between biota and the atmosphere, these kinds of fluxes are rather insignificant on a global scale. Carbon-rich organic matter will begin to build in ecosystems when the level of net primary production is greater than the level of community respiration. These kinds of accumulations have led, from a geological point of view, to the deposition of fossil fuels like coal and oil. In modern peatlands, this still happens, although to a lesser level.

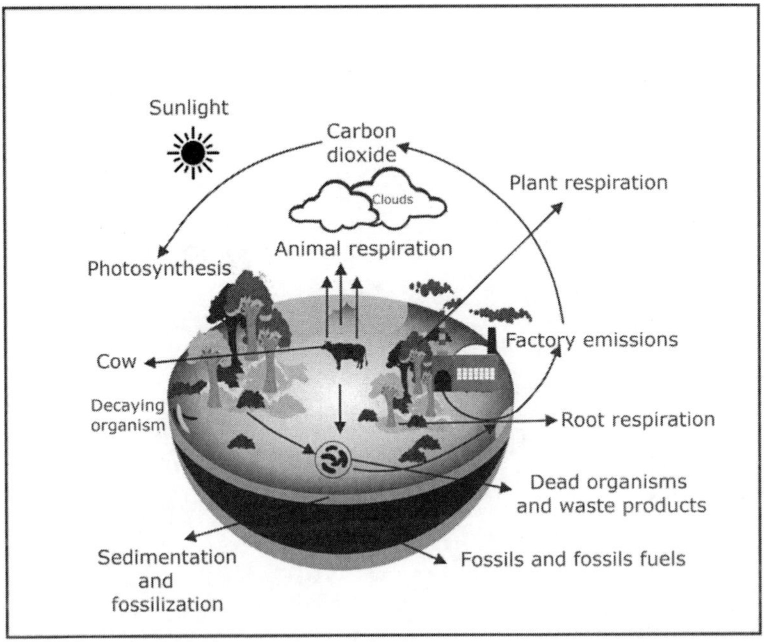

Fig. 6.6: Carbon cycle

Impact of Human activities

The carbon cycle has been altered in the following ways:

a) Huge quantities of carbon dioxide are released into the atmosphere when fossil fuels, such as coal and wood, are burned for energy.

b) We are cutting down trees and other plants (vegetation) that take in carbon dioxide and produce oxygen.
c) The addition of more CO2 to the atmosphere is the cause of the preceding two processes, which cause an increase in the average temperature of the Earth. The natural greenhouse effect, which contributes to the warming of both the lower atmosphere (troposphere) and the surface of the earth, may be amplified by increased amounts of greenhouse gases in the atmosphere, such as carbon dioxide (CO2) and other gases. The effects of global warming may include disruptions to global food supply and habitats for many species of animals, changes to patterns of temperature and precipitation, and an increase in the average sea level in many parts of the globe.

Nitrogen cycle

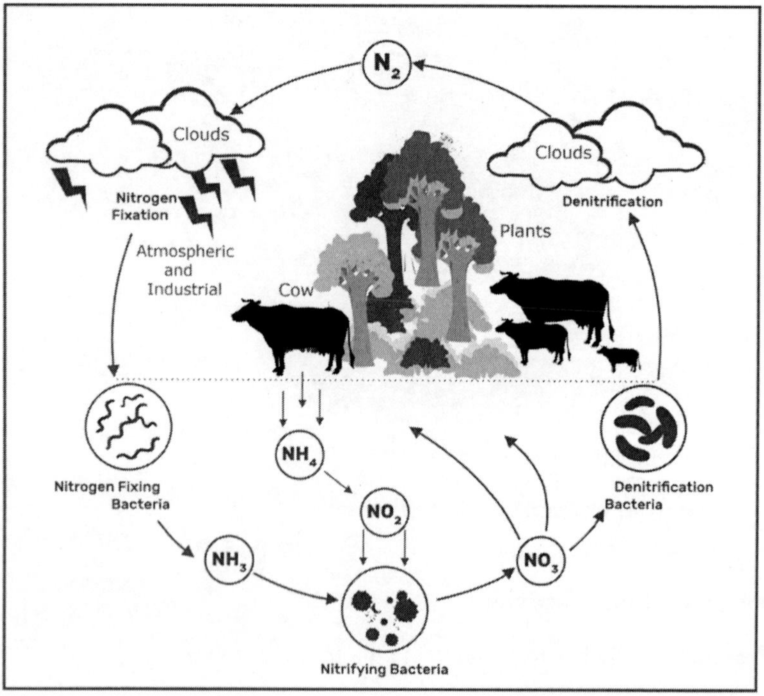

Fig. 6.7: Nitrogen cycle

The Nitrogen Cycle is a biogeochemical process in which nitrogen is converted into different forms before making its way from the atmosphere to the soil to the organism and then back into the atmosphere again (Fig. 6.7). There are a few different processes that are involved, including nitrogen fixation, nitrification, denitrification, degradation, and putrefaction. Both biological and inorganic forms of nitrogen may be discovered in the natural world. Organic nitrogen is a kind of nitrogen that may be found in living things and is passed up the food chain when one living thing consumes another living thing. Inorganic forms of nitrogen may be found at high concentrations in the atmosphere. This nitrogen is made accessible to plants by symbiotic bacteria that are capable of converting inert nitrogen into forms that are helpful to plants, such as nitrites and nitrates. In order to keep the natural balance of the ecosystem intact, nitrogen passes through a number of different changes. This process also occurs in other biomes, with the nitrogen cycle in marine environments being one of the most complex biogeochemical cycles.

Stages of Nitrogen cycles

Nitrogen fixation, Nitrification, Assimilation, Ammonification, and Denitrification are the stages in the Nitrogen Cycle process. These processes occur in phases, which are described below:

1. Nitrogen fixation:

The process begins with this step of the nitrogen cycle. In the course of this process, inert atmospheric nitrogen (N2) is converted into the more beneficial form of ammonia (NH3). During the nitrogen fixation process, the inert form of nitrogen gas is deposited into soils from the atmosphere and surface waters, with precipitation being the primary means of deposition. Later on, the nitrogen will go through a series of transformations that will ultimately result in the separation of two atoms of nitrogen. These nitrogen atoms will then combine with hydrogen to make ammonia, which is denoted by the symbol NH4+. Symbiotic bacteria known as diazotrophs are responsible for the last step in the process of nitrogen fixation. Other significant participants in this process are Azotobacter and Rhizobium. These bacteria have an enzyme called nitrogenase, which allows them to convert hydrogen and gaseous nitrogen into ammonia. Fixation of nitrogen may take place either by air fixation (also known as lightning) or via industrial fixation (producing ammonia at high temperatures and pressures). The production of ammonia and nitrogen-rich fertilisers by industry is one example of a man-made process that may be used to mitigate this issue.

2. Nitrification:

During this stage of the process, the presence of microorganisms in the soil causes ammonia to be converted into nitrate. Nitrosomonas bacteria species are responsible for the production of nitrites via the oxidation of ammonia. The nitrites that are produced are subsequently converted into nitrates by nitrobacter. Because ammonia gas is lethal to plants, it is essential that this reaction takes place. The following is an example of one of the reactions that takes place during the nitrification process:

$$\mathbf{2NH_4^+ + 3O_2 \rightarrow 2NO_2^- + 4H^+ + 2H_2O}$$
$$\mathbf{2NO_2^- + O_2 \rightarrow 2NO_3^-}$$

3. Assimilation:

Plants use their roots to absorb nitrogen molecules from the soil, which are then available to the plant in the form of ammonia, nitrite ions, nitrate ions, or ammonium ions. Nitrogen molecules are necessary for the production of plant and animal proteins and are absorbed by plants through their roots. When the plants are consumed by the main consumers, they make their way into the food chain in this way.

4. Ammonification:

When living things, such as plants or animals, die, the nitrogen that was contained in the organic matter returns to the ground. Decomposers, which may be bacteria or fungus and are often found in the soil, are responsible for converting organic waste back into ammonium. Ammonia is produced as a by-product of this decomposition process and is then put to use in several other biological processes.

5. Denitrification:

The return of nitrogen compounds to the environment occurs via a process known as denitrification. This involves the transformation of nitrate (NO_{3-}) into gaseous nitrogen (N). This is the last stage of the nitrogen cycle, and it takes place when there is no oxygen present. Species of denitrifying bacteria, such as Clostridium and Pseudomonas, are responsible for the process of denitrification. These bacteria consume nitrate in order to get oxygen, and as a consequence, they generate free nitrogen gas.

Impact of Human activities

The Nitrogen cycle is altered in the following ways:

a) We release a significant quantity of nitrogen oxide into the atmosphere whenever we burn any kind of fuel at high temperatures (which helps N_2 and O_2 to combine). It is possible for NO to be converted into NO_2 gas and HNO_3 in the atmosphere, both of which may cause dangerous acid rain to fall to the surface of the planet.
b) N_2O is released into the atmosphere as a result of the activity of anaerobic bacteria on animal dung and commercial inorganic fertilisers that have been applied to soil. It is possible for this gas to warm the atmosphere while at the same time reducing the amount of O_3 in the stratosphere.
c) NO_3^-, which is found in inorganic fertilisers, has the potential to leach through the soil and contaminate ground water, which may be harmful to everyone who drinks it, but especially infants and children.
d) When we cut down large areas of trees, grasslands, and wetlands, we cause massive quantities of nitrogen – found in soils and plants – to be released into the troposphere in the form of gaseous molecules.
e) Before planting crops, we remove nitrogen from the topsoil by harvesting nitrogen-rich crops, irrigating crops, and burning or clearing grasslands and woods. This harms aquatic environments by adding additional nitrates that come from agricultural runoff and urban sewage system discharges.

Sedimentary cycles

The sedimentary cycle is responsible for the transportation of phosphorus, calcium, and magnesium. In most cases, the element that is part of the sedimentary cycle does not cycle through the atmosphere; rather, it adheres to a fundamental pattern of flow through processes such as erosion, sedimentation, mountain building, volcanic activity, and biological movement via the excrement of marine birds.

Phosphorus cycle

The phosphorus cycle is a rather sluggish one in comparison to other biogeochemical cycles, such as the water cycle, the carbon cycle, and the nitrogen cycle. Phosphorus is a nutrient that is necessary for the survival of all creatures that are alive. It plays a crucial role in the phospholipids that make up our cell membranes as well as nucleic acids like DNA. In the form of calcium phosphate, it also makes a contribution to the structural components of our bones. In nature, phosphorus is typically the limiting nutrient, which refers to the nutrient that is in restricted supply and hence inhibits development; this is particularly true for aquatic and freshwater ecosystems (see Fig.6.8). The phosphorus cycle (Fig.6.8) moves at a much slower pace compared to other biogeochemical cycles such as the water cycle, the carbon cycle, and the nitrogen cycle. The most frequent form of phosphorus that may be found in nature is in the form of phosphate ions. Phosphate compounds may be found in sedimentary rocks. As sedimentary rocks weather, or deteriorate over time, the phosphorus that is stored inside those rocks progressively seeps into the surface water and soils. Phosphate may also be found in aerosols, mineral dust, and ash from volcanic eruptions. This is despite the fact that phosphorus, unlike carbon, nitrogen, and sulphur, does not exist in a true gas phase. Plants have the ability to take up phosphate molecules in the soil, which may subsequently be transferred to the bodies of animals that eat those plants. When plants and animals die or produce waste, phosphates may either be eaten up by detritivores or returned to the soil where they were found. Surface runoff has the potential to transport phosphorus-containing substances to bodies of water such as rivers, lakes, and oceans, where aquatic organisms will consume them. New sedimentary layers are produced whenever phosphorus-containing compounds that originate from the bodies of marine creatures or the wastes they produce make their way to the ocean bottom. Phosphorus-rich sedimentary rock may be transported from the ocean onto land through the geological process known as uplift. This process may take an extremely long amount of time. The normal phosphate ion, on the other hand, has an oceanic residence period (the amount of time it remains in the ocean) that ranges from 20,000 to 100,000 years. This makes the process very sluggish.

Impact of Human activities

The Phosphorus cycle is altered in the following ways:

a) In order to manufacture industrial quantities of inorganic fertilisers and detergents, we extract a significant quantity of phosphate rocks from the ground.
b) The quantity of phosphate that is easily available in soils is reduced when we cut down trees.

c) Aquatic ecosystems are disrupted when phosphates, such as those found in animal dung and fertiliser runoff, as well as discharges from sewage treatment systems, are present.

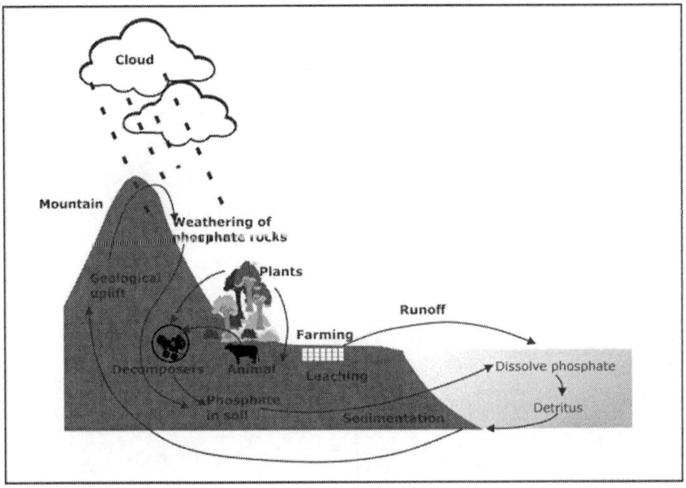

Fig. 6.8: Phosphorus cycle

Sulphur cycle

Sulphur is found everywhere on earth, including the atmosphere, water, and soil, as well as biological things (Fig. 6.9). The sulphur cycle, which can be seen in the Fig., moves sulphur across the biosphere in a circular fashion. A significant portion of the sulphur that exists on the world is locked up in the earth's crust in the form of rocks and minerals, particularly sulphate (SO_4^-) salts that are buried very deeply under ocean deposits. Several different types of natural processes contribute sulphur to the air we breathe. Active volcanoes and the decomposition of organic detritus in flooded swamps, bogs, and tidal flats by anaerobic decomposers are two sources of hydrogen sulphide (H2S), a colourless, very lethal gas with a smell like rotten eggs. H2S is released by both of these sources. Sulphur dioxide (SO2) is an odourless gas that may cause suffocation and is released by volcanoes. Sea spray, dust storms, and forest fires all contribute to the introduction of sulphate (SO_4^{2-}) salt particles into the atmosphere. One example of this is ammonium sulphate. Plant roots are responsible for the absorption of sulphate ions, which are then integrated as an essential component of many proteins. There are several marine algae that create significant quantities of the volatile compound known as dimethyl sulphide, or DMS. The condensation of water vapour into the droplets that make up clouds begins at the nucleus of very minute DMS droplets. Alterations in the levels of DMS emissions may influence cloud cover and the climate in this manner. Sulphur dioxide is the product of the conversion of DMS in the environment. In the atmosphere, sulphur dioxide is converted into sulphur trioxide gas (SO3) and into tiny droplets of sulphuric acid (H2SO4). Additionally, it reacts with other tropospheric chemicals such as ammonia to produce tiny particles of sulphate salts. Sulphur dioxide can come from natural sources or from human activities. These droplets and particles are components of acid deposition, which coupled with other air pollutants may be harmful to plants and aquatic life. Acid deposition falls to the soil like rain or snow.

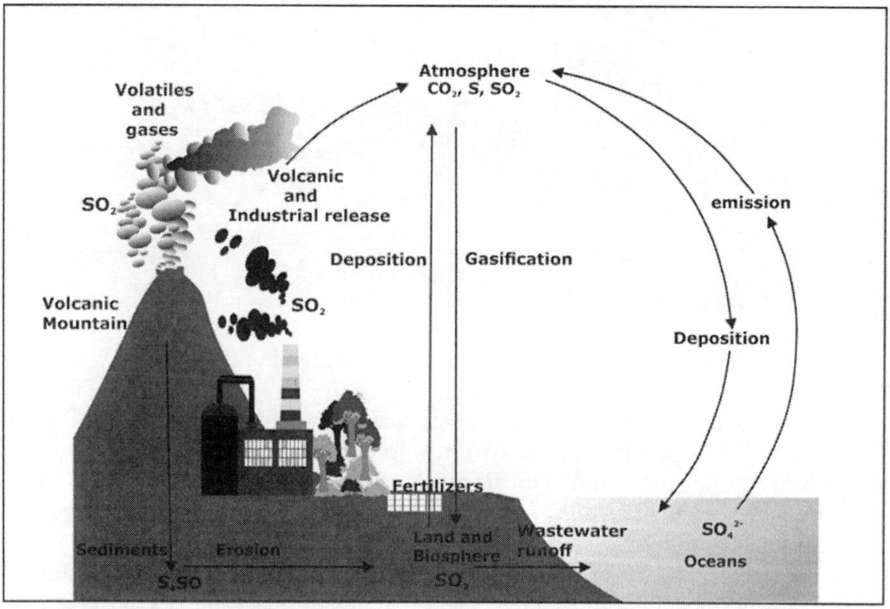

Fig. 6.9: Sulphur cycle

Impact of Human activities

The Sulphur cycle is altered in the following ways:

a) In order to generate electricity, we make use of coal and oil, both of which include sulphur.

b) Sulphur-containing petroleum is refined by our company to produce gasoline, heating oil, and other products.
c) When we transform ores that include metallic minerals that contain sulphur into free metals like copper, lead, and zinc, we produce a significant quantity of sulphide oxide, which is then released into the environment.

CHAPTER 7

TERRESTRIAL ECOSYSTEM

Nature and types of Terrestrial Ecosystem

Up until this point, a comprehensive description of the overall organization and function of an ecosystem as a whole has been provided. As was said before, the biosphere is home to a wide variety of distinct ecosystems that each function as independent, interconnected systems. About 148 million square kilometres, or about 29 percent of the total surface area of the world, are covered by terrestrial ecosystems. The variations in temperature or precipitation that occur in a certain region are what define the types of plant species that are able to thrive there. It is common for height, density, and species diversity to decrease when climate changes from warm and damp to cold and dry. In fact, life form spectra are more similar in situations that are comparable across several continents than they are in climates that are unique to a single continent. Biomes are locations that have climates and key plant varieties that are quite similar to one another. Examples of natural ecosystems include the following: a pond, a lake, a river, a stream, a spring, an estuary, the sea; a desert; a grassland; a forest; a coral reef; and many more types of environments. While all of these ecosystems have a fundamental design that is more or less the same in terms of their general structure and function, there are significant differences between them in terms of the species diversity, production rates, and other factors. As a consequence of this, we will now investigate the distinct organisational patterns of terrestrial ecosystems in more detail, and after that, to some extent, we will assess the relative significance of each.

Tropical forest Biomes

Around the equator are some of the best sites to look for tropical forests. Central and South America are home to around fifty percent of the world's tropical forests. These biomes have a climate that is very consistent, with high yearly rainfall and temperatures that are, for the most part, warm and reasonably constant. The phanerophytes, which include trees, lianas, and epiphytes, are the plants that are now in the majority. Oxisols and Ultisols are the two primary soil types that may be found in tropical rainforests. These soils are rich in iron and aluminium oxides (which give them their red colour), but they are low in natural fertility.

In terms of both the number of different species and the amount of primary products they produce, tropical forests are the most varied and productive biomes on land. It's possible that the net primary productivity is as high as 2–3 kg m-2 per year. It is generally accepted that more than half of the earth's terrestrial species may be found in the biome represented by tropical forests. There are over 170,000 vascular plant species known, and almost half of them are found in tropical biomes. The evergreen rainforest, seasonal deciduous forest, tropical cloud forest, and mangrove forest are all examples of sub-biomes that are included under the umbrella of the tropical forest biome. The changes in precipitation, elevation, and habitat that occur throughout the year give rise to the formation of these sub-biomes.

Savanna Biomes

Savannas are situated to the north and south of tropical forest biomes. Savannas have lower yearly rainfall and longer dry seasons than tropical forests do. These ecotones are characterised by a predominance of grasslands and deciduous trees of a dwarf kind. Savannas cover60 percent of Africa and act as a transition zone between the continent's tropical forests and its deserts. Savannas often have a variety of drought-deciduous trees. There are several distinct varieties of savanna, each of which may be identified by the relative abundance of trees and grass. These characteristics are also connected to the varied patterns of rainfall, water table height, and soil depth. In the last 50,000 years, the African savanna has been ravaged by flames during the dry season on a number of occasions. Savannas need fire to function properly and maintain the right proportion of grasses and trees. When there is a significant amount of time that has passed since the last fire, the number of trees and shrubs in an area will increase.

Because the soil in the savanna is porous, it allows water to flow away from the area fast. It only has a very thin layer of humus, which is the organic element of the soil generated by the incomplete decomposition of plant or animal material and provides nutrients to the plant or animal that grows there. The savannas are grasslands that have a dry season and then a rainy season in succession. In this part of the world, the majority of the land is covered with alfisols and ultisols. These soils are very old yet have very little fertility; nonetheless, since there is a dry season, it is possible that more nutrients will be preserved. Both plants and animals are need to undergo adaptation in order to survive prolonged drought. The acacia tree is one example of a xerophytic plant because of its spiny branches, waxy leaves, and tiny size. Plants also have the ability to store water (as the baobab tree does), or they might have extensive roots that reach down to the water table.

Desert Biomes

Deserts may be found in a band around the planet that runs between 15 and 30 degrees north and south of the equator. They account for between 26 and 35 percent of the terrestrial area of the planet. Desert climates are distinguished by their low levels of precipitation, which are often lower than 250 millimetres per year. There are many distinct types of deserts, some of which are hot, some of which are cold, some of which are high in elevation, and some of which are rain shadow deserts. As a direct consequence of this, the levels of biodiversity, productivity, and the kinds of species that may be found in different sorts of deserts are quite different. Aridisols are the most prevalent kind of soil found in desert environments (dry soil). On the other hand, entisols are a form of soil that may be found in highly dry regions such as the Sahara and the outback of Australia. Entisols are juvenile soils, like sand dunes, that are too dry to create vast soil horizons. Sand dunes are an example of an entisol. The fact that they are able to survive in arid environments and be found in places like deserts is due to the fact that they may be found on floodplains after spring floods. Even though it just seems empty, the desert is expanding all the time. This consists of a live, biological crust that was produced by a collection of lichens, moss, and algae. Aridisols are often quite fertile, yet they only get a trace amount of precipitation each year—not enough to sustain life. Permanent plant life, such as cacti and shrubs, has adapted through millions of years to thrive for extended stretches of time without access to water.

Grassland Biomes

It is possible to identify an area as a grassland if it is more dominated by grasses than by large shrubs or trees. During the Miocene and Pliocene Epochs, which lasted around 25 million years and provided a continental climate suited to grasslands, mountains emerged in western North America. These epochs are known as the "Pliocene Epoch." There was a gradual decline in the number of ancient forests, and grasslands started to dominate the landscape. After the Pleistocene Ice Ages, the global climate began to warm up and become drier, which led to an increase in the amount of territory covered by grasslands. The majority of grassland biomes may be found in the interiors of continents, and they are characterised by the extreme seasonal temperature fluctuations that they experience, with summers that are very hot and winters that are quite cold. The soil in the temperate grasslands is rich and dark, and the top layers are quite productive. It is rich in nutrients as a result of the development and decomposition of extensive grass roots that shoot out in numerous directions. The decomposing roots are responsible for maintaining the structure of the soil and also provide nutrients to the plants that are developing. The quantity of precipitation varies greatly, with a significant increase often occurring during the summer months. Both the kind of grassland community that develops and the amount of grass that may be harvested from grasslands are strongly influenced by the amount of precipitation that the grasslands get. The effect of more rainfall is a tall grass prairie that has a wide variety of grasses as well as shrubs. Lower levels of precipitation lead to the development of short grass prairies and dry grasslands.

The Banni grassland reserve, sometimes known simply as the Banni grassland, is the largest grassland in all of Asia and may be found in Gujarat, close to the Great Rann of Kutch. It has a length of 2,618 kilometres and is responsible for more than 45 percent of the pastures in Gujarat. This grassland is considered to be the best in Asia. There are 192 plant species, 262 bird species, and a wide variety of mammal, reptilian, and amphibian species found in this area. The region is rich in both flora and fauna.

Temperate Deciduous Forest Biomes

Temperature Deciduous forests are often located in the middle latitudes, when the winters are cold, the summers are hot, and there is an abundance of precipitation throughout the year. The amount of litter produced has a significant impact on the net primary productivity, which may vary anywhere from 600 to 1500 g m-2 year-1. The recycling of nutrients is helped significantly by litter, which is an essential source. In these types of forests, it is possible to find an overstory composed of trees with heights of 20–30 metres, an understory consisting of trees and shrubs with heights of 5–10 metres, a shrub layer with heights of 1–2 metres, and a ground layer composed of herbaceous plants.

Soils that are characteristic of temperate deciduous forests are known as mull soils. These soils have a high organic content, particularly near the surface, and the organic matter is well mixed with the mineral matter. Variations in the composition of the soil and the fertility of the land have a significant role in determining the kind of trees that will predominate in the forest.

Mediterranean Climate Biome

The most common kinds of soil found in the Mediterranean region are cambisols (26 percent), calcisols (22 percent), leptosols (20 percent), and luvisols (10 percent). Cambisols are found almost exclusively in European countries, accounting for 26 percent of the total area; they contribute to excellent agricultural land and are considered to be among the most productive soils on the globe. The biomes of Mediterranean forests, woods, and scrub are all intricately connected with the climatic zone that encompasses the Mediterranean. The presence of maquis, also known as sclerophyll shrublands, is indicative of the climate zone, in particular in the Mediterranean Basin. The aquatic ecosystems in this climate zone have adapted through time to the yearly cycle of environmental control that is given by floods and other biotic components. Plants native to the Mediterranean climate zone are forced to undergo a process of adaptation in order to endure the region's blisteringly hot summers and protracted, wet winters. The vegetation consists of evergreen trees, deciduous trees, fruit trees, meadows, and many types of plants. A significant amount of the vegetation has been eradicated in order to create room for agricultural land. In an atmosphere typical of the Mediterranean region, the summers are dry and hot, while the winters are cold and damp. In the western parts of the continent, the regions may be found between 30 and 45 degrees north and south of the equator. The climate zone is associated with the five primary subtropical high-pressure belts that are found in the ocean. These pressure belts include the Azores, the South Atlantic High, the North Pacific High, the South Pacific High, and the Indian Ocean High. Anticyclones are pressure belts that spin in the opposite direction in the Northern and Southern Hemispheres—clockwise andcounter clockwise, respectively.

Coniferous forest Biome

A coniferous forest is a kind of biome that may be found in regions across the globe that have warm summers and mild winters, in addition to a sufficient amount of precipitation to sustain the establishment of the forest.

Trees such as pines, cedars, redwoods, and kauris are examples of coniferous trees that may be found in a coniferous forest biome. The majority of coniferous trees have leaves that are either scale-like or needle-like and they retain their green colour throughout the whole year. The fact that these coniferous tree species belong to the softwood family enables them to thrive in environments with low temperatures and acidic soils. The average temperature in the coniferous forest biome is 10 degreeCelsius. Temperatures as low as -40 and as high as 20 degrees Celsius are usual during the winter, while 10 degrees Celsius is the norm during the summer. The winters are exceptionally lengthy and covered with snow, while the summers are very hot and muggy.

Tundra Biome

The word "tundra" originates from the Finnish word "tunturia," which literally translates to "treeless mountain tract," which is an accurate term for this environment. Of all the biomes on Earth, the tundra is the coldest. Despite its relative peace and quiet, the tundra is home to unique and abundant varieties of plant and animal life. It occupies around 10 percent of the total surface area of the world. According to the research done by scientists, it is the youngest biosphere in the world. It is believed that the tundra biome first appeared around 10,000 years ago. The tundra is a kind of biome that is distinguished by its arid landscape and very cold temperatures. This kind of biome is characterised by a predominance of rocks over vegetation. The density of trees is so low that they are almost negligible. It is not very often that the temperature in the region reaches the point where the ice completely melts away. The tundra is characterised by its very low temperatures, brief growing seasons, little amounts of precipitation, and low levels of nutrient availability in the soil. The tundra biome extends across an area equal to more than 3 million square miles. Included in its territory are the states of Alaska and Greenland's peripheries, as well as Northern Canada, Russia, Northern Siberia, and Northern Scandinavia. This biome may be found in latitudes ranging from 55 to 75 degrees north. Even though the average annual precipitation in the tundra biome is just 10 inches, there are substantial exceptions to this rule. In portions of the tundra that are located in close proximity to the ocean or other large bodies of water. There is a possibility that the annual quantity of precipitation approaches 20 inches. Alpine tundra and arctic tundra are the two subtypes that fall within the tundra biome. Both forms of tundra have many characteristics in common. The amount of life that can be sustained by each of them is the sole distinction between them.

Because the tundra is the ecosystem with the lowest average temperature on Earth. During the winter, the temperature doesn't usually get below -18 degrees Fahrenheit (-8 degrees Celsius). Throughout the month of June, the temperature may reach 50 degrees Fahrenheit at its highest point (10 degrees Celcius). The winters on the tundra are dreary, protracted, and very cold. The sun shines more brightly during the summer months than at any other time of the year. The sun often sets at an earlier time in the autumn, and there is very little available light. In the absence of light, temperatures have the potential to fall to far below -70 degrees Celsius (or -94 degrees Fahrenheit). Even though it's July, there's still a chance of snow showers. Because they have to be able to endure prolonged exposure to cold temperatures, plants suffer when they are in this kind of environment. Cold weather is a challenge for animals because they have to learn to adjust to it while also figuring out how to locate food and shelter for themselves.

Deforestation

The growing need for space is becoming a cause for worry as the population of the globe continues to increase at an unfathomable rate. A direct action known as "deforestation" takes place as a response to a severe need for land for agricultural, industrial, and most crucially, urban requirements to limit cities and their growing populations. In words that are more understandable to a layperson, deforestation is the act of cutting down and removing forest cover or tree plantations in order to create room for agricultural, industrial, or urban use. It comprises the removal of forest cover in order to make land acceptable for residential, commercial, or industrial use. This may be done in order to make property appropriate for all three uses.

The percentage of land covered by forests throughout the world has dropped to an all-time low of around 30 percent during the course of the last century due to extensive deforestation efforts. According to the estimates provided by the United Nations Food and Agriculture Organization, around 18 million acres (7.3 million hectares) of forest are lost every single year.

Causes: The primary causes of deforestation are given below:

a) Agricultural activities: Because of the growing need for food, a significant number of trees are cut down to make room for crop cultivation; subsistence agriculture is responsible for 33 percent of the deforestation that is caused by agricultural activities.

b) Ranching of animals: It is estimated that livestock are responsible for around 14 percent of the total amount of deforestation that occurs on the earth. Farmers typically clean land by cutting down trees and then burning them in order to make room for livestock and produce food. They will continue to exploit the land until the soil has been fully damaged, at which point they will try the same process with fresh forests. Although it will ultimately reforest, it will take many years for it to return to the condition it was in before. Surprisingly, the area covered by forests has shrunk by more than 40 percent in the last 40 years, whilst the area covered by grasslands and the number of cattle have both substantially and rapidly expanded.

c) Illegal logging: Wood-based industries like paper, matchsticks, and furniture need a substantial number of wood supply. Illegal logging is one of the main causes of this problem. Wood is used as a source of energy in a variety of ways, some of which are more obvious than others. As a direct consequence of this demand for resources, trees are being felled to satisfy it. In the form of firewood and charcoal, wood is used both directly and indirectly as a fuel. Some of these firms find that it is advantageous to engage in illegal tree cutting and felling.

d) Urbanization: In a similar fashion, roads are constructed in order to get access to these forests; once again, trees are destroyed in order to construct roadways. There is a clear correlation between overpopulation and loss of forest cover due to the fact that as cities expand, more land is needed for the construction of dwellings and other settlements. As a direct consequence of this, forest land is being cleared.

e) Land degradation: One further factor that contributes to deforestation is one that is partially caused by natural reasons and partly caused by human activity; this factor is land degradation. It occurs as a consequence of inappropriate usage of land, which renders the area unfit for the growth of plants. Many petrochemical companies dispose of their waste in rivers, which leads to soil erosion and renders the land unfit for the cultivation of plants and trees.

f) Mining: Mining for oil and coal requires a significant amount of land that was formerly covered with trees. In addition to that, roads and highways will need to be built so that they can handle trucks and other forms of heavy equipment. Mining waste contaminates the environment and has a detrimental effect on the creatures that live in the area.

g) Forest fires: Each year, hundreds of trees are lost all over the world as a result of forest fires in various parts of the world. It takes place as a consequence of summers that are unusually hot and winters that are unusually mild. Forest fires are a leading cause of catastrophic forest loss, and they can be started either by humans or by natural causes.

h) Paper: According to estimates provided by the Environment Paper Network, 640 million trees are cut down annually for the purpose of producing paper waste. Each year, the United States, China, Canada, and Japan produce more than half of the paper that is used throughout the world, which amounts to 400 million tonnes. It is possible to reduce or eliminate the emission of 27.5 million tonnes of carbon dioxide into the atmosphere through recycling. When we use paper that has been recycled, we are doing our part to ensure that forests continue to support a thriving ecology and provide a habitat for animals.

i) Overpopulation: An increase in the world's population results in a greater demand for residential and community space. As a consequence, there will be a significant rise in the demand for food as well as the amount of land required to cultivate food and raise cattle. In order to facilitate transportation and communication, it is necessary to construct a large number of additional roads and highways, all of which contribute to the clearing of forest land. Logging firms cut down trees so that they may manufacture things like furniture, paper, building materials, and a wide range of other products.

In addition, the growth of the human population is intimately linked to the rate at which forests are being cut down. As a direct consequence of this, it is now almost obligatory to make purchases from environmentally responsible businesses that wage aggressive campaigns against deforestation.

Effects of Deforestation: The various effects are discussed below:

a) Climate Imbalance and Climate Change: Deforestation has a number of different impacts on the environment's temperature and humidity levels. Our planet's lungs are forests. Because trees take in carbon dioxide and release oxygen and water vapour into the atmosphere, tropical rainforests have an exceptionally high relative humidity. In addition, trees generate shade, which contributes to the retention of moisture in the soil. Every single one of them is in jeopardy as a result of the loss of trees. It leads to an imbalance in air temperature, a drier environment, and ecological situations that are more severe, all of which contribute to the shift in climate.

b) Global warming: Trees play a vital part in mitigating the effects of climate change. The trees are responsible for the maintenance of the atmosphere's equilibrium via their absorption of greenhouse gases. As a consequence of ongoing deforestation, the proportion of greenhouse gases in the atmosphere has increased, which has contributed to the problems we are having with global warming.
c) An increase in emissions of greenhouse gases: Forests contribute to a decrease in emissions of carbon dioxide and other potentially damaging greenhouse gases. After being chopped down, burned, or removed in some other way, they become into sources of carbon. It is estimated that deforestation is responsible for around 20 percent of the emissions that contribute to global warming; more specifically, tropical deforestation is responsible for the annual release of 1.5 billion tonnes of carbon into the atmosphere.
d) Erosion of the soil: Trees are also essential to the local water cycle because they are responsible for the return of water vapour to the atmosphere. The soil is able to retain its moisture because rainwater seeps into it over time. The complex root systems of many different layers of trees are responsible for preserving the rich soil in its original location. Now that the tree cover has been destroyed, the ground is instantly exposed to the sun, which will cause it to become dry. When there are no trees around, erosion is more likely to occur, which causes the land to be carried away by the nearby rivers and streams. The trees of the world are nature's water filtration facilities. Erosion of the soil exposes it to a variety of contaminants, which then make their way into the water supply and pollute our drinking water.
e) Flooding: When it rains, trees utilise the roots in their roots to soak up and store a significant quantity of water. This may cause flooding. When they are cut down, the natural flow of water is disrupted, and the ability of the soil to retain water is reduced. It may lead to floods in certain areas, while in others it might lead to a lack of precipitation and drought.
f) Extinction of animal species and destruction of habitat: As a consequence of extensive tree cutting, a number of different animal species have gone extinct. They are forced to leave their native

surroundings and find a new home because of this. Many of them have become extinct as a result of human activity. In the course of the last several decades, many plant and animal species have disappeared from our planet. Research that was carried out in the Brazilian Amazon suggests that as much as 90 percent of the predicted species extinctions will take place over the next 40 years.

g) Ocean acidification: The levels of carbon dioxide in the atmosphere are increasing as a direct consequence of human activities such as deforestation and the use of fossil fuels. As a result, our oceans are getting more acidic. Since the beginning of the Industrial Revolution, there has been a 30 percent increase in the acidity of coastlines, putting marine life and ecosystems at jeopardy.
h) A decrease in quality of life: Millions of people all over the globe depend on woodlands for a variety of different things, including hunting, small-scale agriculture, harvesting, and medicine. Tropical forests provide us with essential resources, like latex, fruit, natural oils, and resins. These woods also offer us with food. Deforestation has a negative impact on the lives of millions of people worldwide. In Southeast Asia, there has been an upsurge in social strife as well as migration as a result of deforestation.
i) Uncertainty about the availability of food in the future: Uncertainty regarding the availability of food may develop in the future as a consequence of deforestation for agricultural purposes. The amount of agricultural land that is affected by moderate or severe soil erosion is equal to 52 percent of the total. In the long term, the absence of good soil may lead to low yields and food insecurity because of its effects.
j) The Considerable Loss of Biodiversity: The process of deforestation leads to a significant loss of biodiversity. The world's tropical rainforests are home to around 80 percent of the planet's total plant and animal species. Not only do forests provide a habitat for a variety of animal species, but they also play an important role in the preservation of plants used in medicine. A significant contribution is made by the forest to the protection of a wide variety of animal and plant species. In addition to this, it wreaks havoc on the population of microorganisms, which is essential for the production of clean water, the elimination of toxins, and the recycling of nutrients.

Solutions to Deforestation: There are various solutions to deforestation and some of them are explain below:

a) Government regulations: The implementation of a system of rules and laws that regulate tree cutting is the most effective method for fighting deforestation.
b) Prohibiting the cutting of trees in their whole inside forests: This will stop the forest cover from being entirely destroyed. It is a sensible and quite achievable course of action to consider.
c) Reforestation and afforestation: Land that has had its tree cover removed to make way for urban expansion should be incentivized to plant new trees in the area so that they may replace the ones that have been cut down. In addition to this, the trees that were chopped down need to be replaced with new trees in order to take the place of the ones that were destroyed. Trees are planted as part of a variety of programmes on an annual basis; nevertheless, the total number of trees planted does not come close to making up for the total number of trees that have been cut down.
d) Decrease the amount of paper that you use: paper for printing, notebooks, napkins, toilet paper, and several other forms of paper are all used on a daily basis. Reduce how much paper you use, cut down on how much paper you throw away, and go for paper products made from recycled paper. Your life will be much simpler if you print on both sides of the paper, use less toilet paper, get rid of paper plates and napkins, and go paperless wherever it is possible.
e) Help educate others: A large number of individuals are still living in full ignorance about the dangers posed by global warming. Inform your friends, family, and neighbourhood about the causes and effects of deforestation, as well as the facts surrounding the issue of deforestation. You have the power to move things in the right direction!
f) Reduce your consumption of meat: Livestock agriculture is one of the leading causes of deforestation. Cut down on the amount of meat you consume. It could be challenging for some individuals to give it a go. However, even if you limit your consumption of meat to just one meal a day, you would still have a sizeable impact on the natural environment.
g) Make purchases from sustainable businesses that are friendly to forests: Whenever it is feasible, make purchases from businesses that are committed to reducing the amount of deforestation that occurs. Asian Pulp and Paper, Disney, Hershey, and Willmar International are some examples of businesses that have made efforts to reduce their contributions to deforestation.

h) Decrease your usage of items that contribute to deforestation: Palm oil may be found in almost every product we come across. Make it a habit to give the ingredients a quick once-over every so often. Another cause of deforestation is the production of soybeans.

Desertification

Desertification is a kind of land degradation in which an already dry piece of land gets much drier over time, leading to the extinction of both flora and fauna as the process continues. There are a number of factors that contribute to climate change, one of which is the acts of humans. Desertification is a significant ecological and environmental problem that affects the whole planet. The United Nations Educational, Scientific, and Cultural Organization (UNESCO) reports that desertification poses a danger to one-third of the land area on the planet. This affects the way of life of millions of people whose livelihoods depend on the ecosystem services given by dry lands. In many places of the globe, desertification is primarily caused by excessive usage of livestock for grazing. Desertification is caused by a number of factors, including urbanisation, climate change, improper use of groundwater, deforestation, natural disasters, and agricultural tillage practises that make soils more susceptible to wind erosion. Urbanization is one of the most significant contributors to desertification. Desertification has repercussions not just for the topsoil but also for the groundwater reserves, surface runoff, and the human, animal, and plant populations. In arid regions, where water is in short supply, there is less opportunity for the production of lumber, crops, and grazing, as well as other ecological services that are beneficial to our society.

Various causes of Desertification

a) Excessive animal grazing: This is a significant problem in many areas that are through the process of transforming into desert biomes. Plants in places that have been overgrazed by an excessive number of animals have a more difficult time recovering, which is detrimental to the ecosystem and causes it to lose its lush, verdant appearance.
b) Deforestation: People are a contributing factor to the issue of desertification because they contribute to the problem of desertification when they wish to live in a specific place or need trees to construct homes or undertake other activities. Without the

existence of plants, the bulk of the biome would not be able to survive (particularly trees).

c) Methods of farming: Many farmers aren't sure how to make the most use of their land in their farming operations. They could eliminate almost all of the land's inhabitants and their belongings before moving on to another section of land. When the nutrients in the soil are removed, there is a greater risk of desertification occurring in the area that is now used for agricultural purposes.
d) Excessive use of fertilisers and pesticides: Excessive use of fertilisers and pesticides, with the goal of increasing crop yields in the short term, usually leads in a significant loss of soil quality. After a few years of intense cultivation, the land will no longer be able to support agricultural production since it has been degraded to an unacceptable degree. In the long term, this region may go from being arable to becoming desert.
e) Excessive pumping of groundwater: Groundwater is a category of freshwater that is located underground, and it is one of the most important sources of water in the world. The process of collecting groundwater at a rate that is higher than the aquifer's equilibrium yield is referred to as over drafting. This may also be understood as the practise of drawing an excessive amount of groundwater from underground aquifers. The depletion of its resources is the root cause of desertification.
f) Urbanization and other forms of land development: As was said before, expansion may result in the destruction of plant life caused by humans. In addition to this, it has the potential to contaminate the soil with pesticides and other potentially hazardous substances. Desertification happens when there are less places for plants to grow, which happens when there is an increase in the amount of urbanisation in a region.
g) Climate change: Desertification is made worse by climate change, which is another factor that contributes to global warming. As the temperature of the days continues to rise and droughts grow more frequent, desertification is becoming more and more apparent. If action is not taken to mitigate the effects of climate change, vast expanses of land will be transformed into deserts, and some of these areas may eventually become inhospitable.
h) Plundering the land's resources: If a certain region of land has natural resources such as natural gas, oil, or minerals, then other people will go there to mine them or extract them. This causes the soil's nutrients to be depleted, which in turn eradicates plant life

and ultimately results in the establishment of a habitat more typical of a desert.

i) Natural disasters: In some circumstances, the land may suffer damage as a result of natural disasters such as drought. In instances like this, the only thing people can really do is labour to help rehabilitate the land after it has been damaged by nature. There isn't much more they can do.

j) The contamination of the soil: Desertification is made worse by the contamination of the soil. The majority of plant species are very responsive to the environment in which they are grown. Desertification is a process that may occur in areas of land that have had their soil polluted as a consequence of human activity such as mining, agriculture, and manufacturing. Over time, the quality of the soil will deteriorate to an increasing degree depending on the level of pollution present.

k) Overpopulation and excessive consumption: As the number of people living on our planet continues to increase at an alarming pace, the demand for food and other material goods is also growing at an alarming rate. Our overall consumption is exhibiting the same pattern of persistent expansion. As a consequence of this, in order for us to satisfy our need, we need to improve the methods that we use in agriculture in order to get even greater crop yields. On the other hand, over-optimizing agricultural practises would be detrimental to the land, and in the long run, it would lead to the destruction of the land.

l) Mining: Mining is another factor that contributes to the progression of desertification. Industries are required to use significant quantities of resources in order to satisfy our want for material goods. For mining to be successful, large sections of land need to be used, which leads to the destruction of forests and the pollution of the surrounding surroundings. When the vast bulk of a region's natural resources have been extracted, and when mining activities are no longer profitable, the soil in that region will have suffered significant damage, the region will have been dry, and desertification will have taken place.

Devastating effects of Desertification: Some of the devastating effects of desertification are mentioned below:

a) Farming would become almost impossible: If an area turned into a desert, it would be very difficult to grow large crops there without the use of specialised technologies. As a result of the huge costs associated with trying to do so, many farmers will be compelled to sell their land and relocate out of the arid regions.

b) A decrease in agricultural production: Desertification is one of the processes that has the most substantial impact on crop yields. Whenever a landscape changes from being fertile to becoming desert, the land is often no longer suitable for agricultural use. Because of this, the livelihoods of a great number of farmers might be put in jeopardy, given that farming is often the major source of income for most farmers. If their land becomes arid, it is possible that they will no longer be able to produce enough food crops to maintain their own existence.
c) Famine: Without farms in these regions, the food that these farms offer would become much scarcer, and the people who live in these areas will be much more likely to turn to extreme means to assuage their hunger. In addition, animals will experience hunger, which will make the current shortage of food much more severe.
d) Flooding: The probability of flooding in an area where there is no plant life is significantly increased. There are certain deserts that are more humid than others, and since there is nothing to prevent water from pooling up and spreading out in these areas, there is a greater risk of flooding. The availability of water may also be compromised in the event of flooding.
e) Deteriorating water quality: When a territory becomes a desert, the water quality deteriorates far more than it would be under normal circumstances. This is due to the fact that plant life plays a significant role in maintaining the cleanliness and clarity of the water; in its absence, maintaining the cleanliness and clarity of the water is a far greater challenge.
f) Over population: An excessive population will cause animals and people to relocate to other places of the world where they have a better chance of surviving if the local environment starts to deteriorate into a desert. Because of this, there is an increase in both the density of population and the number of people living in a given area, both of which contribute to the cycle of desertification that was the original cause of the problem.
g) Poverty: If the issues that have been brought up thus far aren't addressed, they may all contribute to the problem of poverty, which is tied to the issue of desertification. People have a much harder time surviving when they do not have access to food and water, and it takes a very long time for them to get what they need.

h) Loss of biodiversity: The degradation of habitats and the spread of deserts are two factors that may contribute to the loss of biodiversity. There is a possibility that some species may be able to adapt to the altered climatic conditions; nevertheless, the vast majority of species won't be able to do so and will suffer catastrophic declines in population.
i) The extinction of species: Desertification is one of the leading causes of population decreases, which may lead to some species becoming endangered or becoming extinct. This predicament is especially difficult for species that are already in risk of extinction since the little population of animals or plants that is still alive may perish over the course of time, which may result in the species' extinction.
j) The destruction of habitats: As a consequence of desertification, many species of animals and plants are unable to survive since their homes are destroyed. There is a possibility that desertification will have an effect on the living conditions of local flora and fauna, making it more difficult for animals and plants to keep their population levels stable. After desertification, locations confront water shortages as a consequence of climate change, and animals may suffer and die as a result of this. Because water is necessary for all life on our planet, climate change has caused this.

Solutions to desertification: There are various solutions to combat desertification, some of them are mentioned below:

a) Changes in policy related to how people can farm: New policies relating to how frequently people can pasture as well as how much they can farm on particular areas may be implemented in nations where policy change will actually be imposed on people in the country to help minimize the challenges that are typically associated with agriculture and desertification.
b) Changes in policy regarding other types of land use: If people are using land to acquire natural resources or to develop it for people to live on, the rules and regs that govern them must aid the land in thriving rather than allowing them to further harm the land. This would be a change in policy regarding other types of land use. The revisions to the policies might be general or they could be particular to the kind of land use that is in concern.
c) Education: In countries still on the path to economic development, education is a tool that is highly important to have since it teaches people how to make the most of the land they farm on. By

educating people about sustainable practises, a greater amount of land will be saved from being turned into a desert.

d) The development of new technologies: The majority of the problems that we have with the environment may be resolved via research, and the same is true for desertification. In certain situations, it may be difficult to forestall the process of desertification from occurring. In some situations, research and the use of cutting-edge technology are necessary in order to push the limits of what we now know about the factors that contribute to desertification. With the advancement of technology, it is possible that we may be able to devise further measures to stop the issue from developing into an epidemic.
e) Placing restraints on mining practises: Mining operations often necessitate the destruction of large tracts of land. As a consequence of this, governments should exercise control over it in order to maintain the natural habitats of a variety of animals and plants and to preserve nature reserves. As a direct consequence of this, a smaller percentage of land will continue to be characterised by dry conditions, and the issue of desertification will be mitigated to some degree.
f) Reforestation: Lands that have been cleared off in the past should carefully consider the possibility of replanting such forests. It is of the utmost importance to engage in tree planting in such areas because trees serve as natural carbon dioxide storage spaces, which in turn helps to halt the progression of global warming and contributes to the maintenance of natural balance. On the other hand, putting those lands to use for anything other than what they were intended might gradually turn them into dry territory. Because of this, reforesting affected areas with trees not only stops the spread of desertification but also helps mitigate a variety of other environmental problems.
g) Sustainable practises to stop desertification: There are a number of environmentally friendly methods that may be used in the process of addressing the actions that could be causing desertification. By integrating them into what we should be doing with land, we can prevent the drying up of the whole planet and transform it into a desert instead. Desertification is a big problem that requires urgent attention, and if we are able to address it now, we will be able to prevent far more serious problems in the future. By applying an analytical approach to the processes of desertification, which we

now have at our disposal, we will be able to successfully manage these processes.

Status of Indian Desertification:

As per the Indian Desertification and Land Degradation Atlas of 2007, dry regions cover 69.6 percent of the country as a whole. The amount of land in India that has been degraded is 105.48 million hectares, which is equivalent to 32.07 percent of the total land area in the nation. There are 81.45 million hectares in the nation that are affected by desertification, which is the deterioration of land found inside drylands. India served as the host country for the 14th Conference of Parties (COP 14) of the United Nations Convention to Combat Desertification in the month of September 2019(UNCCD). To attain its national objectives of Land Degradation Neutrality (LDN) as well as the rehabilitation of 26 million hectares of deteriorated land by the year 2030, India plans to place emphasis on the efficient and environmentally responsible use of its land resources. India has been a pioneer in the movement to elevate the issue of land degradation to the forefront of important worldwide coalitions for the preservation and protection of the environment. In order to meet its commitments for the restoration of national land, the Indian government has adopted a strategy based on collaboration.

The latest edition of the Desertification and Land Degradation Atlas of India was published on June 17, 2021. ISRO's Space Application Centre in Ahmedabad is responsible for its distribution. This atlas provides data for the years 2018 and 2019 on the total area of degraded land in each state. In addition to that, it offers a change analysis that covers the 15-year term beginning in 2003-2004 and ending in 2018-2019. The land in India has degraded to the degree that it is 29.7 percent, according to a new atlas that was released by ISRO.

Indian Forest types

Out from the tropical rainforests of Kerala in the south to the alpine meadows of Ladakh in the north, and from the deserts of Rajasthan in the west to the evergreen forests of the north-east, India's forests cover a wide variety of climates and terrains (Fig. 7.1). There are a number of essential aspects that go into categorising different kinds of forests, including climate, geography, soil type, and elevation. Forests are distinct not just in their beginnings and structures but also in the environments in which they flourished and the ways in which they interacted with their surroundings.

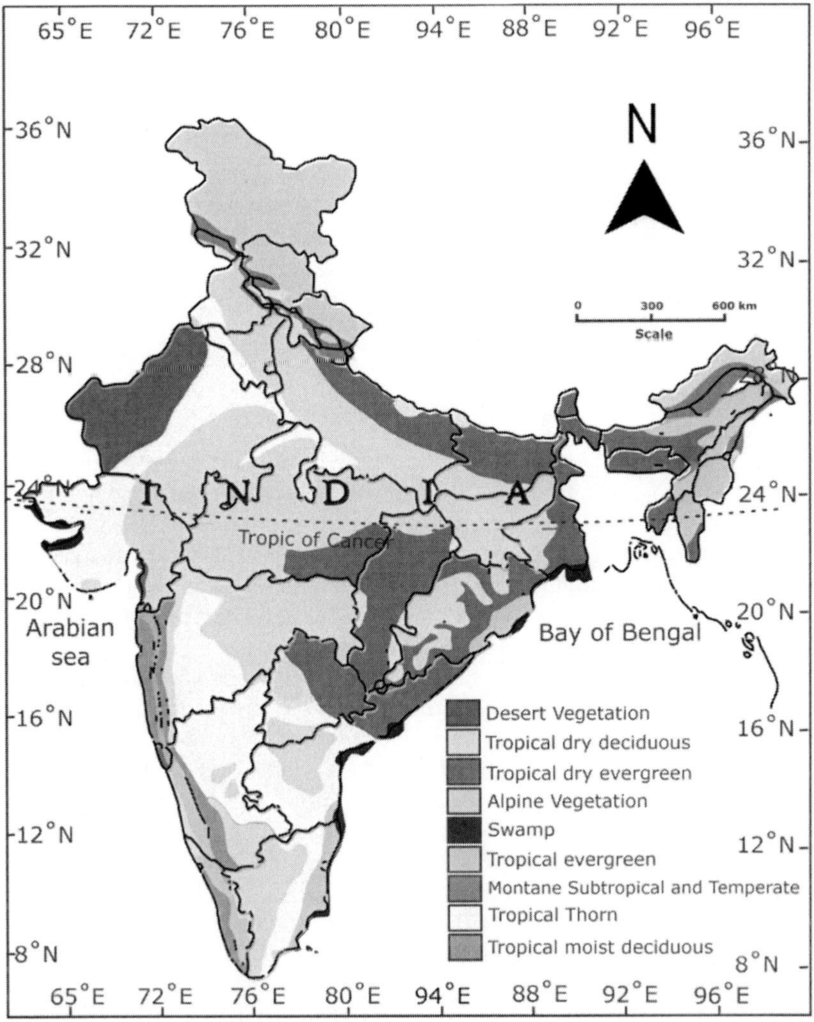

Fig. 7.1: GIS map of Indian forest types

In the Indian Constitution, forests are listed in the Concurrent List, which can be found in the Seventh Schedule. The 42nd Amendment Act of 1976 shifted responsibility for matters pertaining to forests and the protection of wild animals and birds from the state to the concurrent list. In accordance with the provisions of Article 51 A (g) of the Constitution, it is the primary duty of each individual citizen to preserve and improve the natural environment, with a particular focus on forests and animals. According to the Directive Principles of State Policy, Article 48 A, the state is obligated to make every effort to preserve and improve the environment, as well as to protect the nation's wildlife and woodlands. The National Forest Policy of 1988 is what is used to administer India's forests now, and its primary goals are to maintain ecological harmony and protect people's livelihoods. The India State of Forest Report 2019 found that tree and forest cover represented for 24.56 percent (8,07,276 sq km) of the total land area in India. Long-term goals for the nation include increasing the percentage of land that is covered by forest cover to 33 percent.

Champion and Seth classification of forest

Champion and Seth classify Indian forest types into 16 categories:

a) Tropical wet evergreen forest: The Western Ghats, the Nicobar and Andaman Islands, and the whole of the north-eastern region are all covered with wet evergreen forest. Evergreen trees stand tall and erect here, giving it a unique appearance. The jackfruit tree, the betel nut palm, the jamun tree, the mango tree, and the hollock tree are the most common trees that can be found here. The trees in this forest are arranged in tiers, with the lowest layer being covered by shrubs, the next layer having trees with short structures, and the highest layer having tall trees. It thrives in areas that are warm and humid, with an annual mean temperature of more than 22 degrees Celsius and an average precipitation of more than 200 centimetres. The tree trunks are covered with stunning orchids of a wide variety of species, as well as ferns of every imaginable hue.
b) Tropical semi-evergreen forest: The Eastern Himalayas, the Western Ghats, and the Andaman and Nicobar Islands all feature tropical semi-evergreen forests. These woodlands have a mixture of wet evergreen trees and moist deciduous trees throughout their composition. The forest is quite extensive and well-forested, and it has a broad variety of coniferous and deciduous trees. Hollock, white cedar, and kail are the species that are most often seen.
c) Tropical moist deciduous forest: This kind of forest may be found in regions that get rainfall of between 100 and 200 centimetres. There are wet deciduous forests in almost every region of India, with the exception of the western and north-western parts of the nation. The trees are rather tall, with trunks that are both wide and branched, and roots that keep them firmly established in their locations. Some of the taller trees lose their leaves during the dry

season because of the lack of moisture. There is a layer of shorter trees and evergreen plants located in the undergrowth of the area. In addition to mango, bamboo, and rosewood, sal and teak are the most common types of trees found in these forests.

d) Littoral and swamp: You may find littoral and swamp forests in the deltas of the Ganges and Brahmaputra rivers, as well as surrounding the Andaman and Nicobar islands. They have fine tissue roots that provide the plant the ability to breathe when it is submerged in water. India has a total mangrove forest area of 6,740 square kilometres, which is equivalent to 7 percent of the total mangrove forest area on the planet. The woodlands contribute to the shoreline's stability and assist prevent erosion along the coastal zones. The Sunderbans is the largest tidal forest in the world, and it is located in the Ganges delta in India.

e) Tropical dry deciduous forest: Dry deciduous woodlands may be found all across the northern part of the country, with the exception of the north-eastern region. Additionally impacted are the states of Madhya Pradesh, Gujarat, Andhra Pradesh, Karnataka, and Tamil Nadu. The canopy of the trees often does not reach farther than 25 metres in any direction. The sal, the acacia, and the bamboo are the most common types of trees.

f) Tropical thorn forest: Areas with dark soil, such as the northern, western, central, and southern parts of India, are home to this kind of forest. The trees seldom reach heights more than 10 metres in height. The spurge, caper, and cactus that may be found in this region are well-known. The places with annual precipitation of less than 50 centimetres are ideal for the growth of the woods. The plants do not have any visible leaves for the vast majority of the year.

g) Tropical dry evergreen forest: You may find dry evergreens near the shores of Karnataka, Andhra Pradesh, and Tamil Nadu. It is mostly made up of evergreen trees with prickly leaves and fragrant flowers, but there are also some deciduous trees in there for good measure. There are many different kinds, but among of the more common ones include pomegranate, olive, and oleander.

h) Subtropical broad-leaved forest: Broad-leaved woods are located in the Western Ghats and the Silent Valley in the Eastern Himalayas. There is a clear distinction in the kind of plants that may be found in each location. The Silent Valley is home to a wide variety of plants, the most common of which are poonspar, cinnamon, rhododendron, and fragrant grass. The altering agricultural practises and forest fires have had a devastating effect on the plant life of the Eastern Himalayas. These wet woods are dominated by evergreen trees, with just a few deciduous trees scattered throughout for variety. Oak, alder, chestnut, birch, and cherry are some of the tree species that may be found here. Orchids, bamboo, and creepers come in a vast variety of species and varieties.

i) Subtropical pine forest: Pine forests may be found on the high, arid sides of the Shivalik Hills, the Western and Central Himalayas, the Khasi, Naga, and Manipur Hills. These areas are all located in the Indian subcontinent. In these areas, you may find trees such as chir, oak, rhododendron, and pine. The lower regions of these areas include trees such as sal, amla, and laburnum.
j) Subtropical dry evergreen forest: Dry evergreen forests often have a protracted season that is both hot and dry, followed by a season that is characterised by low temperatures. It is characterised by evergreen trees that have shiny, lustrous foliage in most cases. Up to a height of one thousand metres, these types of forests are located in the Shivalik Hills and the foothills of the Himalayas.
k) Montane wet temperate forest: Montane wet temperate forests are found in the area that extends eastward from Nepal into Arunachal Pradesh and receives an average of 2,000 millimetres (mm) of precipitation annually. The northern forests are composed of three distinct layers: the top layer is dominated by coniferous trees, the middle layer is composed of deciduous trees like oak, and the bottom layer is dominated by rhododendron and champa. In the southern part of Kerala, it is possible to locate it at the Niligiri Hills, which are located in the higher altitudes. The northern woodlands are home to a greater variety of plant life than their southern counterparts. There is a possibility of discovering rhododendrons and other types of ground plants in this region.
l) Himalayan moist temperate forest: You may find this kind of forest everywhere in the Himalayas, from the Western to the Eastern Himalayas. The western region is home to a variety of tree species, including broad-leaved oak, brown oak, walnut, rhododendron, and others. The Eastern Himalayas get a larger amount of rainfall than the Western Himalayas, which results in vegetation that is more verdant and dense. There are a number of different species of broad-leaved trees, ferns, and bamboo. Coniferous trees may also be found in this region; however, the species that grow here are distinct from those in the southern regions. Located all over the belt at elevations ranging from a thousand to two thousand metres.

m) Himalayan dry temperate forest: This kind of forest may be found in the Himalayan areas of Lahul, Kinnaur, and Sikkim, among other places. The majority of the trees are coniferous, although there are also some broad-leaved trees including oak, maple, and ash. At higher heights, you may find fir, juniper, and deodar trees, as well as chilgoza.

n) Subalpine forest: The subalpine woods may be found between an elevation of 2900 and 3500 metres and reach from Kashmir to Arunachal Pradesh. Juniper, rhododendron, willow, and black currant are the most common types of plants that may be found in the Western Himalayas' flora. In the eastern parts, you'll find an abundance of trees such as red fir, black juniper, birch, and larch. The timber line in this part of the world is much higher than in the West. This is due to the copious rainfall and high humidity. With these regions, the hillsides are covered in rhododendrons of a wide variety of species.

o) Moist alpine scrub: Moist alpine vegetation may be found all across the Himalayas, particularly on higher peaks that are located close to the border with Myanmar. It is characterised by a thick evergreen forest with low scrub that is predominated by rhododendron and birch trees. The surface of the ground is covered with moss and fern patches. Snowfall totals may be rather high in this region.

p) Dry alpine scrub: Dry alpine scrub may be found at altitudes that range from 3000 to 4900 metres above sea level. The majority of the plants that grow there are dwarf varieties, such as black juniper, drooping juniper, honeysuckle, and willow.

CHAPTER 8

AQUATIC ECOSYSTEM

Divisions of Aquatic Ecosystem

Aquatic ecosystems are defined as those that mostly consist of water as their environment. The salt content that is present in aquatic ecosystems is what defines their category.

a) Ecosystems of fresh water: The amount of salt in fresh bodies of water is exceedingly minimal, and it is never more than 5 ppt (parts per thousand). Examples include lakes, ponds, pools, springs, streams, and rivers. There are three different types of freshwater ecosystems: lentic, lotic, and wetland habitats. Lentic ecosystems have slow-moving water, such as lakes, ponds, and pools. Lotic ecosystems have rapid-moving water, such as streams and rivers. Wetland environments have a combination of both slow- and rapid-moving
b) Marine ecosystems are defined as bodies of water that have salt concentrations that are comparable to or higher than those of sea water (i.e., 35 ppt or above). Consider the contrast between the shallow seas and the vast ocean.
c) Brackish water environments: these types of bodies of water have a salt content that falls between 5 and 35 parts per thousand. wetlands such as estuaries, salt marshes, mangrove swamps, and forests, for instance.

Factors limiting the productivity of Aquatic ecosystem

a) Sunlight: The amount of sunlight that is able to penetrate the water diminishes drastically as one moves deeper into the ocean. The depth at which light is able to penetrate a lake is a factor that influences the amount of plant distribution that occurs. On the basis of the light penetration and the distribution of the plants, they are separated into photic and aphotic zones. The photic zone is exposed to sunlight, which is necessary for the process of photosynthesis to take place. Because conditions in the photoic zone are more favourable for the maintenance of life, a greater variety of organisms may be found there. The part of the ocean where there is no exposure to direct sunlight is known as the aphotic zone. This region cannot be seen with the naked eye at all. Photosynthetic organisms are unable to survive in this zone because there is insufficient exposure to sunlight. There is not a lot of biological variety here, and just a few different kinds of animals call this area home.

b) Dissolve Oxygen: In aquatic ecosystems, oxygen is often found in a dissolved state in the water, and the concentration of this oxygen is continuously shifting as a result of the many factors that influence oxygen intake and output. The concentration of oxygen in a given amount of air is about 150 times higher than the concentration of oxygen in a given volume of fresh water, which has an average dissolved oxygen content of 0.0010 percent (also reported as 10 ppm) by weight. Oxygen is brought into the aquatic environment via the interaction of air and water as well as the photosynthetic process carried out by aquatic plants. Because of this, the rate at which the two processes described above take place is a major factor in determining the total quantity of dissolved oxygen that is present in an environment. Through the interaction of air and water and the respiration of living organisms, dissolved oxygen is expelled from the body of water (fish, decomposers, zooplanktons, etc.). The quantity of oxygen that is still dissolved in water may be affected by the temperature of the water. Oxygen solubility is decreased in water at higher temperatures. Additionally, the activity of decomposers is increased by warm water. As a consequence of this, increasing the temperature of a body of water causes the rate at which oxygen is extracted from the water to pick up speed. If the concentration of dissolved oxygen in the water falls below 3–5 ppm, then many aquatic organisms will die.

c) Transparency: the amount of light that may pass through a material is influenced by its degree of transparency. Clay, silt, phytoplankton, and other forms of suspended particulate matter are the primary contributors to turbidity in water. As a consequence of this, the amount of light that is able to penetrate and the amount of photosynthetic activity are both greatly reduced.

d) Temperature: The temperature of water changes more slowly than the temperature of air because water has a much higher specific heat than air does. This indicates that a greater quantity of heat energy must be added to or removed from water in order to increase or lower its temperature. The temperature range that aquatic organisms can tolerate is less than that of land animals because water temperatures are more stable. Because of this, even very minor shifts in water temperature pose a considerable threat to the continued survival of aquatic life. This is in contrast to the effects of changes in air temperature on terrestrial organisms.

Aquatic organisms: The three most common categories of aquatic life are called plankton, nekton, and benthos. The three distinct categories of organisms are distinguished from one another by the environments in which they live and the modes of travel they use.

a) Plankton: Plankton are microscopic organisms that are only able to live in water and are unable to swim across it. They include phytoplankton in addition to zooplankton in their make-up. Bacteria and algae that create food via the process of photosynthesis are collectively referred to as phytoplankton. Microscopic organisms known as zooplankton are responsible for the consumption of phytoplankton.
b) Nekton: Nekton are creatures that live in water and may move about by swimming or using other methods in addition to swimming. Nekton are classified as aquatic organisms. Nekton may include both invertebrates like shrimp and vertebrates like fish.
c) Benthos: Benthos are marine organisms that are found living on the surface of the sea bed. They make their home in or on the sediments at the bottom. Benthic creatures include things like sponges and clams, as well as a few other types of marine life.

Fresh water Ecosystem

Freshwater ecosystems take into account both flowing and motionless bodies of water, in addition to the creatures that make their homes in those environments. Lentic ecosystems are those that are found in still bodies of water, whereas lotic ecosystems are those that are found in moving bodies of water. The habitat of these ecosystems is actively regulated by almost all ecological parameters, including temperature, sunlight, pH, dissolved gases and salts of water, turbidity, alkalinity, salinity, depth, and spatial distribution. This section will discuss the biological features that are unique to lotic ecosystems, such as rivers and lakes. The following are some of the components that make up the lotic ecosystem:

a) River as an ecosystem: Water is a vital component for supporting life. The locations of human settlements most often selected are those in close proximity to surface water supplies. The vast majority of human civilizations also emerged along water channels, notably in regions around large rivers. The accumulation of raindrops over time forms rivers, which are massive natural channels for the flow of water. Because of gravity, the water at the surface will always flow in a downward direction. There are many different sorts of water courses, which may be categorised according to their size and where they are located using terms like streams, tributaries, brooks, creeks, and springs. The number and length of a river's tributaries, as well as its stage of development, catchment area, and geomorphological qualities all have a role in determining the river's classification as large, medium, or small. Every significant river must have a place of origin upstream, sometimes referred to as the headwaters, and a point of convergence with the sea or another body of water downstream. Both of these points are known as confluence points. Each river has its own distinctive longitudinal profile as well as cross-sectional profile. The cross-section of a river changes as one travels from the headwaters to the mouth of the river. River valleys may be anything from deep gorges and canyons to broad, flat streams as they get closer to the delta. River valleys come in all different shapes and sizes. Geologically speaking, a river is a tremendously powerful agent. It may also erode, move, and deposit sediments in addition to depositing them. River alluvium is another name for these sediments. There are three distinct categories that may be applied to rivers: Because it is a perennial river, the flow of water will remain consistent throughout the year. Flow in intermittent streams is dependent on the season. Ephemeral streams are defined by their inconsistent or temporary flow patterns.

b) Limiting factors and structure: Streams are categorised as either being isothermal or non-isothermal depending on the temperature of their water. The following is a list of the most important abiotic elements that have an effect on lotic ecosystems:

i. The circumstances of the slope and the geomorphology of the area, in addition to the make-up of the substratum.

ii. The water's physicochemical properties and qualities. There are several aspects to take into consideration, including temperature, colour, alkalinity, pH, and dissolved oxygen.

iii. The amount of flow and its velocity.

iv. The kinds and amounts of sedimentsthat are suspended in the water and carried by the bedload.
v. Turbidity.
vi. The amount of light that is able to go through the water and the height of the water column.
vii. Variables associated with the climate, include air temperature, humidity, sunshine hours, evaporation, and wind speed.

c) Characteristics of lotic adaptations: The impacts of continually flowing water, pollution, suspended sediments, floods, and other human activities have an influence on lotic habitats. These effects may be detrimental to lotic ecosystems. Animals and plants that dwell in lotic ecosystems have special adaptations. They are subjected to a broad variety of changing environmental circumstances, such as water currents, pollutants, and sediments that are suspended in the air. The following is a list of some of the distinctive qualities of moving water:

i. River life has a stream-like appearance.
ii. The establishment of a solid link to the substratum. Examples of these organisms include the vast majority of sponges, diatoms, and moss. They make their homes on stones, logs made of wood, and rocky outcroppings.
iii. The swimmers are expected to have hooks or suckers on their feet in order to keep their grasp on the polished surfaces.
iv. In order to secure a steady supply of food, some of them erect nets around themselves.
v. Some of them, like snails and worms, could have sticky bottoms that enable them to move along the base of the structure.
vi. Rheotaxis is a phenomenon that may be seen in rivers. This is the capacity or mechanism that enables fish and other species to swim against currents and quickly moving water. It may also be called "anti-current swimming."
vii. Clinging habitat is another aspect of river life that must be taken into consideration. Certain organisms have a strong desire to get as near as they can to solid substances or materials.

d) Life along rivers: Rivers are home to an astonishing variety of life, including fish, plants, animals, and untold numbers of microscopic organisms that are invisible to the naked eye. In addition to this, trees and shrubs tend to grow along the riverbanks, which provides a safe habitat for a variety of animal and bird species. River waters are home to a plethora of diverse organisms, each of which plays a vital role in the maintenance of the ecosystem as a whole and the provision of food. They are capable of acting as both feeders and collectors, in addition to grazers. They contribute to the breakdown of plant matter that accumulates along streams or is dropped from trees and shrubs that hang above. In order to strengthen their shells, river snails must first digest the calcium that is found in the water. Some trees and plants not only shield other kinds of life from the sun's harmful rays, but they also filter harmful chemicals out of the air and pull trace metals from sediments. Rivers tend to be places where interactions between predators and their prey are more common. When it comes to fish, bigger fish in rivers end up eating the smaller ones, whereas smaller predatory species end up parasitizing the larger ones. To get fresh water for drinking, fishing, cooking food, bathing, cleaning, and living, local and migrating birds, snakes, frogs, and other terrestrial species, including cattle and humans, travel to the river. Other terrestrial creatures also go there.

e) Longitudinal zonation: Rapids and pools are the two primary kinds of ecosystems found in streams. As a direct consequence of this, there are two distinct communities of stream life forms: rapid communities and pool communities. The kind of stream bottom and the population density in a river community are the primary factors that define the nature of the river community. Sand, pebbles, clay, bedrock, or rubble rock are some of the possible components of the riverbed. Torrential fauna is the name given to the population that lives in the rapids because of its susceptibility to the turbulence caused by the currents. A blackfly larva that was discovered on the rock bottom is an illustration of an example that falls into this category. The species that dig burrows and make their homes along the stream banks or bottom make up a significant portion of the population that lives in the pools. For instance, the mayfly nymph and the dragonfly nymph. There is also a zonation present in the communities found in the streams. The species that live in headwaters are unique from those that live in deltas. Changes in temperature, the velocity of water flow, and the quality of the water, including pH, are the root causes of the gradual shifts that occur in community structures.

Lake Ecosystem

Lakes are large, naturally occurring pools of fresh water that stand still and are produced when depressions in the earth's surface are filled with water from precipitation, runoff, and groundwater seepage. Lakes may be quite deep. These depressions may be caused by glaciation (as in the case of North America's Great Lakes) or crustal displacement (as in the case of East Africa's Lake Nyasa), and they can be as vast as miniature seas. There are a lot of similarities between the characteristics of oceans and those of huge lakes. The depth of a body of water as well as its distance from its adjacent shoreline are used to categorise lakes and ponds into one of four "Zones." They have been divided into the following four zones, which may be found in Fig. 8.1:

a) Littoral Zone: The topmost zones along a lake or pond's margin are known as the littoral zones. This Zone is the warmest because it is shallow, it is well-lit, and it can absorb more of the heat that the sun has to offer. It is home to a wide range of organisms, including grazing snails, worms, insects, crustaceans, fishes, and frogs, as well as rooted and floating aquatic plants.
b) Limnetic Zone: The near-surface open water that is bounded on each side by the littoral zone is what the limnetic zone is defined as. This region is characterised by well-lit open surface water that is located at a considerable distance from the shore and extends to depths that may be illuminated by light. This part of the lake is home to phytoplankton, zooplankton, and higher animals. It is also responsible for producing food and oxygen for the bulk of the lake's consumers.
c) The Profundal Zone: The profundal zone is the portion of the lake that is the deepest. This zone is both much colder and thicker than the two that came before it. Only a little amount of light is able to penetrate the limnetic zone and enter the profundal zone. Deep aphotic zones that are devoid of oxygen and too dark for photosynthesis make up this part of the image. In these murky, cold waters, you'll find species of fish that have adapted to the conditions.
d) The Benthic Zone: This zone refers to the bottom of the lake. There are organisms that are able to survive in environments with low quantities of oxygen and low temperatures. The majority of the inhabitants of this zone are decomposers, animals that feed on debris, and fish that migrate from one zone to the next. It is predominantly nourished by detritus that is washed in from the littoral and limnetic zones, in addition to sediments that have been washed into lakes.

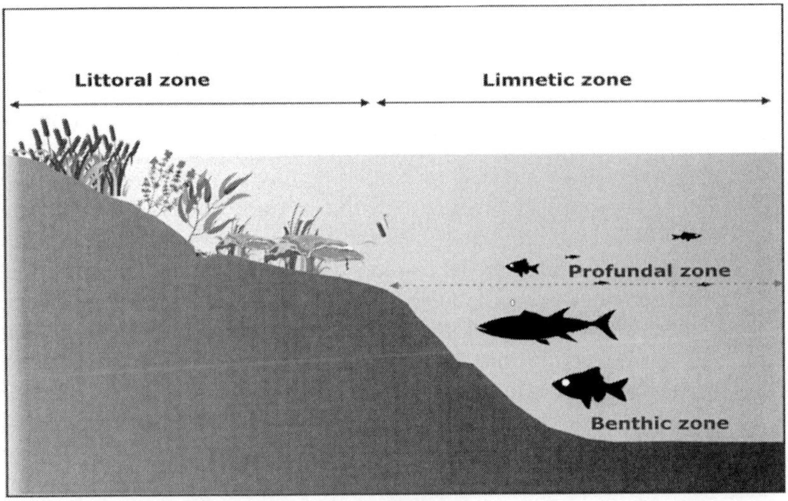

Fig. 8.1: Diagram showing divisions of lake ecosystem

Lakes are often categorised according to the amount of organic matter that they yield. Alternatively, lakes have been separated into the three types that are mentioned below according to the amount of nutrients that they contain or their overall productivity.

a) Oligotrophic: An oligotrophic, also known as an undernourished lake, is a lake that has only been around for a short period of time and has a restricted supply of plant nutrients. A lake like this one is quite deep and has a low temperature. It also has steep banks and a very small surface area in relation to its depth. Due to the lack of nutrients in this lake, there are very few phytoplanktons here. Because these lakes are not highly productive, there is not a significant amount of life to be found in them. The waters of an oligotrophic lake are often very pure, since there are typically very few deposits of organic debris that may be broken down. These kinds of lakes, like Lake Baikal in Siberia, are often found in untouched regions that are free from the influence of humans.
b) Eutrophic: Over time, sediments, organic waste, and organic nutrients wash into the majority of oligotrophic lakes, which encourages the development of both plants and animals. As a result of their breakdown, sediments of various types are produced in the benthos. A lake is said to be eutrophic (well-nourished) when it contains an adequate or an excessive supply of the nutrients

(mostly nitrates and phosphates) that are necessary for producers. These lakes are often on the shallow side, with murky water that is either brown or green and poor visibility overall. Because of the high concentration of organic materials in the benthos, the rates of decomposition are accelerated, which may cause oxygen levels to drop. Eutrophication may be caused by human activities, which can lead to a significant increase in the amount of nutrients entering lakes (called cultural eutrophication).

c) Mesotrophic: Several lakes are considered to be mesotrophic because their levels of nutrient enrichment fall somewhere in the middle of the two extremes. They have a moderate nutrient content and a moderate amount of phytoplankton, and they have a moderately productive rate. These kinds of lakes might be viewed of as a transitional phase between oligotrophic and eutrophic phases of lake development.

Eutrophication

The process of eutrophication describes the addition of nutrients to bodies of water, whether those nutrients are introduced naturally or deliberately, as well as the impact of those nutrients (Fig.8.2). When it results in unfavourable outcomes, eutrophication has the potential to be categorised as a kind of pollution. The term "eutrophication" refers to the process through which inorganic plant fertilisers enrich bodies of fresh water (e.g., nitrate, phosphate). It may happen naturally, but it can also happen as a consequence of human activity, andit is most visible in slow-moving rivers and shallow lakes. Increased sediment deposition may ultimately cause a lake or river bed to rise in level. This will enable land plants to colonise the borders of the body of water, which will finally lead to the transformation of the region into dry land. The natural process of eutrophication takes place over the course of millennia as lakes become older and fill up with sediments. However, human activities have accelerated the rate of eutrophication and expanded its scope. These activities have done so by injecting limiting nutrients, such as nitrogen and phosphorus, into aquatic ecosystems via point-source discharges as well as through non-point loadings. It wasn't until the middle of the twentieth century that eutrophication was recognised as a significant source of pollution in the lakes and reservoirs of Europe and North America. To put it another way, eutrophication is made worse by the increased flow of inorganic nutrients and organic compounds into lakes, which is caused by increased runoff from agriculture and construction, contamination from sewer tanks and sewage, as well as other human-related activities.

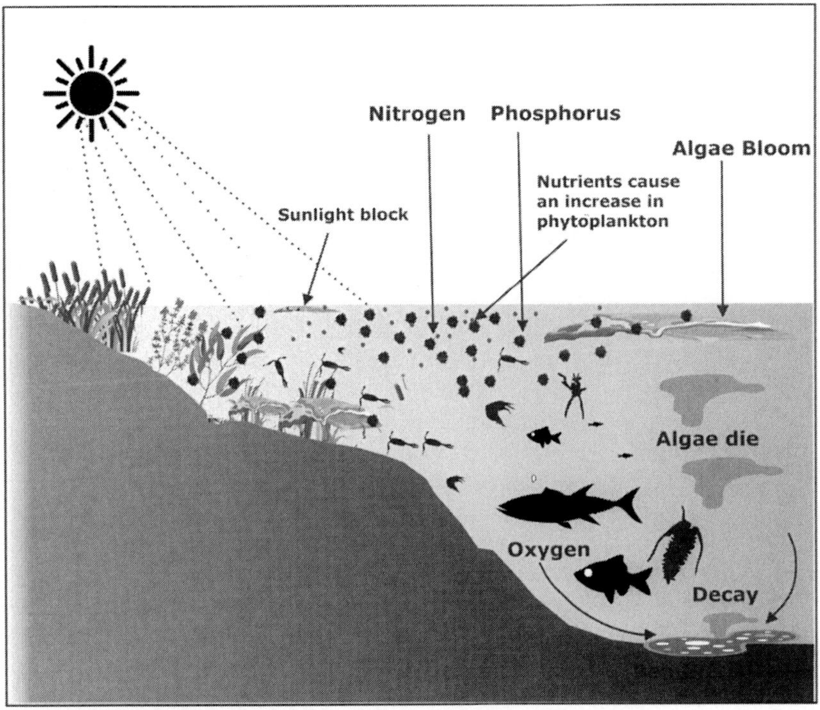

Fig. 8.2: Eutrophication

Effects of Eutrophication

a) Algal Blooms: An algal bloom, also known as a marine bloom, is a rapid increase in the population of algae found in an aquatic environment. Blooms of algae may form in environments consisting of either freshwater or saltwater. Typically, only one or a few species of phytoplankton are active at any one time, and particular blooms may be recognised by the colouring of the water that is generated by a high concentration of pigmented cells. Algae may be considered to be blooming in levels ranging from hundreds to thousands of cells per millilitre, depending on the species that is causing the problem, even though there is no recognised threshold level that exists. It's possible for there to be millions of cells in only one millilitre of water when there's a bloom of algae. It has been observed that the colours green, yellowish-brown, or red are present. There is also the possibility of blossoms with a bright green colour. The presence of blue-green algae is to blame for these. The presence of an excessive amount of nutrients (particularly phosphorus and nitrogen) in water is thought to be the cause of certain algal blooms. Higher concentrations of these nutrients in water are thought to encourage the development of algae and other green plants. This organic material that has begun to decompose provides food for the microbes. Because there is now a greater supply of food, the bacteria are able to grow, which causes them to consume the dissolved oxygen in the water. When there is a decrease in the quantity of dissolved oxygen, it becomes impossible for many species of fish and aquatic invertebrates to survive. As a direct consequence of this, a dead zone has developed. The production of neurotoxins by some species of algae makes algal blooms another potential cause for concern.

b) Harmful algal blooms (HABs): When these toxins reach the high cell densities obtained during certain blooms, they may have detrimental biological impacts on animals. Harmful Algal Blooms, often known as HABs, are blooms of phytoplankton that are recognised for their ability to naturally produce biotoxins. Even though the colour of the sea surface caused by individual species blooms may range from red to brown to yellow to green or even white, these types of blooms are commonly referred to as "red tides" and are often associated with unpleasant or deadly outcomes. HABs are also being found in regions where they had not been seen or known to exist before. In other instances, species of algae that were previously considered harmless or were even unknown have suddenly developed into ones that cause problems. Hazardous algal blooms have the potential to cause the extinction of marine life and damage to aquaculture systems. Diverse organisms are responsible for the production of lethal toxins, which may be passed up the food chain to eventually affect humans in addition to seabirds, marine mammals, and other marine life. Some species are dangerous because they may cause a wide variety of diseases in humans. These illnesses can be acquired by breathing airborne toxins, coming into touch with the skin, or, most often, by eating infected seafood. These poisons have the potential to cause amnesia, stomach aches, nausea, memory loss, and even death in susceptible individuals.

c) Hypoxia: When the dissolved oxygen content in a body of water drops, it is known as hypoxia. It's possible that the algae will obstruct the light after they die. The process of degrading them is carried out by bacteria, which use oxygen in the process. This results in the water becoming anoxic, which is dangerous for aquatic life.

d) Fish and other aquatic animals are exterminated on a massive scale.
e) Water may have a poor flavour, colour, and odour, all of which are detrimental to the environment and the areas around it.
f) It also reduces biodiversity since it leads to the extinction of species.
g) The toxins that are generated by some species of phytoplankton may cause serious symptoms in humans, including diarrhoea, memory loss, paralysis, and even death in the most severe of cases.

Estuary Ecosystem

The area at the mouth of a river or stream where it meets the ocean is known as an estuary. It is a coastal region that is only partially enclosed, has brackish water (with a salinity that ranges from 0 to 35 parts per thousand), has one or more rivers or streams that flow into it, and has an open connection to the ocean. Estuaries are areas where freshwater from the land, which contains fertile silt and runoff, mixes with salty water from the ocean. Estuaries act as ecological transition zones (ecotones) between the habitats of rivers and the ocean. River mouths, coastal bays, tidal marshes, lagoons, and deltas are all types of ecosystems that can be classified as estuaries. Rising sea levels, the movement of sand and sandbars, the activities of glaciers, and other geological processes can all lead to the formation of estuaries. Every plant and animal that lives in an estuary is subjected to varying salinities, to which they have adapted over time. Estuaries are significantly influenced by the activity of the tides. They are washed in sea water once or twice daily, the frequency of which is determined by the number of tides. In certain estuaries with a narrow mouth, tidal bores are very noticeable. Tidal bores have a significant influence on the ecosystems that are found in estuaries. The word "lagoon" refers to a body of salty water that is cut off from the ocean by a narrow strip of land or a reef of coral.

Depending on factors such as their geomorphology, water circulation and stratification, and system energetics, estuaries can be described in a variety of different ways.

Based on geomorphology

a) Drowned river valleys
b) Ford-type estuaries
c) Bar- built estuaries
d) River delta estuaries
e) Tectonic estuaries

Based on water circulation and stratification

a) Highly stratified or salt wedge estuaries
b) Partially mixed or moderately stratified estuaries
c) Completely mixed or vertically homogenous estuaries
d) Hyper saline estuaries

Based on ecosystem energetics proposed by Odum

a) Physically stressed system of wide latitudinal range.
b) Natural arctic ecosystem with ice stress
c) Natural temperate coastal ecosystem with seasonal programming.
d) Natural tropical coastal ecosystem of high diversity.
e) Emerging new systems associated with humans.

Vegetation and Importance

In order to be successful in estuaries, plants and animals must have adapted to the "brackish" conditions of the water. Two of the characteristics that have an effect on the distribution of organisms in an estuary are the salinity and the degree of flooding that occurs there. Estuaries provide a home for many different types of ecosystems, including mangrove forests, salt marshes, beds of seagrass, and mudflats. Because of the constant changes in river flow, tidal range, and sediment distribution, estuaries are among of the most dynamic and productive types of ecosystems. The majority of the components that make up phytoplankton in estuaries are called diatoms, dinoflagellates, green algae, and blue-green algae. Along the seacoasts of the estuaries, there is often a dense growth of both algae and seagrass. Around the mouths of rivers and in deltas are often times where one can find mangrove forests. Many species of plants and animals that live on land call estuaries their home, such as wood storks, pelicans, coniferous and deciduous trees, and butterflies. In addition, estuaries provide a home for a wide variety of aquatic plants and animals, such as sea turtles, sea lions, sea catfish, saltworts, eelgrass, salt grasses, cordgrasses, seagrass, sedge, and bulrush, amongst others.

More than 60 percent of the world's population lives in close proximity to the coast or an estuary. This makes estuaries the most densely populated locations on the entire planet. They are the most productive bodies of water in the world, even more so than wetlands, as a result of the mixing of freshwater and salty water zones where marine species from both ecosystems interact. Ecotone areas (also known as transitional zones) include mangroves, wetlands, estuaries, grasslands, and other similar ecosystems. The production in ecotone areas is significantly higher than that of natural ecosystems, such as forest ecosystems, ocean ecosystems, pond ecosystems, riverine ecosystems, desert ecosystems, and so on. This is as a result of the presence in the ecotone of a wide variety of organisms originating from ecosystems that are adjacent to it. Additionally, due to the fact that estuaries experience low levels of wave activity, they offer a tranquil refuge from the chaos of the open ocean, which makes them ideal for the continued existence of a wide variety of aquatic species. Large mangrove trees that are located on the landward side of an estuary can act as a barrier for the coastal environment and reduce the amount of wind speed that is experienced during storms and high-velocity landward winds. Estuaries act as a barrier between coastal catchments and the marine environment, while also storing, recycling, and trapping sediment. In addition, estuaries store nutrients. In addition to this, they act as a natural water filter by removing impurities from the water via a process of absorption, retention, and purification. As a result of the presence of wetlands, streams, lagoons, mangroves, and seagrass beds, estuaries have an abundance of natural resources, particularly fisheries. Due to the fact that they have the ability to remove carbon from the atmosphere and store it, they are frequently referred to as the blue carbon ecosystem.

Indian estuary

There are a total of 162 minor rivers, 44 medium rivers, and 14 big rivers in the nation, all of which empty into the ocean through a number of estuaries. The Bay of Bengal is home to a number of important estuaries. There are several estuaries that are home to some of the largest seaports in the world. The eastern coast of India is home to the bulk of the country's most important estuaries. On the other hand, the estuaries that are located on the west coast are much smaller. The Mandovi and Zuari estuaries are two instances of typical estuaries that may be found along the western coast. The Hooghly-Matlah estuary system is the largest of the Indian coast's estuaries. It covers the Gangetic delta known as Sundarbans, which is the world's largest delta equipped with the most mangrove forest. Its location in West Bengal (latitude 21° 23' N and longitude 88° 89' E) makes it the largest of the Indian peninsula's estuaries (4,264 kilometres squared).

Issues

a) Changes made to the land around estuaries can cause fluctuations in water flow in the estuaries themselves, which can either be significantly higher than what is required or significantly lower than what is required (e.g.,in Hooghly, Godavari, Pulicat etc.).
b) Pollution resulting from the operation of factories and the discharge of combined sewage from cities.
c) Fishing and boating for recreational purposes.
d) Shipbuilding, harbour maintenance, and navigation.
e) The growth of both urban and rural communities, as well as mining and industry, agricultural production, and the disposal of solid waste.
f) The target fish stock is being overexploited as a result of the growing demand for the fish.
g) Reclaiming the area around the centre for intensive aquaculture in pens.
h) The pathways that fish and prawn use to recruit new members are being blocked (e.g., in Chilka and Pulicat).
i) The reduction in biodiversity that results from the collection of prawn seeds and the use of nets with a fine mesh (e.g., in Hooghly, Chilka and Pulicat).
j) Catchment areas are being inundated as a direct result of the rising water level.

Mangrove Ecosystem

The environment in which mangroves forest thrive is known as a littoral (which means "near the sea"). Evergreen forests are typically found in tropical and subtropical regions, and they tend to form on protected low-lying beaches, estuaries, mudflats, backwaters, marshes, and lagoons. Coastal waters are held back on land in these areas. The high water levels that occur during spring tides are ideal for the growth of mangroves. The best places are either on the upwind side of beaches that have accumulated a lot of sand or where rivers have brought down a lot of silt. Trees in mangrove forests can reach heights of up to 8 or 20 metres, making these ecosystems among the most productive on Earth. They protect the coastline from the devastation that can be caused by storms and tsunamis. Since they are situated in the transition zone between land and sea, mangroves serve as an excellent example of an ecotone. Mangroves are a type of plant or small tree that can be found along the coast and thrives in salty or brackish water. The trees known as mangroves are able to thrive in harsh coastal environments because of their resistance to salt. Mangroves are also sometimes referred to as halophytes. As a result of the growth of mangroves, additional water is lost. The leaves are thick and contain glands that secrete salt into the environment. At the root level, certain plants are able to block salt absorption. They have a complicated salt filtration system in order to survive the constant exposure to salt water and the constant movement of the waves, and their root systems are similarly complicated. They can thrive in the anoxic conditions of wet mud, which have low oxygen levels. They do this by developing pneumatophores, which helps them overcome the difficulty of breathing in anaerobic soil conditions (blind roots). Mangrove forests are widespread across the world's tropics and subtropics, although the majority of them are located between the latitudes of 25 degrees north and south. They need a great deal of sunshine in order for their roots to be able to filter salty water. This goes some way to explaining why mangroves are only found along the coasts of oceans that are tropical or subtropical in temperature. There is a diverse variety of forms and dimensions that mangroves may take. Certain species, like as Rhizophora, have the ability to grow arching prop roots that reach into the water. Others, like Avicennia, stretch their "pneumatophores," often known as their air roots, vertically upward from the mud (Fig. 8.3.). Adventitious roots, also known as stilt roots, are those that grow from the main stem of a tree at a location that is elevated above the surrounding soil. Mangroves reproduce asexually via a process known as viviparity. That is, the seedlings develop inside the body of the tree itself (before falling to the ground). This is a strategy of adaptation that may be used to cope with the challenge of germination in salt water.

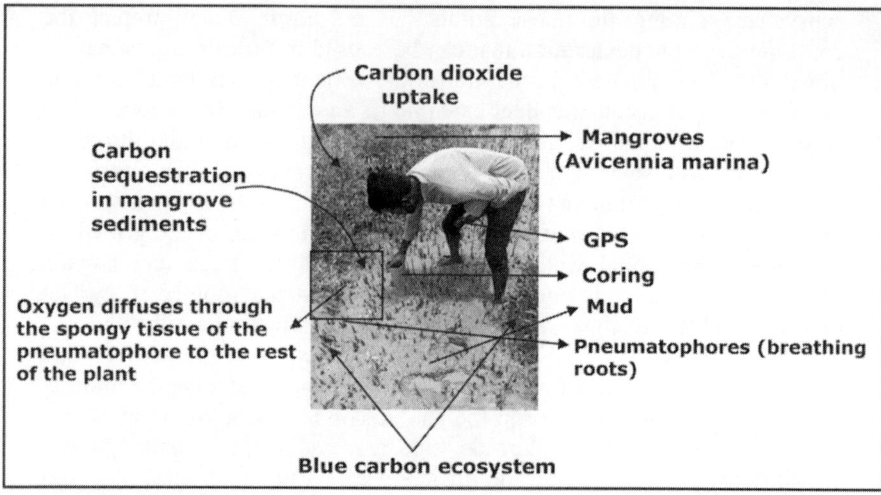

Fig. 8.3: Diagram showing Pneumatophores

Indian Mangroves

The mangroves that are found in the Sundarbans make up the world's biggest single block of tidal halophytic mangroves. Crocodiles and Royal Bengal Tigers both make their homes in this dense mangrove forest. The mangroves at this location are being cut down to make way for agricultural development. The mangroves of Bhitarkanika (which are located in Orissa) are the second biggest on the Indian subcontinent. They have a significant amount of genetic diversity and a high concentration of typical mangrove species. In the Andhra Pradesh deltaic areas that are fed by the Godavari and Krishna rivers, mangrove swamps thrive on the intertidal mudflats that are located on both sides of the creeks. The destruction of mangroves in Pichavaram and Vedaranyam, primarily, has been caused by the construction of aquaculture ponds and salt pans.

On the west coast of India, the states of Maharashtra, Goa, and Karnataka have mangroves that are often overgrown and in poor condition. In the coastal zone of Kerala, the mangrove forests are not very dense and are instead relatively sparse. The Gulf of Kutch and the Kori stream in Gujarat are the most common locations for the mangrove species Avicennia marina, Avicennia officinalis, and Rhizophora mucronata to be found (north-west coast). The health of the mangroves is becoming better, especially in the zone that is located around Kori creek, which is a paleodelta of the Indus river (once upon a time it was part of Indus delta).

Small, bushy areas of mangroves may be found in the Gulf of Kutch, whereas towering stands of mangroves can be seen in the Sunderbans. The tiny tidal estuaries and lagoons found on the Andaman and Nicobar Islands are home to a rich and diversified mangrove forest that has not been damaged (Fig.8.4).

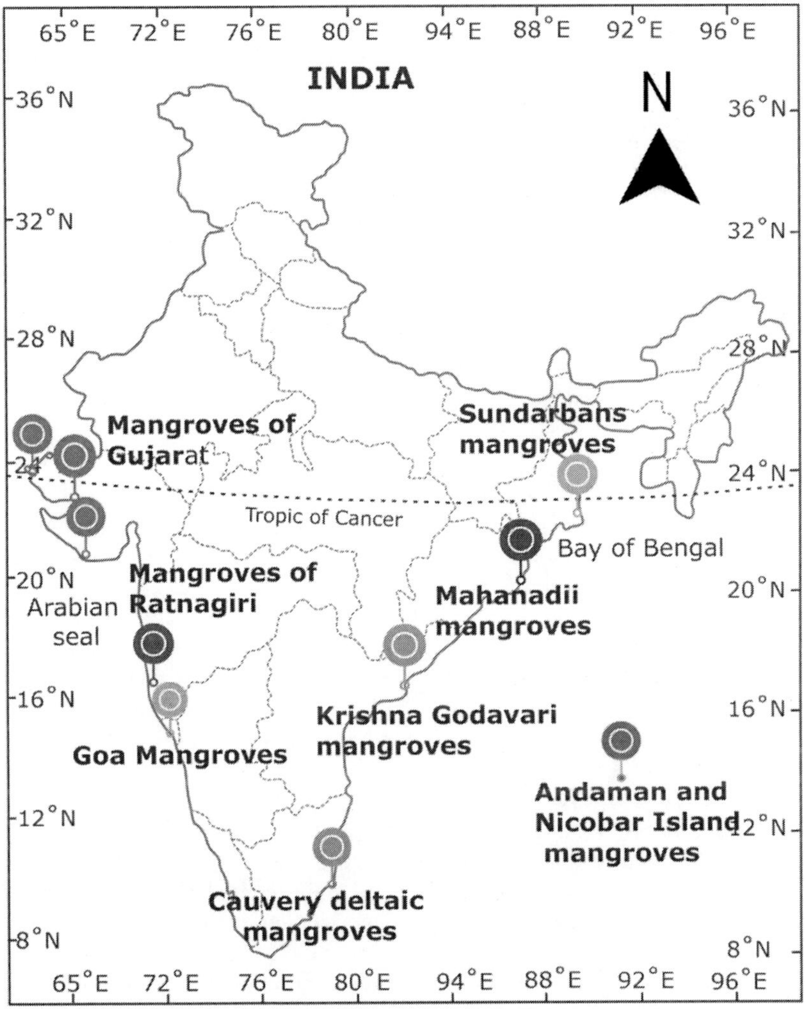

Fig. 8.4: GIS map of Indian mangroves

Importance of Mangroves

Mangrove plants feature (extra) characteristic roots such as prop roots and pneumatophores that aid to limit water flow and, as a result, stimulate silt deposition in locations (where it already exists), stabilise coastal beaches, and offer a breeding spot for fishes. These roots also allow the mangrove plants to create a stable environment for fishes to grow their young. In addition to this, it helps to lessen the severity of monsoon tide floods and protects the lowlands that are located along the shore from being flooded. They prevent the erosion of the coastal soil. They shield populations along the shore from the dangers of tsunamis, storms, and floods. Mangroves are an important part of the natural cycle of recycling nutrients. The mangrove ecosystem is home to a diverse collection of plants, birds, and animals. It is important to ensure that a variety of fish have a comfortable and secure habitat in which to reproduce, spawn, and raise their young.

They provide the locals with materials like as timber and firewood, as well as medicinal and delicious plants and herbs. They provide a plethora of employment alternatives for local individuals, which in turn increases their income.

Threats

a) Urbanization: They are eradicated for the purposes of agriculture, the production of fuel and feed, salinization, mining, oil spills, aquaculture (shrimp farming), and the use of chemical pesticides and fertilisers, as well as for industrial applications.
b) Elevated levels of CO2: Elevated levels of CO2 may be damaging to a range of mangrove species, leading to major changes in the vegetation that occurs along salinity and aridity gradients. On the other hand, it is not understood how exactly CO2 would affect mangroves.
c) Rise in sea level: An increase in the level of the sea is most likely the threat that is of the most concern to coastal mangroves. Recent studies have shown that mangrove ecosystems are vulnerable to extinction when subjected to more significant shifts in the level of the ocean. It has been reported that stronger cyclones cause damage to mangroves by stripping the trees of their leaves, uprooting the trees, and ultimately killing the trees. In addition to the mortality of trees, the character of the sediments in the soil is also changed. It is also possible that this will result in ecological conversion.

d) Rise in temperature: It is anticipated that an increase in temperature will have an effect on the species, and it is possible that an increase in sea level will have a significant effect on mangrove forests. The optimal temperature for photosynthesis in mangrove plants is between 28 and 32 degrees Celsius, which can be significantly decreased when leaf temperatures approach 38 and 40 degrees Celsius, resulting in a decrease in net output.
e) Decreased precipitation: It has been hypothesised that reduced precipitation and increased evaporation will lead to an increase in mangrove salinity, which will in turn lead to decreased production as well as poor seedling and growth survival. Altering the levels of competition between existing species can also result in a reduction in the richness of mangrove zones and a shrinking of the area occupied by mangroves.

Wetland Ecosystem

The presence of water is the fundamental feature that determines both the ecology of a wetland and the plant and animal species that resides inside it. Wetlands are one kind of habitat. They may be found in areas where the water table is at or very close to the surface of the ground, as well as in areas where the land is completely immersed in water. The term wetland refers to lands that are transitory among terrestrial and aquatic eco-systems, with the water table usually at or near the surface or the land covered by shallow water. This definition describes wetlands as lands that are transitory between terrestrial and aquatic eco-systems. These habitats are inundated on a regular basis by deeper settings that are located nearby; as a result, the flora and fauna that inhabit these habitats are uniquely adapted to survive under conditions of shallow flooding or waterlogging. Wetlands are separated from other types of ecosystems by the presence of water-loving plant life (hydrophytes) and hydric soils (inadequate oxygen). There are several different types of natural wetlands in India. These include elevated wetlands in the Himalayas, flood plains of major river basins, saline and momentary wetlands in dry and semi-arid regions, and coastal wetlands such as lagoons, backwaters, estuaries, mangroves, swamps, and coral reefs. World Wetlands Day is celebrated on February 2 each year all across the world. On February 2, 1971, in Ramsar, Iran, the Convention on Wetlands was signed, and this day is set aside to honour that event.

Types of wetlands

a) Coastal wetland: These are areas between land and open water that are not influenced by rivers or landforms such as shorelines, beaches, mangroves, or coral reefs. These wetland areas may be found between the land and the open water. A good example of this kind of wetland is one that has mangrove trees and is located in a protected tropical coastal location.
b) Shallow lakes and ponds: Shallow lakes and ponds are areas of water that are either permanent or semi-permanent but have a restricted amount of flow. Among these are lakes that only appear seasonally, spring pools, salt lakes, and lakes formed in volcanic craters.
c) Marshes: Marshes are characterised by herbaceous (non-woody) plants that are adapted to circumstances of moist soil and are periodically saturated, flooded, or ponded by water. Marshes may be found in wetlands. Marshes can be found in wetlands. Marshes may be either tidal or non-tidal, depending on whether or not they are near a body of water.
d) Swamps: These areas are characterised by the predominance of trees and bushes and get the majority of their water from surface sources. In both freshwater and saltwater floodplains, you may find wetlands in the form of swamps.
e) Bogs: Bogs are moist peatlands that may be found in the depressions of the terrain or in former lake basins. Bogs get almost all of their water from precipitation that falls from the sky.
f) Estuaries: The region where rivers meet the sea and the water changes from fresh to salt may be home to a wide variety of different kinds of flora and fauna. Wetlands may take a variety of forms, including deltas, tidal mudflats, and salt marshes.

Status of wetlands in India

The number of Ramsar sites in India is 49, covering a total area of 10,93,636 hectares as of February 2022. This is the greatest number in all of South Asia. Wetlands cover around 15.26 million hectares of land and account for approximately 4.6 percent of India's total land area. The Ramsar Convention has very specific rules that must be followed in order to conserve any wetlands that have been given that designation. There are over 2400 Ramsar sites located all over the globe, and they have a combined area of around 2.5 million square kilometres. On January 2, 2022, in celebration of World Wetland Day, a Ramsar site in Haryana,

Khijadia Wildlife Sanctuary in Gujarat, and Bakhira Wildlife Sanctuary in Uttar Pradesh were all designated as Ramsar sites. The event took place at Sultanpur National Park. The designation of the Haiderpur Wetland in Uttar Pradesh as the 47th Ramsar Site took place in the month of December 2021. In February of 2020, the Ramsar List was expanded to include the Sunderban Reserve Forest, also known as the Sunderban Wetlands, which is located close to Kolkata in West Bengal.

Importance of wetlands

Wetlands are essential because of the many benefits, also known as "ecosystem services," that they provide to humanity. These ecosystem services range from the provision of freshwater, food and building materials, and biodiversity to the management of floods, the recharge of groundwater, and the reduction of the effects of climate change. Wetlands provide a safe haven for a wide range of aquatic plant and animal species, in addition to resident and migratory bird populations. The preservation of wetland areas is essential to the growth of the tourist industry in the long run. They do so by removing impurities and nutrients from the surface water as well as purifying the water itself. They contribute to the recycling of nutrients, the recharging of groundwater, and the stabilisation of the local climate. The rate of runoff is one of the most critical aspects of flood management must be managed. They act as a storehouse of genetic material for a number of different plant species.

Threats to wetlands

a) Urbanization: Wetlands that surround metropolitan centres are being developed for residential, industrial, and commercial reasons at a rising rate as a direct result of urbanisation. Wetlands in urban areas are very necessary for the protection of public water supplies.
b) Agriculture: Significant portions of the wetland have been flooded and converted into rice fields. The hydrology of the wetlands that were affected was significantly altered as a result of the construction of a large number of reservoirs, canals, and dams for irrigation purposes.
c) Pollution: Since wetland ecosystems act as natural water filters, they help to lower overall levels of pollution. However, they are only capable of removing pesticides and fertilisers from agricultural runoff. They are unable to remove mercury from industrial sources or any other types of contamination. The influence of industrial

pollution on the biological diversity of wetland ecosystems and the water sources that are used for drinking is a major cause of worry.

d) The effects of climate change: shifts in precipitation patterns, an increase in the frequency of storms, droughts, and floods; rising sea levels; and a higher concentration of carbon dioxide in the atmosphere all have the potential to have an effect on wetlands.
e) Dredging: removing debris from a river bed or a wetland is accomplished through the process of dredging. The water table is lowered and the surrounding wetlands are dried out when streams are dredged.
f) Drainage: The process of draining wetlands involves digging canals into the ground, which then collect water and carry it away from the wetland. This leads to a drop in the water table, which in turn causes the wetland to dry up.
g) Non-native species: Wetlands in India are at risk from non-native plant species like water hyacinth and salvinia, which were illegally imported from other countries. They interfere with drainage systems and are a competitor to native plant life.
h) Salinization: The process of salinization is a consequence of the excessive extraction of groundwater.

Efforts towards wetland conservation

Ramsar convention

Key facts:

a) It is a multilateral agreement on the protection and sustainable use of wetland resources.
b) Additionally, it is referred to as the Wetlands Convention.
c) It was called after the city of Ramsar, which is located in Iran.
d) The signing of the Convention took place on February 2, 1971.
e) On February 2 each year, the international holiday known as World Wetlands Day is celebrated.
f) There were 170 parties at the conference (COP).
g) The "appropriate management" of wetland areas is at the core of the Ramsar concept.
h) It champions wise application, which involves preserving the ecological character of an area while yet allowing for long-term growth.

Need

a) Wetlands are necessary because of the many benefits that they provide, which are referred to as "ecosystem services." These services include the provision of freshwater, the production of food and building materials, the maintenance of biodiversity, the prevention of flooding, the recharging of groundwater, and the reduction of the effects of climate change.

b) During the course of the previous century, 64 percent of the world's wetland areas were eradicated.

Ramsar site criteria

a) The first criterion stipulates that "it contains a representative, unusual, or unique example of a natural or near-natural wetland type found within the eligible biogeographic region."

b) The second criterion asks if the solution "protects vulnerable, endangered, or severely endangered species or natural communities."

c) The third criterion states that "it maintains populations of plant and/or animal species vital to the ecological variety of a certain biogeographic area."

d) The fourth criterion asks if the area "supports plant and/or animal species at a vital point in their life cycles or whether it offers refuge under extreme circumstances."

e) The fifth criterion states that "it consistently provides a habitat for 20,000 or more waterbirds."

f) The sixth criterion is that the habitat "regularly supports one percent of the individuals in a population of one species or subspecies of waterbird."

g) According to the seventh criterion, "it supports a substantial number of indigenous fish subspecies, species or families, life-history stages, species interactions, and/or populations that are indicative of wetland advantages and/or values, hence contributing to global biological diversity."

h) The eighth criterion states that "it supplies a vital source of food for fishes, a spawning place for fishes, a nursery for fishes, and/or a migratory path on which fish populations depend, either inside the wetland or outside."

i) The ninth criterion stipulates that "it continuously supports 1 percent of the individuals in a population of one species or subspecies of wetland-dependent non-avian animal species."

Montreux Record:

a) Since it is included on the Ramsar List, its information is regularly reviewed and revised.
b) The Montreux Record is a database of wetland sites that have experienced, are experiencing, or are predicted to suffer changes in ecological character as a consequence of technological developments, pollution, or other human interference. These wetland sites are all included on the List of Wetlands of International Importance.
c) Keoladeo National Park in Rajasthan and Loktak Lake in Manipur are both included on the Montreux Record as wetland locations. The Chilka Lake (Odisha) entry in the record was formerly included but was afterwards removed.

Regulations of wetlands in India

a) The Wetlands (Conservation and Management) Rules of 2017 are the rules that regulate the management of wetlands.
b) The edition of the Rules that was published in 2010 included the establishment of a Central Wetland Regulatory Authority; however, the revised Rules that were published in 2017 replaced this authority with authorities at the state level and established a National Wetland Committee to act as an advisory body.
c) Under the new legislation, the terms "backwaters," "lagoons," "streams," and "estuaries" are no longer included in the definition of "wetlands."
d) Under the regulations that went into effect in 2017, the responsibility for determining how to classify wetlands was given to the individual states.

Wetlands International

a) Wetlands International is a non-governmental organisation (NGO) that operates on a worldwide scale and is committed to the protection and restoration of wetlands and the resources found within them for the benefit of both humans and animals.
b) Research, campaigning, and interaction with governments, companies, and international policy fora and conventions are all part of the work that Wetlands International does.

c) It has a headquarters in Ede, the Netherlands, as well as 20 regional, national, or project offices located on all continents, and its primary emphasis is on the areas of the globe that are still in a state of development. The nongovernmental organisation works to solve problems related to wetland areas in over one hundred countries and on a variety of sizes. It does this with the assistance of dozens of governmental, non-governmental, and corporate donors and partners, and it supports over 80 different initiatives.

National wetland Conservation Programme (NWCP)

a) The first year of operation for the NWCP was 1985–1986.
b) 115 wetland areas have been identified by the Ministry of the Environment as being in need of urgent management and conservation efforts as part of the plan.
c) The criteria used by the Ramsar Convention on Wetlands and the National Wetlands Conservation Program (NWCP) to designate wetlands of national importance are the same.
d) It is the government's obligation to be in charge of the overall coordination of the various wetland conservation programmes.
e) In addition to this, it provides guidance, financial help, and technical support to state governments.
f) Because state governments and the administrations of UT hold the land resources, they are the ones who are responsible for the management of wetland areas.

Aim: Protection of wetlands in order to halt any further degradation and to ensure that they may be used in a sustainable manner, both for the benefit of the surrounding community and for the general preservation of biodiversity.

Objectives

a) The provision of policy guidelines for the protection and administration of wetland areas is the primary objective.
b) To provide monetary assistance in order to facilitate the adoption of stringent conservation measures.
c) In order to monitor the manner in which the programme is being carried out.
d) To build a list of the wetlands found in India.

Key Initiatives to safeguard marine and coastal environments

a) **Coastal ocean monitoring and prediction system:** The levels of marine pollution have been monitored by the Ministry of Earth Sciences of the Government of India, which was previously known as the Ministry of Ocean Development. This monitoring takes place at around 80 different locations along the coastline of the nation. This particular programme goes by the name COMAPS. This program's key objectives are to investigate the current state and trends of the quality of the coastal marine environment over the long term and to bring the attention of government and public institutions to the repercussions of those findings. Since the academic year 1991–1992, the programme has been running. Data on approximately 25 environmental parameters, such as the physical, chemical, biological, and microbiological characteristics of water and sediment, are collected with the assistance of research and development institutes in the 0–10 km sector of these locations.

b) **Land-ocean interaction in the coastal zone (LOICZ):** Since 1993, the LOICZ has functioned as an international research project that brings together experts from all over the globe for the purpose of investigating changes in the chemistry, physics, and biology of the coastal zone. Since 2003, LOICZ has included the human aspects of the coastal zone into its research by broadening the scope of its study disciplines to include the social, political, and economic sciences. The results of the study were put to use in an investigation of the factors that contribute to and are affected by global change on coastal ecosystems, as well as the crucial part that people play in both driving change and adapting to it. This involves the fragility of socio-ecological coastal zone systems as well as their capacity to adapt in response to changing circumstances as well as decreasing or growing land and sea usage choices. The fundamental idea was to provide a scientific basis for more informed decision making, or alternatively, to preserve beaches and coastal goods and services for future generations: "provide the knowledge, insight, and prediction necessary to allow coastal communities to analyse, anticipate, and react to the interaction of global change and local forces that characterise coastal change" was the goal statement for LOICZ.

c) **The Integrated Coastal and Marine Area Management (ICMAM):** The International Development Association (often known as the World Bank) has indicated that it would contribute to the Integrated Coastal and Marine Area Management by implementing the Environment Management Capacity Building Project (ICMAM). Numerous coastal establishments, such as ports and harbours, as well as activities, such as garbage disposal, that are common in inland and neighbouring coastal areas have a detrimental impact on the natural coastal and marine habitats. This effect may be seen in the form of pollution. According to the data that was collected, the coastal and marine ecology is becoming more degraded in a number of locations. These areas include important ecosystems such as mangroves, coral reefs, seagrass beds, biodiversity hotspots, and a variety of other types of habitats. The Indian government has suggested that the concept of ICMAM be implemented in India in order to forestall the further deterioration of the coastal and marine environment and ecosystems in the future. The Ministry of Earth Sciences is actively participating in capacity development programmes in a variety of sectors in order to fulfil its responsibility to protect the marine environment. The creation of knowledge in specialised areas that are connected to the application of the concept of integrated coastal zone management is one of the primary responsibilities of this component. The use of scientific tools and methods such as GIS, remote sensing, and mathematical modelling to analyse issues that are prevalent across coastal marine zones and produce integrated management solutions is one of the subjects that has been proposed for capacity development.

d) **Society of Integrated Coastal Management (SICOM):** Under the auspices of the Ministry of Environment, Forests, and Climate Change of the Government of India, the Society of Integrated Coastal Management (SICOM) was established with the mission of creating a thriving, resilient, and resilient coastal and marine environment that will provide ongoing and improved benefits to the country as a whole as well as to coastal communities. The following is a list of SICOM's primary goals:

1. Nodal organization for strategy development, administration, performance, monitoring, and effective implementation of Integrated Coastal Zone Management (ICZM) practises across the nation, as well as the National Project Management Unit (NPMU) for the Integrated Coastal Zone Management Project and the ENCORE Project in all 13 Coastal States/UTs in the country. Integrated Coastal Zone Management (ICZM) is an

umbrella term that refers to a set of practises that aim to protect and manage coastal areas.

2. To act as a technical secretariat for the National Coastal Zone Management Authority (NCZMA) with regard to the regulatory norms and CRZ categorization of the country's various coastal regions.
3. The coordinating agency and interface for the Enhancing Coastal and Ocean Resource Efficiency Project (ENCORE) on behalf of many ministries of the Government of India, as well as the governments and several line departments of the 13 coastal states/UTs, as well as financing institutions (the World Bank).
4. To assist in the reduction of CRZ violations via the improvement of enforcement made possible by technological advancements, the fortification of existing institutions, and the introduction of new regulatory and legislative measures.
5. SICOM has also started the Pilot Blue Flag Programme, a first in India, as a part of Beach Environment & Aesthetic Management Services (BEAMS), with the intention of creating beaches in India that are on par with those seen in other developed countries.
6. Training for coastal state units and other national and state organisations and agencies participating in the projects on the design, implementation, and administration of integrated coastal zone management programmes.
7. Acting as a bridge between coastal communities, experts, and governments, including the development and dissemination of best practises and standards for the protection of coastal and marine ecosystems, as well as the protection of coastal and island community livelihoods.
8. To perform any other duties or functions that the MOEF&CC may designate at any time in the context of coastal management and other activities that are pertinent to the field.

e) **Institution for coastal management:** The Coastal Regulation Zone (CRZ) Notification, 1991, is an effort to protect the coastal areas of India. Institutional mechanisms in India, such as the National Coastal Zone Management Authority (NCZMA) and the State Coastal Zone Management Authority, have been put in place for the purposes of ensuring compliance with the CRZ Notification as well as monitoring its implementation (SCZMA). These agencies were given the authority to take various actions for the purpose of preserving and improving the quality of the coastal environment, as well as avoiding, mitigating, and regulating environmental pollution in coastal regions, by virtue of Section 5 of the Environmental (Protection) Act of 1986.

f) **Ganga Action Plan:** In 1986, a strategy called the Ganga action plan was established with the goal of lowering the level of pollution in the Ganga River. The government is solely responsible for providing money for this endeavour. This strategy resulted in the formation of the National River Ganga Basin Authority and the designation of the Ganga as India's national river. Rajiv Gandhi was in charge of supervising the Ganga Action Plan. The prime ministers and chief ministers of all the states that are located along the Ganga are responsible for leading the authority. The GAP programme consisted of two distinct stages. The first phase of the project started in 1985 and originally comprised three states: Uttar Pradesh (UP), West Bengal, and Bihar (WB). GAP Phase II was started in 1993, and it is now active in seven states: Uttarakhand, UP, Bihar, Jharkhand, West Bengal, and Haryana. Delhi is also a part of this phase. The second step included the development of an action plan for the Ganga that included all of these tributaries. Under the umbrella of the same programme, the second stage of implementation of the national river conservation plan got underway. In addition to that, it included Ganga tributaries like as the Yamuna, Mahananda, Gomti, and Damodar rivers. The design and execution of GAP are ultimately the duty of the Ministry of Environment and Forests (MoEF), which was granted this role. The Central Ganga Authority, often known as the CGA, came into being as a result of the passage of the Environmental Protection Act in 1986. The Prime Minister of India serves as the CGA's chairman.

Objectives:

1. When the Ganga Action Plan was first conceived in 1985, and its primary objective was to reduce the amount of pollutants that were allowed to flow into the Ganga in order to improve the river's water quality to an acceptable level.
2. The Ganga Action Plan placed a high priority on the elimination of pollutants and the enhancement of water quality. The majority of attention in GAP I is given to sewage collection and treatment infrastructure.
3. It also placed an emphasis on the conservation of biodiversity, the development of an integrated river basin management strategy, the undertaking of extensive research

to further these goals, and the gathering of expertise for the implementation of similar river clean-up initiatives in other polluted rivers throughout India.

4. The Ganga action plan contained a broader objective of restoring biodiversity along the Ganga's path in addition to the more specific goal of reducing pollution.

g) **Namami Ganga Programme:** The Union Government in June 2014 designated the NamamiGange Programme as a "Flagship Programme" in order to achieve the dual goals of effective pollution abatement as well as conservation and restoration of the National River Ganga. The NamamiGange Programme is an Integrated Conservation Mission that was given this designation. The Department of Water Resources, River Development, and Ganga Rejuvenation under the Ministry of Jal Shakti is in charge of its administration. Implementation of the initiative is the responsibility of both the National Mission for Clean Ganga (NMCG) and its state counterpart organisations, which are referred to as State Initiative Management Groups (SPMGs). The NMCG is the implementation department of the National Ganga Council (set in 2016; which replaced the National Ganga River Basin Authority - NGRBA). About 288 projects may be accommodated by it, and it has a total capitalization of 20,000 billion rupees that comes from centralised financing. The following are the program's primary pillars:

1. Facilities for the Treatment of Wastewater
2. River-Front Development
3. The Cleaning of River Surfaces
4. Afforestation
5. Raising Awarenessto the Public
6. The Monitoring of Industrial Effluents
7. Ganga Gram

h) **Indian Antarctic Research Programme (IARP):** The first Indian expedition to the Antarctic was sent out from Goa in December 1981, marking the beginning of the Indian Antarctic Research Programme. The Department of Ocean Development is responsible for ensuring that yearly expeditions to Antarctica are organised and carried out. The first station in India was called Dakshin Gangotri. The second company, Maitri, was established in 1989. In the area of the Larsemann hills, India is now building a third station (named PARAM). As a consequence of India's participation in scientific research projects in Antarctica, the country was granted consultative membership in the Antarctic Treaty.

i) **Ocean Observation and Information Services (OOIS):** The government of India has begun an all-encompassing campaign with the objective of drawing attention to the significance of the information and marine expertise found on OOIS. Ocean Observing System, Ocean Information Services, Ocean Modelling and Dynamics (INDOMOD), and Satellite Coastal Oceanographic Research are the four essential parts that make up this system (SATCORE). The OOIS intends to accomplish the following goals:

1. The formulation of a diverse range of models for the ocean, the atmosphere, and the coasts.
2. Development of techniques for extracting satellite parameters.
3. Extending ocean observations via the use of data taken in-situ and through satellite.
4. The establishment of advice services for the ocean.

CHAPTER 9

ECOLOGICAL SUCCESSION

Concept

Ecological succession is the process by which one population succeeds another within an ecosystem of a particular location on the surface of the earth. This results in a gradual but steady change within the ecosystem. The ecological transitions that take place on abandoned farms are some of the most well-known, and include the change from grass and weeds to trees and forest (Fig. 9.1). Even though the ecology is always changing, the fields are always occupied by the same series of plant communities in the same generally predictable order. Even though surroundings alter most drastically when wet soil develops on dry dunes, identical populations are able to successfully colonise newly formed marshlands and sand dunes. It seems that the driving functions of ecosystem change are supplied by the biomass of increasing plant communities. This biomass affects the physical state of the environment and provides a habitat for animals or decomposers. However, the composition of animal populations may change throughout time. Within the context of their own colonising dynamics, they may cause changes to the plant environment or replace other animal populations. The fact that these ecological succession facts point to ordered growth is essentially what has attracted the attention of ecologists.

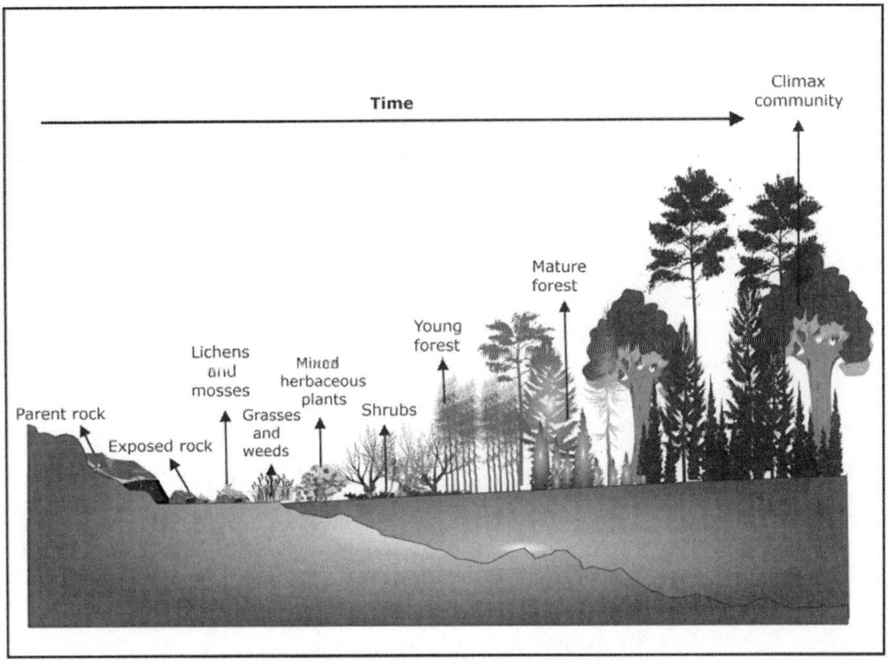

Fig. 9.1: Diagram showing primary succession on a terrestrial site

Causes and Trends of Succession

It is comprehensible that there is not a single reason for succession since it is a process, or more precisely, a collection of sophisticated activities. In general, there are three categories of causes, which are as follows:

a) The initial or initiating causes: These have their roots in both the climate and the biological world. The first category includes things like erosion and deposition, wind, fire, and other things caused by factors like lightning or volcanic activity, while the second category covers the many activities that organisms participate in. Either current populations are wiped out or sterile landscapes are created as a result of these events.
b) Ecesis or continuous causes: These are the processes, such as migration, ecesis, aggregation, competition, reaction, and so on, that result in repeated waves of populations as a consequence of changes in the edaphic features of the region.

c) Causes that contribute to the community's stability: These factors play a role in maintaining the established order. According to Clements, the climate of the area is the fundamental cause for its stable conditions, while other aspects are secondary.

The four tendencies of ecological succession are as follows:

a) Changes in the sorts of plants and animals are continual.
b) A general growth in species variation.
c) An increase in organic matter and biomass, helped along by an accessible flow of energy (but in heterotrophic succession reverse is true).
d) A decrease in the annual yield or the community's overall production.

Process of Succession

The whole main autotrophic succession process is broken down into a number of distinct stages that are carried out one after the other in sequential order. Secondary succession takes place a lot quicker than primary succession does since there is more readily available enriched soil and other factors. In such a sequence, the disturbance is the one and only factor that may slow things down. The technique is essentially the same as the one used for primary, with the exception of the omission of some of the more conventional stages. Both of these things will, in the end, lead to the creation of stable communities that are predominantly composed of K-selected species (species that generally possess relatively stable populations fluctuating near the carrying capacity of the environment). The following is an orderly rundown of the stages:

a) Nudation: It is the growth of an inhospitable environment devoid of any living organisms. In most cases, disturbances are what kick off the process. These factors may be topographic (such as soil erosion and wind action), climatic (such as hail, storms, glaciations, and fire, among other things), or biotic (caused by insects and animals, for example, human activities). In a location like this one, the only creatures that have a chance of surviving are those that are autotrophic and able to use inorganic substrates. The ecological system is now ready to accommodate the introduction of new species. The creation of acidic compounds, for instance, by lichen species is a factor that contributes to the transformation of rocks into soil.

b) Invasion or migration: The introduction of seeds, spores, or other reproductive propagules, which are necessary for the establishment of a species, may be facilitated by invasion or migration. An invasive species is a non-native creature that has the potential to quickly spread across a population. These factors often pose dangers to the natural environment and are the origins of disturbances within communities. Throughout the course of succession, they do, however, make a contribution to changes in the structure and function of the soil. R-selected species are typically the first ones to colonise a new area due to the high reproductive rates and great dispersion capabilities of these species.

c) Ecesis: The establishment of the plant community may now be seen here. The structure of the soil has a role in determining this. This stage is often referred to as the "colonisation" stage. As a result of germination, development, and reproduction, the early colonising species experience an explosion in their population at this period. Allogenic mechanisms are the only ones that may produce ecesis. At this point in time, the pioneer species has been able to withstand the techniques of spread. This process may be stretched out over a longer period of time, allowing for certain pioneer species to be gradually displaced by others. This is because different pioneer species develop at varying rates. This process also modifies the structure of the soil so that it may accommodate species whose seeds were already present in the ecosystem but were unable to germinate because they did not get sufficient support from the community.

d) Aggregation: The term "aggregation" refers to the increase in population of a species that has become established in the area after first settling there. In most successions, little herbs are eventually replaced by bushes. Additionally, this will provide future occupants with a source of food.

e) Competition: As soon as the foundation has been laid by the first few species, both intraspecific and interspecific competition amongst the species will commence. This is the stage in which competition takes place. The majority of the time, competition is for resources such as food, water, and other such things. There is competition between different types of plants and animals. Either resource sharing (also known as resource partitioning) or competitive exclusion will occur as a consequence of the procedure.

f) Reaction: The conditions of the environment are changed as a result of the activities of the creatures that live in the habitat. As a direct consequence of these shifts, some species go extinct while others take their place. The current community won't be able to survive in its current state because of the challenging environment. The most important mechanism at work here is called autogenic succession, and it refers to the process by which plants alter their own environmental conditions. In many cases, the way is prepared for more mature species to follow after early immigrants have established themselves.
g) Stabilization: The term "stabilisation" refers to the process of building the climax community as a whole. The climax community is one that has reached maturity, achieved self-sufficiency, and achieved stability; it is the last step of the succession process. In order to maintain the health of the whole population, the physicochemical conditions are adjusted and then stabilised. The climax communities are the ones that have adapted to the successional zones the best, and it is probable that the community structure will remain in place until another disturbance takes place. This is a state of ecological equilibrium that is always present, and it has a specific composition, structure, and flow of energy.

Basic Types of Succession

The many forms of succession have been classified in various ways based on various factors. However, some fundamental forms of succession are as follows:

a) Primary succession: One kind of succession that may take place in any of the three primary environments (terrestrial, fresh water, and marine) is called primary succession. This type of succession starts with the primordial substratum, which was devoid of any living things in the past. The pioneers, primary community, or main colonisers are the first organisms to establish themselves in a new location.
b) Secondary succession: Another sort of general type of succession, secondary succession starts with substrata that have already been built up and living matter that has previously been established. The current community will disappear if it is subjected to any kind of disruptive external force, such as a sudden change in the local climate, the introduction of a new species, a fire, or any other

similar event. As a consequence of this, the area loses all of its living organisms, but its substratum, rather than remaining undeveloped, begins to be developed. These kinds of successions take a relatively short amount of time.

c) Autogenic succession: Once the process of succession has started, it is often the community itself that has an effect on its own environment as a consequence of the community's responses to the environment, which ultimately leads to the community's own replacement by other communities. Autogenic succession is the name given to this particular sort of succession.
d) Allogenic succession: On the other hand, there are certain situations in which the replacement of the current population is caused primarily by any other external condition rather than by the organisms that are already present. This kind of succession is known as allogenic succession.
e) Autotrophic succession is differentiated from other types of succession by the early and persistent dominance of autotrophic organisms like green plants. It begins in an inorganic environment, and the flow of energy continues indefinitely after that. The quantity of biological matter expands over time thanks to the constant flow of energy.
f) Heterotrophic succession is characterised by the early dominance of heterotrophic organisms such bacteria, actinomycetes, fungi, and animals. This kind of succession may be separated from other types of succession by this trait. It starts out in a mostly organic environment, and the amount of energy that is there steadily diminishes.

There are many other types of succession that are mentioned in ecological literature. These types of succession are determined primarily by the nature of the environment (primarily based on moisture relations), where the process has begun, and so it may be a hydrosere or hydrarch — beginning in regions where there is an abundance of water, such as ponds, lakes, streams, swamp, bog, etc.; a mesarch — beginning where adequate moisture conditions are present; and a xerosere or It is common practise to distinguish between the lithosphere, which originates on rocks, the psammosere, which originates on sand, and the halosere, which originates in salty water or soil.

CHAPTER 10

NATURAL RESOURCES

Concept

All means of meeting human requirements for critical services at a certain time and place are referred to as "resources." Therefore, the term "resources" refers to the many means through which one might improve their own and their community's well-being. Land, water, minerals, forests, animals, energy, and even humans themselves may be considered resources as well as components that contribute to the creation of resources. The question of whether or not the present pace of global population expansion will be sufficient to meet the resource needs of humanity for both survival and comfort has emerged as a prominent concern on the international level. Because of this, there is a battle to find ways to use resources in a sustainable way that can meet the ever-increasing demands of human populations. Recent advances in technology have been of tremendous help in finding solutions to problems caused by resource depletion in a more appropriate fashion.

These are resources that may be found in nature and are generated without the assistance of humans. Natural resources may take many forms and include things like air, sunlight, water, soil, rocks, plants, animals, and fossil fuels. Natural resources are materials that occur naturally and are useful to humans or that have the potential to be used in certain technological, economic, or social contexts. Natural resources also include supplies that are derived from the ground, such as food, materials for construction and clothing, manures, metals, water, and geothermal power, among other things. Historically, the scientific world has been responsible for exploring natural resources.

Types of Resources

There are four major types of resources:

a) Continuous resources: These resources are those that can be accessed on a continual basis and with little effort. These resources are still accessible, and with the possible exception of solar energy, the distribution of which may be restricted by air pollution, they cannot be degraded, even if they are managed in a very inefficient manner. There are many different types of renewable energy sources; some examples are wind energy, gravity, solar energy, tidal energy, and geothermal energy.
b) Renewable resources, also known as flow resources: Flow resources are any materials that are capable of being used in the production of energy. These are the kind of resources that, given enough time, have the potential to spontaneously transform back into things that can be put to human use. If their capacity to regenerate is not hindered by natural calamities or by the acts of humans, then these resources are seen to be renewable, and it is possible that they will always be available. If a renewable resource has been degraded to the degree that it has passed a certain threshold, there is a possibility that it may never recover. Some examples of this are clean water, vegetation and fauna, soil, and air.
c) Non-renewable resources, also known as stock resources: These are either accessible in limited quantities or the rate of regeneration is so slow that they must be regarded only obtainable in fixed quantities. Non-renewable resources are also known as fixed-quantity assets. Minerals and underground water are two examples in this category.
d) Extrinsic resources: Extrinsic resources are natural resources that have been modified or improved by humans in order to provide additional advantages, most prominently for recreational purposes. These are perishable resources that are prone to degradation or breakdown, but if they are handled properly, they have the potential to be continuous resources. Examples include things like human capabilities, institutional capacities, managerial skills, and so on.

Status of Major resources

The most important components of the world's resources are its water, land, forests, energy, and minerals. The existence of such resources is essential to the continuation of life on Earth. It is general knowledge that early man's needs were quite limited; but, as industry advanced, man's demand significantly expanded. This contrasts with the fact that early man's needs were relatively limited. The resources that a nation has, as well as its capacity to make use of those resources and to safeguard them, are the primary factors that determine the extent of that nation's power on any scale, whether it social, economic, or political. Developing nations are responsible for a significant amount of the world's raw material production, yet they only control a tiny part of the global commerce in completed goods. Their natural riches have been consistently taken advantage of by the market procedures of the industrialised nations. The fact that less developed nations are still residing in deep poverty due to the under-utilization of their assets and are prompted to live in an environment contaminated by waste-generating economies is perhaps the most striking aspect of the situation. While the over-utilization of natural resources by developed countries has caused a negative shift in the world's ecological balance, less developed nations are still residing in desperate poverty due to the under-utilization of their resources.

a) Water: Water is the primary most essential resource for human existence. About 97.2 percent of the water on our planet is made up of salt water, 2.15 percent is made up of ice, and the remaining 0.65 percent is made up of fresh water that can be found either on the surface or in the ground (Fig. 10.1). There is a significant lack of available fresh water resources. The details of water distribution on Earth are outlined in Tables 10.1 and 10.2, which also detail the process of water loss and the frequency with which water is regenerated. As a result of rapid expansion in population, agricultural production, and industrial activity, the demand for potable water is rising and will continue to do so in the foreseeable future. As a direct consequence of this, the supply of clean water is diminishing on a regular basis. Because the amount of clean water required by each individual is around 2.7 litres per day, the amount of water that is required on a daily basis all across the globe is approximately 16.5 billion litres. Agriculture is another sector that consumes a significant amount of freshwater resources.

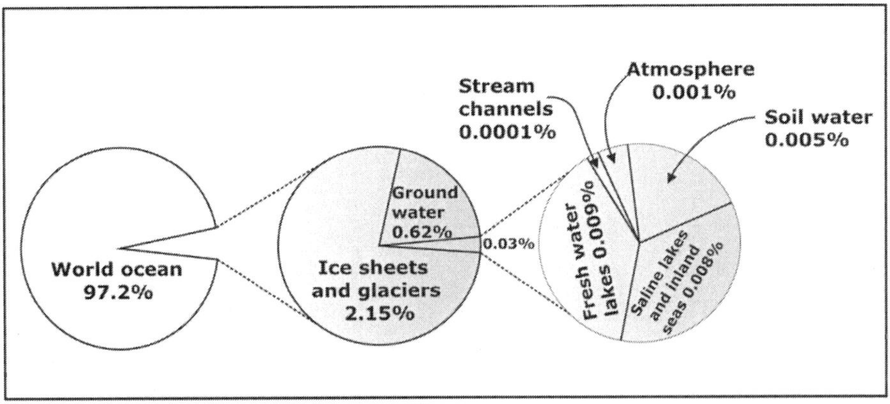

Fig. 10.1: Distribution of water on the Earth's surface

Table 10.1: Distribution of Water on the Earth

	Volume, km^3	Mass, g
Sea water	1.37×10^9	1.41×10^{24}
Fresh water lakes	1.3×10^5	1.3×10^{20}
Saline lakes and Inland seas	1.0×10^5	1.0×10^{20}
Rivers	1.3×10^3	1.3×10^{18}
Groundwater	8.4×10^6	8.4×10^{21}
Icecaps	2.9×10^7	2.9×10^{22}
Water vapour	1.3×10^4	1.3×10^{19}

Table 10.2: World's Water Resources

Resources	Volume (W), Thousands of km^3	Annual rate removal (Q) (in Thousands of km^3)	Process of removal	Renewal period (T) (T=W/Q)
Total water on Earth	1,460,000	520	Evaporation	2,800 years
Total water in the Ocean	1,370,000	449	Evaporation	3,100 years
Free, gravitational waters in the Earth's crust (to a depth of 5 km)	60,000	13	Underground runoff	4,600 years
Amount of water present in the zone of active water exchange	4,000	13	Underground runoff	300 years
Lakes	750		Underground runoff	16,000 years
Glaciers and permanent snow	29,000	1.8	Runoff	
Soil and subsoil moisture	65	8.5	Evaporation and underground runoff	
Atmospheric moisture	14	520	Precipitation	9 days
River waters	1.2	36.3	Runoff	12 (20) days

Source: UNESCO

India receives about 3 trillion m3 of water from rainfall each year, potentially the most in the world. In spite of the fact that it is home to 4 percent of the world's freshwater resources, India entered a state of water scarcity in 2011. Research that was conducted in 2019 by NITI Aayog found that India is now going through the most severe water crisis in its entire history, with approximately 600 million people lacking access to clean drinking water. The report predicted that by the year 2021, the groundwater supplies of 21 cities, including Bangalore, Delhi, Hyderabad, and Chennai, would most likely have been drained. In rural areas of India, groundwater resources satisfy between 80 and 90 percent of the demand for drinking water. The levels of groundwater in India have dropped by more than 60 percent between 2007 and 2017, which is cause for concern. Over ninety percent of the water that is removed is used for agricultural purposes. Research conducted in 2014 by the BharatiyaAgro Industries Foundation found that rain-fed agriculture makes about 70 percent of all acreage in India. Despite this, the sea absorbs 65 percent of the entire amount of rainwater that falls. Another kind of water stress that may lead to healthcare-related external losses is pollution of the water. Cities in India are home to 36 percent of the country's population yet are responsible for 70 percent of the country's water pollution, according to research published by the Central Pollution Control Board. There are 14 major river systems that are responsible for 83 percent of the drainage basin, 85 percent of the surface water flow, and provide water for 80 percent of the entire population of the nation. In addition, there are 44 medium rivers and 55 minor rivers. The vast majority of these rivers have their beginnings in the coastal highlands, and the monsoon season accounts for eighty percent of their annual flow. Although there are a large number of marshes and wetlands in the nation, which are essential for a healthy ecosystem, it should be noted that these areas are rapidly disappearing as a result of human modification. India is rapidly approaching the point when it will be forced to confront its water resource issue. As a consequence of widespread disruptions to the hydrological cycle, this phenomenon is becoming more widespread. As a consequence of this, having a national water policy is very necessary in order to ensure the availability and use of water in a sustainable manner. The Government of India announced the "National Water Policy" in April of 2002. Within this policy, a greater emphasis was placed on the collection, storage, and rational utilisation of surface water. This was done because ground water levels were decreasing very rapidly in various parts of the country.

b) Land: Land is a valuable resource because it is used to cultivate crops and other biological resources that are harvested for use in the production of food, livestock, pharmaceuticals, textiles, and other goods. The amount of production is dependent on the size of the cultivable land and its potential yield, both of which are subject to change in accordance with changing human attitudes. The limitations of cultivable land come in the form of physical factors such as temperature, precipitation, topography, and soil. It is also impacted by human social patterns, particularly patterns of energy usage, which have an impact on both its availability and its cultivability. India has a population that is more than 17 percent of the world's total yet only has 2.4 percent of the world's land area. Plains make up 43 percent of India's land resources, whereas mountains make up 30 percent, and plateaus account for 27 percent. The majority of the land in the plains is used for agricultural purposes, and there are also some industrial areas. Mountains not only attract visitors but also ensure that rivers continue to flow continuously. Minerals, fuels, and forests are found in large abundance on plateaus. The many types of land resources that may be found in India include forests, areas that are infertile or otherwise unfit for agricultural use, fallow land, agricultural fields, and so on.

1. Forests: Forests encompass around 22.8 percent of the India's land area.
2. Unavailable land for cultivation: 8.61 percent of the land is made up of terrain that is inhospitable to agricultural use, including terrain that is rocky, arid, and desert-like. This includes territory that is barren and wasteland. In addition, up to 5.57 percent of the total is comprised of land that is not utilised for agricultural purposes, such as that which is used for building, roads, industry, and settlements. Other uncultivated areas include land that is used for permanent pastures and grazing (3.38 percent), land that is covered by a variety of tree crop groves (1.11 percent), and culturable waste land (4.17 percent).
3. Fallow land: The overall portion of fallow land that is included in land resources is 8.13 percent, with 4.76 percent covering the existing fallow. The term "current fallow" refers to lands that have been uncultivated for a period of time that is less than one agricultural year.

4. Net Sown Area (NSA): The net sown area, also known as NSA, is the area that is sown more than once in the course of an agricultural year. This is in addition to the gross cropped area or the net sown area. In India, the proportion of total land resources that is comprised of net sown area is 46.24 percent. As an example, 80 percent of the entire land in Punjab and Harvana is used for agricultural purposes. On the other hand, the Andaman and Nicobar Islands, Arunachal Pradesh, Mizoram, and Manipur each have a net sown area that is less than 10 percent of their entire land area.

c) Biological resources: Even though our understanding of the organisms that inhabit the earth, also known as the planet's biological resources, is still in its infancy, there is no question that the abundance and variety of living things provide a great number of benefits and contribute to the allure and charm of our home planet. Unfortunately, the world's rapidly increasing human population as well as the use of resources are currently posing a danger to three different types of biodiversity. These threats include the variety of biological communities, genetic variation within species, and species diversity. We now believe that there are more than 1.7 million species presently known, which is just a tiny percentage of the total number of species that were anticipated to exist on Earth by the conclusion of the great exploration period that occurred throughout the nineteenth century. There are many different factors to blame for the depletion or destruction of biological resources. The actions of humans, whether deliberate or accidental, may have a negative impact on ecosystems and lead to the extinction of species. Changes that occur naturally in the ecosystem may wipe out species that were formerly abundant or limit their population sizes to just a few isolated pockets. As of right now, there are three different sorts of species losses: a decline in the abundance of a species that was once plentiful, the extinction of a species, and ecological disruption. There is a gradient upwards in terms of the severity of these losses. Depletion of biological resources by humans has a long and storied history. In a number of different ways, humans have contributed to the decline of biological abundance or to the extinction of species (Fig. 10.2).

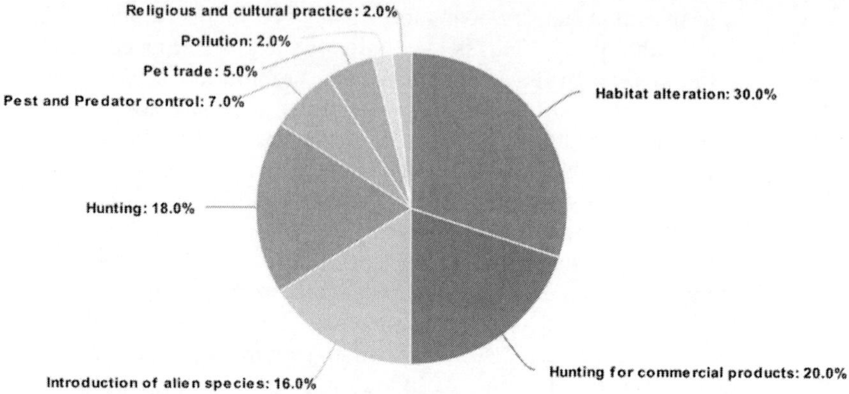

Fig. 10.2: Estimated contribution of various human activities to biological resource extinction

Forests play a vital part in the management of the temperature, the control of water runoff, the provision of habitat and food for animals, and the cleansing of the air. They are responsible for the production of necessary goods, such as wood and paper pulp, which are used by all of us. The overall forest cover in India increased by 0.4 percent, taking it from 21.67 percent to 21.71 percent, which demonstrates that the country's vegetation is continuing to expand.

A total of 721 square kilometres have been added to India's forest cover thanks to recent planting efforts. The State of Forests Report is compiled by the Forest Survey of India (FSI), which is responsible for doing comprehensive research, collecting data from around the nation, and compiling the results. It is not FSI but rather the Union Ministry of Environment, Forests, and Climate Change that is responsible for releasing the report (MoEFCC). It is distributed after every two years, with the report for 2019 arriving prior to the report for 2021. The evaluation is focused on a survey of three different kinds of forests, which are as follows:

1. Intensely Forested Areas (canopy density of more than 70 percent).
2. Forests with a moderate density (canopy density between 40 and 70 percent).
3. Open forest (canopy density between 10 and 40 percent). Even though the canopy density is less than 10 percent, the region is nonetheless surveyed and classed as "scrubs."

New Features of India State of Forest Report, 2021

1. This was the first time that the research and survey encompassed the Gir Forest (the sole habitat for Asiatic Lions), Tiger Reserves and Tiger Corridors.
2. There has been an increase in forest cover in 20 of the 32 tiger reserves over the last decade, while the other 10 have witnessed a decrease.
3. Tiger corridor forest cover has expanded, whereas the forest area in tiger reserves has dropped dramatically.
4. Anamalai Tiger Reserve (Tamil Nadu), Buxa Tiger Reserve (West Bengal), and Indravati Tiger Reserve (Uttar Pradesh) have seen a rise in forest cover (Chhattisgarh).
5. The following tiger reserves have lost their forest cover: Kawal Tiger Reserve (Telangana), Bhadra Tiger Reserve (Karnataka), and Sunderbans Tiger Reserve (West Bengal)
6. The Pakke Tiger Reserve in Arunachal Pradesh holds the record for the highest percentage of total forest cover relative to its size, with a whopping 97 percent forest cover.

d) Minerals: The human race has been responsible for the mining of a diverse variety of minerals, including both metals and non-metals, throughout the course of history. The crust of the planet is where the vast majority of metals that are utilised in industry may be discovered (i.e., 10–70 km thickness). These metals are possible to be mined at a low cost in areas where there are considerable explorable deposits. There are three categories of metals found in the crust of the earth: structural metals like iron and aluminium; rare metals like copper, lead, zinc, magnesium, and mercury; and valuable metals like gold, platinum, and silver.
e) In addition to being deposited as crustal layers, metals may also be found in the form of polymetallic nodules in the ocean floor. There are significant mineral deposits in a number of countries throughout Asia and Europe. A significant proportion had been harvested for

human use at an earlier time. One of the nations that is quite rich in minerals is India. Mineral deposits may be found in states like Rajasthan, Bihar, Orissa, Sikkim, West Bengal, Assam, Andhra Pradesh, Madhya Pradesh, Maharastra, and Karnataka. The Geological Survey of India estimates that India is home to around 22 distinct types of minerals, each of which may be found in large amounts. The majority of them have already had some of their mineral content mined out. It is possible that our mining activities will be wound down in the not-too-distant future if the present rate of exploitation is maintained and there is no discovery of additional deposits. The vast majority of minerals are extracted from deposits on land, as opposed to deposits along the shore or in the sea bottom. The following are some of India's most important mineral resources: coal (with the fourth largest reserves in the world), iron ore, manganese ore (with the seventh largest reserve in the world as of 2013), mica, bauxite (with the fifth largest reserve in the world as of 2013), chromite, natural gas, diamonds, limestone, and thorium. The old crystalline rocks of the peninsular plateau contain the vast majority of the metallic minerals found there. Nearly 97 percent of the world's coal deposits are located in the basins of the Damodar, Sone, Mahanadi, and Godavari basins. In the high sedimentary basins of Assam, Gujarat, and Mumbai there is potential for the discovery of petroleum reserves. Both the Krishna-Godavari and the Kaveri basins have yielded fresh discoveries of deposits. Mineral deposits are mostly found in India's three most important regions. It's possible that here and there you may find a few of scattered deposits.

1. Plateau Area, North-Eastern India: This region encompasses Chhotanagpur (Jharkhand), the Odisha Plateau, sections of Chhattisgarh, and West Bengal. It has an abundant supply of minerals, including iron ore, coal, manganese, bauxite, and mica, among others.
2. The South-Western Plateau: is comprised of the states of Karnataka and Goa in addition to the higher elevations of Tamil Nadu and Kerala. This region has a significant amount of bauxite as well as ferrous metals. In spite of this, there is high-quality iron ore, manganese, and limestone present. This area is dominated almost entirely by coal reserves; the only exception being the Neyveli lignite. The mineral deposits in this part of the country do not have the same breadth of variety as those in the north-eastern strip. In addition to bauxite clay, monazite and

thorium may be found in abundance in Kerala. Iron ore reserves may be found in the state of Goa.

3. The North-Western area: Minerals belonging to the Dharwar system of rocks may be found along the Aravali mountains. In this region, both copper and zinc have historically been considered to be precious minerals. Sandstone, granite, and marble are just few of the many construction materials that can be found in Rajasthan. In addition to that, there are substantial amounts of gypsum and fuller's earth. The cement industry makes use of both the dolomite and the limestone that is found in this area as raw materials. It is well knowledge that Gujarat has enormous oil and gas reserves. Both Gujarat and Rajasthan are home to substantial salt deposits.

f) Energy: The proportion of the overall energy mix that is comprised of oil continues to be the greatest (31.2 percent). In the year 2020, coal was the second most important fuel, accounting for 27.2 percent of total primary energy consumption, which is an increase from the 27.1 percent it accounted for in the previous year. Natural gas production achieved a new all-time high of 24.7 percent, while renewable energy production reached a new all-time high of 5.7 percent. The share of renewable energy in total energy production has already overtaken that of nuclear energy, which only accounts for 4.3 percent of the total. The energy contribution of hydro increased by 0.4 percentage points to 6.9 percent in the previous year, marking the first gain in this sector since 2014. The India Energy Outlook 2021 Report was recently released by the International Energy Agency (IEA). This report examines the potential and problems that India will encounter as it works toward the goal of providing dependable, affordable, and sustainable energy to a population that is continuing to increase. The India Energy Outlook 2021 is a new special publication that is part of the World Energy Outlook series published by the International Energy Agency. The most important conclusions are as follows:

1. In the next two decades, India will be responsible for a quarter of all new energy consumption, and by the year 2030, it will have surpassed the European Union to become the third biggest energy consumer in the world.
2. After China, the United States of America, and the European Union, India is presently the country with the fourth-highest energy consumption in the world.

3. Based on the current state of national policy, it is anticipated that India's energy consumption would almost quadruple by the year 2040, when the country's GDP is projected to reach an estimated 8.6 trillion US dollars.
4. Prior to the global pandemic, it was anticipated that India's energy consumption would increase between 2019 and 2030 by more than 50 percent; however, it is now anticipated that growth would be closer to 35 percent.
5. It is anticipated that by the year 2040, India would contribute for about 20 percent of the increase in global industrial value-added and will be first in the world in terms of industrial final energy consumption, notably in the production of steel.
6. If policies remain the same, it is anticipated that India's demand for oil would increase by 74 percent by the year 2040, reaching 8.7 million barrels per day. If the number of cars owned per person in India increased by a factor of five, the country would vault to the forefront of the world's rising need for oil.
7. By the year 2040, India will have more than doubled its demand for natural gas, making it the market with the highest rate of growth. The percentage of dependency on natural gas imports increased from 20 percent in 2010 to roughly 50 percent in 2019, and it is predicted to increase to more than 60 percent by the year 2040.
8. Coal is the primary source of electricity generation in India, accounting for more than 70 percent of the country's total output. Demand for coal is anticipated to increase from its current level of 590 million tonnes to 772 million tonnes in the year 2040.

g) Ocean: The ocean is the most valuable asset on Earth. It has a wide variety of biological resources, mineral resources, and sources of energy in the form of tides and waves. One such resource is the biological diversity of the ocean, which may be broken down into its component parts: phytoplanktons, dinoflagellates, zooplanktons, meroplanktons, fishes, and other benthic flora and fauna. Since the beginning of human civilization, people have used the seas for a variety of purposes, including as a source of food and tools, as a place for commerce and colonisation, and even as a source of welfare. Despite the vastness of its living resources, just a fraction of them have been used thus far. When it comes to making effective use of and growing upon these resources throughout the course of time, there is a vast amount of potential for improvement.

It has been determined that the resource potential of conventional marine fisheries may reach up to one hundred metric tonnes per year. It is possible that in the future, aquaculture practises that are more suitable for coastal areas may greatly increase productivity. Coral reefs and other coastal resources, such as mangroves, seaweed, and sea grass, are not being used to their full potential. Many maritime resources are in fast decline over the course of time for a variety of reasons, including over-exploitation, pollution, and the loss of habitat. To adequately preserve the coastal and marine ecosystem from such a state, protective measures need to be taken. Oceanic resources are relatively numerous in India, especially along the coast line. Oceanic resources in India account for 46 percent of the total living resources that are harvested from the Indian Ocean. Even while our capacity for exploitation is now somewhat restricted, there is still reason to be optimistic about the future. The entire amount of India's hydrocarbon reserves is more than 12.70 billion tonnes, with an estimated 8.70 billion tonnes coming from the sedimentary offshore basin. The country's hydrocarbon reserves are also rather big. Approximately 6.3 percent of the coastline of Peninsular India is covered by mangrove forests. There is a coral reef off roughly 7 percent of the coastline. An ever-increasing number of biological forms, found in coastal and marine environments, have been recognised as potential new drug sources. As a consequence of this, we need to restrict our activities in coastal areas in order to protect the abundant biological richness that is available for human use.

CHAPTER 11

ENVIRONMENTAL POLLUTION

Concept of Environmental pollution

The presence of undesired substances in any segment of the environment, generally as a result of human activities discharging by-products, waste products, or hazardous secondary products that are damaging to man and other species, is referred to as pollution. However, it should be noted that no environmental component, even in its natural condition, is pure in the scientific sense; there are admixtures and impurities, but they are little and do not cause substantial harm. Clean air is a natural mixture of many gases, moistures, and particulate matters, but when foreign substances, such as oxides of nitrogen and sulphur, or lead-containing fumes or fluorides, are added in significant quantities, air becomes harmful to plants, animals, and humans, either directly or indirectly through the food chain and thus becomes polluted.

Sources of Pollution

Environmental contamination is today regarded as a worldwide phenomenon that draws a great amount of human attention due to its serious long-term implications. Various causes of pollution have been recognised during the last several decades, altering the makeup of the earth's water, air, and soil. It is commonly said that industrialization is the primary source of pollution in the environment. However, many such unknown sources were previously ignored. So far, six significant sources of contamination have been identified. Fig. 11.1 depicts the specifics. Depending on the site-specific circumstances, the relative worth of all sources was analysed. According to experts, anthropogenic sources of pollutants are the most important agents in urban environments, whereas agricultural sources are the most important agents in rural agricultural districts.

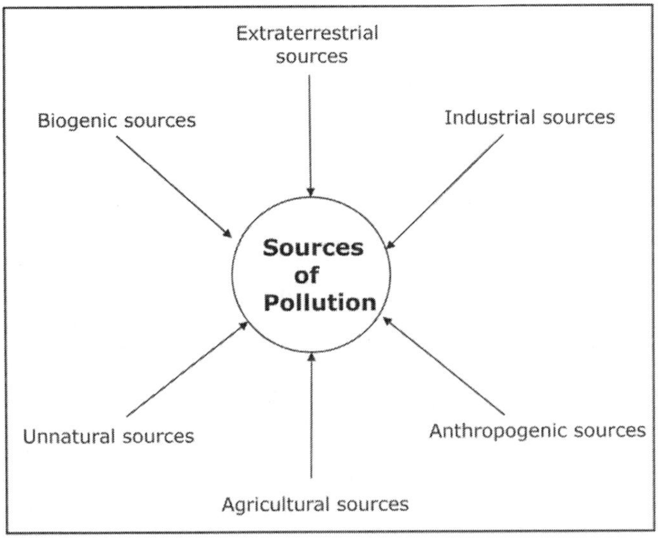

Fig. 11.1: Sources of pollution

Nature of Pollutants

A pollutant is any chemical (toxic metal, radionuclide, organophosphorus compound, Bases), geochemical substance (dust, sediment), biological organism or product, or physical substance (heat radiations, sound wave) that is intentionally or unintentionally released by man into the environment, with actual or potential adverse, harmful, unpleasant, or inconvenient effects. Such unfavourable consequences may be direct (affecting man) or indirect (affecting his resource organisms or climate changes). There are four basic groups of pollutants based on their nature: physical, chemical, biological, and pollutants of unknown types (Table 11.1). Again, different agents are detailed in depth for each group (Fig. 11.2). Polluting may be classified into six broad categories based on the type of the pollutants and the consequent pollution of environmental components:

a) Water Pollution
b) Soil/Land pollution
c) Noise Pollution
d) Air Pollution
e) Radioactive Pollution
f) Thermal Pollution

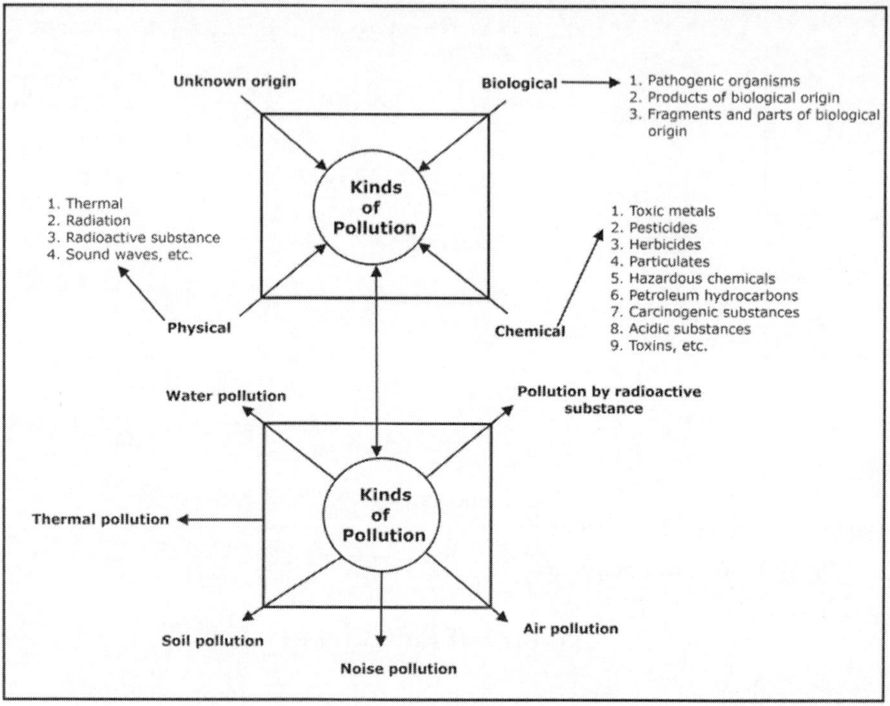

Fig. 11.2: Diagram showing kinds of pollutants and pollution

Table11.1: Pollutants

On the basis of	Type of Pollutants	Description
Form in which they exist	Primary pollutant	That is immediately emitted from the source. Sulphur and nitrogen oxides are two examples.
	Secondary pollutant	This is formed when contaminants react with one another. As examples, consider ground ozone and acid rain.
Nature of Disposal	Biodegradable pollutants	Microbial activity can degrade these contaminants. Sewage is one example.
	Non-biodegradable pollutants	These contaminants cannot be decomposed and hence persist in the environment for an extended length of time. Plastic, for example, and glass.
Natural or Artificial	Natural Pollutants	They are already prevalent in nature, but become pollutants when their concentration exceeds a certain threshold. Carbon dioxide is one example.
	Artificial Pollutants	These are manufactured by humans. Herbicides, pesticides, and DDT are examples.
Physical form of Existence	Particulate pollutants	Examples includelead andfly ash.
	Gaseous Pollutants	Examples include: Carbon monoxide (CO) Carbon dioxide (CO2) Chlorofluorocarbons (CFCs) Ozone (O3) Nitrogen oxide (NOx) Sulphur dioxide (SO2

Environmental pollution is becoming more prevalent. Water is polluted by four types of substances: organic waste, industrial waste, chemical agents for fertilisers and pesticides for crop protection, and silt from deteriorated catchments. While municipal garbage is expected to create three-fourths of waste water by volume, industrial waste though tiny in volume provides more than half of total pollution load, with the majority of this coming from large and medium companies. Trends in ambient air quality in major cities demonstrate that levels of Suspended Particulate Matters (SPM) are higher than authorised standards or limits, especially during the summer months. NOx levels in cities are growing as automobile emissions rise. Environmental concerns are becoming increasingly prevalent. Every year, the chemical industry creates more material, which has a detrimental influence on vital components of the environment, soil, and water. In industrial high-density zones, we face harm to the environment's social and economic functions in addition to the effects on local health and the environment. The problem of noise pollution in metropolitan areas, in particular, appears to be becoming worse by the day. The environment's status continues to deteriorate. The advancement of scientific and technological understanding has enabled the usage of an ever-increasing amount of natural resources. The growing population puts additional strain on the ecosystem. Forest decline has been followed by an increase in pollutants harming the atmosphere, soil, and water.

Air Pollution

The Earth's atmosphere is 99.9 percent nitrogen, oxygen, and argon, with the remaining 0.1 percent (approximately half of the remaining 0.04 percent) composed of the three transition gases CO2, Ne, and He. The remaining 0.66 percent of the atmosphere is a complex mixture of hundreds of trace gases, the most abundant of which being methane. Many gases at parts per trillion are wholly man-made, yet they can still have a substantial impact on the ecosystem. Some trace gases influence or govern the Earth's temperature and stability. Long-lived gases that are accumulating at significant rates as a result of human activity are of special contemporary concern since they eventually lead to stratospheric ozone depletion, global warming, and chemical disruptions that many fear will be damaging to human existence. The introduction of contaminants, organic molecules, or other hazardous chemicals into the Earth's atmosphere is referred to as air pollution. This can take the form of surplus gases such as carbon dioxide and other vapours that cannot be adequately eliminated by natural cycles such as the carbon or nitrogen cycles. The

Environment (Protection) Act, 1986 (two years after the Bhopal disaster) was passed by the Government of India to safeguard and enhance the quality of our air, water, and soil in order to reduce environmental pollution.

Major air pollutants and their sources

Carbon monoxide (CO)

a) CO is a colourless, odourless, tasteless, and very poisonous gas that is slightly denser than air. It is only present in the atmosphere for a few months.
b) Carbon monoxide is created by internal combustion engine exhaust and incomplete combustion of several other fuels. As a by-product of iron smelting, carbon monoxide is produced.
c) It occurs when there is insufficient oxygen to make carbon dioxide ($CO2$).
d) Carbon dioxide is produced when carbon monoxide burns with a blue flame in the presence of oxygen.
e) Natural photochemical processes in the troposphere are the world's greatest producer of carbon monoxide.
f) Volcanoes, forest fires, and other kinds of combustion are also natural sources of CO.
g) The most prevalent kind of lethal air poisoning is carbon monoxide poisoning (poor ventilation and heat management in laptops and other electrons can cause the release of CO).
h) When present in amounts greater than 35 ppm, it is hazardous to haemoglobin animals (including humans).
i) It combines with haemoglobin to form carboxyhaemoglobin, which takes over the oxygen-carrying area in haemoglobin.

Carbon dioxide ($CO2$)

a) Gas that is colourless and odourless. It weighs more than air.
b) Volcanoes, hot springs, and geysers are natural sources, and it is liberated from carbonate rocks by dissolving in water and acids.
c) Because carbon dioxide dissolves in water, it may be found in groundwater, rivers and lakes, ice caps and glaciers, and ocean.
d) $CO2$ is a suffocating gas (asphyxia: a condition arising when the body is deprived of oxygen, causing unconsciousness or death).

e) Even in the presence of enough oxygen, concentrations of 7 percent can cause asphyxia, appearing as dizziness, headache, and unconsciousness.
f) Carbon dioxide is a major contributor to global warming.
g) Since the industrial revolution, the use of carbon-based fuels has contributed to global warming.
h) Because it dissolves, it contributes significantly to ocean acidification.

Chloroflorocarbons (CFC)

a) Refrigerators, air conditioners, and aerosol sprays all contain chlorofluorocarbons (CFCs).
b) When CFCs are released into the atmosphere, they ascend to the stratosphere and interact with few other gases, resulting in a decrease in the ozone layer, which shields the planet from the sun's damaging UV radiation.
c) Because of their harmful effects on the ozone layer, CFCs have been severely restricted since the late 1970s.
d) The Montreal Protocol on Compounds that Deplete the Ozone Layer (a protocol to the Vienna Convention for the Protection of the Ozone Layer) is an international convention that aims to safeguard the ozone layer by phasing out the manufacturing of a variety of ozone-depleting substances, notably CFCs.

Lead

a) One of the most dangerous heavy metals is lead.
b) Lead may cause substantial harm to the human body, including: damage to the nervous system, digestive problems, kidney disease, and an influence on intelligence.
c) As a result, lead was prohibited as an addition to fuels and other items.
d) Lead poisoning may occur when lead is combined with water and food.
e) It has long-term consequences for children since it reduces intellect.

Ozone

a) It occurs naturally in the stratosphere and absorbs the sun's damaging UV radiation.
b) However, it is a pollutant (Greenhouse gas) with very harmful effects at the ground level.
c) Ground-level ozone emissions are mostly caused by vehicles and industry.
d) Carbon monoxide and nitrogen dioxide are important in the conversion of O2 to O3.
e) Ozone causes itching, watery eyes. It reduces our resistance to the common cold and pneumonia.

Particulate pollutants

a) Particulate pollutants are particles that contaminate the air by being suspended.
b) These particles are produced by anthropogenic processes such as automobiles, factories, construction sites/activities, and so on, or by natural sources such as pollen, volcanic eruptions, natural gaseous precursors, and so on.
c) Their diameters range from 0.001 to 500 micrometres (μm).
d) Particulate contaminants can cause significant harm to the human respiratory system.
e) The Central Pollution Control Board has designated PM 2.5 particles (2.5 (μm or less) as one of the most dangerous particulate pollutants (CPCB). They are so small that they can only be seen using an electron microscope.

Heavy particulate matter	Suspended Particulate matter	Nano Particulate matter
More than 10 μm	Less than 10 μm	Less than 0.02 μm
Settles down after a point	Floats and move freely	Form aerosols, very light and powerful

Sulphur dioxide (SO2)

a) It is a poisonous gas with a strong, unpleasant odour. Acid rain is caused by it.
b) Volcanic activity naturally releases it. It is abundantly accessible in Venus' atmosphere.

c) Sulphur dioxide is primarily manufactured for the production of sulphuric acid.
d) Sulphur dioxide inhalation has been linked to increased respiratory symptoms and untimely mortality.
e) It also impairs the function of certain nerves.
f) It is generated by burning coal in thermal power plants and diesel fuels, various industrial operations such as paper manufacturing and metal smelting, reactions involving Hydrogen Sulphide (H2S) and oxygen, Sulphide ores such as pyrite, sphalerite, and cinnabar are roasted (mercury sulphide).

Smog

Smog is caused by air pollution and decreases visibility. The term "smog" was coined in the early twentieth century as a contraction of the words smoke and fog to describe smoky fog because of its opacity and odour. The term was originally meant to relate to pea soup fog, a common and major phenomenon in London from the nineteenth to the mid-twentieth centuries. Nitrogen oxides, sulphur oxides, ozone, smoke, and other particles contribute to visual air pollution. Coal combustion emissions, automotive emissions, industrial emissions, forest and agricultural fires, and photochemical reactions to these pollutants all contribute to man-made haze. Smog is caused by the reaction of pollutants from cars, factories, and industries with the atmosphere and sunlight.

Smog Types and formation

Smog formation is caused by a variety of factors. As a result, they are divided into two categories.

a) Photochemical Smog (Also called Los Angeles Smog): The most frequent type of smog is photochemical smog. When sunlight combines with nitrogen oxides and at least one volatile organic compound (VOC) in the atmosphere, photochemical smog is formed. This type of smog does not require smoke or fog. Automobiles, power plants, and manufacturing emissions all produce nitrogen oxides into the environment. Paints, gasoline, and cleaning solvents all emit volatile organic chemicals into the environment.
b) Sulphurous smog (Also called London Smog): Sulphurous smog is caused by excessive levels of sulphur oxides in the environment. This is mainly caused by the usage of fossil fuels such as coal.

Typically seen in large cities with a high concentration of automobiles and industry. Because the smog becomes trapped and cannot be moved away by the wind, it is seen in towns located in basins surrounded by mountains.

How to control Smog

a) Using renewable energy sources.
b) Reducing the number of automobiles, an effective public transportation system with last-mile connection is required.
c) Improved energy efficiency.
d) Smog towers have been utilised effectively in China.

Harmful effects of Smog

a) Wheezing and coughing.
b) A burning feeling affecting the eyes and throat.
c) The possibility of developing major heart disease.
d) There is a risk of significant lung disease.
e) For asthmatics, this is quite dangerous.
f) Plants can also be killed by smog.

Fly ash

a) Fly ash consists of oxide and other heavy metal particles. The bulk of this includes substantial levels of aluminium silicate, silicon dioxide (SiO2), and calcium oxide (CaO).
b) Thermal power stations are a significant source of fly ash pollution.
c) Its deposition in agricultural areas has the potential to contaminate crops and vegetables with heavy metals.
d) The Ministry of Environment and Forests has made the use of fly ash-based products mandatory in all construction projects, road embankment works, and low lying landfilling works within a 100-kilometer radius of Thermal Power Stations, as well as mine-filling activities within a 50kilometer radius of Thermal Power Stations.

Effects of Air Pollution

a) Effects on vegetation: Plants can be affected by air pollution to varied degrees. There is no effect at the lowest levels, i.e., below the 'threshold,' such as apparent damage, cumulative chronic effects, genetic consequences, or even slow changes in the makeup of the plant community. Even at this level, however, air pollutants can be retained in plants and transferred into the food chain, damaging animals that consume the plants. Air pollutants, particularly acidic air pollutants, can enter plants directly by gaseous diffusion or through polluted soil. Gaseous air pollutants such as SOx, NO, CO2, and others may enter directly through leaf stomata. However, solid particles are adsorbed on the surface. Contaminants, in general, induce harm to sensitive plants at fatal concentrations, whereas tolerant plant species are capable to sinking pollutants to a significant level without injury. As a result, different air contaminants cause different sorts of damage to exposed sensitive plants. Suspended particles on foliagecause harm to leaf functions, in the way of:

1) Changes in the amount of sunlight falling on the leaf surface, influencing energy exchange owing to dust layer.
2) Reducing chlorophyll content.
3) Interruptions in gaseous exchange caused by dust particles obstructing stomata.
4) Dust deposition alters the soil characteristics that enable plant development.

b) Effects on animals: For a long time, air contaminants have been recognised to have an indirect influence on animals. It is most common around nonferrous smelters and companies such as phosphate fertiliser operations, brick kilns, and aluminium smelters, where fluorides are discharged and accumulated in the grasses and adjacent fields. Similarly, lead compounds from automotive exhausts are deposited near roadways, however the amounts detected in vegetation are much lower than those found near smelters and have so far been assessed as being below the acceptable threshold for animal toxicity. However, over time, such metals accumulate in animal bodies via the food chain, posing a threat to animal health. Heavy metal poisoning symptoms include diarrhoea, anaemia, and stiffness. Fluorosis is a more prevalent issue that affects livestock, notably dairy cows.

c) Effects on man and health: Air pollution enters the nose along with inhaling air, where fine hair filters out most particles larger than around 10 micrometres in diameter. The air is then warmed and humidified before being transmitted down the windpipe into the bronchial tubes, which divide the air stream and carry it into the lungs, where it is divided into many air sacs. Gaseous contaminants enter blood streams by capillary diffusion from the air sac, and particles settle in the alveolar sacs. The hazards of some microscopic particles, such as silica and asbestos, which are common in mines, quarries, and some industrial sites, have long been recognised. They cause occupational diseases such as silicosis and asbestosis.

d) Effects on Materials: Pollutants in the air have a negative impact on materials such as stone, paintwork, stained glass, fibre materials, and others. Particulate soiling is seen in industrial cities, where buildings made of light-coloured stones and bricks quickly take on a black appearance.
e) Global climate changes: Over the last several decades, it has become clear that air pollution have a significant impact on global climates and associated processes, such as greenhouse effects, ozone depletion, acid rain, and El niño impacts, among others.

India Status

State of global air 2020 report

a) The State of Global Air 2020 study was recently issued. Two cautions are noted in the study for India. For starters, India had the world's greatest annual average PM 2.5 concentration exposure in 2019. Second, over the last decade, the country has had the highest levels of PM 2.5 in the world.
b) According to it, India has the world's greatest per capita pollution exposure (83.2 μg/cubic metre). Over 116,000 newborns perished in India within a month after birth in 2019 as a result of severe air pollution.
c) Despite several government initiatives to reduce air pollution, the Indo-Gangetic plains pollution load has remained high. Furthermore, air pollution is currently the leading cause of mortality among all health concerns.
d) As a result, there is an urgent need for a concerted effort to address this public issue in order to lessen the negative impact of air pollution on people's health.

WHO

a) According to the WHO, toxic air is now the leading environmental cause of premature mortality, accounting for one in every nine deaths.

b) It kills 7 million people per year, which is significantly more than HIV, TB, and malaria combined.
c) Ambient air pollution is responsible for an estimated 4.2 million premature deaths worldwide, primarily from heart disease, stroke, chronic obstructive pulmonary disease, lung cancer, and acute respiratory infections in children.

World Bank

a) According to a 2016 World Bank report, the cost of lost lives and bad health is enormous: $225 billion in lost labour income in 2013, or $5.11 trillion per year (about $1 million every minute) if welfare losses are factored in.
b) Almost all of India's 1.4 billion inhabitants are exposed to harmful levels of ambient pollution of PM 2.5.

Government initiatives

a) National Air Quality Monitoring Programme (NAMP): The Central Pollution Control Board is in charge of the National Air Quality Monitoring Programme, which is a nationwide programme for monitoring ambient air quality (NAMP). The network comprises of 804 working stations that serve 344 cities/towns throughout 28 states and six union territories. The objectives of the N.A.M.P. are to determine the status and trends of ambient air quality; to determine whether the prescribed ambient air quality standards are violated; to identify non-attainment cities; to obtain the knowledge and understanding required for developing preventive and corrective measures; and to comprehend the natural cleansing process that occurs in the environment through pollution dilution, dispersion, wind-based movement, dry deposition, and precipitation. Under NAMP, four air pollutants have been designated for regular monitoring at all locations: Sulphur Dioxide (SO2), Nitrogen Oxides as NO2, Respirable Suspended Particulate Matter (RSPM / PM10), and Fine Particulate Matter (PM2.5). Meteorological data such as wind speed and direction, relative humidity (RH), and temperature were also monitored in conjunction with air quality. Pollutant monitoring is carried out for 24 hours (4-hourly sampling for gaseous pollutants and 8-hourly sampling for particulate matter) twice a week, for a total of 104 observations each year. Monitoring is carried out with the assistance of the Central Pollution Control

Board, State Pollution Control Boards, Pollution Control Committees, and the National Environmental Engineering Research Institute (NEERI), Nagpur. The CPCB works with these agencies to maintain the uniformity and consistency of air quality data, and it offers technical and financial assistance to them in order for them to operate the monitoring stations. N.A.M.P. is managed by a number of monitoring organisations. A large number of people and machines are involved in sampling, chemical analysis, data reporting, and so on.

b) National Ambient Air Quality Standards (NAAQS): National Ambient Air Quality Standards (NAAQS) are standards for air quality that are set by the Central Pollution Control Board (CPCB) that are applicable all over the country. These criteria are critical for the creation of effective ambient air quality management. The first ambient air quality criteria were established in 1982 in accordance with the Air Act. These standards were later amended in 1994 and 1998. The most recent adjustment to the NAAQS was completed in 2009, and this is the version that is being used. The 2009 regulations reduced the maximum allowable levels for pollutants even more and made the requirements uniform across the country. Prior to now, industrial zones had less severe criteria than residential areas. The National Air Quality Monitoring Programme monitors compliance with the NAAQS (NAMP). The CPCB is in charge of implementing NAMP. The current limits (2009) include the following pollutants:
1) Particulate Matter 10 (PM10)
2) Particulate Matter 2.5 (PM2.5)
3) Nitrogen Dioxide (NO2)
4) Sulphur Dioxide (SO2)
5) Carbon Monoxide (CO)
6) Ozone (O3)
7) Ammonia (NH3)
8) Lead (Pb)
9) Benzene
10) Benzopyrene
11) Arsenic
12) Nickel

c) National Air Quality Index: Government entities are responsible for informing the public about health dangers and how to avoid them. The National Air Quality Index was launched in September 2014 in New Delhi as part of the Swachh Bharat Abhiyan to monitor India's air quality index. The Central Pollution Control Board (CPCB) has assembled an expert panel comprised of medical professionals, air quality specialists, academics, advocacy organisations, and SPCBs. IIT Kanpur was also given a technical study. The IIT Kanpur and the expert committee suggested an AQI plan for India in 2014. Six Indian cities have a continuous monitoring system that captures data in real time: New Delhi, Kolkata, Mumbai, Pune, and Ahmedabad.

The Air Quality Index is calculated using the average air pollutant concentration over a specific period collected from an air sensor or model. The AQI rises due to an increase in the release of air pollutants; for example, a high AQI is recorded during heavy traffic or a forest fire. Furthermore, because to anticyclones or temperature inversions, stagnant air contains significant amounts of contaminants. As a result of chemical interactions between air contaminants and foggy circumstances, the AQI rises. Among the instruments used to measure air quality are: PCE-RCM 05, PCE-RCM 8 and PCE-HFX 100. The National Air Quality Index categorises air pollution into six levels: good, satisfactory, moderately polluted, poor, very poor, and severe. As shown in the table 11.2 below, each category outlines the health consequences connected with it:

Table 11.2: National Air Quality Index

AQI	Category	Health Impacts
0-50	Good	Minimal Influence
51-100	Satisfactory	Sensitive persons may encounter modest breathing difficulties.
101-200	Moderately polluted	Patients with lung illness may encounter breathing difficulties, as may those with heart disease, youngsters, and the elderly.
201-300	Poor	Breathing difficulty with extended exposure, as well as pain for people with heart disease
301-400	Very Poor	Prolonged exposure might result in respiratory disease. People with lung and heart disorders are particularly vulnerable.
401-500	Severe	While healthy people may encounter respiratory issues, persons with lung or heart illness may suffer serious health consequences. People may suffer health consequences even while engaging in mild physical activity.

d) National Clean Air Programme: The programme is a pollution management strategy with the primary objective of lowering the concentration of coarse and fine particulate matter in the atmosphere by at least 20 percent by 2024. The NCAP is a collaborative effort between the Ministry of Road Transport and Highways, Ministry of New and Renewable Energy, Ministry of Petroleum and Natural Gas, Ministry of Heavy Industry, Ministry of Health, Ministry of Housing and Urban Affairs, Ministry of Agriculture, Central Pollution Control Board and NITI Aayog. The program's objectives are:
   1) To extend the nationwide network of air quality monitoring stations.
   2) To increase capability for managing air pollution
   3) To increase public understanding of the dangers of air pollution
   4) The NCAP also aspires to establish a workable strategy for air pollution prevention, management, and control.
   5) The initiative will be implemented at the national level by an apex committee at the Environment Ministry. Committees at the Chief Secretary level will monitor the scheme's implementation at the state level.
   6) The Central Government's Smart Cities Mission will also be used to launch clean air initiatives in 43 of the 102 cities.

Delhi Air Pollution Control Measures:

1) Reduce dieselization.
2) Expand integrated public transportation.
3) Encourage walking and cycling.
4) Tax polluting modes.
5) Decide to adopt Bharat Stage IV countrywide in 2015 and Euro VI in 2020.
6) Put controls on other pollution sources.

Water Pollution

Water is an essential component of the life support system. It is a great chemical media with the unique capabilities of dissolving and suspending a wide range of chemicals. As a result, it is readily polluted. Impurities from diverse causes are frequently found in natural surface water bodies. Impurities might include suspended particles, colloidal materials, and dissolved cationic and anionic substances. Various natural and man-made activities—industrial, household, agricultural, and others are causing water contamination on a daily basis, particularly in fresh water systems. The time has come to collaborate on a war footing to battle water contamination concerns with new safeguards in order to provide all of our children the gift of clean and safe water in the twenty-first century.

Water pollution is defined as the addition or presence of unwanted chemicals to/in water, such as organic, inorganic, biological, radiological, or heat, which reduces the quality of water to the point that it is unsuitable for use. Soil erosion, leaching of minerals from rocks (owing to natural solubility and solubility caused by acid rain), and organic matter decomposition are all natural sources of water contamination.

Sources

1) Point sources: Point source pollution occurs when pollutants are discharged from a single area, such as a drain pipe carrying industrial effluents dumped directly into a water body.
2) Diffuse or non-point sources: Non-point sources, on the other hand, include pollutant discharge from scattered sources or from a greater region, such as runoff from agricultural fields, grazing grounds, construction sites, abandoned mines and pits, and so on.

Causes of Water pollution

a) Sewage Water: Sewage water is wastewater from a community's residents, i.e., water discharged from residences after usage for different reasons such as dishwashing, laundry, and toilet flushing, hence the term wastewater. The used water is removed from the dwellings via pipes constructed during the plumbing process. The sewage water then flows through sewers, which are either built by the house owner or installed by the municipality. Grey and black water make up the majority of sewage water. Grey water is waste water from bathing, cleaning dishes, or doing laundry. Toilet waste is referred to as black water. Due to the chemical nature of the various waste elements, it is distinguished by detritus such as paper wrappings, sanitary items, soap residues, and dirt. Furthermore, sewage water stinks. The problem is that sewage contamination has come from overcrowding in metropolitan areas without sufficient planning, posing a hazard not just to the environment but also to human health. It also has an impact on biodiversity, aquatic life, and agriculture, and it contributes significantly to eutrophication and an increase in Biological Oxygen Demand (BOD).

1) Biological Oxygen Demand (BOD): Organic waste water contamination is quantified in terms of Biochemical Oxygen Demand (BOD). BOD is the quantity of dissolved oxygen required by microorganisms to decompose organic waste in water. It is measured in milligrammes per litre of water. A greater BOD value suggests that the water has a low DO concentration. BOD is not a valid way of evaluating water contamination since it is confined to biodegradable materials.
2) Dissolve Oxygen (DO): Water with a DO concentration less than 8.0 mg/L may be deemed polluted. Water with a DO concentration less than 4.0 mg/L is considered extremely contaminated. The DO concentration of water is critical for aquatic life viability. Surface turbulence, photosynthetic activity, O2 consumption by organisms, and organic matter breakdown are all variables that influence the quantity of DO in water. The higher the amount of waste, the faster it decomposes and consumes O2, lowering the DO level of water.
3) Chemical Oxygen Demand (COD): COD is a somewhat better method for measuring pollution load in water. COD quantifies the amount of oxygen required to oxidise organic (biodegradable and non-biodegradable) and oxidizable inorganic chemicals in a water sample in parts per million. The Chemical Oxygen Demand (COD) test determines the quantity of oxygen required to chemically oxidise organic and inorganic nutrients in water, such as ammonia and nitrate.

b) Industrial Wastes: Discharge of wastewater from industries such as petroleum, paper manufacturing, metal extraction and processing, chemical manufacturing, and others that frequently contain toxic substances, particularly heavy metals (defined as elements with densities greater than 5 g/cm3 such as mercury, cadmium, copper, lead, and arsenic) and a variety of organic compounds.
c) Agricultural Sources: Dissolved salts such as nitrates, phosphates, ammonia, and other nutrients, as well as hazardous metal ions and organic compounds, are found in agricultural runoff. Fertilizers are rich in plant nutrients including nitrogen, phosphorus, and potassium. Excess fertilisers may leak into groundwater or be combined with surface water. Pesticides include insecticides, fungicides, herbicides, and other chemicals. They include a variety of compounds, including chlorinated hydrocarbons (CHCs, such as

DDT and Endosulfan), organophosphates, metallic salts, carbonates, and others. Many pesticides are non-biodegradable, and their residues last for a long time. Runoff carries waste from chicken farms, piggeries, and slaughterhouses into the river and causes pollution.

d) Thermal and Radiation Pollution: Thermal and nuclear power plants, as well as chemical and other industries, consume a lot of water for cooling, and the utilised hot water is discharged into rivers, streams, or seas. The discharge of hot water may raise the temperature of the receiving water by 10 to 15 degrees Celsius over the ambient temperature. This is called thermal pollution. The dissolved oxygen in water diminishes as water temperature rises. Aquatic creatures, unlike terrestrial organisms, are suited to a constant, homogeneous temperature in their habitat. Temperature spikes harm fish and other aquatic species. One of the most effective ways to reduce thermal pollution is to keep hot water in cooling ponds and allow it to cool before releasing it into any receiving water body. Mutations in the DNA of marine species are caused by radiation exposure. If the mutations are not fixed, the cell may develop cancer. The thyroid gland absorbs radioactive iodine, which can lead to thyroid cancer.

e) Marine Pollution: All natural and man-made pollution eventually end up in the oceans. Coastal cities' wastewater and garbage are also discharged into the sea. Other causes of maritime contamination include oil, grease, detergents, sewage, rubbish, and radioactive waste discharged during navigation, offshore oil mining, and oil spills. Leaking during sea transport and leakage from subterranean storage tanks are the two most prevalent causes of oil spills. An oil leak might also occur during offshore oil production. Because oil is lighter than water, it forms a thin layer on the water's surface, cutting off oxygen to floating plants and other producers. Fish, shellfish, and plankton die within hours after an oil spill owing to asphyxia (a condition arising when the body is deprived of oxygen, causing unconsciousness or death by suffocation.) and metabolic problems.

f) Invasive species: Water hyacinth plants are the world's most troublesome aquatic weed, commonly known as the 'Terror of Bengal.' They proliferate in eutrophic water basins, causing an imbalance in the environment. They cause havoc by causing the stagnation of dirty water due to their rapid development.

g) Underground pollution: In general, ground water is chemically and microbiologically unpolluted, making it suitable for drinking and cooking, as well as agricultural and industrial purposes. However, there have been several cases of ground water pollution with arsenic, fluoride, or nitrate in recent decades. These contaminants pose substantial health risks in the long run. It is generally understood that toxic metal contamination of ground water occurs mostly as a result of leaching from hazardous waste dumps or from the crustal layer of the earth via biotransformation. Agricultural pollutants have also leached into higher aquifers, contaminating ground water.

1) Nitrate: Nitrate in ground water is generally caused by the mineralization of soil organic matter or the application of excessive nitrogen fertiliser. The nitrate concentration of source water ranges from less than 1 mg NO3 per lit to around 50 mg NO3 per lit. The initial restriction of 50 mg NO3 per lit was designed to protect babies from methanoglobinaemia. This is a condition in which more than 10 percent of the haemoglobin in the blood is converted to methanoglobin. It is reversible, however mortality can occur only if more than 40 percent of the haemoglobin in the blood is converted due to decreased oxygen.

2) Arsenic: Ground water arsenic pollution has been documented in numerous places of the world throughout the years. However, the biggest occurrence in the Indian subcontinent has been seen in West Bengal and Bangladesh, where a large tract is affected by the arsenic tragedy. Thousands of individuals are suffering from chronic arsenicosis as a result of consuming polluted ground water. Chronic arsenic exposure causes black foot disease. It can also induce diarrhoea, as well as lung and skin cancer.

3) Fluoride: Fluoride contamination, like arsenic contamination, has been documented in recent years in various regions of West Bengal and Assam. There have previously been instances of fluoride contamination of surface water (dug wells) in Andhra Pradesh. The exposed community suffers greatly as a result of such pollution, both in the short and long term. There have been instances of bone deformities and issues with locomotory organs. Children are more vulnerable than adults. Identification of polluted regions is critical, as is the delivery of clean, non-contaminated water for cooking and drinking. Fluoride in excess in drinking water causes neuromuscular issues, gastrointestinal

difficulties, tooth malformation, bone hardening, and stiff and painful joints (skeletal fluorosis). Knock-Knee syndrome is characterised by pain in the bones and joints as well as outward bending of the legs from the knees. Fluorosis is a widespread condition in numerous states owing to the use of water with a high fluoride content.

Effects of Water pollution

a) Effects on Human Health: Domestic and hospital sewage contain numerous dangerous germs, and dumping it into water without sufficient treatment can lead to an epidemic of deadly illnesses like typhoid and cholera. Metals in industrial wastewaters such as lead, zinc, arsenic, copper, mercury, and cadmium are harmful to people and other creatures. Consumption of arsenic-polluted water causes arsenic build-up in body components such as blood, nails, and hair, resulting in skin lesions, rough skin, dryness, and thickening of the skin, and ultimately skin cancer. Bacterial activity converts mercury compounds in wastewater into very poisonous methyl mercury, which can induce numbness of the limbs, lips, and tongue, hearing, blurring of eyesight, and mental instability. Mercury pollution of bodies of water produces Minamata (neurological syndrome) sickness in people. Lead poisoning is caused by lead (Lead interferes with a variety of body processes and is toxic to many organs and tissues). Lead compounds induce anaemia, headaches, muscular weakness, and a blue line around the gums. Water polluted with cadmium can induce itaiitai illness, also known as ouch-ouch disease (a painful bone and joint condition), as well as lung and liver cancer.
b) Effects on Aquatic Ecosystems: Water pollution reduces the presence of dissolved oxygen (DO) and therefore eliminates sensitive creatures such as plankton, mollusks, and fish. Some tolerant species, such as Tubifex (annelid worm) and insect larvae, may be able to thrive in extremely contaminated water with low DO levels. These species have been identified as contaminated water indicator species. Heavy metals, biocides, and polychlorinated biphenyls (PCBs) all kill vulnerable aquatic creatures directly. When hot fluids from industry are dumped into bodies of water, the DO concentration is reduced. Such dangerous experiments and natural processes also lead to eutrophication and algal blooms.
c) Effects on Environment: Microorganisms engaged in the biodegradation of organic materials in sewage waste use a lot of

oxygen, resulting in low oxygen levels in the water, which kills fish and other aquatic species. Algal bloom (excessive development of planktonic algae) occurs when huge levels of nutrients are present in water. Lakes age as a result of this. In the aquatic food chain, a few hazardous chemicals that are often found in industrial wastewaters can be biologically magnified (Biomagnified). This is a well-known phenomenon with mercury and DDT. DDT at high levels disrupts calcium metabolism in birds, causing eggshell thinning and premature breakage, eventually leading to bird population decrease.

Control Measures

a) Buffered riparian areas
b) Sewage water and industrial effluents should be treated before being released into bodies of water.
c) Hot water from power plants should be cooled before being released.
d) Domestic cleaning in drinking water tanks, streams, and rivers should be forbidden.
e) Use of fertilisers and pesticides in excess should be avoided.
f) Organic farming and effective fertiliser utilisation of animal waste.
g) Water hyacinth (an aquatic weed) may cleanse water by removing certain hazardous compounds as well as a variety of heavy metals.
h) Oil spills in water may be cleansed with bregoli, a paper industry by-product that includes saw dust, oil zappers, and microorganisms.

Pollution of Indian lakes: Lakes and reservoirs are the most valuable water resources among surface water bodies. These surface water bodies are now under severe pollution danger, not just in India but also internationally, and this is seen as a major concern. National and worldwide programmes on lake water quality monitoring and management in the context of conservation have been tried throughout the last few decades. Significant research has been conducted in Himalayan lakes in J&K (Dal & Nagin Lake), Kumaun (Nainital and Bhimtal Lake), Madhya Pradesh (Bhopal Lake), Orissa (Chilka Lake), Manipore (Loktak Lake), and others. In virtually all situations, lake water is heavily contaminated with silts, agrochemicals, fertiliser, organic waste, and industrial discharges, among other things. As a result, significant biotic alterations have been seen, as well as a decrease in output. Most limnological investigations of the lake system thus found that most of the lakes gradually turned into swampy marsh without any adequate conservation management. The Ministry of Environment and Forest, Government of India, has had a national lake protection policy in place since the 1990s.

River Pollution in India: Most Indian rivers and their tributaries, including the Ganges, Yamuna, Godavari, Krishna, Sone, Cauvery, Damodar, and Brahmaputra, have been reported to be severely polluted as a result of untreated sewage disposal and industrial effluent discharge straight into the rivers. For example, the river Ganges alone gets sewage from 29 class I cities on its banks, as well as industrial effluents from around 300 small, medium, and large industrial units across its entire 2,525 kilometre journey. Similarly, another important river, the Yamuna, has been endangered by pollution in the Delhi and Gaziabad areas. Every day, around 5,15,000 kilolitres of sewage waste water are estimated to be released into the Yamuna River. Similarly, several additional rivers have been assessed for contamination during the last three decades. Aside from home and industrial discharge into rivers, there was also surface run-off from agricultural regions, mining, and even cremation on river banks. Since the 1980s, the Indian government has taken significant measures to reduce water pollution in the holy river Ganga and other important rivers in India. The main effort is the Ganga Action Plan, which has been in place for decades to reduce and control the causes of river pollution. One of these noteworthy developments is the construction of Farakka Barrage.

Soil Pollution

Soil pollution is a type of land pollution in which the top layer of the soil is affected. Chemical fertiliser overuse, soil deterioration caused by running water, and various pest control procedures, among other factors, contribute to the loss of agricultural land, forest cover, and grazing pastures. Soil gets polluted by a number of ways. The major kinds of soil pollution are:

a) Acidification: The process of soil acidification occurs when the pH of the soil falls over time. The concentration of hydrogen ions in the soil rises, causing soil acidification. Agricultural output quickens this process, which can influence both the top soil and the subsoil. Some factors that contribute to soil acidity include:
   1) the use of ammonium-based nitrogen fertilisers at high levels in naturally acidic soils;
   2) nitrogen nitrate leaching, originally used in ammonium-based fertilisers; and

3) plant material harvesting (plant material is alkaline so when it is removed the soil is more acidic than if the plant material had been returned to the soil).

b) Salinisation and sodification: The accumulation of water-soluble salts in the soil is known as salinisation. Potassium (K^+), magnesium (Mg^{2+}), calcium (Ca^{2+}), chloride (Cl^-), sulphate (SO_4^{2-}), carbonate (CO_3^{2-}), bicarbonate (HCO_3^-), and sodium (Na^+) are among these salts. Sodification refers to the accumulation of sodium. With water, salts dissolve and travel around. The salts are left behind after the water evaporates. One of the primary physiological risks to ecosystems is the accumulation of salts (especially sodium salts). Plant development is hampered by salt, which restricts nutrient absorption and lowers the quality of water supplied to the plant. It disrupts the metabolism of soil organisms, resulting in a significant reduction in soil fertility. Because of the rise in osmotic pressure and the harmful effects of salts, high levels of salinity in soil cause plant withering. An overabundance of sodium destroys the soil structure, rendering it incapable of supporting plant or animal life owing to a lack of oxygen. The impermeability of deep soil layers is increased by salinisation, making it hard to cultivate the area.

c) Agrochemical Pollution: Inorganic fertilisers have become increasingly popular in recent decades, at the cost of more traditional organic nutrient treatments. Between 1952 and 1985, worldwide fertiliser usage climbed by about 900 percent, from 14 million tonnes to 125 million tonnes. Because the nutrients are in a more easily available form and are released quickly after application, inorganic fertilisers are preferred over organic treatments. Because typical nutrient cycles between plants, soil, and animals are significantly interrupted as animal production rises, synthetic herbicides, insecticides, bactericides, and fertilisers are widely used, all of which contribute to pollution.

d) Urban and Industrial pollution: Intensive urbanisation will increase poverty by making it impossible for local governments to provide services to all citizens. Air pollution caused by concentrated energy use has a direct impact on human health. Because of automotive emissions, lead levels in metropolitan air are increased. Radioactive waste from nuclear research stations and power facilities, as well as radioactive fallout from nuclear explosions, pollute the earth. Because radioactive elements have a long half-life, they can remain for lengthy periods of time in the soil.

Effects of Land pollution

The various effects of land pollution are:

a) Climate change: Land pollution, such as that caused by mining, farming, and manufacturing, may allow dangerous chemicals to infiltrate the soil and water, contributing to climate change. These toxins have the ability to damage animals and plants, causing the food chain to be disrupted. Methane, a greenhouse gas that contributes to global warming, is released by landfills.
b) Acid rain: Acid rain and fog affect forests, particularly those at higher elevations. Acid deposits decrease essential elements such as calcium and enable aluminium to leak into the soil, making it harder for trees to absorb water. Acids can also harm tree leaves and needles.
c) Deterioration of fields: As a result of soil pollution, a chain reaction ensues. It diminishes soil filtering ability, decreases soil biodiversity, and decreases soil organic matter. It also pollutes the water in the soil and groundwater, causing nutritional imbalances in the soil.
d) Respiratory health problems: Shortness of breath, coughing, wheezing, asthma attacks, and chest discomfort are all symptoms of air pollution. Air pollution exposure is linked to lung cancer, heart attacks, strokes, and, in the worst-case scenario, early deaths.

Solutions

Minimize, reuse, and recycle to reduce land emissions. It is critical to engage in reforestation and afforestation activities. Farmers can utilise organic fertilisers, an integrated pest management strategy, and crop rotation. Recycling practises are one of the most significant strategies to help decrease landfill trash, maintain natural resources, preserve wildlife, reduce noise, reduce energy use, and slow global warming. By decreasing surface erosion and protecting the fertile topsoil, reforestation prevents river and lake silting. It keeps the soil from drying out and reduces the quantity of precipitation that flows off. Biodegradable plastics create substantially less trash than petroleum-based plastics. Biodegradable polymers disintegrate into benign, harmless components as they decay. They create just 32 percent of the greenhouse emissions released by petroleum-based polymers.

Marine Pollution

Waste and toxins that reach the ocean from terrestrial sources are referred to as marine pollution. This pollution has far-reaching consequences for marine life and economic systems that rely on maritime infrastructure. Marine pollution happens when contaminants from industries, agriculture, and homes, among other things, enter the ocean. Because oceans are home to a diverse range of marine creatures and plants, it is everyone's responsibility to do their part in keeping them clean so that marine species can survive for extended periods of time.

Laws and policies:

a) In 1948, President Harry Truman signed the Federal Water Pollution Jurisdiction Act, which gave the federal government control over marine pollution in the United States.
b) The Council on Environmental Quality passed the Marine Protection, Research, and Sanctuaries Act of 1972, which regulates ocean dumping.
c) MARPOL was a convention formed between 1973 and 1978 to regulate maritime pollution, particularly oil pollution. The MARPOL pact was officially enforced in 1983 by the International Convention for the Prevention of Pollution from Ships.
d) The United Nations Convention on the Law of the Sea (UNCLOS) was founded in 1982 to safeguard the marine environment by requiring governments to regulate pollutants into the sea. It imposed worldwide limits on the quantity of poisons and pollutants released by all ships.
e) FAO launched the BOBLME (Bay of Bengal Large Marine Ecosystem) in 2009. It is supported by the Global Environment Facility. It includes all eight Indian Ocean nations. The goal is to encourage sustainable fishing, minimise pollution in the ocean, and enhance coastal livelihoods.
f) For a West Indian Ocean Region with healthy rivers, beaches, and seas, the Nairobi Convention was established. It is a collaboration between the government, civil society organisations, the private sector, and UNEP.

Causes of Marine pollution

a) Plastics: Between 4 and 12 million metric tonnes of plastic end up in the ocean. Every year, India produces 62 million metric tonnes of trash, of which 10-12 percent is plastic waste, including single-use plastics.

b) Sewage: Pollution can immediately reach the ocean. Sewers or contaminating chemicals run straight into the ocean through sewage, rivers, or drainages. In India, 80 percent of municipal sewage is collected, but only 20 percent of it is cleaned before being dumped straight into the seas.
c) Toxic chemicals from the industries: Another prominent type of trash that is directly thrown into the seas, causing contamination, is industrial and agricultural waste. Toxic substances dumped in the water have a direct impact on marine life.
d) Land Runoff: Another cause of pollution in the water is runoff from the land. When water infiltrates the soil to its maximum capacity and extra water from rain, floods, or melting runs over the land and into the ocean, this occurs. This water frequently picks up dangerous man-made chemicals that end up in the ocean, such as fertilisers, petroleum, pesticides, and other soil toxins. Plastic garbage accounts for 80 percent of land discharge into the sea.
e) Large scale Oil Spills: Ship pollution is a major cause of ocean pollution, with oil spills being the most severe consequence. Crude oil may linger in the water for years and is exceedingly hazardous to marine life, frequently smothering them to death if entrapped. Crude oil is also incredibly difficult to clean up, which means that once it has been divided, it generally stays split.
f) Ocean mining: Another source of contamination in the water is deep sea mining. Ocean mining sites that drill for silver, gold, copper, cobalt, and zinc develop sulphide deposits up to 3,000 metres below the surface.
g) Littering: Believe it or not, pollution from the atmosphere is a major source of ocean contamination. This happens when things from far inland are blown large distances by the wind and end up in the ocean. These particles can range from natural materials such as dust and sand to man-made materials like garbage and rubbish. The majority of waste, particularly plastic litter, is incapable of decomposition and stays floating in the ocean's circulation for years.

Effects of Marine pollution

a) Toxic Wastes Effect on Marine Animals: The oil spill poses a threat to marine life in a number of ways. The oil spilt in the ocean might get into marine animals' gills and feathers, making it harder for them to move, fly, or feed their young. Cancer, reproductive system failure, behavioural abnormalities, and even death can be long-term effects on marine life.

b) Eutrophication occurs when a body of water gets too loaded with minerals and nutrients, causing excessive algae growth or an algal bloom. The water body loses oxygen as a result of this process.
c) Human Health: Animals from the damaged food chain are consumed by humans, which has an influence on their health since chemicals from these polluted animals are deposited in human tissues, potentially leading to cancer, birth abnormalities, or long-term health problems.
d) Disruption to the cycle and food chain: The oil spill floats on the water's surface, blocking sunlight from reaching marine plants and interfering with photosynthesis. Long-term effects on marine life include skin irritation, eye discomfort, lung and liver disorders. Chemicals used in industry and agriculture wash into rivers and are transferred into the seas from there. These compounds do not dissolve and sink to the ocean's bottom. Small animals consume these pollutants, which are ultimately consumed by larger species, affecting the entire food chain.

Solutions

a) One large-scale option is tighter government control of industry and manufacturing.
b) Limit offshore drilling by using renewable energy sources like wind and solar electricity.
c) Pesticides in agriculture should be limited, and organic farming and environmentally friendly pesticides should be encouraged.
d) Sewage treatment and investigation of environmentally friendly wastewater treatment methods.
e) Reduce industrial and manufacturing waste, and keep landfills from spilling into the ocean.

Noise Pollution

The term 'noise' comes from the Latin word 'nausea,' which denotes an illness that makes you want to vomit. Noise is an unpleasant and unwelcome sound that causes discomfort in people. A sound's intensity is measured in decibels (dB), and noise pollution and its levels are assessed using a decibel scale. A whisper is defined as a sound intensity of less than 20 decibels. Sound levels less than 70 dB are not harmful to living beings, according to the World Health Organization (WHO), regardless of how long or constant the exposure is. Continuous noise above 85 dB for more than 8 hours might be dangerous. Noise pollution has become an issue as a result of increased noise surrounding civilizations. Vehicles, aeroplanes, industrial machinery, loudspeakers, and firecrackers are all important contributors. Other equipment, such as televisions, transistors, and radios, add to noise pollution when they are utilised at excessive volumes.

Causes of Noise pollution

Because of the increased noise around us, which is mostly created by automobiles, aeroplanes, industrial machinery, loudspeakers, firecrackers, and other sources, noise pollution has been a source of worry. The following are some of the most significant sources of noise pollution:

a) Growing industries in metropolitan areas are a major source of noise pollution these days, since they use a variety of machinery capable of producing a huge quantity of noise.
b) Improper urban area planning: Due to packed housing, tiny spaces, insufficient parking facilities, and frequent battles over basic services, improper and bad urban planning plays a big part in causing noise pollution, especially in developing nations.
c) Social gatherings: At various social events, individuals frequently play songs at full intensity, causing noise pollution and deteriorating living conditions. Loudspeakers are used to play music at weddings and other events, resulting in unpleasant noise in the neighbourhood.
d) Cars and transportation: One of the causes of noise pollution is an increase in the number of vehicles on the road. Heavy noise is produced by traffic jams, subway trains, aeroplanes, and other sources, which can contribute to hearing loss.
e) Construction sites: Various construction activities such as mine, bridge, dam, and building construction, among others, contribute significantly to noise pollution.

Effects of Noise pollution

Loud music, vehicles, construction, electrical generators, and people are the primary contributors of noise pollution in residential areas. Noise pollution has a variety of negative consequences, including exhaustion, lifelong hearing loss, and mental illnesses.

a) Longer exposure to loud noise causes raised blood pressure, which can lead to hypertension in humans.
b) Hearing loss: Constant exposure to loud noise that exceeds the usual sound intensity range can damage the eardrums, resulting in hearing loss.
c) Sleeping problems: Noise pollution can disrupt an individual's sleep cycle, resulting in sleeping disorders, poor energy, and exhaustion.
d) Loud sounds also induce an increase in normal blood pressure, which can lead to a variety of cardiovascular illnesses in healthy people.

Control of Noise pollution

a) Highway traffic should be redirected by bypasses and overpasses, rather than passing through towns and cities.
b) Machines that produce a lot of noise should have soundproof chambers constructed.
c) Workers in diverse sectors and construction sites should be provided with protection gear such as ear muffs or cotton plugs.
d) Distancing human populations from industrial sites, aerodromes, and railway stations to impose acoustic zoning.
e) Educational institutions, hospitals, and key offices should all have silence zones.

Laws related to Noise pollution in India

a) The Noise Pollution (Regulation and Control) Rules, 2000 control sources of noise pollution individually. The Noise Pollution (Control and Regulation) Rules of 2000 establish the following ambient noise levels for specific areas (Table 11.3):

Table 11.3: Showing ambient noise levels for specific areas

Category of Area/ Zone	Limits in dB(A) $_{Leq}$	
	Day time 6 a.m. to 10 p.m.	Night time 10 p.m. to 6 a.m.
Industrial Area	75	70
Commercial Area	65	55
Residential Area	55	45
Silent Zone	50	40

b) Previously, the Air (Prevention and Control of Pollution) Act of 1981 addressed noise pollution and its causes.
c) The Environment (Protection) Rules, 1986 also set noise limitations for automobiles, air conditioners, refrigerators, diesel generators, and some types of construction equipment.
d) Under the Air (Prevention and Control of Pollution) Act, 1981, State Pollution Control Boards / Pollution Control Committees (SPCBs / PCCs) for states / Union territories control noise from industry.
e) The Indian government launched a Real-Time Ambient Noise Monitoring Network in March 2011. In phase I of this network, five Remote Noise Monitoring Terminals were placed in distinct noise zones throughout seven metros (Delhi, Hyderabad, Kolkata, Mumbai, Bangalore, Chennai and Lucknow). Another 35 monitoring stations will be installed in the same seven cities during Phase II. Phase III will see 90 stations installed in 18 additional locations. Kanpur, Pune, Surat, Ahmedabad, Nagpur, Jaipur, Indore, Bhopal, Ludhiana, Guwahati, Dehradun, Thiruvananthpuram, Bhubaneswar, Patna, Gandhinagar, Ranchi, Amritsar, and Raipur are the cities in Phase III. A Silence Zone is a 100-meter radius around hospitals, educational institutions, courts, religious sites, or any other area designated as such by a competent authority.

Plastic Pollution

Plastic pollution happens when a large amount of plastic has accumulated in a region, severely impacting the natural environment and causing difficulties for plants, wildlife, and even humans. This involves destroying plant life and endangering local wildlife. Although plastic is a very useful material, it is not biodegradable.

Causes of Plastic pollution

a) Plastic is one of the most readily available and misused materials in the world today since it is less costly.
b) Cheap plastics are in more demand as a result of rapid urbanisation and population rise.
c) Because it is a cost-effective and long-lasting material, it is used in everything from packaging to plastic bottles and containers, straws, and carry bags.

d) Decomposition of plastic takes 400 years or more. Plastic disintegration takes anywhere from 500 to 600 years, depending on the material.
e) Floating Fishing Nets.
f) Plastic waste is frequently mishandled and ends up in landfills.
g) Burning plastic is extremely hazardous to the environment and can result in severe infections.

Effects of Plastic pollution

a) Negative effects on Human health: Microplastics entering the human body through direct ingestion or inhalation can cause a variety of health effects, including inflammation, genotoxicity, oxidative stress, apoptosis, and necrosis, all of which have been linked to a variety of negative health outcomes, such as cancer and cardiovascular disease.
b) Plastic contaminated sea food: Microplastics have been detected in 114 marine species, with almost one-third of them ending up on our plates, according to scientists.
c) The Food Chain is Upset: Polluting plastics, which come in a variety of sizes, have an impact on even the tiniest species on the planet, such as plankton.
d) Plastics are responsible for the majority of trash and pollution in the world's seas and groundwater.
e) Land pollution: When plastic is thrown away in landfills, it reacts with water, forming harmful compounds. These contaminants impair the water quality as they seep beneath. The wind transports and discharges plastic from one location to another, increasing the amount of trash on the ground.
f) Air pollution: Burning plastic in the open air pollutes the ecosystem by releasing harmful chemicals.
g) Economic Impacts: Plastic pollution causes an estimated $13 billion in economic harm to marine ecosystems each year. This includes lost revenue from fishing and tourism, as well as the expense of beach clean-up. Clean-up operations, litter disposal, the repair and replacement of damaged vessels and gear, lower fishing catches, and a drop in coastal tourism and its influence on allied sectors are all examples of economic consequences.
h) Marine life: Ingestion, suffocation, and entanglement of hundreds of marine species are the most apparent and unpleasant effects of marine plastics. Seabirds, whales, fish, and turtles mistake plastic

garbage for food, and the majority of them starve to death as their bellies fill with plastic debris.

Micro plastics

Plastic waste particles smaller than 5mm in length, or roughly the size of a sesame seed, are referred to as micro-plastics. They can arise from a number of places, one of which is when bigger bits of plastic disintegrate into tiny, harder to detect ones.

Micro plastic pollution

a) The durability of plastic, which means that depending on the type of plastic and where it was deposited, it might take hundreds to thousands of years for it to degrade.
b) Plastic pollution in the oceans has an influence on marine life, ocean health, coastal tourism, and even human health.
c) Various news reports in recent years have revealed that marine species such as whales, seagulls, and turtles accidentally swallow plastic and suffocate as a result.
d) Marine plastic pollution is also dangerous to people if it enters the food chain. Micro-plastics, for example, have been discovered in tap water, beer, and even salt.
e) According to one of the first studies to assess plastic pollution in human consumption, an average individual consumes at least 50,000 microplastic particles each year.

Government Initiatives

a) On July 2022, India banned single-use plastics.
b) Single-use plastic goods have been banned in all federal and state government offices, as well as major PSUs.
c) Imports of solid plastic garbage have been prohibited in India.
d) The Plastic Waste Management Rules of 2016 were approved in India, introducing Extended Producer Responsibility. Its goal is to raise the minimum thickness of plastic carry bags from 40 microns to 50 microns.

The Strategy of the 3R's +E:

a) Reduce: In order to effectively reduce plastic pollution, we must reduce our use of plastic.

b) Reuse: Many plastic goods may be reused or repurposed for a variety of applications. Before tossing away plastic things, think about how they may be repurposed.
c) Recycle: To limit the amount of plastic in the waste stream, plastic recycling involves collecting garbage and turning it into new goods.
d) Educate: Education is another critical method for increasing awareness and behavioural change.

Thermal Pollution

The rise or reduction in temperature of a natural aquatic environment induced by human intervention is known as thermal pollution. Due to the expanding call of globalisation everywhere, this has become an increasing and current pollution. Thermal pollution is generated by dumping hot water from factories and power plants, or by removing trees and vegetation that shade streams, allowing sunshine to raise the temperature of these waters, and then releasing cold water to cool them down. Thermal pollution, like other types of water pollution, is pervasive, impacting numerous lakes and a large number of streams and rivers across the world. The term "thermal pollution" has been used to describe the negative impacts of different power stations' heated effluent discharge. It refers to the degradation of the aquatic and terrestrial environment caused by various industrial plants such as thermal, atomic, nuclear, and coal-fired facilities, as well as oil field generators, factories, and mills.

Sources: The released effluents from these sources are hotter than the intake water, which lowers the concentration of oxygen in the water and has a negative impact on the marine ecology.

a) Nuclear Power Station
b) Coal fired power plant
c) Industrial Effluents
d) Domestic Sewage
e) Thermal Power Plant
f) Thermal Power Plant

Harmful effects of Thermal pollution

a) Reduction in dissolve oxygen: When pollutants from various industrial facilities are heated, the concentration of oxygen falls as the temperature of the water rises.

b) Changes in water properties: The reduction in density, viscosity, and solubility of gases in water accelerates the settling speed of suspended particles, threatening aquatic organisms' food supply.
c) Increase in Toxicity: The concentrated pollutant raises the temperature of the water, increasing the toxicity of the toxin contained in the water. Water poisoning will raise the mortality rate of marine life.
d) Disruption of Biological activities: Temperature fluctuations damage the whole marine environment since temperature variations affect an aquatic organism's physiology, metabolism, and biological processes such as respiration rate, digesting, excretion, and development.
e) Damage of biotic organisms: Aquatic creatures that travel through screens and condenser cooling systems, such as juvenile fish, plankton, fish, eggs, larva, algae, and protozoa, are very sensitive to temperature fluctuations. They are accustomed to warmer water and may unexpectedly meet an increase or drop in the temperature of water bodies, causing them to perish as a result of the sudden changes in water temperature.

Thermal Pollution prevention

a) Heated water from industry can be treated before being discharged directly into bodies of water.
b) The building of cooling ponds and cooling towers can treat heated water from factories.
c) Industrial treated water can be reused for household or industrial heating purposes.
d) Through manmade lakes: Industries can dump their used or heated water at one end of the lake, while water for cooling purposes can be retrieved from the other. Heat is finally dispersed via evaporation.
e) Cogeneration is a method in which surplus thermal energy from electricity generation is utilised in another manufacturing process that requires such energy. Waste hot water can be utilised to heat residences or other structures near industrial plants, a technique common in Scandinavian towns and cities and proposed for use in China.
f) The prescription for preventing thermal pollution caused by devegetation is simple: do not devegetate and leave strips of trees and vegetation along streams and shorelines.

Radioactive Pollution

The contamination of water and air caused by radioactive elements is known as radioactive pollution. It is possible to cause life-threatening contamination if radioactive waste is not properly disposed of. For example, the first atomic bomb, which exploded during World War II, had far-reaching consequences that the world will never forget. Flora and wildlife were entirely annihilated. The radiation generated by the bomb burnt or killed about 30 percent of the population, and another 30 percent was gravely affected. Even despite these findings, the nuclear race between various nations continues, resulting in radioactive contamination. Radioactivity is the ability of some elements (such as radium, thorium, and uranium) to spontaneously emit protons (alpha particles), electrons (beta particles), and gamma rays (short-wave electromagnetic waves) through the disintegration of their atomic nuclei (nuclides). Each radioactive material decays at a constant rate. The half-life is the amount of time it takes for half of its atoms to decay. The half-life can range between a fraction of a second and thousands of years. Radionuclides with extended half-lives are the primary cause of radioactive contamination in the environment. Radiations are classified into two types: non-ionizing radiations and ionising radiations.

a) Non- Ionizing radiation: Non-ionizing radiations are made up of electromagnetic waves of longer wavelengths, ranging from near infrared rays to radio waves (includes higher wavelength ultraviolet rays, microwaves). These waves contain enough energy to excite the atoms and molecules in the medium they pass through, causing them to vibrate faster, but not enough to ionise them. They may cause eye injury due to reflections from coastal sand, snow (snow blindness), or directly looking at the sun during an eclipse. They cause sunburns by injuring skin cells and blood capillaries, resulting in blisters and reddening.
b) Ionizing radiation: Ionisation is the process by which an atom or molecule gains or loses a negative or positive charge by acquiring or losing electrons to create ions, which is frequently combined with other chemical changes. Ionising radiations ionise (pull one or more electrons from an atom's outer shells) the atoms and molecules in the medium through which they pass. Electromagnetic radiations such as short-wavelength ultraviolet (UV), X-rays, and gamma rays, as well as energetic particles produced in nuclear processes, electrically charged particles such as alpha and beta particles produced in radioactive decay, and neutrons produced in

nuclear fission, are extremely harmful to living organisms. Electrically charged particles created by nuclear reactions can have enough energy to knock electrons out of the medium's atoms or molecules, resulting in ions. Ions generated by water molecules, for example, can cause processes that break bonds in proteins and other essential molecules. When a gamma ray travels through a cell, the water molecules around the DNA may get ionised, and the ions may react with the DNA, causing it to break. They can also create chemical changes by disrupting chemical bonds, causing harm to biological cells. Burns, decreased metabolism, dead tissues, and organism death are among short-term impacts. Long-term impacts include increased tumour and cancer incidence, life-span shortening, and developmental abnormalities.

Types of radiation particles:

a) Alpha particles can be stopped by a sheet of paper and human skin.
b) Beta particles can penetrate through skin but are prevented by some glass and metal objects.
c) Gamma rays may readily penetrate human skin and harm cells as they travel, and can only be stopped by a very thick, strong, large piece of concrete.

Sources:

a) Artificial sources:
   1) Nuclear power plant accidents and nuclear waste
   2) Explosion and testing of nuclear weapons (Nuclear fallout). The fall Out contains radioactive elements such strontium-90, caesium-137, and iodine-131.
   3) Mining for uranium and other radioactive materials such as thorium (monazite is the ore of thorium).
   4) Radiation treatment and direct radiation exposure for diagnostic reasons (e.g., X-rays), chemotherapy, and so on.
   5) Slow nuclear radiations can come from a number of sources, including nuclear reactors, labs, and so on.
b) Natural sources: Cosmic rays from space and terrestrial radiations from radionuclides found in the earth's crust, such as radium-224, uranium-238, thorium-232, potassium-40, carbon-14, and others, are examples.

Impacts

a) Every mobile phone tower antenna emits electromagnetic radiation (power). The radiation emitted by mobile towers is non-ionizing radiation.
b) One mobile phone tower is shared by several operators; the more the number of antennas, the greater the power intensity in the surrounding region.
c) The power level is stronger near towers and decreases as we walk away.
d) EMR may cause physiological and psychological alterations in humans owing to thermal effects caused by microwave radiation absorption.
e) Exposure can cause genetic flaws, affects on reproduction and development, Central Nervous System behaviour, and other problems.
f) Non-thermal effects can also be generated by radio frequency fields at levels too low to cause considerable heating and are caused by the migration of calcium and other ions across cell membranes.
g) This type of exposure is known to cause tiredness, nausea, irritability, headaches, lack of appetite, and other psychiatric issues.
h) The present exposure safety limits are solely focused on thermal effects, with just a few pieces of information from non-thermal impacts being considered.
i) A bird's surface area is bigger than its body weight in compared to the human body, hence it absorbs more radiation.
j) Also, because the bird's body contains less fluid due to its little weight, it heats up quickly.
k) The magnetic field from the towers disrupts birds' navigation ability, causing them to get disoriented and fly in all directions when exposed to EMR.
l) Every year, a considerable number of birds are killed by collisions with telecommunications poles.

Control measures: Because there is no treatment for radiation harm, prevention is the best control approach.

a) All safety precautions must be properly observed. The age of radioactive elements should be thoroughly examined.
b) Radioactive waste disposal that is safe.
c) Monitoring on a regular basis is accomplished by frequent sampling and quantitative analysis.

d) Nuclear accident prevention methods
e) Nuclear explosions and the use of nuclear weapons should be outright prohibited.
f) Appropriate precautions should be taken to avoid occupational exposure.

Acid Rain

Acidification of natural water and soils has become an issue as global industry has increased. This acidity is caused by the transportation and subsequent deposition of SO2, NOX, and other acid oxidation products. The phenomenon is known as acid rain. As a result, great worry was expressed on the international forum concerning future methods for reducing industrial pollution, both regionally and locally. The pH relationship of several substances is represented on the following scale(Fig.11.3).

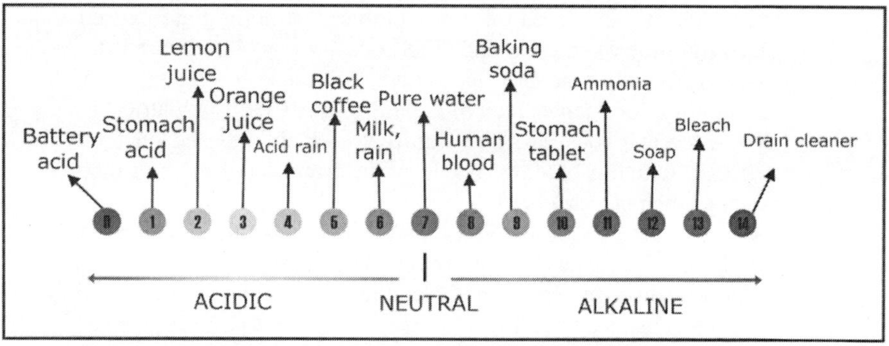

Fig. 11.3: The pH scale

The pH scale:

a) The pH scale determines whether a solution is acidic or basic (alkaline).
b) It has a scale of 0 to 14. A pH of 7 is considered neutral.
c) A pH less than 7 indicates acidity, whereas a pH greater than 7 indicates basicity.
d) It is based on the concentration of hydrogen ions in an aqueous solution.
e) As hydrogen ion levels rise, pH values fall.

f) A pH 4 solution is ten times more acidic than a pH 5 solution, and a hundred times more acidic than a pH 6 solution.
g) While the standard pH range is 0 to 14, lesser and higher values are potentially feasible.

Acid rain is any precipitation (rain, fog, mist, or snow) that has a lower pH than usual (less than 5.6). Acid rain is produced by atmospheric pollution caused by acidic gases such as sulphur dioxide and nitrogen oxides released by the combustion of fossil fuels.Acidic smog, fog, and mist are also known to migrate out of the sky and fall on dust particles, which then build on plants as acid depositions. Acid dews arise when rain falls on these depositions of acid.

Causes of Acid Rain

Acid rain is caused by Sulphur and Nitrogen particles that interact with the moist components of rain. Sulphur and nitrogen particles that combine with water can be obtained in two ways: man-made (emissions from industry) or natural (lightning strikes in the atmosphere releasing nitrogen oxides and volcanic eruptions releasing sulphur oxide). According to the Royal Society of Chemistry, who deems him the "father of acid rain," Scottish scientist Robert Angus Smith coined the term "acid rain" in 1852. The typical clean rain we feel, even if it is not clean, i.e., water and carbon dioxide combine to generate mild carbonic acid, which is not highly hazardous on its own. The resulting response is:

$$H_2O\ (l) + CO_2\ (g) \rightleftharpoons H_2CO_3\ (aq)$$

Regular rainfall has a pH of roughly 5.7, indicating that it is acidic. The nitrogen and sulphur oxides, as well as dust particles, are carried away by the wind. They settle on the earth's surface after falling as precipitation. Acid rain is mostly caused by human activities that produce nitrogen and sulphur oxides into the atmosphere. Examples include the use of fossil fuels and unethical waste disposal practices. Sulphur dioxide and nitrogen dioxide are oxidised before reacting with water to generate sulphuric acid and nitric acid, respectively. The acid generation reaction will be clarified by the following reaction:

$$2SO_2\ (g) + O_2\ (g) + 2H_2O\ (l) \rightarrow 2H_2SO_4\ (aq)$$
$$4NO_2\ (g) + O_2\ (g) + 2H_2O\ (l) \rightarrow 4HNO_3\ (aq)$$

Acid rain and Geology

The environmental impact of acid rain is determined not only by the quantity of acidity in the rain, but also by the type of the ecosystem itself. Areas underlain by granitic or quarzatic bedrock, for example, are more vulnerable because soils and water are already acidic and lack the capacity to buffer or neutralise further acidity from precipitation. As a result, acid levels rise, the environmental equilibrium is upset, and catastrophic ecological damage is unavoidable. Areas that are geologically basic, such as those underlain by limestone or chalk, are far less susceptible and may even benefit from the added acidity. These places' strongly alkaline soils and water guarantee that the acid contributed to the environment by rain is efficiently neutralised. The sensitivity of the environment to acid rain damage in places covered by glacial drift or other unconsolidated deposit will be governed by the character of the surface material rather than the composition of the bedrock. In principle, it is critical to establish baseline values of acidity or alkalinity in order to evaluate the environment's sensitivity to acidification.

Types of Acid deposition

a) Wet Deposition: Acid compounds in the air can fall to the ground as rain, snow, fog, or mist if they are blown into places where the weather is wet. This acidic water impacts a range of plants and animals as it travels over and through the earth(fig.11.4).
b) Dry Deposition: Acid chemicals may become absorbed into dust or smoke and fall to the ground by dry deposition in locations where the weather is dry, clinging to the ground, buildings, vegetation, automobiles, and so on. Rainstorms can wash away dry deposited gases and particles from these surfaces via runoff. This discharge water raises the acidity of the resultant combination. Dry deposition returns around half of the acidity in the atmosphere to the earth.

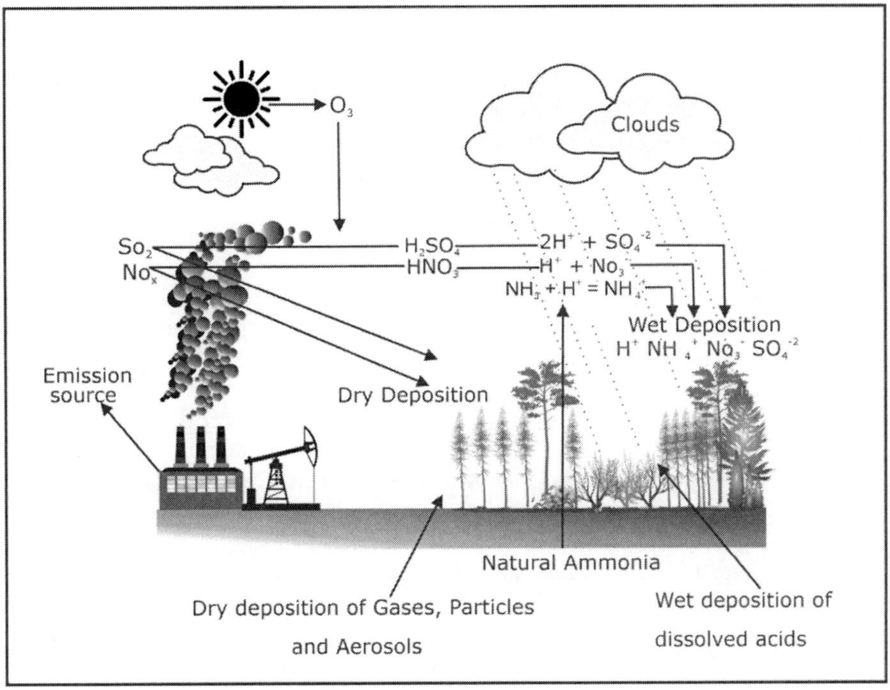

Fig. 11.4: Diagrams showing types of acid rain deposition

Chemistry of acid rain

The formation of acid rain involves six main steps:

a) Sulphur and nitrogen oxides enter the atmosphere from both natural and man-made sources.
b) Some of these oxides fall back to the earth as dry deposition, either near the source or some distance away.
c) Sunlight promotes the production of photo-oxidants in the atmosphere, such as ozone.
d) These photo-oxidants undergo oxidation to produce sulfuric acid (H2SO4) and nitric acid (HNO3) by interacting with the sulphate and nitrate oxides of nitrogen as well as other gases (such as NH3).
e) Wet deposition of acid rain comprising ions of sulphate, nitrate, ammonium, and hydrogen.

Harmful effects of acid rain

a) Effects on Human: Bad odours, impaired vision, skin, eye, and respiratory tract irritation Chronic bronchitis, lung emphysema, and cancer are some of the direct impacts.
b) Effects on Soil: The interchange of hydrogen ions with nutrient cations like potassium and magnesium in the soil causes nutrient leaching, rendering the soil barren. An increase in ammonia in the soil as a result of a loss in other nutrients slows decomposition. The soil's nitrate level has also been observed to drop. Acid rain has a lower impact on soil in India since Indian soils are typically alkaline and have a good buffering capacity.
c) Aquatic life: Fish, frogs, and other aquatic creatures' eggs and sperms are sensitive to pH fluctuations. Acid rain destroys their gametes, disrupting life cycles and production (ecosystem imbalances). Acidic lake waters can destroy bacteria and render them ineffective. Metals bound on soils can be released into the aquatic environment as a result of acid rain.
d) Terrestrial life: Acid rain damages plant cuticles and decreases photosynthesis. Acidic environments encourage the leaching of heavy metals such as aluminium, lead, and mercury. When such metals seep into ground water, they have an impact on soil microflora and fauna. The loss or change of food and habitat resources is another indirect consequence of acid rain on wildlife.
e) Microorganisms: The proliferation of any microbial species is determined by pH. Most bacteria and protozoa like a pH close to neutral. The majority of fungus enjoy an acidic atmosphere. The vast majority of blue-green bacteria prefer an alkaline environment. As a result, microbial species in soil and water change from bacteria to fungus. This creates a delay in the breakdown of organic substances in the soil.
f) Effects on buildings, monuments and materals: Acid rain has a negative impact on many old, historical, and antique structures, as well as pieces of art and textiles. Acid rain degrades limestone and marble. Such things are covered by smoke and soot. Because of acid vapours in the air, they progressively dissolve/flake away from the surfaces.

Real life examples:

a) Taj Mahal: The first report of acid rain in India came from Bombay in 1974. Acid rain has a significant impact on one of the world's seven wonders. Agra has various factories that produce sulphur and nitrogen oxides into the environment. People continue to utilise low-quality coal and firewood as residential fuel, exacerbating the situation. Acid rain reacts with marble (calcium carbonate) as follows:

CaCO3(s) + H2SO4(l) → CaSO4(s) + H2O(l) + CO2(g)

The deterioration of this wonderful monument is caused by the production of calcium sulphate.

b) Statue of Liberty: The copper Statue of Liberty has also been harmed by the cumulative impact of acid rain and oxidation for over 30 years and is thus becoming green.

c) Leshan Giant Buddha, Mount Emei (China): The Leshan Buddha has succumbed to pollution caused by the region's unrestrained growth. The culprit in this case has been identified as the expanding number of coal-fired power stations located near the Giant Buddha, notably the poisonous fumes emitted by their smokestacks, which eventually return to the soil as acid rain. The Buddha's nose has become black over time, and his hair curls have begun to fall off his head.

d) Acropolis OF Athens (Greece): Pollutants and high car emissions from the developing contemporary metropolis of Athens have contributed to acid rain in the region in recent decades, as Greece has undergone significant economic expansion and development. Marble, the ancient Greeks' preferred monumental and sculptural stone, is very sensitive to surface damage from even modest quantities of acid rain. Acid rain, for example, has chemically changed the Parthenon's exquisite marble relief frieze panels into mushy gypsum.

e) Dampier Rock art Complex (Australia): The Dampier Archipelago, which stretches into the Indian Ocean from the Burrup Peninsula in northern Australia, is home to a beautiful collection of Aboriginal petroglyphs engraved on rock faces and outcroppings. The National Trust of Australia has categorised the Burrup Peninsula's rock art sites as endangered, although industrial growth across more than 25 percent of the rock art region since 1963 has created serious dangers to the site. Much of the heaviest industry (mine and petrochemical) is placed very next to some of the most delicate art collections. Acid rain has begun to destroy many of the meticulously, but sometimes shallowly, etched rock surfaces, and archaeologists and geologists have predicted that most of the rock art would vanish by the middle of the twenty-first century.

Control measures:

a) One of the important control strategies is buffering, which is the technique of adding a neutralising chemical to acidified water to raise the pH. Lime in the form of calcium oxide and calcium carbonate is commonly utilised.
b) Reducing SO2 emissions from power plants by using less fossil fuel, employing alternative energy sources such as tidal, wind, and hydropower, and so on.
c) using low-sulphur fuel
d) desulphurization
e) reducing NOx emissions from power plants and
f) Engine customization.
g) SOx emissions can be minimised by converting them to sulphuric acid, elemental sulphur, and neutalizing them for use in other goods.

CHAPTER 12

SOLID WASTE

Introduction and Waste Generation

The term "solid waste" refers to a variety of garbage materials generated by animal and human activities that are thrown as undesired and worthless. Solid waste is created in a specific region by industrial, residential, and commercial activity and can be treated in a variety of ways. As a result, landfills are classed as sanitary, municipal, building and demolition, or industrial waste sites. Garbage can be classified according to its composition, such as plastic, paper, glass, metal, and organic waste. Hazard potential, such as radioactive, flammable, infectious, poisonous, or non-toxic wastes, can also be used to categorise waste. The origin of the garbage, whether industrial, home, commercial, institutional, or building and demolition, may also be classified. Solid waste, regardless of its origin, substance, or hazard potential, must be managed methodically to guarantee environmental best practises. Because solid waste management is such an important part of environmental hygiene, it must be included in environmental planning.

Effects

a) The ground becomes unsightly and unsanitary as a result of plastic bag waste.
b) Traditional plastics have been linked to reproductive issues in both people and wildlife.
c) One of the toxins thought to be passed on to nursing infants through breast milk is dioxin (a highly carcinogenic and poisonous by-product of the manufacturing process).
d) When plastics, particularly PVC, are burned, dioxin and furan are released into the atmosphere. Thus, ordinary plastics constitute a big environmental hazard from their creation through their disposal.
e) Plastic bags can potentially contaminate food owing to harmful colour leaching and disease translocation. Careless disposal of

plastic bags clogs drains, limits soil porosity, and impedes groundwater recharge.

f) Plastic disrupts the activities of soil microbes. Plastic waste is misinterpreted as food by terrestrial and aquatic creatures, who ingest it and perish.
g) Plastic bags degrade soil fertility because they become part of manure and linger in the soil for years.
h) These bags clogging the city drainage system causes annoyance, difficulty in maintenance, and produces an unsanitary environment, posing a health risk and spreading water-borne illnesses.
i) The demand for eco-friendly, biodegradable polymers is urgent.

Agricultural wastes and Sewage Sludges

Agricultural waste includes animal slurries, silage effluents, spray residues and tank cleaning after pesticide usage, and discarded plastic containers. The treatment of industrial and household wastes results in the production of sewage sludges.

The major components of agricultural waste are animal manures and silage effluents. These, like sewage sludges, are a slurry of fine-grained, organic-rich particles. Depending on the source of effluent and the efficacy of treatment, sewage sludges may be polluted with heavy metals, water-soluble organic compounds, greases, and oils.

Municipal Wastes

Household garbage, construction and demolition debris, sanitation residue, and street waste comprise municipal solid waste. With expanding urbanisation and changes in lifestyle and eating habits, the volume and content of municipal solid trash has been constantly increasing and changing. In 1947, cities and towns in India created an estimated 6 million tonnes of solid trash; by 1997, the figure had risen to around 48 million tonnes. Over a quarter of municipal solid garbage is not collected at all. Around 70 percent of Indian cities lack appropriate transportation capacity, and there are no hygienic landfills to dispose of trash. Existing landfills are neither well-equipped nor appropriately lined to prevent soil and groundwater pollution. The consumer market has risen quickly in recent years, resulting in products being packed in cans, aluminium foils, plastics, and other nonbiodegradable materials that cause immeasurable harm to the environment.

Industrial and Mining Wastes

Industrial process wastes might comprise everything from normal industry trash to packing materials, organic wastes, acids, alkalis, and metalliferous sludges. Mining wastes are produced as by-products of the extraction process and may contain topsoil, rock, and dirt, as well as trace amounts of metals and coal. The most notable characteristic of industrial and mining wastes is that a considerable fraction is considered hazardous and, as such, requires specific treatment and disposal.

Hazardous Wastes

Because harmful compounds are present in industrial and hospital waste, it is deemed dangerous. Hazardous wastes can be very hazardous to humans, animals, and plants, as well as caustic, highly flammable, or explosive.

Every year, India produces around 7 million tonnes of hazardous waste, the majority of which is concentrated in four states: Andhra Pradesh, Bihar, Uttar Pradesh, and Tamil Nadu. Outdated batteries, shoe polish, paint tins, old pharmaceuticals, and medicine bottles are examples of dangerous household garbage. Metal, chemical, paper, pesticide, dye, refining, and rubber products sectors are important sources of hazardous waste in the industrial sector. Direct exposure to hazardous waste compounds such as mercury and cyanide can be dangerous.

Radioactive Wastes

Radioactive wastes are wastes produced by a nuclear power plant or a medical laboratory. They include radioactive elements that are dangerous to most living things and the environment. The wastes degrade over time;therefore, they would be kept in a secure location until they lost their radioactivity and no longer posed a hazard to the environment. The time frame previously indicated is dependent on the kind of waste and radioactive isotopes.

Because cosmic rays are involved and the earth's surface is reached from space, natural radiations are also known as background reactions. Radioactive elements such as uranium, thorium, radon, carbon, and radium are examples of these. These elements can be found in soil, water, and rock. Man-made radiations include the refinement of plutonium, thorium, and mining, as well as the explosion and creation of nuclear power plants, radioactive isotopes, and nuclear fuels.

Biomedical Wastes

Biomedical waste encompasses any waste created in health care institutions and medical labs that requires specific processing, such as anatomical waste, pathological waste, infectious waste, hazardous waste, and other garbage. Pathological and institutional wastes were historically used to refer to what is now called biomedical waste. Hospital waste is produced during the diagnosis, treatment, or vaccination of humans or animals, as well as during research and the manufacturing or testing of biologicals. These compounds include formaldehyde and phenols, which are employed as disinfectants, as well as mercury, which is utilised in thermometers and blood pressure measuring devices. Soiled waste, disposables, anatomical waste, cultures, abandoned pharmaceuticals, chemical wastes, disposable syringes, swabs, bandages, bodily fluids, human excreta, and other wastes may be included. These are extremely contagious and can pose a major hazard to human health if not controlled scientifically and judiciously. According to several surveys, health care organisations in India are not paying enough attention to waste management. Following the publication of the Bio-Medical Waste (Handling and Management) Rules, 1998, these institutions are gradually streamlining the waste segregation, collection, treatment, and disposal processes.

Classifications

The Ministry of Environment and Forests of the Government of India has categorised biomedical waste into the categories shown in Table 12.1. To simplify identification, processing, storage, decontamination, and transportation, the Ministry advises that biomedical waste be separated and stored in leak-proof, colour-coded plastic bags.

Table 12.1: Ministry classification and colour coding for Biomedical waste

Classification	Description	Colour Code
Type A class 1	Human anatomical	Red
Type A class 2	Animal anatomical, Infected	Orange
Type A class 3	Non anatomical infected	Yellow
	Laboratory waste	yellow
	Wastes from DNA works	Yellow
Type B class 1	Animal anatomical (Non infected)	Blue

Solid Waste Management

Solid waste management is defined as the discipline associated with the control of solid waste generation, storage, collection, transport or transfer, processing, and disposal in a manner that best addresses a variety of public health, conservation, economic, aesthetic, engineering, and other environmental considerations. Solid waste management include planning, administration, finance, engineering, and legal duties. Complex inter-disciplinary relationships among subjects such as public health, city and regional planning, political science, geography, sociology, economics, communication and conservation, demography, engineering, and material sciences might be used to provide solutions.

Residential and industrial producers, urban and rural locales, and developed and developing countries all have different solid waste management techniques. Local government bodies are responsible for the management of non-hazardous garbage in urban regions. Hazardous waste management, on the other hand, is often the duty of those who create it, as subject to municipal, national, and even worldwide authorities.

Objectives

The major purpose of solid waste management is to reduce and eliminate the negative effects of waste products on human health and the environment in order to promote economic development and a high quality of life. This must be done in the most effective way feasible in order to keep expenses low and waste building to a minimum.

Waste Management System Functional Elements

a) Waste generation: This includes all operations that entail identifying items that are no longer useable and are either gathered for systematic disposal or discarded.
b) Onsite handling, storage and processing: This pertains to operations at the place of trash creation that make collection simpler. Rubbish bins, for example, are put in locations where there is a considerable amount of waste.
c) Waste collection: This is an important phase of waste management since it comprises tasks such as putting waste collection bins, collecting garbage from those bins, and gathering trash in the area where collection trucks are emptied. Although transportation is used during the collecting phase, it is not the primary mode of trash transportation.
d) Waste transfer and transport: These are the operations involved in transporting rubbish in big waste transport vehicles from local waste collecting locations to regional waste disposal sites.
e) Waste processing and recovery: This term refers to the facilities, equipment, and procedures used to recover reusable or recyclable materials from waste streams and increase the efficacy of other waste management functional aspects.
f) Disposal: The ultimate stage of trash disposal. It entails efforts targeted at the methodical disposal of waste products in places such as landfills or waste-to-energy plants.

Solid Waste Management Rules 2016

a) These regulations, which supersede the Municipal Solid Wastes (Management and Handling) Rules, 2000, now apply to locations other than municipalities, such as urban agglomerations, census towns, notified industrial townships, and so on.
b) Waste Segregation is Critical: Under the new rules, all waste generators must begin segregating their waste into three categories: biodegradables, dry waste (plastic, paper, metal, wood), and domestic hazardous waste (diapers, napkins, mosquito repellents, cleaning agents) before handing it over to collectors.
c) Collection and Disposal of Sanitary Napkins: In order to efficiently manage sanitary waste such as diapers and sanitary pads, the new standards require producers to offer a bag or wrapper for disposal whenever they sell their products to the consumer.

d) Introduction Of New Things – User Fees and Spot Fines: Garbage generators will be required to pay a fee to waste collectors for tasks such as waste collection, disposal, and processing under the new laws. The 'User Fees' are decided by the local bodies across India. The rules also stipulate a zero tolerance for throwing, burning, or burying solid waste generated on streets, open public spaces, and water bodies. "Spot Fines" are also introduced under these new rules, which means that if someone is found littering or disobeying the guidelines, local governments have the authority to fine them.
e) Collect Back System for Non-biodegradable Packaging Waste; According to the new guidelines, brand owners who sell or market their products in packaging materials that are not biodegradable or environmentally friendly must put in place a system to collect back the packaging waste generated as a result of their production.
f) A Bin is Required for Street Vendors: All street vendors shall retain appropriate containers or bins for storing garbage created by them, such as food waste, disposable plates, cups, cans, wrappers, coconut shells, leftover food, vegetables, fruits, and so on. They are also responsible for their own garbage and must deposit it in a waste storage station, container, or vehicle as designated by the local authorities.
g) Guidelines for Overburdened Landfills: Non-recyclable garbage with a calorific value of 1500 K/cal/kg or above shall not be disposed of in landfills. That garbage may either be utilised to generate energy or prepared into refuse-derived fuel. It can also be utilised in cement or thermal power plants as a co-processor. The landfill site should be 100 metres away from a river, 200 metres away from a pond, 500 metres away from roads, habitations, public parks, and water supply wells, and 20 kilometres away from airports/airbases, according to the new laws.
h) Waste Processing and Treatment Guidelines: According to the new rules, biodegradable waste should be processed, treated, and disposed of within the premises as much as possible through composting or bio-methanation, and residual waste should be given to waste collectors or agencies as directed by the local authority. Developers of Special Economic Zones, industrial estates, and industrial parks will also be required to set aside at least 5 percent of the total area of the plot or a minimum of 5 plots/sheds for recovery and recycling facilities. The laws also require bioremediation or capping of old and abandoned waste sites to be completed within five years.

i) Ragpickers' Involvement: The new guidelines assist the state government in integrating ragpickers, garbage pickers, and kabadiwalas from the informal to the official sectors.
j) Guidelines for Municipal Authorities: All local bodies are expected to establish a few by-laws concerning trash management and garbage segregation within their community. They must also put in place a system that allows all waste generators to follow the waste segregation procedure smoothly. They must also encourage composting, garbage segregation, and waste management through various educational efforts.
k) The Monitoring Process: The government has also formed a Central Monitoring Committee, chaired by the Secretary, MoEF&CC, to properly oversee the overall execution of the regulations. It is recommended that the committee, which would be comprised of diverse stakeholders from the national and state governments, meet once a year to debate and oversee the implementation of these laws.

Bioremediation

Bioremediation is a biotechnological method that reduces or eliminates pollution. It is a waste management strategy that employs organisms to remove or consume contaminants from contaminated areas. There are various methods for purifying polluted water or solids, including chemical treatment, burning, and burial in a landfill. Other waste management techniques include solid waste management, nuclear waste management, and so on. Bioremediation is distinct in that it does not employ hazardous chemicals.

When it comes to conducting the Bioremediation process, microorganisms such as Bacteria and Fungi play a key role. Bacteria are the most important bacteria in this process because they convert waste into nutrients and organic stuff. Even though this is an effective waste management procedure, bioremediation cannot eliminate all toxins. Bacteria can quickly digest toxins such as chlorinated insecticides and clean up oil spills, however microorganisms are incapable of destroying heavy metals such as lead and cadmium.

Bioremediation strategies

a) In- situ: It involves treating the polluted material on-site.
   1) Bioventing: Bioventing is a potential new method that promotes the spontaneous in-situ biodegradation of any aerobically degradable substance by supplying oxygen to the soil microorganisms already there. It employs low air flow rates to deliver just enough oxygen to keep microbial activity going.
   2) Biosparging: In order to induce in-situ aerobic biological activity, biosparging involves pumping pressurised air or gas into a polluted zone. This remediation approach targets chemical substances that can be biodegraded under aerobic conditions and applies to dissolve and residual contaminants in the saturated zone.
   3) Bioaugmentation: Bioaugmentation entails introducing microorganisms into a polluted area in order to speed up the degradation process.
b) Ex-situ: It involves the removal of the contaminated material to be treated elsewhere.
   1) Landfarming: polluted soil is dug and put over a prepared bed, then tilled on a regular basis until contaminants are degraded. The purpose is to promote the aerobic breakdown of pollutants by indigenous bio- degradative bacteria.
   2) Biopiles: it is a hybrid of landfarming and composting. Essentially, engineered cells are constructed as aerated composted piles. Typically used for treatment of surface contamination with petroleum hydrocarbons.
   3) Bioreactors: The processing of contaminated solid material (soil, silt, sludge) or water via a designed containment system.
   4) Composting: Compost bioremediation is the utilisation of a mature, cured compost's biological system of microorganisms to sequester or break down pollutants in water or soil.

Genetic Engineering approaches

a) Phytoremediation: Phytoremediation is a low-cost plant-based technique that takes use of plants' ability to concentrate elements and chemicals from their surroundings and metabolise different molecules in their tissues. It refers to certain plants' natural capacity to bioaccumulate, digest, or render harmless pollutants in soil, water, or air. Phytoremediation targets toxic heavy metals and organic contaminants.

1) Phytoextraction / Phytoaccumulation: This is the process through which plants acquire pollutants in their roots and aboveground shoots.
2) Phytotransformation / Phytodegradation: The absorption of organic pollutants from soil, sediments, or water and their transformation to a more stable, less hazardous, and less mobile form is referred to as phytotransformation or phytodegradation.
3) Phytostabilization: Plants are used in phytostabilization to reduce the movement and migration of polluted soil. Leachable elements are absorbed and bonded into the plant structure, forming an unstable plant mass from which pollutants do not re-enter the environment.
4) Rhizodegradation: Rhizodegradation is the breakdown of pollutants caused by rhizosphere activity. The presence of proteins and enzymes generated by plants or soil organisms such as bacteria, yeast, and fungi is responsible for this action.
5) Rhizofiltration: Rhizofiltration is a water treatment process that includes plant roots absorbing impurities. Rhizofiltration is a technique used to minimise pollution in natural wetlands and estuaries.

b) Mycoremediation: Mycoremediation is a type of bioremediation that involves the use of fungus to breakdown or isolate pollutants in soil.
c) Mycofiltration: Mycofiltration is a unique procedure that employs fungal mycelia to filter hazardous waste and microorganisms from soil water.

Advantages

a) Useful for completely destroying a wide range of pollutants.
b) It is possible to completely destroy target contaminants.
c) Cost-effective
d) Environmentally conscious

Disadvantages

a) Bioremediation is restricted to biodegradable compounds. Not all compounds are capable of a full and quick breakdown.
b) Biological processes are frequently quite specialized.
c) Extrapolating from bench and pilot size investigations to full scale field operations is tough.

d) Bioremediation usually takes longer than other treatment methods.

Treatment and Disposal of Solid Waste

a) Open Dumps: Open dumps are exposed regions where various types of solid waste are dumped. The garbage is not processed, is not separated, and is not segregated. It serves as a breeding place for flies, rodents, and other disease-carrying insects. Rainwater runoff from these dumps pollutes adjacent land and water, promoting illness. The use of open dumps is being phased away.
b) Landfills: Landfills are typically found in metropolitan areas. It is a hole excavated in the earth. Every day, the waste is deposited and the hole is filled with earth, preventing the reproduction of flies and rats. As a result, waste is deposited and sealed every day. When the landfill is full, the area is covered with a thick layer of mud and can then be developed as a parking lot or park. Problems - All forms of garbage are put in landfills, and when water seeps through them, it becomes contaminated, polluting the environment. Leaching is the polluting of groundwater and soil caused by landfills.
c) Incineration Plants: Incineration is the process of burning garbage in massive furnaces at high temperatures. The recyclable material is separated at these factories, while the remainder of the material is burned and ash is produced. Garbage burning is a dirty procedure that pollutes the air and water and creates tonnes of poisonous ash. A considerable portion of the garbage burned here may be collected and repurposed. In reality, at the moment, incineration is reserved as a last resort and is mostly utilised to remove infectious garbage.
d) Pyrolysis: It is a process of combustion in the absence of oxygen or the material burned in an oxygen-controlled environment. It is an alternative to burning. The resulting gas and liquid can be utilised as fuels. Pyrolysis of carbonaceous wastes such as firewood, coconut, palm trash, corn combs, cashew shell, rice husk, paddy straw, and saw dust produces charcoal as well as tar, methyl alcohol, acetic acid, acetone, and a fuel gas.
e) Composting: Composting is a biological process in which degradable organic waste is decomposed by microorganisms, mostly fungus and bacteria, into humus-like material in the presence of oxygen. This completed product, which resembles dirt, is abundant in carbon and nitrogen and serves as an ideal substrate for plant growth. It improves the soil's ability to retain water and

makes it easier to grow. It aids in the retention of plant nutrients in the soil.

f) Vermiculture: It is sometimes referred to as earthworm farming. Earth worms are introduced to the compost in this manner. These worms break down garbage, and their excreta enriches the compost with nutrients.

g) The Strategy of the 3R's +E:
   1) Reduce: In order to effectively reduce pollution, we must reduce our use of products that contributes to more solid wastes.
   2) Reuse: Many plastic goods may be reused or repurposed for a variety of applications. Before tossing away plastic things, think about how they may be repurposed.
   3) Recycle: To limit the amount of plastic in the waste stream, plastic recycling involves collecting garbage and turning it into new goods.
   4) Educate: Education is another critical method for increasing awareness and behavioural change.

CHAPTER 13

ENERGY

Non-renewable energy use has caused more environmental damage than any other human activity. Electricity produced from fossil fuels has resulted in excessive levels of hazardous substances in the environment. As a result, numerous environmental and health issues are now being addressed. As a result, alternative energy sources have become increasingly vital and useful. Renewable energy is energy that is created from naturally regenerated resources. This comprises solar energy, geothermal heat, wind, tides, water, and many biomass types. This energy is never depleted and is continually replenished. They are a feasible source of clean, infinite energy that emits less pollution and is readily available locally. When compared to non-renewable energy, the usage of renewable energy significantly reduces all types of pollution. The majority of renewable energy sources are low-polluting and deemed clean. However, while being a renewable resource, biomass is a major cause of indoor pollution. Electricity is the flow of energy or current and is one of the most extensively used kinds of energy worldwide. The two energy sources are categorised as follows:

Conventional Fuels

Apart from hydroelectric power, conventional sources of energy are also known as non-renewable forms of energy and are accessible in finite quantities. The majority of the world's energy is derived from fossil fuels. Carbon-based fuels are vast reservoirs of high-quality solar energy that have been amassed over millions of years. Nuclear fuels, on the other hand, are derived from the concentration of elements found in geologic strata since the earth's origin.

a) Oil: Crude oil is a complex mixture of liquid hydrocarbon molecules that may be discovered in sedimentary rocks. Its composition includes carbon, 82.2–87.1 percent by weight; hydrogen 11.7–14.7 percent; oxygen 0.1–4.5 percent; nitrogen 0.1–1.5 percent; and sulphur 0.1–5.5 percent. Products manufactured from crude oil are given different names based on the quantity of carbon atoms in their molecules. They are referred to as gasoline (C4–C10), kerosene (C11–C13), diesel fuel (C14–C18), heavy gas oil (C19–C25), lubricating oil (C26–C40), and waxes (Over C40). Oil is formed from organic materials in sedimentary rocks at depths ranging from 800 to 5000 metres and temperatures ranging from 66 to 150 degrees Celsius. Its primary source material is most likely marine organisms. India imports 85 percent of its crude oil and over half of its natural gas. While crude oil is converted into fuels such as gasoline and diesel, gas is utilised as compressed natural gas (CNG) in cars and fuel in factories. The oil and gas sector is one of India's eight key industries, and it has a significant impact on all other critical sectors of the economy. Because India's economic growth is strongly tied to its energy needs, the demand for oil and gas is expected to rise further, making the industry attractive for investment. According to the IEA (India Energy Outlook 2021), primary energy consumption is predicted to nearly double to 1,123 million tonnes of oil equivalent by 2040, as the country's GDP rises to USD 8.6 trillion.

Government Initiatives

1) The government unveiled important oil and gas projects in Assam in February 2021, including the INDMAX Unit at Indian Oil's Bongaigaon Refinery, Oil India Limited's secondary tank farm at Madhuban, Dibrugarh, and a 'Gas Compressor Station' in Hebeda Village, Makum, and Tinsukia, distant from Dhemaji.
2) The government unveiled important oil and gas projects in February 2021, including the Ramanathapuram – Thoothukudi natural gas pipeline and the Gasoline Desulphurisation Unit at Chennai Petroleum Corporation Limited, Manali.
3) Indian Oil Corp. Ltd. signed a letter of intent with Greenstat Hydrogen India Pvt. Ltd. in February 2021 to build a centre of excellence for the hydrogen value chain and other associated technologies such as hydrogen storage, fuel cells, and so on.
4) The Minister for Road Transport and Highways has inaugurated India's first liquefied natural gas (LNG) facility plant in Nagpur, Maharashtra, in July 2021.
5) India announced its first cargo from Guyana in July 2021, diversifying its oil buying. This step also implies a potential roadmap for the expanded oil and gas partnership with Guyana.

6) To increase hydrocarbon output, the government stated in June 2021 that it will sell off unmonetized big oil and gas assets controlled by state-owned ONGC and OIL.
7) The Prime Minister of India said in February 2021 that the government expects to invest Rs. 7.5 trillion (US$ 102.49 billion) on oil and gas infrastructure over the following five years. The government allotted Rs. 12,480 crore (US$ 1.71 billion) in Union Budget 2021 for direct benefit transfer of LPG (liquefied petroleum gas) and Rs. 1,078 crore (US$ 147.31 million) for feedstock subsidy to BCPL/Assam Gas Cracker Complex.
8) The Finance Minister said in the Union Budget 2021 that 1 crore new LPG connections will be provided under the Pradhan Mantri Ujjwala Yojana (PMUY) plan.

b) Coal: Coal is a complex organic substance made up of fused carbon rings that are bound together by various hydrocarbons and their atomic (O,N,S) connections. Its typical chemical formula is C10H8O. (This ratio of 10 carbon atoms to 8 of hydrogen can be contrasted with the ratio of 10 carbons to 17.5 hydrogen in crude oil). Coal was first mined in India in 1774 by the East India Company at Raniganj Coalfield on the western bank of the Damodar River. India now possesses the world's sixth greatest coal reserves. After China, India is the world's second largest producer of coal. Coal is formed when dead plant matter decays into peat and is transformed into coal over millions of years by the heat and pressure of deep burial. Coal was formed around 300 million years ago when the planet was covered in swampy (marshy) forests. Some of the plants perished and dropped into the marsh waters as they developed. New plants sprang to take their place, and as they perished, others sprouted. The wetland eventually developed a thick covering of decomposing dead vegetation. The earth's surface altered, and water and dirt rushed in, halting the degrading process. More plants sprouted, but they, too, perished and fell, forming distinct strata. Many strata had accumulated over millions of years, one on top of the other. The weight of the top layers, as well as the water and mud, compacted the plant matter's bottom levels. Heat and pressure caused chemical and physical changes in the plant layers, forcing oxygen out and leaving behind rich carbon deposits. The deposited material eventually turned into coal. Peat is transformed into lignite, lignite is transformed into sub-bituminous coal, sub-bituminous coal is transformed into bituminous coal, and bituminous coal is transformed into anthracite. As a result of exploration up to a maximum depth of 1200 m, a total of 319.02 billion tonnes of geological coal resources have been estimated in the nation as of April 2018. Hard coal deposits, which are scattered throughout 27 major coalfields, are mostly concentrated in the country's eastern and south-central regions. The lignite deposits are estimated to be over 36 billion tonnes, with 90 percent being in the southern state of Tamil Nadu. Jharkhand is the top state in India in terms of total coal reserves, followed by Odisha, Chhattisgarh, West Bengal, and Madhya Pradesh.

c) Natural Gas: Natural gas (also known as fossil gas; often simply gas) is a naturally occurring hydrocarbon gas mixture predominantly composed of methane, but sometimes including different proportions of other higher alkanes and, on occasion, a trace of carbon dioxide, nitrogen, hydrogen sulphide, or helium. Natural gas is a fossil fuel that is found in the same geological structure as petroleum. Natural gas pressure can sometimes drive oils to the surface. This type of natural gas is referred to as related gas or wet gas. Some reservoirs contain just natural gas and no oil. This is known as non-associated gas or dry gas. Natural gases frequently include significant amounts of hydrogen sulphide or other organic sulphur compounds. In this situation, the gas is referred to as "sour gas." Because it contains no hydrogen sulphide, coalbed methane is referred to as "sweet gas." Natural gas is typically bought and sold on the market by calorific value rather than volume. Natural gas purchases are often represented in MMBTUs (millions of British thermal units (BTU or Btu)) = 1,000 cubic feet of natural gas. Natural gas is odourless and colourless; the slightly sour stench we associate with gas coming from a burner is caused by an odorization process that adds mercaptan compounds to the end-use gas (for safety and leak detection). Natural gas commercial production in India began considerably later (since 1961). The discovery of gas reserves in Assam and Gujarat in the 1960s launched India's natural gas sector. Natural gas is created over millions of years when layers of decaying plant and animal waste are exposed to tremendous heat and pressure under the Earth's surface. Currently, around 40 percent of natural gas is utilised in the manufacturing of fertilisers. About 30 percent is utilised in electricity generation, while 10 percent is used in LPG. Natural gas production and all of these industries have grown in tandem. Natural gas output has increased dramatically, particularly after 1971. Gas-fired power plants provided about 10 percent of India's electricity. Gulf of Kutch, Gulf of Khambhat, Bassein field, Bombay High, Barmer in Rajasthan, KG basin, Tamil Nadu's Cuddalore region, Odisha, Assam, Tripura, and others. India's economically feasible natural gas reservoir is 541 BCM (on-shore, Assam, and Gujarat), with an additional 190 BCM offshore in the Bay of Cambay and 190 BCM (billion cubic metres) in the Bombay High.

A 400 BCM deposit was recently revealed in the Tripura Basin. Aside from these, 72 BCM is in the Rava structure, and a massive reserve has been reported near Andaman and Nicobar island. The reserves in Andaman and Nicobar are expected to be approximately 1700 BCM based on remote sensing data. Its economic feasibility has yet to be demonstrated, which is why manufacturing has yet to begin. This reserve will address India's problem or requirement for the next 100 years. Eastern India may see an economic revolution as a result of this.

d) Uranium: Because of its unique nuclear capabilities, uranium, which has a silvery grey metallic look, is mostly employed in nuclear power plants. Depleted uranium is also utilised as a radiation shield in medical operations that involve radiation treatment, as well as for transporting radioactive materials. Despite being radioactive, uranium's high density makes it good at stopping radiation. Because of its high density, it is also helpful as a counterbalance in aircraft and industrial gear. India possesses vast quantities of atomic minerals. Uranium and Thorium are the primary atomic minerals to which beryllium, lithium, and zirconium may be added. Thorium is extracted from monazite, which includes 10 percent thoria and 0.3 percenturania. Thorianite is another mineral that contains thorium. India's nuclear power facilities employ nuclear fuel that has been processed in three stages. The first stage includes Uranium, which is being used in India. The second stage nuclear fuel is made up of products derived from processed nuclear fuel, whereas the third stage nuclear fuel is made up of Thorium. India possesses the world's greatest thorium resources and is working hard to advance to the third stage of nuclear fuel consumption in order to become self-sufficient in nuclear fuel supply. India generates around 2 percent of the world's uranium. The overall uranium reserves are estimated to be 30,480 tonnes. Uranium deposits are found in crystalline rocks in

India. Jharkhand is home to 70 percent of Uranium reserves. Uranium deposits may be found in the Jharkhand districts of Singhbhum and Hazaribagh, the Bihar district of Gaya, and the sedimentary rocks of Uttar Pradesh's Saharanpur area. Atomic mineral mines are listed below in table 13.1:

Table 13.1: List of atomic mineral mines in India

Mines	States
Coastal Sands	Kerala
Doimasiat, Khasi hills	Meghalaya
Dilwara-Kilovari	Rajasthan
Rohil, Sikar	Rajasthan
Jadugoda, Turamdih, Narwapahar	Jharkhand
Tumallapalle	Andhra pradesh
Lambapurpedagatu	Andhra pradesh
Gogi	Karnataka

Centres of Atomic Energy in India

a) Tarapur Nuclear Reactor: The Tarapur Nuclear Reactor, located in Maharashtra, Western India, is the country's oldest nuclear reactor, having begun commercial operations in 1969.
b) The Rana Pratap Sagar Plant (Rajasthan) is located in the Kota Industrial Complex in eastern Rajasthan.
c) The Narora Nuclear Reactor in Uttar Pradesh, Northern India, is home to two PHWR with a combined capacity of 440MW. Unit 1 was erected in January 1991, followed by Unit 2 in July 1992.
d) Kalapakkam Nuclear Power Plant: The Kalapakkam Nuclear Power Plant in Tamil Nadu commenced operations in 1984 and now has two 235MW reactors, with two more 500MW and 600MW reactors to be constructed at a later date.
e) The Kakarapar Atomic Power Plant in Gujarat, Western India, contains two PHWR reactors with a total installed capacity of 440MW. The two reactors were finished in May and September of 1993, respectively.

Non-Conventional Fuels

Non-conventional energy sources are often known as renewable energy sources. Solar energy, bioenergy, tidal energy, and wind energy are examples of non-conventional energy sources.

a) Solar Energy: Solar energy is energy derived from the sun. The Sun provides enough energy to suit our power needs if we could only harness it properly. The energy absorbed by the earth is used to dry clothing, by plants in the process of photosynthesis, by the ocean, where the heat forms wind and heat currents, and for heating houses, among other things. India is one of the few nations in the world that has lengthy days and an abundance of sunshine. We can generate power from sunlight in two ways:

   1) Photo Voltaic Electricity: Photovoltaic power is generated by solar photovoltaic (SPV) cells, which convert solar radiation (sunlight) into electricity. A solar cell is a semi-conducting device constructed of silicon and/or other materials that generates energy when exposed to sunshine. Solar panels are mounted on an aluminium framework. PV cells are made up of at least two semiconductor layers, one with a positive charge and one with a negative charge. Photons are reflected, passed through, or absorbed by a PV cell when it is exposed to sunlight. When enough photons are absorbed by the photovoltaic cell's negative layer, electrons are liberated from the negative semiconductor material. These liberated electrons travel to the positive layer, resulting in a voltage difference. When the two layers are linked to a load, electrons travel through the circuit, producing electricity. The produced electricity is direct current (DC), which is transformed to alternative current (AC) using inverters.
   2) Solar Thermal Energy: Solar Thermal energy is generated via a solar collector with a mirrored surface that reflects sunlight onto a receiver, which warms a liquid. This heated liquid is used to generate steam, which generates energy. It uses concentrated sunlight to generate high-temperature heat. This heat is then sent via a traditional generator to generate power. Solar collectors gather and concentrate sunlight to heat a fluid, which produces electricity. The collectors are available in a variety of shapes. Parabolic troughs are the most usually utilised. A parabolic trough power plant uses a curved, mirrored trough to reflect direct solar radiation onto a glass tube carrying a fluid,

which heats up due to the concentrated solar radiation and is utilised to crank the turbine to create energy. Fluids that are often employed include synthetic oil, molten salt, and pressured steam. The produced electricity is direct current (DC), which is transformed to alternative current (AC) using inverters.

Solar Energy in India

1) India, being in the tropical region, benefits from peak solar radiation for 300 days, equating to 2300–3,000 hours of sunshine equivalent to around 5,000 trillion kWh.
2) Solar power in India is a rapidly growing business that is a component of renewable energy in India. As of August 31, 2021, the country's installed solar capacity was 44.3 GW.
3) Nearly 42 solar parks have been built in India to make land available to solar plant developers.
4) India's Intended Nationally Determined Contributions (INDCs) include a promise to 100 GW of solar power out of a total of 175 GW of renewable energy by 2022.
5) The sector also has enormous potential for employment creation; one gigatonne of solar manufacturing capacity produces around 4000 direct and indirect jobs.
6) Furthermore, solar deployment, operation, and maintenance generate new recurring jobs in the industry.
7) Storage advancements are ongoing, with the potential to revolutionise this industry globally; but, until then, dependency on fossil fuels may be lessened by steadily increasing the use of renewables.
8) By 2035, India is estimated to account for 8 percent of worldwide solar capacity. With a projected potential capacity of 363 Gigawatts (GW), India has the potential to be a worldwide leader in leveraging energy sector benefits.
9) With India struggling to meet its energy demands, solar energy can play a vital role in ensuring energy security.
10) Non-fossil fuel sources will account for 40 percent of total installed capacity for power.
11) To lower the GDP's emissions intensity by around one-third. India has committed an extra carbon sink (a method of absorbing carbon dioxide from the atmosphere) of 2.5 to 3 billion tonnes of carbon dioxide equivalent by 2030 through increased forest and tree cover. Below in table 13.2 is the list of major solar parks in India

Table 13.2: Major solar parks in India	
Solar Power Parks	State
Bhadla Solar Park (world's largest)	Rajasthan
Pavagada Solar Park	Karnataka
Kurnool Ultra Mega Solar Park	Andhra Pradesh
NP Kunta	Andhra Pradesh
Rewa Ultra Mega Solar Park	Madhya Pradesh
Charanka Solar Park	Gujarat
Kamuthi Solar Power Project	Tamil Nadu
Ananthapuramu-II	Andhra Pradesh
Galiveedu Solar Park	Andhra Pradesh
Mandsaur Solar Farm	Madhya Pradesh

International Solar Alliance

The International Solar Alliance (ISA) was founded on November 30th during the CoP21 Climate Conference in Paris as a particular forum for mutual collaboration among 121 solar resource rich nations located entirely or partly between the Tropics of Cancer and Capricorn. The alliance is geared to meet the ISA member countries' unique energy demands. The legal name of the International Solar Alliance is the International Agency for Solar Policy and Application (IASPA). The ISA secretariat is headquartered in the National Institute of Solar Energy in Gurgaon. The Prime Minister coined the term "Surya Putra" to refer to those nations that lay between the Tropics of Cancer and Capricorn and have been asked to join the alliance. These countries are often known as "Sunshine Countries."

Objectives

1) To drive down prices by increasing demand.
2) To standardized solar technology.
3) To promote research and development

Luminescent Solar Concentrator

A luminescent solar concentrator (LSC) is a device that traps solar radiation across a vast area before directing the energy (through luminous emission) to cells positioned on the material layer's thin edges. The thin sheet of material is often made of a polymer (such as polymethylmethacrylate (PMMA)) that has been doped with luminous species like organic dyes, quantum dots, or rare earth complexes. The major reason for using LSCs is to replace a huge area of costly solar cells in a standard flat-plate PV panel with a less expensive alternative. As a result, both the module cost (£/W) and the solar electricity generated (£/kWh) are reduced. LSCs have a significant advantage over traditional concentrating systems in that they can gather both direct and diffuse solar light. As a result, solar tracking is not necessary. LSCs are ideal for building integrated photovoltaics (BIPV) as well as cloudier northern regions.

b) Wind Energy: Wind energy is the kinetic energy associated with atmospheric air movement. Wind turbines convert wind energy into mechanical power, which is then converted to electric power to create electricity. Wind energy is a major renewable energy source in India. Many of the largest onshore wind farms in operation are in the United States, India, and China. Germany, the United States, Denmark, Spain, and India account for 80 percent of the world's installed wind energy capacity. Wind energy conversion equipment, such as wind turbines, transform wind energy into mechanical energy. Wind turbines are made up of a few sails, vanes, or blades that radiate from a central axis. The blades or vanes revolve about the axis as the wind blows against them. This circular motion is used to accomplish some valuable job. Wind energy may be transformed into electric energy by connecting the wind turbine to an electric generator. When compared to a heat engine, it is highly efficient.

Wind Farm / Park

A wind farm or wind park, also known as a wind power station or wind power plant, is a set of wind turbines that produce energy in the same place. Wind turbines create power by harnessing the energy of flowing air. Wind farms range in size from a few turbines to hundreds of turbines spread across a large region. The two types of Wind farms are:

1) Onshore: Onshore wind turbines are those that are positioned on land.
2) Offshore: Offshore turbines are positioned either at sea or in freshwater. A fixed-foundation turbine is constructed in shallow water, but a floating wind turbine is constructed in deeper water, with its foundation anchored in the seabed. Floating wind farms are still in their early stages. Offshore wind farms must be at least 200

nautical miles from the coast and 50 feet below sea level. Electricity is generated by offshore wind turbines and returned to land via cables buried on the ocean floor.

Wind Turbine Working principle

Wind turbines transform wind kinetic energy into mechanical energy. This mechanical power can be employed for specific activities (such as grinding grain or pumping water), or it can be converted into energy via a generator. Most turbines are equipped with three aerodynamically designed blades. Wind energy rotates two or three propeller-like blades around a rotor linked to the main shaft, spinning a generator to generate power. Wind turbines are positioned atop a tower to maximise energy extraction. They may take advantage of the quicker and less turbulent wind at 100 feet (30 metres) or higher above land. Three major factors influence how much power a turbine can generate:

1) Wind Speed: Higher wind speeds generate more energy. Wind turbines create electricity at speeds ranging from 4 to 25 metres per second.
2) Blade radius: The bigger the radius of the blades, the more energy is generated. Doubling the radius of the blade can result in four times the power.
3) Air density: Heavier air lifts a rotor. The density of air varies with height, temperature, and pressure. Because high altitude areas have low air pressure and lighter air, they are less productive for turbines. The dense heavy air around sea level propels rotors quicker and hence more efficiently.

Types of Wind Turbines

1) Horizontal axis turbines: Two or three blades rotate upwind of the tower in a horizontal-axis arrangement. The blades of a horizontal axis machine rotate on an axis parallel to the ground.
2) Vertical axis turbines: Vertical-axis turbines are distinguished by vertical blades that revolve in and out of the wind. The blades of a vertical axis turbine rotate on an axis perpendicular to the ground. This drag-type turbine rotates slowly but produces a lot of torque. It can grind grain, pump water, and perform a variety of other duties, but its modest rotating rates make it unsuitable for generating energy. Vertical-axis turbines, unlike horizontal-axis turbines, do

not benefit from increasing wind speeds at higher heights (100 feet and above).

Status of Wind Energy in India

1) The indigenous wind power industry leads the country's wind energy sector, which has made steady development. The wind industry's growth has resulted in a robust ecosystem, project operating capabilities, and a manufacturing base of around 10,000 MW per year.
2) The country presently boasts the world's fourth-highest wind installed capacity, with a total installed capacity of 39.25 GW (as of March 31, 2021), and has generated around 60.149 billionunits during the period 2020–2021.
3) Between 2010 and 2020, the compound annual growth rate for wind power was 11.39 percent, while it was 8.78 percent for installed capacity.
4) According to the Ministry of New and Renewable Energy (MNRE), India's 7,600 km of coastline can create 127 GW of offshore wind energy. The entire wind energy potential at a 100-meter hub height, according to the National Institute of Wind Energy (NIWE), is 302 GW.
5) Andhra Pradesh, Gujarat, Karnataka, Madhya Pradesh, Maharashtra, Rajasthan, and Tamil Nadu account for more than 95 percent of economically exploitable resources. Tamil Nadu is the state having the most installed wind power producing capacity in India. In 2018, wind power accounted for around 28 percent of total electricity generation. At the end of 2018, overall wind capacity was at 8,631 MW, while total installed power generating capacity stood at 30,447 MW. Gujarat has the country's second-largest installed wind power producing capacity. In 2018, wind power accounted for around 19 percent of total electricity generation.
6) The Union Ministry of New and Renewable Energy (MNRE) has set a target of 5 GW of offshore wind energy by 2022, with a goal of 30 GW by 2030.

Policies related to Wind Energy

1) The major goal of the 2018 National Wind-Solar Hybrid Policy is to create a framework for the development of large grid-connected wind-solar PV hybrid systems for the optimal and efficient

exploitation of wind and solar resources, transmission infrastructure, and land.

2) The National Offshore Wind Energy Policy was established in October 2015, with the goal of developing offshore wind energy in the Indian Exclusive Economic Zone (EEZ) along the Indian coastline of 7600 km.

National Offshore wind energy policy 2015

1) In order to establish offshore wind farms, the government of India has chosen to implement a strategy that would allow for the most efficient use of offshore wind energy in the best interests of the nation and achieve the following goals.
2) Exploration and promotion of offshore wind farm deployment in the country's Exclusive Economic Zones (EEZ), including those under Public Private Partnership.
3) To minimise carbon emissions, offshore wind energy farms will be used to harness the wind and create power on a huge scale.
4) To encourage energy infrastructure investment and attain energy security.
5) Through appropriate incentives, to encourage spatial planning and management of the country's offshore renewable energy resources in the exclusive economic zone.
6) To promote the indigenous development of offshore wind energy technologies.
7) To encourage offshore wind energy research and development.
8) To develop a qualified workforce and create jobs in the offshore wind energy sector.
9) To assist the development of offshore wind project EPC (Engineering, Procurement, and Construction) and operation and maintenance.
10) Develop and maintain coastal infrastructure and supply chain to enable heavy construction and fabrication activity, as well as operation and maintenance (O&M) operations in the offshore wind energy industry.

c) Hydro power: Hydroelectric power generates electricity by using the kinetic energy of flowing water. The use of flowing water to generate electricity is known as hydroelectricity or hydel power. It is less expensive than thermal or nuclear electricity. Dams are constructed to hold water at a higher level, which is then released to drive turbines that create energy. One of the most significant advantages of hydropower is that it is a reasonably cheap and clean source of energy after the dam is finished and the turbines are operational. Hydropower has certain drawbacks as well. Dam construction severely disrupts and ruins natural ecosystems, and some are lost permanently. There are 197 hydropower plants in India. The end of the nineteenth century saw the rise of authority in India. Darjeeling received electricity in 1897, and a hydropower station at Sivasamudram, Karnataka, was inaugurated in 1902.

Types of Hydro power stations

Hydropower plants are classified into three types: impoundment, diversion, and pumped storage. Some hydroelectric plants utilise dams, whereas others do not.

1) Impoundment: An impoundment facility is the most prevalent form of hydroelectric power plant. A dam is used in an impoundment facility, which is often a big hydroelectric system, to hold river water in a reservoir. The reservoir's discharged water runs through a turbine, turning it and activating a generator, which produces power.
2) Diversion: A diversion, also known as a run-of-river facility, conducts a section of a river through a canal or penstock and then through a turbine, spinning it and activating a generator to generate energy. It may not be necessary to use a dam.
3) Pumped storage: It functions similarly to a battery, storing electricity supplied by various power sources such as solar, wind, and nuclear for later use. When power demand is low, a pumped storage plant stores energy by pumping water from a lower reservoir to an upper reservoir. When there is a significant demand for energy, the water is discharged back into the lower reservoir and turns a turbine, creating electricity.

Classification of Hydro Projects based on Installed Capacity

Hydro Projects are classified depending on their installed capacity. Hydropower projects are typically divided into two categories: small hydro and large hydro. Small Hydro Power (SHP) projects in India are hydro projects with station capacity of up to 25 MW.

1) Micro (up to 100 KW):

2) Mini: 101KW to 2 MW
3) Small: from 2 to 25 MW
4) Mega: Hydropower plants with installed capacities greater than 500 MW
5) Thermal projects with installed capacity more than 1500 MW

Small Hydro Power

Any hydropower plant with an installed capacity of less than 25 MW is considered a small hydro project. It is mainly run-of-river, with a minor dam or barrage, usually simply a weir and little or no water held. As a result, run-of-river installations have less of an impact on the local ecosystem than large-scale hydro projects. Small hydropower facilities may provide energy to isolated rural regions on their own. India and China are the biggest participants in the SHP industry, with the most installed projects.

India Status

1) India ranks fifth in the world in terms of installed hydroelectric generating capacity.
2) India's installed utility-scale hydroelectric capacity was 46,000 MW as of March 31, 2020, accounting for 12.3 percent of total utility power production capacity.
3) Hydropower potential is concentrated mostly in the north and north-eastern areas.
4) Arunachal Pradesh has the most untapped hydropower potential (47 GW), followed by Uttarakhand (12 GW).
5) Unutilized potential is concentrated mostly along three river systems: the Indus, Ganges, and Brahmaputra.
6) Over 90 GW of potential pumped storage capacity exists in India, with 63 locations identified and recognised in national energy plans for their significant grid services.
7) India has a hydropower potential of 1,45,320 MW, excluding small hydro plants (SHPs), which have a potential of 20 GW.
8) India's hilly states, namely Arunachal Pradesh, Himachal Pradesh, Jammu and Kashmir, and Uttarakhand, account for over half of this potential. Maharashtra, Chhattisgarh, Karnataka, and Kerala are other possible states.
9) Bhutan also exports excess hydroelectric electricity to India.
10) The public sector generates 92.5 percent of India's hydroelectric electricity. The development of hydroelectric electricity in the

Himalayan mountain ranges and in north-east India is also projected to boost the private sector.

d) Ocean Thermal Energy: The oceans and seas retain a large quantity of solar energy. The tropical waters receive solar energy equivalent to the heat content of 245 billion barrels of oil on average, covering 60 million square kilometres. OTEC is the method of harvesting this energy (ocean thermal energy conversion). It uses temperature variations between the ocean's surface and depths of around 1000 metres to power a heat engine that generates electricity.

Wave Energy

Waves are produced by the interaction of the wind with the water's surface and signify a transfer of energy from the wind to the sea. The first wave energy facility, with a capacity of 150MW, has been established at Vizhinjam, near Trivandrum.

Tidal Energy

Tides may be used to create energy by establishing a reservoir or basin behind a barrage and then sending tidal waters through turbines in the barrier. Tidal power offers enormous promise in locations such as Gujarat's Gulf of Kutch, the Gulf of Cambay, and West Bengal's Sunderban area, where the tide height is adequate for the building and economic operation of tidal power plants. A huge tidal wave power plant for Rs.5000 crores is proposed for the Hanthal Creek in Gujarat's Gulf of Kutch.

Current Energy

It resembles the wind over the oceans. Underwater turbines, which are enormous propellers attached to the seabed, create power by moving with the marine currents. Given the magnitude of open ocean currents, the Intergovernmental Panel on Climate Change (IPCC) predicts major project scale development when technologies capture lower-velocity currents.

e) Geo Thermal Energy: Geothermal energy is natural heat from the earth's interior that may be exploited to create power and heat buildings. A layer of hot, molten rock known as magma exists under the earth's crust. This layer is constantly producing heat, mostly from the decay of naturally radioactive elements such as uranium and potassium. The quantity of heat within 10,000 metres

(33,000 feet) of the earth's surface holds 50,000 times more energy than all of the world's oil and natural gas resources combined. The following are some natural instances of geothermal energy. Hot Springs, Geysers, Lava Fountain etc. There are three types of geothermal resources:

1) Geo pressurized zones
2) Hot rock zones
3) Hydro thermal convection zones

How is it captured

Geothermal systems can be found in areas with a normal or slightly above normal geothermal gradient (the geothermal gradient is a progressive shift in temperature that represents the increase in temperature with depth in the earth's crust. The typical geothermal gradient is around 2.5-3 degrees Celsius/100 m), and notably in places near plate boundaries, where geothermal gradients can be much greater. The most popular contemporary method of harnessing geothermal energy is to tap into naturally existing "hydrothermal convection" systems in which colder water penetrates into the earth's crust, heats up, and then rises to the surface. When hot water is driven to the surface, steam may be easily captured and used to power electric generators.

Challenges

1) The Resource and location: Geothermal energy exists beneath the whole earth's surface, but not all of it can be utilised. Only a limited fraction of land is located above sufficient pockets of water and steam that may heat houses or power electrical facilities, restricting the possibilities of installing geothermal power plants. Many of the locations that are suited for supplying significant amounts of geothermal energy that may be turned into power are also quite tectonically active. When there is a persistent risk of earthquakes or volcanic activity, firms are hesitant to build large-scale power plants.
2) Infrastructure: Apart from a lack of suitable resources, one of the reasons geothermal power is not widely used in the United States is a lack of infrastructure. By definition, geothermal energy can only be utilised to generate baseline power for an electrical grid, which poses challenges in and of itself. Drilling wells and constructing power plants are both extremely expensive, and training individuals to work in a geothermal power plant is both time-consuming and

costly. There is also a limit to where geothermal energy may be employed. Once the energy is recovered from the subterranean wells, it cannot be transferred to another facility whose grid is in greater demand; it must be used as it is extracted.

3) High cost: Geothermal energy is a costly resource to access, with prices ranging between $2 and $7 million for a 1 megawatt facility. However, where the upfront expenses are substantial, the investment can be recouped over time.
4) Location restricted: The most significant drawback of geothermal energy is that it is site dependent. Geothermal plants must be developed in regions where energy is available, which implies that certain localities are unable to utilise this resource. Of course, if you live in an area where geothermal energy is easily accessible, such as Iceland, this is not an issue.
5) Earthquakes: Geothermal energy also has the potential to cause earthquakes. This is due to changes in the Earth's structure caused by digging. This issue is particularly widespread in enhanced geothermal power plants, which drive water into the Earth's crust to open fissures for better resource extraction. However, because most geothermal facilities are located far from population centres, the effects of these earthquakes are negligible.
6) Sustainability: To ensure the long-term viability of geothermal energy, fluid must be injected back into subsurface reservoirs quicker than it is depleted. This means that geothermal energy must be carefully handled in order to be sustainable. It is critical for industry to evaluate the benefits and drawbacks of geothermal energy in order to capitalise on the benefits while limiting any potential drawbacks.
7) Environmental side effects: Although geothermal energy does not generally emit greenhouse gases, many of these gases are stored under the Earth's surface and are released into the atmosphere during excavation. Although these gases are naturally discharged into the environment, the rate rises around geothermal installations. These gas emissions, however, are significantly lower than those connected with fossil fuels.

Geothermal Energy in India

1) Exploration and research of geothermal fields began in India in 1970.

2) The Geological Survey of India has discovered over 350 geothermal energy sites in the nation. The most promising of them is in Ladakh's Puga valley.
3) India features seven geothermal provinces [Himalayas, Sohana, West Coast, Cambay (Gujarat), Godavari, Mahanadi, and Son-Narmada-Tapi (SONATA)] as well as many geothermal springs.
4) According to the Ministry of New and Renewable Energy, geothermal resources in India have been identified, and in general estimates that there might be 10 Giga tonnes of geothermal power potential.
5) The Chhattisgarh government planned in 2013 to build the country's first geothermal power plant in Tattapani in the Balrampur district.
6) A deal to build Ladakh's first geothermal power plant was reached in 2021.
7) Satellites such as the IRS-1 have been important in finding geothermal zones by taking infrared photos.
8) The government intends to grant a capital subsidy of up to 30 percent for industrial projects.
9) The first geothermal power plant in Chhattisgarh will be built through a collaboration between NTPC and the Chhattisgarh Renewable Energy Development Agency (CREDA). Tattapani geothermal field is located in the SONATA geothermal province.
10) The Ministry of New and Renewable Energy (MNRE) offers significant incentives and subsidies for geothermal energy research, design, development, and demonstration (RDD&D) in India.

The Ministry of Renewable Energy intends to create up to 1000 MW of geothermal energy by 2022. The Major Sites for Geothermal Energy in India is listed in Table 13.3.

Table 13.3: List of major sites for geothermal energy in India

Sites	State / Union Territory
Puga geothermal field	Ladakh
Chhumathang geothermal field	Ladakh
Manikaran geothermal field	Himachal pradesh
Beas valley	Himachal pradesh
Satluj and spiti valley	Himachal Pradesh
Tapoban geothermal field, Chamoli, Uttarakhand and Alaknanda Valley	Uttarakhand

Tatapani geothermal field	Chhatisgarh
Lasundra geothermal province	Gujarat
Western thermal province	Maharashtra
Damodar Valley basins	West Bengal
Tuwa and Chhabsar geothermal fields	Gujarat
Sohna thermal region	Himachal Pradesh
Cambay geothermal region	Gujarat
Godavari Graben Konkan geothermal provinces	Andhra Pradesh
Andaman-Nicobar region	Andaman and Nicobar Island
Anhoni- Samoni area	Madhya Pradesh
Unkeshwar	Maharashtra
Salbardi region	Madhya Pradesh

f) Fuel Cells: Fuel cells are electrochemical devices that transform the chemical energy of a fuel directly and efficiently into direct current (DC) and heat, eliminating the need for combustion. The best fuel for such cells is hydrogen or a combination of hydrogen-containing molecules. An electrolyte is placed between two electrodes in a fuel cell. When oxygen flows over one electrode and hydrogen passes over the other, they react electrochemically to produce electricity, water, and heat.

Working of Fuel cell

This fuel cell worked by transferring hydrogen and oxygen via carbon electrodes into a concentrated solution of sodium hydroxide. The cell response is expressed as follows:

Cathode Reaction: $O2 + 2H2O + 4e- \rightarrow 4OH-$

Anode Reaction: $2H2 + 4OH- \rightarrow 4H2O + 4e-$

Net Cell Reaction: $2H2 + O2 \rightarrow 2H2O$

This electrochemical process, however, has a very slow response rate. This problem is solved by using a catalyst such as platinum or palladium. The catalyst is coarsely split before being introduced into the electrodes to maximise the effective surface area.

International Renewable Energy Agency (IRENA)

The International Renewable Energy Agency (IRENA) is an international agency tasked with facilitating collaboration, advancing understanding, and encouraging the adoption and long-term usage of renewable energy. It is the first worldwide organisation to focus only on renewable energy, serving both developed and developing countries. It was established in 2009, and its law went into effect on July 8, 2010. Masdar City, Abu Dhabi, serves as the agency's headquarters with 150 member nations. In the pursuit of sustainable development, energy access, energy security, and low-carbon economic growth and prosperity, IRENA encourages the broad adoption and sustainable use of all sources of renewable energy, including bioenergy, geothermal, hydropower, ocean, solar, and wind energy.

REN 21

REN21 (Renewable Energy Policy Network for the Twenty-First Century) is a renewable energy policy think tank and multistakeholder governance organisation. The purpose of REN21 is to promote policy formation, knowledge sharing, and collaborative action toward a rapid worldwide transition to renewable energy. REN21 brings together governments, non-governmental organisations, research and academic institutions, international organisations, and business to share knowledge and accelerate renewable energy adoption. REN21 provides knowledge about what is happening now in the renewable energy market and how the newest trends will affect future developments to aid policy decision-making.

REN21 supports the collecting of renewable energy data. This is accomplished through the following six products: the Renewables Global Status Report (GSR), regional status reports, global futures reports (GFR), theme reports, the REN21 Renewables Academy, and the International Renewable Energy Conference (IREC) series. The REN21 Secretariat is situated at UN Environment in Paris, France, and is a non-profit organisation recognised under German law. As of 2021, the organisation had more than 80 member organisations.

Biological Energy

Biomass is a type of sustainable biological material derived from plants and animals. Energy generated or produced by living or once-living creatures is referred to as biomass energy. Biomass is a significant fuel in many nations, particularly in underdeveloped countries for cooking and heating. Many industrialised nations are boosting their use of biomass fuels for transportation and power generation in order to reduce carbon dioxide emissions from fossil fuel consumption. It contains chemical energy from the sun that has been stored. Photosynthesis is the process through which plants generate biomass. Biomass can be directly burnt for heat or transformed to renewable liquid and gaseous fuels via a variety of processes. Biomass sources for energy include:

1) Wood and wood processing wastes: firewood, wood pellets, and wood chips, lumber and furniture mill sawdust and waste, and black liquor from pulp and paper mills.
2) Agricultural crops and waste materials: corn, soybeans, sugar cane, switchgrass, woody plants, and algae, as well as agricultural and food processing residues
3) Biogenic materials in municipal solid waste: paper, cotton, and wool products, as well as food, yard, and wood waste.
4) Human sewage and animal manure.

Converting Biomass to Energy

Various methods are used to convert biomass to energy, including:

1) To generate heat, direct combustion (burning) is used: The most prevalent technique for converting biomass to useable energy is direct burning. All biomass may be burnt directly to heat buildings and water, as well as to generate power in steam turbines.
2) Solid, gaseous, and liquid fuels are produced by thermochemical conversion: Pyrolysis and gasification are two types of thermochemical conversion of biomass. Both are thermal degradation processes in which biomass feedstock materials are cooked to high temperatures in confined, pressure containers known as gassifiers.
3) Chemical conversion to produce liquid fuels: Transesterification is a chemical process that converts vegetable oils, animal fats, and greases into fatty acid methyl esters (FAME), which are used to make biodiesel.
4) Biological conversion to produce liquid and gaseous fuels: Fermentation to convert biomass into ethanol and anaerobic digestion to create sustainable natural gas are examples of biological conversion. Ethanol is a type of motor fuel. Renewable natural gas, commonly known as biogas or biomethane, is generated in anaerobic digesters in sewage treatment plants, as well

as dairy and cattle operations. It can also develop in and be collected from solid waste dumps.

5) Anaerobic digestion, often known as biomethanation or methanogenesis: This is a scientific process in which anaerobic microbes degrade biodegradable waste to produce methane-rich biogas and effluent. Hydrolysis, acidogenesis, and methanogenesis are the three sequential functions.
6) Cogeneration: Co-generation is the process of generating two types of energy from a single source. One type of energy must always be heat, whereas the other might be electrical or mechanical energy. Fuel is burned in a boiler to produce a high-pressure system in a traditional power station. The steam is used to power a turbine, which generates electricity. The exhaust steam is usually condensed to water and returned to the boiler. Because low-pressure steam loses a substantial amount of heat during the condensing process, typical power plants have an efficiency of only around 35 percent. In a cogeneration facility, the low-pressure exhaust steam from the turbine is not condensed but utilised for heating purposes in industries or households, allowing for very high efficiency levels in the region of 75-90 percent. Because co-generation can cover both power and heat demands, it provides additional benefits such as considerable cost savings for the plant and lower pollutant emissions owing to reduced fuel usage.

Bio energy role meeting India's energy demands

1) Energy need: Bioenergy can assist to fulfil the country's expanding energy demand, particularly in rural regions. Almost 25 percent of its primary energy comes from biomass resources, and nearly 70 percent of the rural population relies on biomass to satisfy their daily energy demands. Biomass can also assist address rural energy demands.
2) Climate change mitigation: Bioenergy has significant advantages over fossil fuels, particularly in terms of GHG emissions. Biomass recycles carbon from the atmosphere while avoiding the usage of fossil fuels, limiting the amount of extra fossil carbon released into the atmosphere from the earth.
3) Market expansion: The market for renewable energy systems in rural and urban India is expected to rise at an exponential rate. Despite this, bioenergy is categorised as 'non-commercial' energy in most energy studies. Plants such as Jatropha, Neem, and other

wild plants have been discovered as possible sources of biodiesel in India.

4) Waste to energy: Biofuels can help turn waste into wealth. As a by-product of renewable biomass resources such as plastic, municipal solid waste, forestry residues, agricultural wastes, surplus food grains, and so on, it has enormous potential to assist the government in meeting its renewable energy goal of 175 GW.
5) Revenue generation: Using biofuels as an alternative energy source may greatly increase farmers' income, create job possibilities, and so on.
6) Reduce imports: Imports meet about 46.13 percent of India's total primary energy consumption. Bioenergy can assist reduce these imports while also increasing India's energy security and self-reliance.

Various Government Initiatives

1) National objective of 10 GW: The Ministry of New and Renewable Energy (MNRE) has established a national target of 10 GW of installed biomass power by 2022.
2) National Biofuel Policy: The programme aims to advance the suggested objective of 20 percent biofuel blending with fossil-based fuels by 2025.
3) Policy for biomass and bagasse cogeneration: The MNRE has created a policy for biomass and bagasse cogeneration to assist satisfy India's energy demands. Financial incentives and subsidies are provided for both biomass plants and sugar mills that utilise this technology.
4) Fiscal Incentives: The government provides a ten-year income tax vacation. Exemption from customs and excise duties on machinery and components used in the early setup of biomass power plants. Certain states provide general sales tax exemption.
5) Waste to energy projects: Waste to energy projects are being established to generate electricity from urban, industrial, and agricultural waste, such as vegetable and other market trash, slaughterhouse waste, agricultural leftovers, and industrial wastes and effluents.
6) MNRE also intends to establish a 'National Biomass Repository' through a nationwide assessment programme, which will assist secure the supply of biofuels generated from domestic feedstock.

Advantages

1) Versatile: Biomass energy may be used for a variety of purposes, including cooking gas and power generation with minimal environmental impact.
2) Renewable: It is a renewable source of energy that can be abundantly supplied by nature.
3) (Ideally) no net CO2 emissions: Biomass energy emits considerably too little CO2 than other conventional energy sources.
4) Less SO2 and NOx emissions than fossil fuels: The sulphur and nitrogen pollution produced by biomass use are negligible.

Disadvantages

1) Low energy density/yield: Some biofuels, such as ethanol, are inefficient when compared to gasoline. To boost its efficiency, it must be strengthened with fossil fuels. Some processes (for example, corn-derived bioethanol) may provide no net energy.
2) Land conversion: Land required to create biomass may be in demand for other uses such as conservation, housing, or agriculture usage, thus leading to a loss in agricultural food output. It may also result in the loss of biodiversity.
3) Aside from the aforementioned issues, intensive agriculture is frequently related with the pollution of nutrients, depletion of soil, erosion of soil and other issues with water contamination.

Energy from Waste

Due to urbanisation, industrialization, and changes in living patterns that are hazardous to the environment, there are rising amounts of garbage in today's period. In recent years, technological advancements have aided in the reduction of garbage for safe disposal and the generation of electricity from it. Garbage-to-energy offers the ability to remove waste from landfills while also producing sustainable electricity without emitting damaging greenhouse gases. This considerably minimises the amount of garbage that must be disposed of and has the potential to generate power. Apart from incineration and biomethanation, pyrolysis and gasification are developing technologies.

Potential of Waste to energy

1) In India, there is an estimated potential of around 225 MW from all sewage and approximately 1460 MW from Municipal Solid Waste (MSW), totalling approximately 1700 MW of electricity.
2) There is now the ability to recover 1,300 MW of electricity from industrial waste, with the capacity expected to grow to 2,000 MW by 2017.
3) The overall installed capacity of grid interactive electricity from waste to energy is 99.08 MW grid power and about 115.07 MW off-grid power.
4) MNRE actively promotes energy generation from garbage by offering incentives and subsidies for projects.

Major problems faced by the Indian waste to energy sector

1) Choice of technology: Waste-to-Energy technology is still a new concept in India. The majority of proven and commercial waste management technology must be imported.
2) High expenses: The costs of projects, particularly those based on biomethanation technology, are high since crucial equipment for a project must be imported.
3) Improper segregation: India lacks a source separated waste stream due to Municipal Solid Waste (MSW) Rules 2000 compliance by Municipal Corporations/ Urban Local Bodies. Organic garbage is combined with other forms of trash. As a result, the operations of waste to energy approaches are hampered, and the endeavours are short-lived due to a lack of smoothness.
4) Lack of policy support: A lack of suitable policy guidelines from state governments in terms of land allocation, waste supply, and power purchase facilities.

India Energy use and pattern

Rural energy consumption is steadily increasing. India is the world's fourth largest energy user. The majority of energy is currently consumed in the areas of home cooking and lighting, agriculture, transportation, and industry. India's energy basket includes a mix of all available resources, including renewables. Coal is the most abundant energy source, followed by petroleum and traditional biomass. According to household-level statistics from the 2011 Census, just 55.3 percent of rural houses utilised electricity as their major source of lighting. Energy access, with around one-fourth of the population without access to electricity, and energy security, with the country reliant on imports for a significant portion of its energy usage, notably crude petroleum, are critical difficulties that the country faces in terms of energy. A wide gap in energy consumption patterns between haves and have-nots, urban and rural, and men and women is another cause for worry.

Present status

1) In India, 83.3 percent of the 121 million people reside in rural regions, while 37.7 percent dwell in urban areas. Energy availability, accessibility, and affordability are critical if our country is to maintain its current rate of growth. In India, 99.92 percent of villages are fully electrified.
2) India's per capita power usage in 2018-19 is 872.87 kilowatt-hours (Kwh). According to official data, the country's total installed capacity stands at 3,70,348 MW as of April 2020. Thermal power plants account for 62.8 percent of installed capacity, while hydropower accounts for 12.4 percent. The nation has 21 nuclear power reactors with a total installed capacity of 6780 MW.
3) The highest power shortfall in December 2019 was 0.5 percent. The total number of active LPG household connections (as of March 2020) was 27.87 crore. According to the NSO's 76th round survey on 'Drinking Water, Sanitation, Hygiene, and Housing Condition,' around 93.9 percent of rural families and approximately 99.1 percent of urban households have electricity for residential use. LPG was utilised as a cooking fuel by about 48.3 percent of families in rural regions and approximately 86.6 percent of households in urban areas.

CHAPTER 14

ENVIRONMENT IMPACT AND AUDIT

Introduction

For a long time, it was recognised that reasonable criteria for environmental quality evaluation and monitoring were required. As a result, numerous criteria have been developed in various nations to establish connections between the intensity, duration, and frequency of pollution exposure and the risk or degree of an undesirable consequence happening in the environment. Each standard has a limit for three separate aspects, namely:

a) A limit set for a pollutant taken up by an organism or a population.
b) A limit set for a pollutant present in a specific environment medium or product (food of consumer products).
c) A limit set for discharges or emissions from polluting sources.

Concept

Environmental Impact Assessment (EIA) refers to a formalised method for examining, analysing, and assessing proposed actions in order to ensure ecologically sound and sustainable development. Environmental impact assessment (EIA) is one of the methods accessible to planners for balancing development activities with environmental concerns. EIA incorporates environmental issues into development operations as early as the preliminary feasibility report is initiated. As a result, it may be possible to include environmental concerns and mitigation strategies into project planning. EIA can frequently predict future liabilities or costly changes in project design. To avoid indiscriminate exploitation of natural resources and to encourage the incorporation of environmental considerations in development projects, the Ministry of Environment and Forests (MoE&F) has made many policy measures and established environmental and pollution control laws. Under the rules of the Environment (Protection) Act, 1986, one such endeavour is the Notification on Environmental

Impact Assessment (EIA) of Development Projects 1994. The following are the primary aims of EIA practises:

a) To establish that before the responsible authority makes a decision to begin a project, the environmental impacts of such actions should be fully considered.
b) To encourage the implementation of acceptable procedures that are compatible with national laws and decision-making processes in all nations.
c) Encourage the creation of reciprocal processes for information sharing, notification, and consultation among states where planned activities are anticipated to have major transboundary effects on those states' environments.

History of Environmental Impact Assessment in India

a) The Indian experience with Environmental Impact Assessment dates back more than 20 years. It began in 1976-77, when the Planning Commission ordered the Department of Science and Technology to conduct an environmental assessment of the river-valley projects.
b) Previously, development projects were done with no regard for the environmental effects.
c) Given the massive environmental damage, governments and the general public are now worried about the environmental consequences of development operations. Thus, the EIA method was developed to examine the environmental implications.
d) The Environmental Impact Assessment (EIA) was first used in India in 1978 for river valley development. Later, the EIA Act was expanded to cover additional parts on development.
e) Under the requirements of the Environment (Protection) Act, 1986, EIA is referred to as the Notification on Environmental Impact Assessment of Development Projects 1994.
f) Aside from EIA, the Government of India published a number of additional notifications relevant to environmental impact assessment under the Environment (Protection) Act 1986.
g) The Impact Assessment Agency of the Ministry of Environment and Forests, Government of India, grants environmental clearance or the "go-ahead" signal. In September 2006, the Ministry of Environment, Forests, and Climate Change (MoEFCC) announced new EIA regulations. The notification requires environmental

approval for a variety of projects, including mining, thermal power plants, river valley infrastructure (road, highway, ports, harbours, and airports), and industries, including extremely small electroplating or foundry units.

h) Unlike the EIA Notification of 1994, however, the new legislation has placed the onus of project clearance on the state government, depending on the size/capacity of the project.
i) EIA is presently required for over 30 types of projects, and these projects receive Environmental Clearance (EC) only once the EIA standards are met.

Environmental Components of EIA

The EIA procedure investigates the following environmental components.

a) Air environment
1) Quality of ambient air present and predicted.
2) Meteorological data: Wind speed, direction, humidity, etc.
3) Quantity of emission likely from the project.
4) Impact of the emission on the area.
5) Pollution control desires/air quality standards.

b) Noise
1) Noise levels present and forecast.
2) Noise pollution reduction strategies.

c) Water environment
1) Existing ground and surface water resources in the zone, as well as their quality and quantity.
2) Water resource impact of a proposed project.

d) Biological environment
1) Flora and wildlife in the zone of impact.
2) Potential (probable) project-related harm owing to effluents, pollutants, and landscaping.
3) Biological suffering (prediction).

e) Land environment
1) Soil properties, land usage, and drainage pattern analysis, as well as the project's potential negative impact.
2) The effect on historical monuments and heritage sites.

Process and Methodology

The following stages are included in EIA. However, the EIA cycle repeats, with many parts interacting with one another.

a) Screening: The project plan is reviewed to determine the size of the investment, the location and kind of development, and if the project requires legislative approval.
b) Scoping: The project's possible implications, zone of influence, mitigating options, and monitoring requirements.
c) Baseline data collection: Baseline data is the environmental state of the studied region.
d) Impact prediction: Positive and negative, reversible and irreversible, temporary and permanent consequences must be projected, which requires the assessment agency to have a thorough grasp of the project.
e) Mitigation measures and EIA report: The EIA report should include activities and methods for preventing, limiting, or mitigating consequences, as well as the degree of compensation for potential environmental harm or loss.
f) Public hearing: After the EIA report is completed, the public and environmental groups residing near the project site may be informed and consulted.
g) Making a decision: The Impact Assessment Authority consults with the project manager and a consultant to make a final decision while keeping EIA and EMP in mind (Environment Management Plan).
h) Monitoring and execution of the environmental management plan: The project's different phases of implementation are monitored.
i) Alternatives Evaluation, Mitigation Measures Delineation, and Environmental Impact Assessment Report: Alternatives should be found and environmental features analysed for each project. Alternatives should include project location as well as process technology. Following the analysis of options, a mitigation strategy for the chosen option should be developed, which should be supported by an Environmental Management Plan (EMP) to assist the proponent toward environmental improvements.
j) Risk assessment: EIA methods include a risk assessment, inventory analysis, and hazard probability and index. The generalized EIA process flowchart is given below (Fig 14.1).

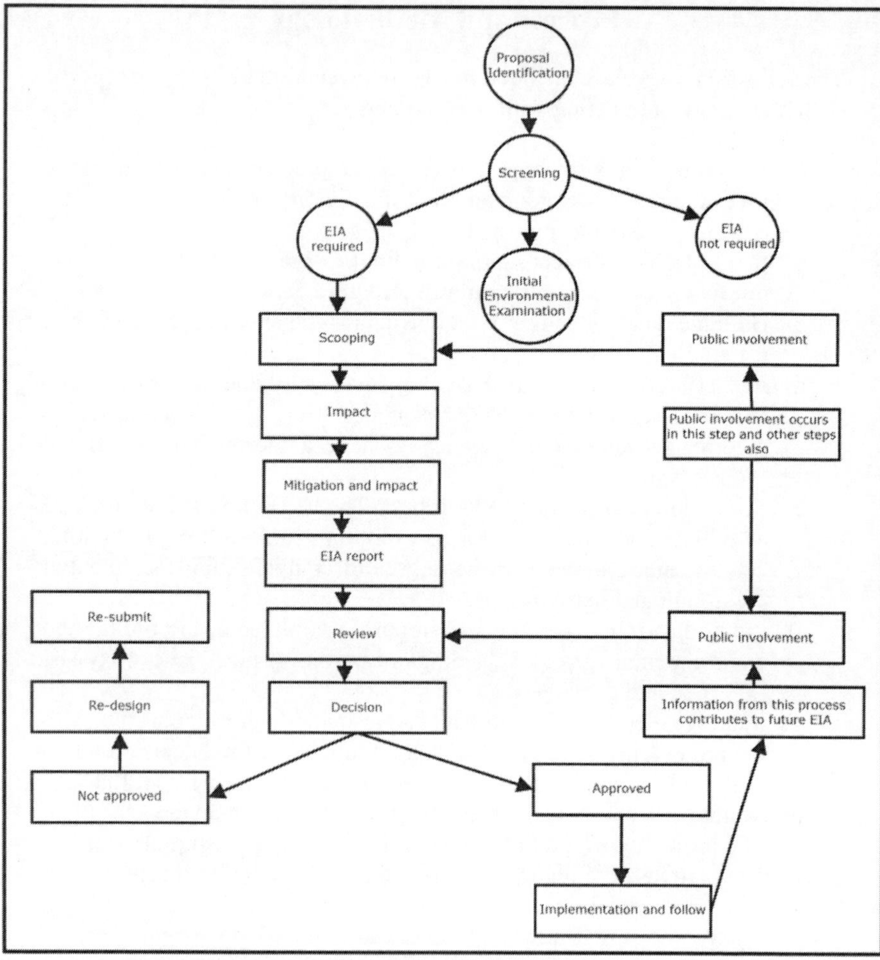

Fig. 14.1: Diagram showing a generalized methodology for environment impact assessment

Steps in preparation of EIA report:

a) Data collection from primary and secondary sources.
b) Impact prediction based on previous experience and quantitative models.
c) Impact evolution vs net cost-benefit analysis.

d) Preparation of environmental management plans to minimise the effects.
e) Estimate the financial cost of the monitoring strategy and mitigating actions quantitatively.
f) Environment management plan: Delineation of mitigation strategies for each environmental component, including prevention and control, as well as a restoration and resettlement strategy.
g) Environment appraisal: The Ministry of Environment and Forests will form an Appraisal Committee to examine a project based on the data submitted by the project authority. If required, the MoEF may undertake discussions with investors and experts on specific concerns. After taking into account all aspects of the project, environmental clearance is granted subject to the application of the specified environmental safeguards. A decision is made within 90 days in the event of projects when the project proponents have presented comprehensive material. The Ministry's six regional offices in Shillong, Bhubaneshwar, Chandigarh, Bangalore, Lucknow, and Bhopal are responsible for project monitoring.
h) EIA of coasts: Coastal Zone Management Plans (CZMPs) are developed by coastal states or union territories in accordance with the regulations established by the CRZ notice of 1991. CZMPs are developed based on the identification and classification of coastal regions for various activities, and they are subsequently presented to the MoEF for approval. The ministry then creates a task team to investigate their ideas.
i) Single window clearance: Environmental and forestry clearances are required. When a project requires both environmental clearance and permission under the Forest (Conservation) Act of 1980, plans for both must be submitted to the Ministry's respective divisions at the same time. The processing is done in parallel for approval or rejection. If the project does not require the diversion of forestland, the case is solely reviewed for environmental reasons.

Stakeholders in the EIA Process

EIA is applicable to both public and private sectors. The six important players are as follows:

a) Those who propose the project.
b) The environmental consultant who prepare EIA on behalf of the project proponent.

c) Pollution Control Board (State or National).
d) The public has the right to express their opinion.
e) The Impact Assessment Agency.
f) Regional centre of the MoEFCC.

EIA expert committee composition

The Committees will be made up of specialists from the following fields:

a) Eco-system management.
b) Air/water pollution control.
c) Water resource management.
d) Flora/fauna conservation and management.
e) Land use planning.
f) Social Sciences/Rehabilitation.
g) Project appraisal.
h) Ecology.
i) Environmental health.
j) Subject area specialists.
k) Representatives of non-governmental organisations (NGOs) and individuals concerned about environmental concerns.
l) The Chairman shall be a distinguished and experienced ecological, environmentalist, or technical practitioner with extensive managerial experience in related development.
m) The Impact Assessment Agency's representative will serve as Member-Secretary.
n) Except for those officially selected as representatives, the chairman and members will serve in their own capacities.
o) A committee's membership cannot exceed 15 people.

Importance of EIA

a) EIA connects the environment with development in order to achieve ecologically safe and sustainable development.
b) EIA is a low-cost way to remove or reduce the negative effects of development initiatives.
c) EIA allows decision-makers to assess the environmental impact of development activities long before the project is completed.
d) EIA advocates the incorporation of mitigation techniques into the development strategy.

e) EIA ensures that the development plan is ecologically sound and within the boundaries of the ecosystem's capacity for absorption and regeneration.

EIA Process Shortcomings

a) Applicability: Several projects with major environmental implications are excluded from notice because they are not mentioned in Schedule I or their investments are less than what is specified in the notification.
b) Expert committees and standards composition: It was discovered that the team created to undertake EIA assessments lacked competence in a variety of sectors, including environmentalists, wildlife specialists, anthropologists, and social scientists.
c) Public hearing: Because public views are not taken into account at an early stage, they can cause disagreement at a later stage of project clearance. Several projects with substantial environmental and social consequences have been exempted from the obligatory public hearing process. Local people's indigenous knowledge is not respected by data collectors.
d) Quality of EIA: One of the most serious issues about the environmental clearance procedure is the quality of the EIA report that is being prepared.
e) Lack of credibility: There have been several incidents of fraudulent EIA studies in which incorrect data was utilised, the same facts were used for two completely different locations, and so on.

Salient Features of 2006 Amendments to EIA Notification

a) The Environment Impact Assessment Notification of 2006 decentralised environmental clearance initiatives by dividing development projects into two categories: Category A (national level assessment) and Category B (local level appraisal) (state-level appraisal).
   1) The Impact Assessment Agency (IAA) and Expert Appraisal Committee (EAC) assess Category A projects at the national level, whereas Category B projects are reviewed at the state level.
   2) To give authorization to the Category B procedure, the State Level Environment Impact Assessment Authority (SEIAA) and

the State Level Expert Appraisal Committee (SEAC) are formed.

b) Following the 2006 Amendment, an EIA cycle consists of four stages: screening, Scoping, Public hearing and Appraisal.
c) Because Category A projects require statutory environmental approval, they do not go through the screening procedure.
d) Category B projects go through a screening procedure and are divided into two groups. Projects in Category B1 (Mandatorily requires EIA) and projects in Category B2 (Do not require EIA).
e) Thus, Category A and Category B projects are subject to the whole EIA procedure, but Category B2 projects are not subject to the full EIA process.

PARIVESH

PARIVESH stands for Pro-Active and Responsive facilitation by Interactive and Virtuous Environmental Single-window Hub. It is a one-stop shop for environmental, forest, wildlife, and CRZ clearances. This Single-Window Integrated Environmental Management System was created in the spirit of the Prime Minister's 'Digital India,' embodying the principles of Minimum Government and Maximum Governance.

Key features:

a) PARIVESH is a workflow-based application built on the web architecture idea. It has been made available for online submission, monitoring, and administration of applications submitted by Project Proponents to the Ministry of Environment, Forest, and Climate Change (MOEFCC) and State Level Environmental Impact Assessment Authorities (SEIAA).
b) It aims to get several sorts of clearances (for example, Environment, Forest, Wildlife, and Coastal Regulation Zone Clearances) from Central, State, and District-level agencies.
c) The Ministry of Environment, Forests, and Climate Change conceived, built, and hosted the system, with technical assistance from the National Informatics Centre (NIC).
d) It offers a single registration and sign-in for all forms of clearances (including Environment, Forest, Wildlife, and CRZ), a unique-ID for all types of clearances necessary for a certain project, and a single Window interface for the proponent to submit applications for all categories of clearances (i.e., Environment, Forests, Wildlife and CRZ clearances).

Environment Protection standards in India

Long before environmental law was enacted in India, the Bureau of Indian Standards (previously Indian Standards Institution) took the lead in developing standards for environmental protection. Water pollution research began in the early 1970s, while solid waste management research began in the late 1970s. Based on global experience, it is believed that two sorts of standards will produce the best outcomes. One of these is classified as an environmental quality standard, while the other is classified as a discharge standard. The environmental quality standards establish the objectives of pollution control programmes. These are directly connected to the highest level of tolerance of human, plant, animal, and aquatic life to a certain pollution. At any given moment, these restrictions are defined by existing data, and no relaxation is permitted to protect environmental quality, regardless of technological or economic disadvantages. Much work has been done in the subject of environmental quality standards in developed nations, particularly Europe, and data in this respect is easily available, however in India, no systematic effort has been made so far in this regard.

In 1963, the Indian Standard 2490 was issued as the first set of standards. These regulations establish the tolerance limits for industrial wastewater released into inland surface waterways. Following that, two further standards were developed: the IS 3306 for discharge into public sewers and the IS 3307 for discharge on land for irrigation purposes. The IS 2490 was then amended in 1974 as industry specific standards, with Part I giving generic limitations and Part II onwards giving limits for individual industries. IS 7968 was developed in 1976 to regulate the discharge restrictions into marine coastal zones. Following that, a variety of standards in the fields of water, air, noise pollution, and solid waste management were issued. Each standard contains the following characteristics:

a) Methods for test, terminology.
b) Guideline for waste treatment.
c) Effluent/emission limits.
d) Receiving courses/ambient air quality limits.

Concept of Audit and Audit Programme

Environmental auditing is a management technique used to offer environmental performance information to the appropriate people at the right time. There are no regulatory obligations mandating a corporation to produce an environmental audit at the time of writing, nor are there any Codes of Practice or even widely recognised recommendations on what should be included in an audit or how it should be carried out. An environmental audit is a method of presenting environmental management information in a systematic manner:

a) to all management levels; and
b) for various objectives.

As a result, the word "environmental audit" refers to a variety of distinct information and evaluation operations. These are classified as follows:

a) A technical examination: An audit of this type will entail the methodical collecting of information on the current and potential environmental effect of the organization's activities; it will typically address compliance with pollution control and waste management legislation. Management techniques will not be covered.
b) A management review: This will concentrate on management practises and records, as well as information on legislative compliance. It may also conduct an evaluation of procedures in light of business policies, programmes, and other needs. It will not conduct a technical examination of the operation's current or potential impact on the surrounding environment.
c) Due diligence review: This will analyse the potential cost of adopting pollution control and site clean-up initiatives while taking current and future regulations into consideration. Such liability analyses are typically conducted in the context of mergers, acquisitions, and long-term business planning.

An environmental audit is therefore a management tool that consists of a systematic, documented, periodic, and objective assessment of the organization's to preserve the environment by:

a) Supporting managerial oversight of actions that may have an environmental effect.
b) Determining compliance with the company's environmental policy.

According to the Regulation, an audit must involve at least the following steps:

a) Planning audit operations, including the assignment of audit roles.
b) Examine the company's environmental protection programme.
c) Evaluation of the organisation, management, and equipment.
d) Data collection and information gathering.
e) Overall performance evaluation.
f) Identifying opportunities for improvement.
g) Internal reporting to upper management.

Setting up an Audit programme

Developing an audit programme necessitates the following decisions:

a) The audit's scope: what information will be gathered.
b) How frequently each site (or issue) will be audited.
c) Who will do the audit.
d) What information, if any, will be made accessible to the public.

Benefits

An organisation can profit from an environmental audit strategy that is effectively implemented. They are as follows:

a) It establishes a framework for assessing (and thereby controlling) environmental performance.
b) It enhances accountability for the business's environmental dimension: the audit process demands management to be clear about their duties and how they are being carried out.
c) It promotes awareness within the organisation of the necessity of professional environment management.
d) To the degree that external reporting is essential, it offers a solid foundation that allows organisations to be certain that the information contained in the reports is reliable.
e) It contains useful information for future planning. This involves future process and product design, as well as inputs into future budgetary projections for pollution control and other investments.
f) It provides senior management with the assurance that all environmental components of their firm are being competently managed: it does not eliminate the chance of environmental incidents, but it does lessen their likelihood.

Environmental auditing programme in India

The notion of environmental audits in industrial facilities appears to have initially gained traction in India in the early 1990s. Before this programme could become required, efforts were made to assess its feasibility. This procedure culminated in the publication of a gazette notification on March 13, 1992, making the filing of environmental audit reports mandatory. Since 1993, companies have been required to submit audit reports to the relevant state pollution control bodies on or before the 15th of May each year. A large number of polluting industries have been recognised for regular environmental audit report submission. They are as follows:

a) Cement factories (above 200 tonnes per day production capacity)
b) Thermal power plants
c) Sugar factories
d) Fertilizers and sulphuric acid plants
e) Oil refineries
f) Caustic soda plants
g) Petrochemical plants
h) Zinc smelting industries
i) Copper smelting industries
j) Lead smelting industries
k) Integrated iron and steel plants

Life cycle concepts and stages

The early 1990s saw the formalisation of life cycle assessment (LCA). It arose from a belief that it was critical to undertake "cradle-to-grave" evaluations of goods, packaging, processes, and activities. The Eco balance research technique established in Europe, emphasising waste reduction to reduce water and air pollution, and Resource and Environmental Profile Analysis work conducted in the United States in the 1980s were forerunners of LCA. However, the fundamental concept for LCA was created during a workshop organised by the Society of Environmental Toxicology and Chemistry (SETAC) in the early 1990s in the United States (SETAC, 1991, 1993, 1994). LCA can be divided into four stages or phases:

a) Initiation: This step is quite similar to scoping in the context of impact assessment. The primary goals are to specify the objectives and specifics of an assessment, to define the "system" to be assessed, and to identify the types of data required.
b) Inventory: The primary focus is on gathering data on raw materials required for, inputs such as energy and water, and wastes created as outputs during the process and at the end of the useful life of the product, process, packaging, or associated activity.

c) Impact: The primary focus is on gathering data on raw materials required for, inputs such as energy and water, and wastes created as outputs during the process and at the end of the useful life of the product, process, packaging, or associated activity.
d) Improvement: This is the cognitive or prescriptive stage, in which attention is devoted on potential ways to eliminate or alleviate any negative affects identified in the preceding stage.

PART 3:

BIODIVERSITY

CHAPTER 15

BIODIVERSITY

Concept and Origin

Our biosphere has enormous diversity (or heterogeneity) not just at the species level, but also at all levels of biological organisation, from macromolecules within cells to biomes. Edward Wilson, a socio-biologist, popularised the word "biodiversity" to represent the cumulative diversity at all levels of biological structure. The term "biodiversity" refers to "the variation among living creatures from all origins, including terrestrial, marine, and other aquatic habitats, as well as the ecological complexes to which they belong; this includes diversity within species, between species, and within ecosystems" (UNES 1992). As a result, it may be characterized in terms of genes, species, and ecosystems, which represents the three fundamental levels of biological organisation:

a) Genetic diversity: Genetic diversity refers to the total genetic information recorded in the DNA of individuals of plants, animals, and microorganisms. A single species may demonstrate significant genetic variation over its range. There are around 50,000 genetically diverse rice strains in India, as well as 1,000 mango varieties.
b) Species diversity: Species diversity signifies population variability in terms of each individual's reproductive distinctiveness. It is a taxonomic unit in the biological world. The Western Ghats, for example, contain a larger diversity of amphibian species than the Eastern Ghats.
c) Ecological diversity: Ecosystem diversity refers to the variety of habitats, biotic communities, and ecological processes found across the biosphere, as well as diversity within ecosystems. India, for instance, has greater biological variety than a country in the Nordic region like Norway because of its deserts, rain forests, mangroves, coral reefs, marshes, estuaries, and alpine meadows.

There are two factors that contribute to population diversity. First, new genotypes are continually appearing in a population as a result of mutation, recombination, and other genetic events, as well as immigration of people, gametes, or propagules. Second, diversity is the population loss due to natural selection and emigration of individuals. Every genetic variety, from gene mutations to whole species, will eventually perish. This loss might happen quickly, or the variations can live for a long time. Horseshoe crabs, which have been around for 200 million years, and cockroaches, which originated even earlier (350 million years) in the carboniferous period, are examples of long-living species.

The biological diversity of the system is determined by the pace at which new variations emerge in comparison to the rate at which they are eliminated. Thus, variety is the consequence of two opposing processes: those that generate new genotypes, variations, and species, and those that eliminate mutations, variants, and species from the system. Natural selection is the primary cause of biodiversity loss; it operates through differential reproduction and differential death.

Measurement of Biodiversity

Two primary components are used to assess biodiversity:

a) Species richness: It refers to the number of species found in a given community and consists of:
   1) Alpha diversity: It refers to the diversity within a certain region or ecosystem, and is generally measured in terms of the number of species (species richness).
   2) Beta diversity: It is an assessment of ecological diversity, commonly assessed as the difference in the number of species between ecosystems.
   3) Gamma diversity: It is a measurement of an area's overall ecological diversity.

b) Species evenness: It quantifies the percentage of species in a specific location; for example, a low evenness suggests that a few species dominate the locality.

Values of Biodiversity

Biological variety is so commonly acknowledged as important that the reasons for this are rarely thoroughly examined. Though not all degrees of variance are equally beneficial for all causes, the values can be accounted for the following reasons:

Value categories	Description	Level of operation
Commodity value	The utility of an organism to humans as food, medicine, and other things. Non-exploited populations may have genes that increase the value of the exploited population.	Population and species diversity
Amenity value	Humans love visiting, observing, and learning about natural communities.	Species and Ecosystem diversity
Ecological integrity value	The significance of sustaining all ecosystem's functionally vital components.	Functional diversity
Ethical value	Humans have a moral imperative not to push other species to extinction.	Species and Ecosystem diversity
Option value	The future evolutionary potential of a species, as well as its potential for future exploitation, is measured in commodities value.	Genetic and population value

It is anticipated that expanding this to 10 percent of the terrestrial area, adequately protected and with recompense to indigenous people, would cost roughly $30 billion each year. It is also stated that 1 percent of global GWP is required to ensure the continued provision of ecosystem services. Now is the moment to analyse the value of biodiversity benefits from a utilitarian standpoint. Currently, only 120 kinds of plants are responsible for 25 percent of the medications on drugstore shelves. However, there are around 25,000 species of traditional medicinal plants utilised by indigenous peoples across the world (about 10 percent of the total number of plant species).

It is commonly believed that species diversity is essential for the development of community-level traits and the efficient operation of communities. A large number of unique DNA-encoded enzymes are needed for a complex organism to operate properly.

Loss of Biodiversity

Natural processes such as speciation and extinction occur. Our planet's biological richness is rapidly deteriorating, and the finger of blame is firmly pointed at human actions. There have been 27 species extinctions in the previous two decades alone. Currently, extinction threatens 12 percent of all bird species, 23 percent of all mammal species, 32 percent of all amphibian species, and 31 percent of all gymnosperm species on the planet. In general, a loss of biodiversity in a region might result in (a) decreased plant productivity, (b) less tolerance to environmental perturbations like drought, and (c) greater unpredictability in ecosystem processes including plant productivity, water usage, and pest and disease cycles.

Causes of Biodiversity loss

a) Habitat loss and fragmentation: The most critical factor pushing animals and plants to extinction is habitat loss and fragmentation. Tropical rain forests are the most severe examples of habitat loss. Rain forests formerly covered more than 14 percent of the earth's land surface, but currently only cover around 6 percent. Mammals, birds, and certain migratory species that need large territories are badly affected, leading to population declines, when large ecosystems are broken up into tiny pieces as a consequence of multiple human activities.
b) Overexploitation: Humans have always relied on nature for food and shelter, but when "need" becomes "greed," natural resources are overexploited. Many species extinctions in the last 500 years have been caused by human overexploitation (for example, Steller's sea cow andthe passenger pigeon).
c) Invasions of alien species: When foreign species are unintentionally or deliberately introduced for whatever reason, some of them become invasive and result in the decline or extinction of native species. The Nile perch, which was introduced into Lake Victoria in east Africa, finally wiped out an ecologically unique assemblage of over 200 species of cichlid fish.
d) Co-extinctions: When a species becomes extinct, all of the plant and animal species that are linked with it must also go extinct.

When a host fish species goes extinct, its unique parasite assemblage goes extinct with it.

e) Natural causes: Biodiversity can also get affected by natural calamities like volcanoes, floods, earthquakes, landslides etc.

Need for Biodiversity Conservation

The diversity of plants and animals on Earth is incredible, and the high degree of species richness in tropical forests and coral reefs is unbelievable. However, hundreds of different bacteria, plants, and animals thrive in habitats with few species, such as the ocean floor. Nature's incredible prodigality has astounded and pleased humanity, and is ultimately the source of their nutrition. Humans, however, threaten this vast diversity of species by utilising plants and animals for food and clothing, building dwellings, and medicine. The present rate of loss of species diversity and decrease in genetic diversity of crops and wild animals may have major consequences for human welfare.

Strategies for Biodiversity Conservation

The tremendous biodiversity of the Earth is essential to humanity's existence. Aside from direct advantages (food, fibre, firewood, medicines, and so on), ecosystem services such as pollination, pest management, temperature moderation, and flood control provide numerous indirect benefits. We also have a moral obligation to preserve the earth's biodiversity and pass it on in good condition to future generations.

Thus, conservation is defined as the establishment and observance of economically, socially, and politically appropriate norms, standards, patterns, or models of behaviour in a particular society's use of natural resources. Conservation is the intentional management of natural resources in order to maintain nature's balance and variety. It also includes judicious use of natural resources in order to meet the requirements of the current generation while leaving enough for future generations. The conservation of biodiversity is crucial to avoid the loss of a species' genetic variety, saving a species from extinction, and protecting ecosystems from damage and deterioration. As a result, conservation activities may be divided into two categories:

a) In-situ: Plant and animal conservation in their native habitats or protected areas is an example of in-situ conservation. Protected areas are areas of land or water devoted to the conservation and preservation of biodiversity. The in-situ strategy stresses the conservation of overall biodiversity of genes, populations, species, communities, and ecological processes through protecting whole ecosystems. The in-situ method comprises the conservation of a set of typical ecosystems through a network of protected areas designated by the United Nations Environment Programme and the World Conservation Union (IUCN). In-situ biodiversity conservation is helpful since it is a low-cost and convenient strategy that relies on people's cooperation. It keeps all species at different trophic levels alive, from producers to top consumers like predators. Organisms not only survive and proliferate in the natural environment, but they also evolve and preserve their capacity to withstand diverse environmental stressors such as drought, snow, temperature swings, extreme rain, flood, fires, infections, and so on. In-situ conservation involves just the removal of conditions that are adverse to the species' survival, allowing a greater number of species to develop concurrently and prosper in their native environment, where they have been growing for a long time. The sole downside of in-situ conservation is that it necessitates greater acreage and reduces room for restraining the human population, which is rapidly expanding. Examples include national parks, sanctuaries, biosphere reserves, reserved forests, and protected forests.

b) Ex-situ: Plant and animal conservation outside of their natural environments. These include botanical gardens, zoos, and gene banks, as well as seed banks, tissue culture, and cryopreservation. For example, the extinct Gangetic gharial has been reintroduced into rivers in Uttar Pradesh, Madhya Pradesh, and Rajasthan.

Botanical gardens

A botanic garden is a carefully organised collection of living trees, shrubs, herbs, climbers, and other plants from all over the world. The main purposes of botanical gardens are given below:

1) To study plant taxonomy and growth.
2) To investigate the introduction and acclimation of alien plants.
3) It serves as a germplasm repository.
4) It aids in the generation of novel hybrids.
5) It contributes to the conservation of rare and endangered species.

Zoos

A zoo is a stationary or mobile establishment where captive animals are maintained for public exhibition and includes a circus and rescue centres, but does not include an establishment of a licenced dealer in captive animals. Zoos began as places for pleasure; yet, over time, they have evolved into places for wildlife conservation and environmental education. Zoos serve an important role in species conservation in addition to preserving individual animals (through captive breeding).

Biopiracy

Biopiracy is the commercial exploitation of naturally occurring biochemicals or genetic materials. Indigenous peoples' traditional cognition often consists mostly of biological traits and genetic variety of the natural environment passed down from generation to generation. Few traditional knowledge items crucial to global survival include the following:

a) Farming or Agriculture.
b) Medicinal Plants.
c) Varieties of Food crops.

Examples of Biopiracy

a) African super-sweet berry biopiracy: Pentadiplandrabrazzein is a plant found in western South Africa. It is an important source of a protein known as Brazzein. It is used as a low-calorie sweetener there. It is thought to be approximately two thousand times sweeter than sugar. Recent advances include the isolation of the gene encoding brazzein, which has been sequenced and patented in the United States.
b) Azadirachta indica – Neem Patent: Since ancient times, Neem has shown to be beneficial in a variety of ways. Indians have spread their neem expertise all over the world. The United States Department of Agriculture and an American business, W.R. Grace, won a European patent in 1994 for different techniques of managing fungal diseases in plants by utilising a neem-derived compound.
c) The rosy periwinkle: The rose periwinkle is a kind of periwinkle that was discovered in Madagascar. It is now being introduced to a number of other tropical countries throughout the world. This

enables researchers to gather information from one country and plant samples from another.

Impact

Bio-piracy has global health effects such as a lack of access to medications, a failure to compensate for vital traditional knowledge, and the depletion of biological resources needed by indigenous groups for ethnomedicine and health care. Western firms' appropriation of indigenous ingenuity robs intellectual contributions from another nation, resulting in intellectual deprivation. Patents based on biopiracy drain biological resources away from local communities. Scarcity and resource poverty are caused by global markets. Bio-piracy allows pirates to gain market domination and exclude others.

Bio-prospecting

Bio-prospecting (also known as biodiversity prospecting) is the search for natural sources of small molecules, macromolecules, and biochemical and genetic data that may be used to create products with a market value in the fields of agriculture, aquaculture, bioremediation, cosmetics, nanotechnology, and pharmaceuticals. For example, over one-third of all small-molecule medications authorised by the United States Food and Drug Administration (FDA) between 1981 and 2014 were either natural products or compounds derived from natural products.

Benefits

a) The therapeutic qualities of plants and other organisms are the most important advantage. Bioprospecting encourages innovation by assisting countries in developing new medicinal goods. Our indigenous medical system is also strongly reliant on bioprospecting.
b) It also favours job possibilities with natural products.
c) The economic worth of the plant and creatures benefits both the local community and corporate companies. The 1992 Convention on Biological Diversity assures that the benefits obtained are not misused by businesses.
d) It provides an incentive to monitor and conserve biodiversity in order to prevent losing economic possibilities to rivals or extinction.
e) Bioprospecting promotes technology and information transfer between countries.

f) Conservation of biodiversity is aided when local communities become more aware of the potential economic worth of natural ecosystems.
g) By rediscovering historic native customs, traditional culture and habits are preserved.

Extinctions

When a species goes extinct, it loses its whole genetic legacy. To adapt to environmental changes or changes in genetic legacy, organisms develop into new species. Over 99 percent of all species that have lived on Earth, totalling over five billion species, are thought to be extinct. According to estimates, there are between 10 and 14 million extant species, of which more than 1.2 million have been researched and more than 86 percent are still to be identified. This might occur naturally as a result of climate change, or as a result of human activity like as overhunting or habitat loss. Regardless of the uncertainty, extinction includes three key components:

a) Species may go extinct if fatal stress is so rare that it is beyond their experience and so beyond the grasp of natural selection.
b) The mass extinctions would pave the way for a major biosphere restructure in which a few thriving groups would be eliminated to make room for lesser groups to flourish.
c) There is some evidence that Darwin favours extinction in certain circumstances. These days, predicting which species will go extinct next in an extinction event is challenging.

Causes of Extinction

The main causes of extinction include:

a) Asteroid strike: A meteor strike on the Yucatan peninsula in Mexico millions of years ago wiped off the dinosaurs. Such events led to the majority of major extinctions, including the KT-extinction and the Permian-Triassic extinction. Astronomers are always on the watch for comets or meteors that might bring about the demise of civilisation.
b) Climate change: Climate change is another factor that might lead to the extinction of terrestrial animals. At the end of the last ice age, the majority of the megafauna were unable to adapt to the warm climate. Due to food scarcity and early humans' hunting, they died.

Even modern civilisation is on the verge of destruction due to global warming.

c) Disease: A significant population of humans and animals on Earth had been wiped off by various plagues. During the Middle Ages, the Black Death killed off one-third of Europe's population.

d) Loss of habitat: Every animal has a safe place where it may reproduce and nurture its young. A bird, for example, is only at ease on a tree limb. Because of the rise of human civilization and industrialisation, the woods that are home to the majority of animals have been devastated. Many species' numbers have been reduced due to a shortage of space and, eventually, food.

e) Lack of genetic diversity: As the number of species decreases, so does the gene pool of that species. There will eventually be a shortage of genetic variety. For example, due to habitat degradation, African cheetahs have significantly low genetic diversity.

f) Better adapted competition: Better-adapted populations triumph over those that lag behind. Prehistoric mammals, for example, were more suited than dinosaurs. The well-adapted ones survive, while the rest become extinct.

g) Pollution: Pollution from factories and cars has resulted in a significant decrease in oxygen levels in the atmosphere and water. This has resulted in the extinction of most aquatic and terrestrial species.

Examples of Extinct animals

The following are some notable instances of extinct animals:

a) Passenger Pigeon: The passenger pigeon (Ectopistesmigratorius) is an extinct species of pigeon native to North America. Because of the species' migratory habits, its common name is taken from the French term passager, which means "passing by."

b) Dodo: The extinct dodo (Raphuscucullatus), a flightless bird, was a native of Mauritius, an island in the Indian Ocean to the east of Madagascar.

c) Tasmanian Tiger: Because of its striped lower back, it is sometimes known as the Tasmanian tiger or Tasmanian wolf (because of its canid-like characteristics). Wilf Batty, a farmer from Mawbanna in the state's north-west, shot the last recorded thylacine in the wild in 1930.

d) Baiji White Dolphin: The baiji is a freshwater dolphin that may be extinct. It is considered to be the first dolphin species to go extinct as a result of human impact. The use of fishing nets with hooks that snare and drown dolphins as bycatch is the most likely reason of the baiji's decrease. Other possible factors include habitat deterioration.

e) Pyrenean Ibex: The Cantabrian Mountains, Southern France, and the northern Pyrenees were the most prevalent habitats for Pyrenean ibex. Pyrenean Ibex DNA tests revealed evidence that, following a demographic boom around 20,000 years ago, its population experienced a bottleneck caused by hunting, inbreeding, and other reasons, eventually leading to extinction.
f) Steller's Sea Cow: The sea cow (Hydrodamalis gigas), sometimes known as Steller's sea cow, was a gigantic extinct aquatic animal that lived around the Komandor Islands in the Bering Sea. Hunters eradicated Steller's sea cows in the eighteenth century, fewer than 30 years after they were discovered by Arctic explorers.
g) Great Auk: The great auk (Pinguinus impennis) is a flightless alcid that went extinct in the mid-nineteenth century. It was the only modern species in the genus Pinguinus.
h) Mammoth: Mammoth (genus Mammuthus) refers to any fossilised member of an extinct group of elephants found in Pleistocene deposits on all continents, with the exception of Australia and South America, as well as in early Holocene deposits in North America. 11,700 years ago marked the end of the Pleistocene Epoch, which lasted 2.6 million years. The Holocene Epoch started 11,700 years ago and is still in effect now.
i) Sabre-toothed Cat: The ice age giants, such as elephant-sized sloths and sabre-toothed tigers, were driven extinct within 100 years by a combination of warming and human hunting. They originally inhabited the windswept plains of Patagonia, South America's southernmost tip, alongside humans.

Conservation Strategies and protected areas

One of the tactics utilised in natural resource management is the construction of protected areas. However, with environmental philosophy, terms, and approaches changing dramatically over the last decade, the international community involved in protected area policy, planning, and management felt it was necessary to revisit the definition of a protected area to ensure that it encompassed the broad range of areas under or requiring protection. This concept was developed during the 4th World Congress on National Parks and Protected Areas, which was held in

Caracas, Venezuela, in February 1992. As a result, a protected area is described as "an area of land and/or sea specifically committed to the conservation and maintenance of biological variety, natural and associated cultural resources, and maintained by legal or other effective measures" (IUCN, 1994). The following are the primary reasons for establishing and managing protected areas:

a) Scientific reasearch
b) Wilderness protection
c) Preservation of species and genetic diversity
d) Maintenance of environmental services
e) Protection of specific natural and cultural features
f) Tourism and recreation
g) Education
h) Sustainable use of resources from natural ecosystems
i) Maintenance of cultural and traditional attributes

Protected areas have been classified into six categories based on the probable mix of priorities provided by the management objectives listed above:

a) Category I: Strict Nature Reserve/ Wilderness Area – Managed primarily for scientific research or wilderness preservation (includes two sub-categories; Strict Nature Reserve - Ia, and Wilderness Area - Ib).
b) Category II: Managed for the sake of ecosystem conservation and recreation
c) Category III: Natural Monument – Anatural feature that is managed primarily for conservation purposes.
d) Category IV: Management Area for Habitat/Species – Managed primarily for conservation through management intervention.
e) Category V: Landscape/Seascape Protection – Managed primarily for landscape/seascape conservation and enjoyment.
f) Category VI: Managed Resource Protected Area – Minly managed for the long-term usage of natural systems.

There are several ways to create protected zones. The two most typical approaches are government intervention (often at the national level, but often at the regional or municipal level) and property purchases by private conservation organisations. A relationship between the governments of developing nations in the tropics and international conservation organisations, multinational banks, and governments of wealthy countries is becoming more widespread. They give finance, training, and scientific and managerial skills to help poor countries build protected areas. The ICUN has devised the following classification system for protected areas, ranging from minimal to extensive human usage of the habitat:

a) Scientific reserve and strict nature reserve: They are strictly protected places that are kept in order for scientific research, teaching, and environmental monitoring. Populations and environmental processes are left alone to the greatest extent practicable.
b) National parks: They are vast regions of scenic and national beauty that are preserved for scientific, educational, and recreational purposes. They are not often employed for commercial resource extraction.
c) National monuments and landmarks: They are frequently smaller zones created to protect unique areas of exceptional national interest.
d) Managed wildlife sanctuaries and nature reserves: They are comparable to strict nature reserves, but some human intervention may be required to maintain the community's traits, and some restricted harvesting may be permitted.
e) Protected landscape: They allow residents to use the environment in a non-destructive manner and give chances for tourism and enjoyment.
f) Resource reserves: These are locations where resource usage is regulated in ways that are consistent with national policies.
g) National biotic areas and anthropological reserves: They enable traditional civilizations to continue to exist without outside intervention. These individuals frequently search and extract materials for their personal needs.
h) Multiple use management areas: They enable the long-term development of natural resources such as water, wildlife, cattle grazing, wood, tourism, fishing, and so on. These actions are frequently compatible with the maintenance of ecological populations.

IUCN Red list Categories

The IUCN, currently known as the World Conservation Union, is the leading coordinating organisation for international conservation initiatives, with headquarters in Gland, Switzerland. It creates directories of people knowledgeable in captive breeding programmes and other conservation issues. Various authorities, as well as certain private organisations, list endangered species. The Red Data Book is the most often mentioned of these listings. It is a loose-leaf book with details on the status of many species. This volume is updated on a regular basis and is published by the International Union for Conservation of Nature (IUCN) in Morges, Switzerland. "Red" represents the threat that both plant and animals are currently facing across the world. The Red Data Book was first published in 1966 as a guide for the development, protection, and management of listed species by the IUCN's Special Survival Commission. The information in this book is more thorough for endangered mammals and birds than for other types of animals and plants, although coverage is also given to less famous organisms facing extinction. The highly endangered species are featured on the pink pages of this magazine. New pages are given to subscribers as the current position of the species changes. Green pages are reserved for species that were formerly endangered but have since recovered to the point where they are no longer endangered. The number of pink pages continues to grow over time. There are a dismal number of green pages.

IUCN Classification For Conservation Priority

The IUCN designated the following main conservation categories (see Fig.15.1) to emphasise the legal position of rare species for the purpose of conservation:

a) Extinct (EX): A taxon is said to be extinct when there is no doubt that its last member has passed away. A taxon's historic range is covered by extensive surveys that fail to find an individual throughout key times (diurnal, seasonal, annual) in known and/or expected habitat.
b) Extinct in the wild (EW): When a taxon is only known to have survived in captivity, in cultivation, or as a naturalised population (or populations) far from its former distribution, it is said to be extinct in the wild. Extensive surveys in known and/or anticipated habitat, at relevant periods (diurnal, seasonal, yearly), throughout the historic range of a taxon have failed to record an individual, at which point the taxon is believed to be extinct in the wild.
c) Critically Endangered (CR): When the best available data indicates that a taxon is critically endangered, this means that there is evidence that it fits any of the Critically Endangered criteria. The

IUCN redlist has given wild species the highest risk category (Fig 15.1).

Criteria:

1) Population decline (almost 90 percent in the previous ten years);
2) Population size (number less than 50 mature individuals); or
3) Quantitative analysis indicating a 50 percent chance of extinction in the wild in the next ten years).

As a result, it is regarded to be in grave danger of extinction in the wild.

d) Endangered (EN): When the best obtainable evidence suggests that a taxon fits any of the Endangered criteria, it is classified as Endangered.

Criteria:

1) Population size decrease (70 percent over the last 10 years);
2) There are less than 250 adult persons in the population; or
3) The chance of extinction in the wild is at least 20 percent within 20 years, according to a quantitative estimate.

As a result, it is thought to be in grave danger of extinction in the wild.

e) Vulnerable (VU): When the best available evidence suggests that a taxon fits any of the criteria for Vulnerable, it is classified as Vulnerable.

Criteria:

1) Population decline (more than 50 percent in the previous ten years);
2) There are less than 10,000 adult persons in the population; or
3) Within 100 years, there is a 10 percent chance of extinction in the wild.

As a result, it is thought to be in danger of extinction in the wild.

f) Near Threatened (NT): When a taxon has been assessed against the criteria but does not currently qualify for Critically Endangered, Endangered, or Vulnerable status, it is considered Near Threatened because it is close to or likely to qualify for a threatened category in the near future.
g) Least Concern (LC): When a taxon is classified as Least Concern, it does not meet the criteria for Critically Endangered, Endangered,

Vulnerable, or Near Threatened. This section includes taxa that are widely distributed and plentiful.

h) Data Deficient (DD): A taxon is deemed Data Deficient if there are inadequate data to establish a direct or indirect assessment of the taxon's risk of extinction based on its range and/or population condition. The biology of a taxon in this category may have undergone extensive study, yet there is a lack of data on its abundance and/or dispersion. As a result, Data Deficient is not a danger category. Taxa that are listed in this category imply that additional information is needed and that future study may reveal that threatened categorization is justified.

i) Not Evaluated (NE): When a taxon has not yet been reviewed against the criteria, it is classified as Not Evaluated.

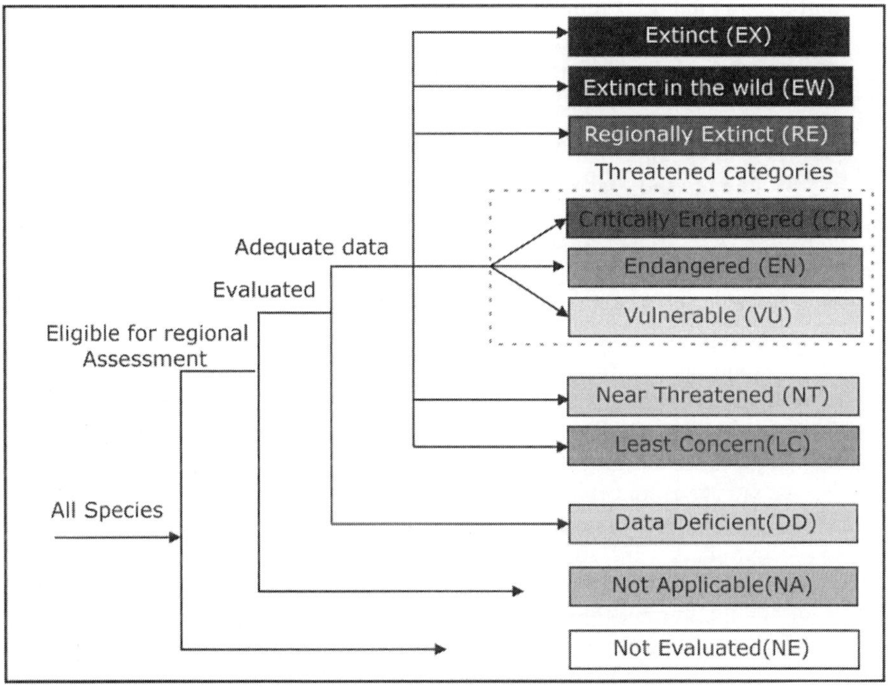

Fig. 15.1: IUCN Classification for Conservation Priority

CHAPTER 16

BIODIVERSITY OF INDIA

Soils of India

Soil is the top layer of the continental crust that contains weathered rock fragments. The soils of India are the result of both physical and human processes. Simply said, soil is a combination of tiny rock particles/debris and organic materials/humus that forms on the earth's surface and supports plant development. VasilyDokuchaev made the first scientific categorization of soil. Soils in India are divided into eight categories by the Indian Council of Agricultural Research (ICAR).

a) Aluvial soils: The Indo-Gangetic plain, which includes the states of Punjab and Haryana in the north, Uttar Pradesh and Bihar in the north, and Bengal and portions of Meghalaya and Orissa in the north east, is home to these. This soil may also be found in peninsular India's east coast deltas and terrace, deltaic, and lagoon alluvium. In the Punjab and western Ganga plains, the soil is rich in loams and clay components, with the loam component rising and sand decreasing in the central Ganga plains, where the calcareous kankar is common. The soil of the Ganga Brahamputra plains is typically alkaline or neutral in response. It has extremely fine particles ranging from loam to very fine silt clay. Normally, the nitrogen concentration is minimal. Potash, phosphoric acid, and alkalies are present in enough amounts. Iron oxide and lime amounts vary substantially. They are largely flat and consistent soils, making them ideal for agriculture. They are ideal for irrigation and respond well to canal and well/tube-well irrigation. Rice, wheat, sugarcane, tobacco, cotton, jute, maize, oilseeds, vegetables, and fruits are abundant.
b) Black soils: This type of soil is abundant in the Deccan traps, including Maharashtra, Mysore, and Madhya Pradesh, but it is also found in Tamil Nadu's Krishna Tungabhadra basin. These are black-cotton soils-regur in the western Deccan, but medium light black in the eastern half. The soils of Vindhyan and the

surrounding hills are also known as "brown soils." The majority of black soils are clay, with areas of clay loamsand sand loams. The black colour is caused by the presence of titaniferous magnetite or iron and black constituents of the parent rock. A typical black soil has a high clay component, 62 percent or more, and is highly argillaceous. In general, black soils in the uplands are poor in fertility, but those in the lowlands are quite productive.

c) Red soils: They cover huge regions in the Peninsula's south and north-east. Andhra Pradesh, Tamil Nadu, and sections of Bihar, Orissa, Uttar Pradesh, and West Bengal have such soils. Their red colour is caused by a high amount of iron constituents. These are primarily sandy to loam in texture, with gravels on the top slopes, sandy soils, deeper loamy soils on the lower slopes, and loams or clay in the valley bottoms. Furthermore, clay of true and low laterites may be found in this zone in the Western Ghats and Kerala, respectively. Furthermore, some sections of South-west India have a variety of red, yellowish, and white clays. This soil formed on Archean granite covers the country's second greatest area. These soils are generally lacking in phosphate, lime, magnesia, humus, and nitrogen. Rainfall is quite erratic. As a result, the soil has formed three kinds.
   1) Red and Yellow soil (200cm): North-east India – Nagaland, Mizoram, Manipur Hills, parts of Malabar coast, quick drainage is needed
   2) Red Sandy Soil (40-60cm): Drier plateaus like Karnataka, Tamil Nadu and Telangana.
   3) RedAlluvial Soil: Along river valleys and has good fertility.

d) Skeletal (Mountain) soils: It has thin layers and poorly defined profiles and horizons. Because of the rapid drainage, it has been prone to soil erosion. It has a high organic content and a good humus content, but it lacks other nutrients. When sand, silt, and clay are mixed, the result is a loamy soil. These are often found over 900 metres in elevation. They are found in the north-western hills of the Aravallis, where they are rocky hillfoot fans and slope colluvium, as well as in the humid south and east of the Himalayas and Meghalaya, where they are clayey in texture.

e) Desert soils: This soil is mostly found in dry and semi-arid areas such as Rajasthan, west of the Aravallis, northern Gujarat, Saurashtra, Kachchh, western sections of Haryana, and southern Punjab. It is deficient in moisture content. The humus level is lower, and the nitrogen concentration is initially low, although part of it is accessible in the form of nitrates. They are sandy and devoid of organic matter. The content of living microorganisms is minimal. It has a high iron concentration. The phosphorus concentration is roughly enough, and the soil is rich in lime and bases. It has minimal soluble salts and moisture retention capability. This soil has a good agricultural return when watered. These are appropriate for crops that require less water, such as Bajra, pulses, feed, and guar.

f) Laterite soils (Latosols): Laterites are high in iron and aluminium. Alternating dry and wet periods are more conducive to the formation of laterite soils. These are found in the Western Ghats, the northern part of the Eastern Ghats, the eastern edges of the Chota Nagpur plateau, Meghalaya, a few patches in Kathiawar, and two locations in the Peninsula's centre, north of Bangalore and west of Hyderabad. These soils have porous clay that is rich in iron and aluminium hydroxides. Such soils are suited for paddy cultivation at low elevations, but at higher elevations, they favour the development of coffee, rubber, tea, and Cinchona.

g) Saline and Alkaline soil: NaCl is abundant in alkali soil. Humus production is nearly nonexistent due to a lack of moisture, humus, and live microbes. The soil is infertile. These are also known as Reh, Usar, Kallar, Rakar, Thur, and Chopan. These are mostly found in Rajasthan, Haryana, Punjab, Uttar Pradesh, Bihar, and Maharashtra. This soil contains sodium chloride and sodium sulphate. It is useful for leguminous crops.

h) Peaty and Marshy soils: This soil comes from places where proper drainage isn't possible. It is high in salinity and contains a lot of organic debris. Potash and phosphate deficiencies exist. Clay and mud predominate, making it heavy. It is good for jute and rice in the Bengal delta, and spices, rubber, large rice, and other crops in Malabar. It has been beneficial to India's Mangrove forests to some extent (Fig. 16.1).

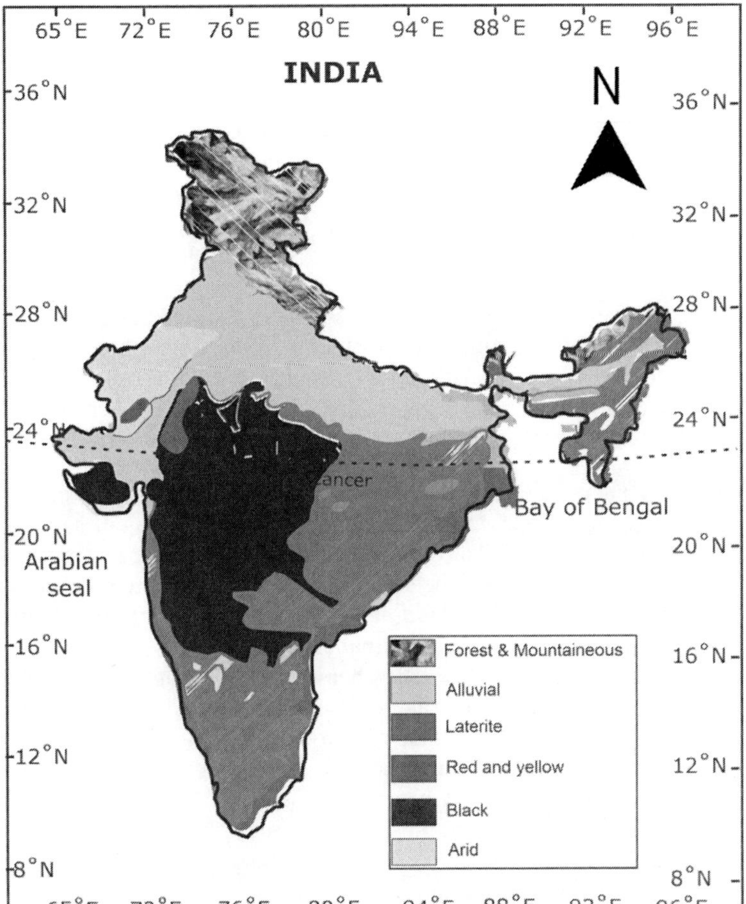

Fig. 16.1: GIS map showing major soil types in India

Climate and Climate region of India

Between latitudes 8° and 36° N, the Indian subcontinent lies fully to the north of the equator. Although India is considered a tropical country, the northern half of the Gangetic Plain is located in the world's northern temperature zone. The physiography of various hills and ranges – such as the Himalayan range, the mountain ranges of Meghalaya and Arunachal, and the Western Ghats with the Nilgiris and other hills in the south – all impact the country's climate. India's climate is monsoon-dominated. According to the Meteorological Department of the Government of India, India has four seasons:

a) The 'cool season', from mid-December to February.
b) The 'hot dry season', from March to mid-June.
c) The 'wet season', from mid-June to mid-September.
d) The 'season of retreating monsoon', from mid-September to mid-December.

India has a variety of geographical climate types as a result of a variety of climatic elements functioning in distinct sorts of physiographic conditions. Because India is a tropical nation, the primary criterion for climatic types has been rainfall rather than temperature. The nation has been classified into the four climate areas below based on yearly precipitation.

a) Wet zones: Rainfall totals more than 200 cm. This zone comprises the western Ghats and hills of Meghalaya and Bengal, as well as the tarai areas of Bihar and Uttar Pradesh. The northern part of the western slopes has extended dry seasons, whereas the southern half gets short ones. The landscape consists primarily of evergreen woods, and paddy is the principal crop.
b) Intermediate zones: Rainfall is heavy, ranging from 100 to 200 cm. This zone encompasses parts of West Bengal, Bihar, Orissa, eastern Madhya Pradesh, the Himalayan foothills, northern Uttar Pradesh, Himachal Pradesh, Jammu, north-east Andhra Pradesh, the eastern slopes of the Western Ghats, and east Tamil Nadu. The dominant vegetation is deciduous forests, with most trees losing their leaves during warmer weather.
c) Dry zones: Rainfall is moderate, ranging from 50 to 100 cm. This zone comprises Carnatic, north-west Deccan, upper Ganges plain, and higher Punjab plain, and encompasses Punjab, U.P., Delhi, western M.P., Gujarat, Maharashtra, Karnataka, southern Andhra Pradesh, and Tamilnadu. The predominant vegetation is of the dry, deciduous, and thorn scrub variety. Dry deciduous woods of the sort found in the intermediate zone can be found in some areas. Wheat and millets are the most important agricultural crops.
d) Arid zones: The average annual rainfall is less than 50 cm. This zone encompasses the north-west lowland and the north-west plateau, including the whole western region of the nation, including the south-western parts of Punjab, Rajasthan, and north-west

Gujarat. The native vegetation is thorn forest with extensive desert and semi-desert regions.

Bio-Geographic Zones of India

India is recognised as a mega-diverse country, with abundant biodiversity and accompanying traditional knowledge. Despite having only 2.4 percent of the geographical area, India is home to about 7 percent of the documented species and nearly 18 percent of the human population. India ranks seventh in mammals, tenth in birds, and sixth in reptiles in terms of species richness. The diverse Edaphic, Climatic, and Topographic circumstances have resulted in a diverse range of ecosystems and habitats in India, including forests, grasslands, wetlands, coastal and marine ecosystems, and deserts. The fact that India has 10 biogeographic zones (Fig. 16.2), two realms (vast geographical areas within which ecosystems share a roughly similar biota), five biomes, and 26 biotic provinces (Table 16.1) demonstrates its biological abundance.

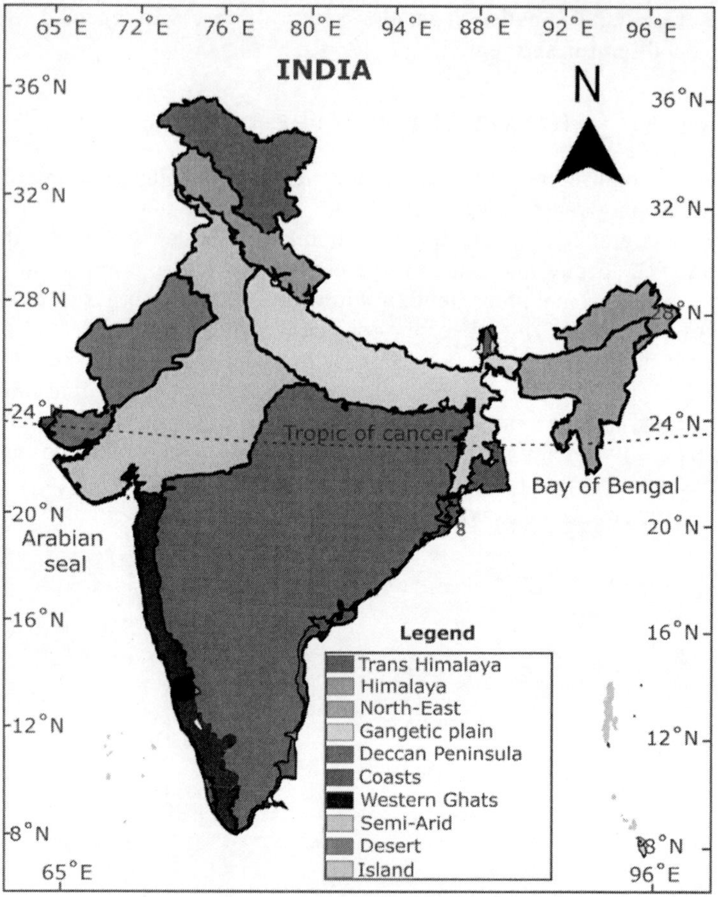

Fig. 16.2: GIS map showing the Bio-geographic zones in India

Table 16.1: Bio geographic zone's biological characterization

No.	Bio geographic zone (ZSI, 1997)	Biotic province	Total Area (km^2)
1.	Trans- Himalaya	Upper Region	186,200
2.	Himalaya	North-West Himalaya	69,000
		West Himalaya	72,000
		Central Himalaya	123,000
		East Himalaya	83,000
3.	Gangetic Plain	Upper Gangetic Plain	206,400
		Lower Gangetic Plain	153,000
4.	Indian Desert	Kutch	45,000
		Thar	180,000
		Ladakh (cold)	----
5.	The Semi- Arid	Central India	107,600
		Gujarat-Rajwara	400,400
6.	Deccan Peninsula	Deccan Plateau south	378,000
		Central Plateau	341,000
		Eastern Plateau	198,000
		Chotta Nagpur	217,000
		Central Highlands	287,000
7.	Western Ghats	Malabar Coast	59,700
		Western Ghats Mountains	99,300
8.	North-East India	Brahmaputra valley	65,200
		North-Eastern Hills	106,200
9.	The Islands	Andaman Islands	6,397
		Nicobar Islands	1,930
		Lakshadweep Islands	180
10.	Coasts	West Coast	6500
		East Coast	6500

a) Trans-Himalayas: A continuation of the Tibetan plateau with high-altitude cold desert in Laddakh (J&K) and LahaulSpiti (H.P), accounting for 5.7 percent of the country's area.
b) Himalayas: The whole mountain chain stretching from north-western to north-eastern India, encompassing a diverse range of biotic provinces and biomes and accounting for 7.2 percent of the country's area.

c) Desert: The severely dry region west of the Aravalli hill range, which includes both Gujarat's saltwater desert and Rajasthan's sand desert. This makes up 6.9 percent of the country's total land area.
d) Semi-arid: The area between the desert and the Deccan plateau, which includes the Aravalli mountain range. This accounts for 15.6 percent of the total land area of the nation.
e) The Western Ghats: The hill ranges and plains that extend along the western coast, south of the Tapti River, cover a wide variety of biotic provinces and biomes. This covers 5.8 percent of the country's total land area.
f) The Deccan Peninsula: The most extensive of the zones, spanning much of the southern and southcentral plateaus with deciduous vegetation. This accounts for 4.3 percent of the country's total land area.
g) The Gangetic plain: These plains, defined by the Ganges river system, are largely homogeneous. This makes up 11 percent of the country's total land area.
h) North-east India: North-eastern India's lowlands and non-Himalayan hill ranges, featuring a diverse spectrum of plants. 5.2 percent of the country's total land area
i) Islands: The biomes of the Andaman and Nicobar Islands in the Bay of Bengal are quite varied. This covers 0.03 percent of the country's total land area.
j) Coasts. A wide coastline that is scattered both west and east, with obvious contrasts between the two; the Lakshadeep islands are included, although their percentage area is minimal.

Vegetation of India

For a long time, human influences have had the greatest impact on this subcontinent. As a result, climax formations are significantly changed and destroyed for agricultural and other comparable uses. The current vegetation has suffered greatly as a result of the influences of plants, animals, soil, climate, and man. As a result, the vegetation of the country that we see around us is greatly impacted. Rainfall, temperature, biotic effects, and life forms are the most essential aspects in vegetation classification. In India, the two most prevalent plant forms are forest and grassland.

a) Forest Vegetation: Indian forests are traditionally categorised into four primary kinds based on temperature:

1) Tropical forests: Common throughout the warmer plains, ranging from thick, multistory forests of various trees, shrubs, and lianas in high rainfall areas to dry, scrub jungles of prickly plants in isolated patches in dry places. As a result, they are classified into two types: moist tropical forests and dry tropical forests.
2) Montane subtropical forests: These are found in the hills of south India, such as Nilgiri, Mahabaleshwar, and Pachmarhi, at elevations ranging from 3,000 to 5,600 feet. These are colder than tropical woods but warmer than temperate woodlands. In the south, common trees include Eugenia, Actinodaphne, Canthium, Mangifcra, and Ficus, while climbers include Piper trichostachyon, Gnetum scandens, Smilax macrophylla, and others. The trees in the northern sections are rather tall. Bamboos, various epiphytes, including orchids and ferns, become plentiful in the eastern Himalayas as humidity levels rise. The majority of the trees are evergreen.
3) Temperate forests: They are found over 5,300 feet in elevation, primarily on Himalayan and Nilgiri ranges. Oaks and conifers abound throughout the Himalayas. Their geographical distribution has already been described. Oaks grow in reasonably stable pure evergreen stands. The sholas at Ootacamund, Nilgiri hills in Tamilnadu, are prominent examples of southern temperate vegetation. The woodlands are densely forested, with considerable development of grasses and tall evergreen trees such as Balanocarpus utilis, Hopea parviflora, Artocarpus hirsuta, Salmaliamalbaricum, and others. Mosses, various woody climbers, ferns, and other epiphytes cover their branches.
4) Alpine forests: They are sometimes classified as sub-alpine forests, alpine scrubs, moist alpine scrubs, and dry alpine scrubs. They are widespread in the Himalayas over 11,000 feet. With rising altitude, the tree height decreases, eventually being replaced by sparse development of tiny plants such as Sedum, Primula, Saxifraga, and lichens, among others. The description of the Himalayas already includes information on the vegetation.

b) Grassland: Natural grasslands are few in India, and the majority of them are kept in their current seral phases owing to biotic effects. Thus, grasslands are not climax formations, but have emerged as a result of forest removal. Edaphic and biotic variables are the two key elements that resulted in their secondary development. India's grasslands are classified into three types:

1) Xerophilous: These are found in semi-desert areas of north-western India.
2) Mesophilous: These wide grass flats, sometimes known as savannahs, are typical of Uttar Pradesh's wet deciduous forests.
3) Hygrophilous: These are also known as wet Savannahs. All of these are influenced by biotic factors.

Whyte et al. (1954) categorised Indian grasslands into eight primary categories, which are represented in Table 16.2.

Table 16.2: Showing grassland type in India

Grassland Type	Environmental conditions	Distribution
SehimaDichanthium	Black soils	Maharashtra, Madhya Pradesh, south-west Uttar Pradesh and parts of Tamil Nadu and Karnataka
Dichanthium Cenchrus	Sandy loam soils	Plains of Punjab, Haryana, Delhi, Rajasthan, Saurashtra, East Madhya Pradesh, Coastal Maharashtra and Tamil Nadu
Phragmites Saccharum	Marshy areas	Tarai areas of North Uttar Pradesh, Bihar, Bengal, Assam and Tamil Nadu
Bothriochloa	Paddy tracts with heavy rainfall	Lonavala tracts of Maharashtra
Cymbopogon	Low hills	Western Ghats, Vindhyas, Satpura, Aravalli, Chota Nagpur
Arundinella	High hills	Western Ghats, Nilgiris, Lower Himalayas (Assam to Kashmir)

DeyeuxiaArundinella	Temperate climate	Upper Himalayas between 6,200 ft to 10,000 ft from Assam, Bengal, through Uttar Pradesh to Punjab and Himachal Pradesh
DeschampsiaDeyeuxia	Temperate alpine with thin soils	Himalayas (above 8,300 ft), Kashmir

Biodiversity Hotspots

Biodiversity hotspots are characterised as areas "with unusual densities of endemic species suffering extreme habitat degradation." Biodiversity hotspots are areas with a high species richness and endemism. Norman Myers introduced the notion of biodiversity hotspots in 1988 after seeing that the tropical forest was losing plant species as well as habitat. Biodiversity hotspots are designated in 36 locations throughout the world. These hotspots account for only 2.3 percent of the total land surface, yet they are home to more than half of the world's plant species and 42 percent of all terrestrial animal species.

Criteria to qualify for Biodiversity Hotspot

To qualify as a hotspot, a location must meet the following two requirements, according to Conservation International:

a) It must have at least 1,500 endemic vascular plants, which means a high percentage of plant life found nowhere else on the earth. In other words, a hotspot is irreplaceable.
b) It must retain 30 percent or less of its original natural vegetation (i.e., it has to have lost at least 70 percent of its original habitat). To put it another way, it must be endangered.

Indian Hotspots

India has four major biodiversity hotspots:

a) Himalaya: The entire Himalayan area of India (and that falling in Pakistan, Tibet, Nepal, Bhutan, China, and Myanmar).
b) Indo-Burma: Includes all of northern India excluding Assam and the Andaman & Nicobar Islands (and Myanmar, Thailand, Vietnam, Laos, Cambodia, and southern China).

c) Western Ghats and Sri Lanka: The whole Western Ghats are included (and Sri Lanka).
d) Sundalands: Includes the Nicobar Islands (and Indonesia, Malaysia, Singapore, Brunei, Philippines).

Hope Spots and Cool Spots

A Hope Spot is an area of the ocean that deserves special preservation due to the presence of species and important underwater ecosystems. The IUCN and Mission Blue, an ocean research group, have selected the Andaman and Nicobar Islands and Lakshadweep Islands as new "hope spots."Andaman and Nicobar Island are home to some of the world's most unusual bird and plant species. In the case of Lakshadweep, the coral reefs are extremely vulnerable to the ocean environment and must be protected.

"Cool-spots" are the world's final refuges, where a large number of vulnerable species survive. Cool-spots may exist as a consequence of protection or as a result of intact habitat that has not yet been destroyed. Higher levels of protection, conservation management, and restoration would benefit all cold areas.

Fauna

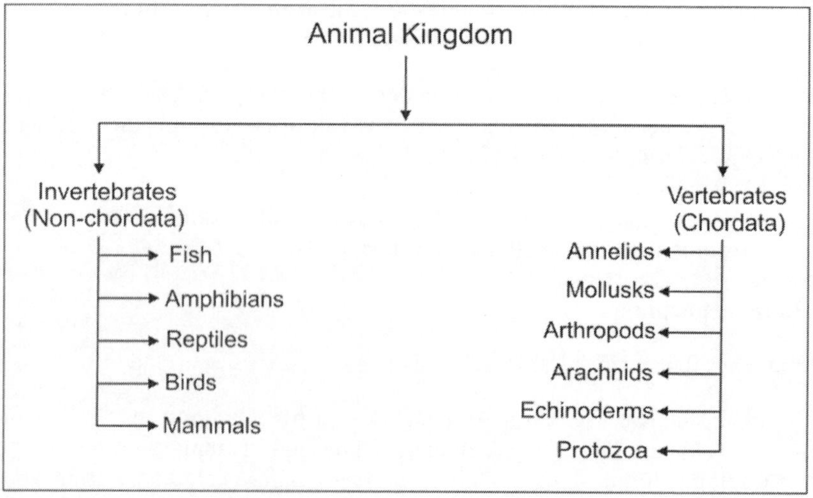

Fig. 16.3: Animal kingdom classification

The Animal Kingdom is classified into:

a) Vertebrates: Vertebrates are creatures that have backbones and spinal cords. Vertebrates are the most evolved creatures on the planet. Although vertebrates make up a very small proportion of all species, their size and mobility frequently allow them to dominate their surroundings.
   1) Fishes:
      a. Cold blooded
      b. Live in water
      c. Breathe under water using gills, not lungs
      d. Have scales and fins
      e. Lay eggs
   2) Amphibians:
      a. Cold blooded
      b. Live on land & water
      c. Webbed feet
      d. Breathe with lungs and gills
      e. Moist smooth skin
      f. Four legs (sometimes none) Lay eggs
   3) Reptiles:
      a. Cold blooded
      b. Have scales
      c. Have dry skin
      d. Usually lay eggs
      e. Ear holes instead of ears
      f. 4 legs or no legs
   4) Aves:
      a. Warm blooded
      b. Have feathers and wings
      c. Lay eggs
      d. Have 2 legs
      e. Ear holes instead of ears
   5) Mammals:
      a. Warm blooded
      b. Have hair or fur
      c. Give birth to live young ones
      d. Mammal mother nurse their young one with milk
      e. Breathe with lungs
      f. Mammals live on land have 4 legs (or 2 legs & 2 arms), and ears that stick out.

b) Invertebrates: Invertebrates lack a backbone. Invertebrates account for more than 98 percent of all animal species on the planet. Invertebrates lack a bone-based internal skeleton. Many invertebrates, such as jellyfish and worms, have a fluid-filled, hydrostatic skeleton. Others, such as insects and crustaceans, have a hard outer shell.
   1) Annelids: The bodies of annelids are separated into segments. Their interior organs are well developed. It can be found practically anywhere on the planet. They don't have any limbs. For example, earthworms, leeches, roundworms, and so on.
   2) Mollusks: Most mollusks have a soft, skin-like organ surrounded by a hard shell. Some mollusks, such as the snail and slug, live on land. Other water-dwelling mollusks include the oyster, mussel, clam, squid, and octopus.
   3) Echinoderms: Echinoderms are sea creatures that dwell in the ocean. The majority of echinoderms have arms or spines that radiate from their centre of mass. The sea star, sea urchin, sand dollar, and sea cucumber are all examples of echinoderms.
   4) Protozoa: Protozoa are single-celled creatures. They are the tiniest of all creatures. Most protozoa are tiny. They do, in fact, breathe, move, and reproduce like multicelled creatures. Amoeba's, flagellates, and so on.
   5) Arthropods: Arthropods have joints in their limbs that allow them to move. In addition, they have an exoskeleton, which is a hard, exterior skeleton. Arthropods include crustaceans, insects, and arachnids.

Flora

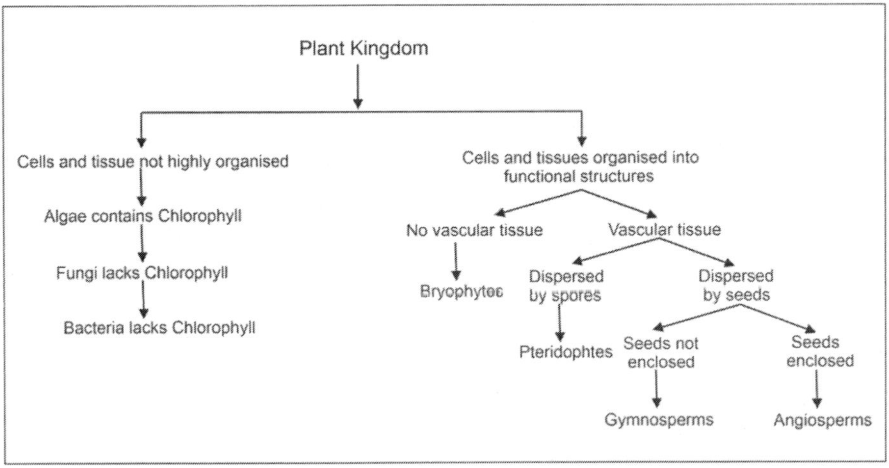

Fig. 16.4: Plant kingdom classification

The following are the descriptions of important flower groupings found in India:

a) Algae: Algae are green non-differentiated plants (non-differentiated into organs such as root, stem, and leaf) that contain chlorophyll. They often develop in water or damp environments. Freshwater algae are green or blue-green in colour, whereas marine algae are red or brown. These are autotrophic plants, which can produce their own nourishment.
b) Fungi: Fungi are non-green, non-differentiated plants that lack chlorophyll completely. They either develop as saprophytes on dead, decaying organic materials or as parasites on other living organisms known as hosts. Saprophytic fungi are most commonly associated with moulds and mushrooms. The Western Ghats have the most fungal variety, followed by the eastern Himalaya and the western Himalaya.
c) Bacteria: Non-chlorophyllous microorganisms that live in a saprophytic or parasitic environment. Many of them are harmful; saprophytic bacteria, on the other hand, are helpful. They are soil-borne, and many are used in industry.
d) Lichens: A lichen is a strange hybrid of an alga and a fungus, with the two living in harmony. They are a type of greyish green plant

that grows on rocks, tree trunks, rotting wood, and other surfaces. Algae produces food that the fungus can use, and also absorbs and retains water, keeping the algal cells wet. A symbiotic relationship. They are most frequent in wetlands, uncommon in rivers and streams, and absent from ground water.

e) Bryophytes: The plant body is divided into a short stalk and simple leaves, but there are no actual roots. They thrive in wet environments. For example, liverworts and mosses. They are India's second biggest group of green plants, mostly found in the Eastern Himalaya, North-eastern India, Western Himalaya, and the Western Ghats. Mosses are the most abundant component of Indian bryoflora, followed by liverworts and hornworts.

f) Pteridophytes: Pteridophytes contain distinct plant bodies made up of roots, stems, and leaves. Furthermore, they have vascular bundles. The majority of them are terrestrial plants that thrive in wet, shaded environments, while others are aquatic. This category contains vascular cryptogams such as clubmosses, horsetails, and ferns, which are found all over the world. The north-eastern area (containing the Eastern Himalaya) has the greatest diversity of pteridophytes, followed by south India (including the Eastern and Western Ghats) and north India (including Western Himalaya).

g) Gymnosperms: Gymnosperms (gymnos=naked, sperma=seed) are plants with bare seeds.

   Flowers are exceedingly basic, with no supplementary whorls, and stamens and carpels stay aggregated in cones. Ovules are found on the surface of the carpels and are pollinated directly by pollen grains. There is nothing resembling an ovary, style, or stigma, and there is, of course, no fruit. Cycas, Pinus, and Gnetum, for example. The pine family is the biggest. Gnetum and Cycas are found mostly in the North Eastern area, the Eastern and Western Ghats, and the Andaman and Nicobar Islands.

h) Angiosperms: Closed-seeded plants are known as angiosperms. These are the most developed plants, bearing blooms with prominent accessory and essential whorls. The ovary, style, and stigma are all found in carpels. The ovary normally develops into fruit and the ovules into seeds in response to fertilisation. As a result, the seeds stay within the fruits. India is home to more than 7 percent of the world's blooming flora.

Natural World Heritage Sites in India

a) Great Himalaya National Park Conservation Area (2014): This park is widely renowned for its high alpine peaks, alpine meadows, and riverine forests, and is located in the western section of the Himalayan highlands in the state of Himachal Pradesh. It also encompasses the catchment region as well as the glacial and snow meltwater sources of multiple rivers. It is a biodiversity hotspot, with 25 different types of woods home to a diverse range of faunal species, some of which are endangered.

b) Kaziranga National Park (1985): Kaziranga National Park is located in the districts of Golaghat and Nagaon in the Indian state of Assam. The park is a UNESCO World Heritage Site because it is home to two-thirds of the world's giant one-horned rhinoceroses. The rhino population in Kaziranga National Park is 2,413, according to a census done in March 2018 by the Forest Department of the Government of Assam and certain recognised wildlife NGOs. The rhino population was 2401 in 2015. In 2006, Kaziranga National Park was designated as a Tiger Reserve. Elephants, wild water buffalo, and swamp deer all have high breeding numbers in the park. On February 11, 1974, the federal government granted the 430-square-kilometer park official status. For its unique natural habitat, UNESCO designated Kaziranga as a World Heritage Site in 1985.

c) Keoladeo National Park (1985): This wetland, located in Rajasthan, was used as a duck hunting reserve until the end of the nineteenth century. The hunting, however, came to an end soon after, and the region was designated as a national park in 1982. This National Park is home to 375 bird species as well as a variety of other animals. It also provides as a wintering site for migrating waterfowl from the Palaearctic, the critically endangered Siberian Crane, and the internationally vulnerable Greater Spotted Eagle and Imperial Eagle. It is well-known for its resident breeding population of non-migratory birds.

d) Manas Wildlife Sanctuary (1985): Assam's Manas Wildlife Sanctuary is a biodiversity hotspot. It is located near the Manas River and is part of the Manas Tiger Reserve. The site's spectacular beauty and tranquil surroundings are due to a variety of wooded hills, alluvial meadows, and tropical evergreen woods. It also supports a variety of endangered species, including the tiger, larger one-horned rhino, swamp deer, pygmy hog, and Bengal florican.

e) Nanda Devi National Park (1988) and Valley of Flowers in 2005 as an extension of Nanda Devi: Both of these National Parks are high-altitude West Himalayan landscapes with extraordinary beauty, and they are located inside the state of Uttarakhand. The rough and high-mountain environment of Nanda Devi National Park is dominated by India's second-highest mountain, the peak of Nanda Devi. The Valley of Flowers, on the other hand, boasts aesthetically appealing alpine flower meadows. These parks are home to a diverse range of floral and faunal species, as well as a major population of internationally vulnerable species like as the Snow Leopard and Himalayan Musk Deer.
f) Sundarban National Park (1987): The Sundarbans National Park in West Bengal, India, is a national park, tiger reserve, and biosphere reserve. It's part of the Sundarbans in the Ganges Delta, and it's close to Bangladesh's Sundarban Reserve Forest. It is located in Bangladesh's south-west corner. The delta is thickly forested, and it is one of the Bengal tiger's most important habitats. A variety of bird, reptile, and invertebrate species, including the salt-water crocodile, call it home. It was designated as a national park on May 4, 1984. It was designated as a UNESCO World Heritage Site in 1987, and it has been a Ramsar site since 2019. Since 1989, it has been designated as a World Network of Biosphere Reserves (Man and Biosphere Reserve).
g) Western Ghats (2012): The Western Ghats are a series of mountains that extend parallel to India's Western Coast, spanning through Kerala, Maharashtra, Goa, Gujarat, Tamil Nadu, and Karnataka. They span over 1600 kilometres and are only broken once by a 30kilometer Palghat gap at roughly 11 degrees North. They also have an impact on the Indian monsoon weather patterns, which help to moderate the region's warm tropical temperature and act as a barrier to rainy monsoon winds from the southwest. Tropical evergreen woods, as well as 325 internationally endangered species, may be found in the Western Ghats.

Biosphere Reserves

UNESCO established the Biosphere Reserve Programme in 1971 as part of its "Man and the Biosphere" Programme (MAB). Biosphere Reserve (BR) is an international designation by UNESCO for representative elements of natural and cultural landscapes spanning extensive areas of terrestrial or coastal/marine ecosystems, or a mix of the two. Biosphere

Reserves are exceptional places for both people and nature, and they are live examples of how humans and nature can coexist while respecting each other's needs. The goal of establishing a biosphere reserve was to preserve all kinds of life, as well as their supporting ecosystems, in their entirety. The Biosphere Reserve Programme (BRP) has the following objectives:

a) The goal is to save ecological samples that are typical of the entire ecosystem.
b) To ensure long-term in-situ genetic diversity conservation.
c) To encourage and make fundamental and applied research and monitoring more accessible.
d) To give educational and training opportunitics.
e) To encourage prudent, long-term stewardship of living resources.
f) Disseminate the knowledge gained in order to foster sustainable growth in other areas.
g) To encourage the implementation of the World Conservation Strategy. In 1983, the inaugural Biosphere Reserve Congress was convened in Minsk, Belarus, and 226 biosphere reserves from 62 nations across the world were recorded. In 1985, there were 243 biosphere reserves spread over 65 nations, encompassing over 115 million hectares. Only 90 of 193 geographical provinces (belonging to eight kingdoms) possessed biosphere reserves, with the remaining 103 awaiting inclusion in BRP. In 1995, the Second Congress was convened in Seville, and 324 biosphere reserves from 82 nations were reported. Currently, there are over 500 biosphere reserves distributed throughout approximately 100 nations.

The world's first biosphere reserve was founded in 1979. Since then, the network of biosphere reserves has grown to around 500, distributed across about 100 nations. The first biosphere reserve in India was established in 1986. This was the Nilgiri biosphere reserve, which spanned 5520 kilometres. The state governments of Karnataka, Kerala, and Tamil Nadu were given monies to undertake various projects for the Reserve. The Nanda Devi Biosphere Reserve, encompassing an area of 1560 square kilometres, was formally proclaimed on January 18, 1988. The third reserve, Nokrek (Thura Range), was established in 1988, and the fourth, Great Nicobar, in 1989. The relevant state governments additionally agreed to the planned biosphere reserves at Sunderbans (West Bengal), Gulf of Mannar (Tamil Nadu), and Manas (Assam). The following 18 biosphere reserves were established in the nation till August of 2011, as stated in the Table 16.3 below:

Table 16.3: Showing the biosphere reserves in India

Name	State / union Territory	Date of establishment	Area (km^2)
Nilgiri	Tamil Nadu. Kerala and Karnataka	1, Aug, 1986	5520 (Core 1240; Buffer 4280)
Nanda Devi	Uttarakhand	18, Jan, 1988	5860.69 (Core 712.12; Buffer 5,148.570; Transition 546.34)
Nokrek	Meghalaya	01, Sept, 1988	82 (Core 47.48; Buffer 34.52)
Manas	Assam	14, Mar, 1989	2837 (Core 391; Buffer 2,446)
Sunderbans	West Bengal	29, Mar, 1989	9630 (Core 1700; Buffer 7900)
Gulf of Mannar	Tamil Nadu	18, Feb, 1989	10,500
Great Nicobar	Andaman and Nicobar Island	06, Jan, 1989	885 (Core 705; Buffer 180)
Simlipal	Orissa	21, Jun, 1994	4374 (Core 845; Buffer 2129; Transition 1400)
Dibru-Saikhowa	Assam	28, July, 1997	765 (Core 340; Buffer 425)
Dehang-Debang	Andhra Pradesh	02, Sept, 1998	5111.50 (Core 4094.80; Buffer 1016.70)
Pachmarhi	Madhya Pradesh	03, Mar, 1999	4926
Khangchendzonga	Sikkim	07, Feb, 2000	2619.92 (Core 1819.34; Buffer 835.92)
Agasthiayamalai	Tamil Nadu and Kerala	12, Nov, 2001	1701
Achanakamar-Amarkantak	Madhya Pradesh and Chattisgarh	2005	3835.51 (Core 551.55; Buffer 3283.86)
Kachchh	Gujarat	29, Jan, 2008	12,454
Cold Desert	Himachal Pradesh	28, Aug, 2009	7770
Seshachalam	Andhra Pradesh	29, Sept, 2010	4755.997
Panna	Madhya Pradesh	25, Aug, 2011	2998.98

National Biosphere Reserve Programme

In 1986, the national Biosphere Reserve Programme was established. To enable local residents' participation in effective management and to propose methods of enhancing local residents' livelihoods via sustainable use.

Objectives

The following are the goals of the Biosphere Reserve programme, as defined by the Core Group of Experts:

a) To preserve the diversity and integrity of plants and animals in natural habitats;
b) To protect the genetic variety of species, which is essential for their continued evolution;
c) To provide spaces for a variety of study and monitoring;
d) To provide educational and training facilities; and
e) To ensure the sustainable use of natural resources with the most suitable technology in order to increase the local people's economic well-being.

Criteria for Selection of Biosphere Reserves

The criteria for selecting Biosphere Reserves sites established by the Core Group of Experts in 1979 are as follows:

a) Primary Criteria: A site that must have a successfully protected and little disturbed core area of conservation importance, as well as adjacent land and water appropriate for study and demonstration of sustainable research and management practises. The core region should be representative of a biogeographical unit and large enough to support sustainable populations from all tropic levels in the ecosystem.
b) Secondary Criteria: This includes rare and endangered species habitats Areas with a variety of soil and microclimatic conditions, as well as indigenous biota. Areas with the ability to preserve traditional tribal or rural patterns of existence for the harmonious use of the environment.

Structure of Biosphere Reserves

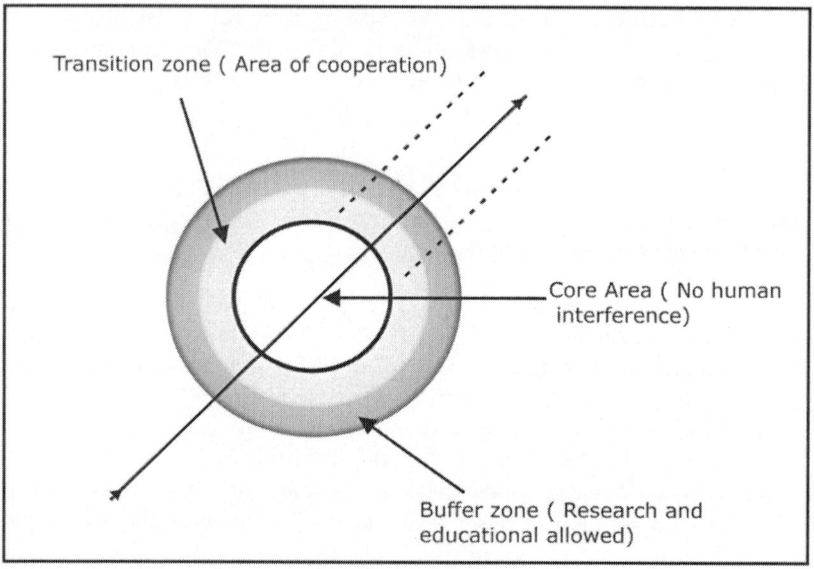

Fig. 16.5: Diagram showing the structure of Biosphere reserve

Biosphere Reserves are divided into three interconnected zones in order to perform complementary efforts of bio-diversity conservation and development of sustainable management aspects (Fig. 16.5).

a) The Core zone: The core zone should be left alone at all times. It must provide sufficient habitat for a diverse range of plant and animal species, including higher order predators, and may contain endemism centres. A core zone provides legal protection and allows for management and research activities that do not interfere with natural processes or animals. Core regions of BR are strict natural reserves and wilderness areas in the area. All human pressures external to the system must be kept out of the core zone.
b) The Buffer zone: The buffer zone is next to or surrounds the core zone. Its usage and activities are governed in such a way that the core zone is protected. These uses and activities include restoration, demonstration sites for increasing the value of the resources, restricted recreation, tourism, fishing, and grazing, all of which are authorised to decrease their impact on the core zone. It is necessary to encourage research and educational initiatives. Human activities,

if they are natural within BR, are likely to be allowed to continue as long as they do not adversely harm biological variety.

c) The Transition zone: The Transition Zone is the furthest reaches of a Biosphere Reserve. This is a zone of collaboration where conservation, knowledge, and management skills are used and uses are managed in accordance with the goal of the Biosphere Reserve. This comprises communities, farming lands, controlled woods, and areas for intense recreation, as well as other ecological uses unique to the region.

CBD Implementation Process in India

Two significant advances in India have been the drafting of the Biodiversity Act and the compilation of a national status report, both of which India failed to present during the COP summit in Bratislava. The proposed legislation is titled "Biological Diversity Act, 1998." The Act proposes the formation of a National Biodiversity Authority, which will include a chairperson, representatives from the Ministry of Environment and Forests, the Department of Agricultural Research and Education, the Department of Biotechnology, the Department of Ocean Development, five non-official members who are experts and scientists familiar with biodiversity issues and CBD provisions, and representatives from Indian industry.

Functionally, the Authority will examine issues of conservation, intellectual property rights, and traditional knowledge, and will provide standards for access and benefit-sharing. State/Union Territory Biodiversity Boards and Biodiversity Management Committees in Local Bodies must be established in order for the Act to be implemented. These entities will advise the appropriate people on biodiversity concerns and assist in the implementation of the Act. Other key provisions of the Act include:

a) Non-Indians are not permitted to transmit biological material or associated information without the prior consent of the National Authority.
b) The requirement that anybody seeking a patent or any other type of IPR have prior approval from the Authority.
c) Fees and royalties may be levied on such applications.
d) Provision for honouring and safeguarding local people's knowledge by measures such as registration of such knowledge at the local, state, and national levels, as well as the construction and

adjustment of the sui-generis system for intellectual property protection of such knowledge.

e) Provision for ensuring that people who have prior informed permission will abide by the terms and conditions that will ensure a fair and equitable benefit-sharing mechanism. These procedures will be made public.
f) State Boards' recommendations for creating Biological Diversity Heritage Sites in conjunction with local units of self-government.
g) Proposes the establishment of National, State, and Local Biodiversity Funds to be funded by fees, royalties, and monetary advantages arising from National Authority approvals.
h) Those who violate the terms of the Act face heavy penalties, including a fine of up to Rs. 10 lakhs or imprisonment for up to 5 years.

Coverage of Biodiversity conservation and use in India

Each country must consider the legal protection and use of biodiversity inside its borders. A similar evaluation has been conducted for India (Kothari and Singh 1992). This examination first gives a conceptual framework within which to study each legislation, and it addresses not just traditional conservation laws, but also laws pertaining to resource development, use, and distribution. Kothari and Singh (1992) classified biodiversity components into five categories for the purposes of the analysis: wild fauna, wild flora, domesticated fauna (livestock), domesticated flora (crops and other cultivated plants), and genetic materials (seeds, eggs, germplasm, semen). They also identified 11 different characteristics of biodiversity:

a) Identification: The process of finding biodiversity components and then determining their identification, distribution, occurrence, status, and value. Both wild and domesticated wildlife and plants are affected.
b) Protection in-situ: Efforts to safeguard and conserve biodiversity, particularly the ecosystems where biodiversity components live and evolve. This applies particularly to wild animals and flora, but also to domesticated fauna and flora in their usage locations, where they may need to be safeguarded from disease/infestation, displacement by new species, and damage from human activity.
c) Protection ex-situ: Attempts to preserve living species or genetic material in gene banks, zoos, botanical gardens, and other locations

other than their native environments. All biodiversity components are affected.

d) Extraction: Controls on the kind and amount of access to wild animals and flora, ex-situ collections, and genetic material; as well as the ways of extraction of these components, particularly for in-situ stock.

e) Use: Efforts to use biodiversity for subsistence, economic, scientific, or other reasons. All biological components are affected.

f) Breeding: Artificial or induced methods of renewing flora-fauna communities under captivity or confinement. Both wild and domestic wildlife and plants are affected.

g) Trade: Activities including the barter, sale, import, export, and other types of biodiversity component trade.

h) Augmentation: Measures to intentionally fill an ecosystem or region with species that did not previously exist there (introduction), where they have dropped or must be increased in number (augmentation), or where they have died out (reintroduction). Relevant to both wild and domesticated wildlife and plants, as well as genetic material employed for the goal (e.g., seeds).

i) Release: Disposal of biodiversity components without the particular goal of introducing or increasing them into the ecosystem. All components of biodiversity are affected.

j) Movement: Measures to transmit biodiversity components by human means from one area to another, particularly in circumstances of transit across ecosystems or bioregions, when components may be accidentally released. Natural or voluntary movement of animals, as well as natural movement of seeds or other floral elements, are not included. All components of biodiversity are affected.

k) Intellectual property rights: Measures connected to state, organisation, and individual rights over biological and biotechnological information, such as patents, royalties, intellectual property rights, farmers' rights, and breeders' rights.

CHAPTER 17

WILDLIFE AND WILDLIFE MANAGEMENT

Wildlife Protection Act, 1972

The Wildlife (Protection) Act 1972, adopted by the Indian Parliament in 1972, provides for the conservation and protection of the country's wildlife (flora and fauna). India is the first country in the world to have environmental protection provisions in its constitution. The topic of the environment was first raised as a problem during the United Nations Conference on Human Environment in Stockholm on June 5, 1972, and since then, June 5th has been observed as World Environment Day. Several acts were passed shortly after, including the Wildlife (Protection) Act of 1972, the Water (Prevention and Control of Pollution) Act of 1974, the Forest (Conservation) Act of 1980, the Air (Prevention and Control of Pollution) Act of 1981, and the Environment (Protection) Act of 1986. The Wild Life Protection Act, 1972 is an Act of the Indian Parliament created to conserve plant and animal species. Prior to 1972, India only had five recognised national parks. It has been fractionate into six schedules with varying levels of protection. Schedule I and Part II of Schedule II guarantee absolute protection; crimes committed under them are subject to the harshest punishments. Schedule III and IV species are likewise protected, although the penalties are substantially lesser. Schedule V lists the animals that can be hunted. Cultivation and planting of the indigenous plants listed in Schedule VI is forbidden. The Wildlife Protection Act has six schedules. They are discussed further below:

a) Schedule I: Threatened species are covered under this Schedule. These species need to be protected rigorously, hence this Schedule lists the worst penalties for breaking the law. Except in cases where they endanger human life, hunting is prohibited in India for the species listed in this Schedule. These species are completely protected. These animals cannot be traded. Examples of animals include the tiger, blackbuck, Himalayan brown bear, brow-antlered deer, blue whale, common dolphin, cheetah, clouded leopard, hornbills, and Indian gazelle.

b) Schedule II: Animals on this list are also given special protection. Their business is illegal. They cannot be hunted unless there is a threat to human life. Examples include the insect Kohinoor, the Assamese Macaque, the Bengal Hanuman Langur, the Large Indian Civet, the Indian Fox, the Larger Kashmir Flying Squirrel, the Kashmir Fox, and other species.
c) Schedule III and IV: Only species that are not endangered are included on this list. This also applies to protected species, however the penalties for violating them are less severe than in the first two schedules. Hunting is prohibited. Examples given here include the hyena, Himalayan rat, porcupine, flying fox, Malabar tree toad, and others.
d) Schedule V: This schedule includes huntable animals (vermin). Mice, rats, common crows, fruit bats, and other animals are examples.
e) Schedule VI: Cultivation and planting of the indigenous plants listed in Schedule VI is forbidden. Pitcher plant, blue vanda, red vanda, kuth, and more examples

Salient Features of the Act

a) The Act establishes wildlife advisory boards and wildlife wardens, as well as their authorities and tasks.
b) It made it easier for India to ratify the Convention on International Trade in Endangered Species of Wild Fauna and Flora (CITES). A global agreement called CITES aims to preserve plants and animals that are in risk of extinction. The Washington Convention is another name for it; it was accepted as a result of an assembly of IUCN members.
c) The country's endangered species were thoroughly inventoried for the first time.
d) The Act made it illegal to hunt endangered species.
e) The terms of the Act make it illegal to trade scheduled animals.
f) The Act authorises the sale, transfer, and ownership of certain animal species in exchange for a licence.
g) It calls for the creation of wildlife refuges, national parks, and other protected areas.
h) Its regulations laid the groundwork for the establishment of the Central Zoo Authority. This is the central organisation in charge of zoo monitoring in India. It was founded in 1992.
i) The Act included six schedules that offered various levels of protection to various flora and fauna groupings. The harshest

penalties are applied to offences in Schedules I and II (Part II), which are completely protected. The schedules also provide a list of the species that may be hunted.

j) Under the terms of this Act, the National Board for Wildlife was established as a statutory institution. The Prime Minister presides over it. It is composed of 47 members, including the Prime Minister. There are 19 ex-officio members among these. Other members include three Members of Parliament (two from the Lok Sabha and one from the Rajya Sabha), five non-governmental organisations (NGOs), and ten notable ecologists, conservationists, and environmentalists.
k) The Act also established the National Tiger Conservation Authority. It is a statutory entity of the Ministry of Environment, Forests, and Climate Change with overall supervision and coordinating responsibilities, as defined by the Act. Its mandate is to improve tiger conservation in India. It grants formal legitimacy to Project Tiger, which was begun in 1973 and has set the endangered tiger on a guaranteed road to resurrection by safeguarding it from extinction.

Wildlife Conservation in India

It combines ex-situ and in-situ conservation, which are complementary to one another. Due to the constant growth in the number of endangered species of flora and fauna, initiatives have been taken to safeguard and manage the country's wildlife. To conserve wildlife, non-governmental volunteer organisations (NGOs) as well as governmental organisations at the state and federal levels have been established. Organizationally, forestry and wildlife have been generally under the jurisdiction of state governments. However, the issue was recently elevated to the top of the priority list, and a separate Ministry of Environment and Forests was established to oversee environmental protection. The Department of Environment, Forests, and Wildlife was established within this Ministry in order to facilitate cooperation between states and the centre, as well as the prompt and faithful implementation of the actions to be followed in the country's wildlife management programme.

The following goals are pursued by wildlife management:

a) Natural habitat conservation through restricted, limited species exploitation.

b) Conservation of a viable number of species in protected regions (national park, sanctuary, biosphere reserve etc.).
c) Establishment of Biosphere Reserves for plant and animal species.
d) Legislative protection.

Wildlife can also be protected by doing the following:

a) Enhancing existing protected areas such as sanctuaries, national parks, and so on.
b) Putting export restrictions on rare plant and animal species and their goods.
c) Educating public for environmental protection at all levels of education.

Non-Governmental Organizations

There are several non-governmental, volunteer, national, and worldwide organisations committed to wildlife conservation. The primary organisations are:

a) The Bombay Natural History Society: It was established in 1883 to collect information and specimens of animals and plants from India, Burma (Myanmar) and Ceylon.
b) The Wildlife Preservation Society of India, Dehradun: It was established in 1958 with numerous wildlife management goals in mind.
c) The World Wide Fund for Nature, India: The World Wildlife Fund, Indian National Appeal was established in India in 1969, during the XIIth General Assembly of the International Union for Conservation of Nature and Natural Resources, which was held in Delhi. WWF-International was founded in 1961, with headquarters in Glands, Switzerland, and is governed by an international board of Trustees. It has brought about National Appeals in a number of nations. The WWF in India was founded with an eight-trustee board and maintains its headquarters in Mumbai. It has backed the well-known "Project Tiger" and other similar initiatives.

Government Organizations

State and Union governments have passed a number of Wildlife Acts, and carried out a number of actions, to promote wildlife conservation. Among these are:

a) The Madras Wild Elephant Preservation Act, 1873.
b) The All India Elephant Preservation Act, 1879.
c) The Wild Birds and Animals Protection Act, 1912.
d) The Bengal Rhinoceros Preservation Act, 1932.
e) The Assam Rhinoceros Preservation Act, 1954.
f) The Indian Board for Wildlife (IBWL) 1952.
g) The Wildlife (Protection) Act, 1972, amended in 1983, 1986, 1991.
h) The establishment of National Parks, Sancturaries and Zoological Gardens.
i) In 1976, India became a party to CITES (Convention of International Trade in Endangered Species) (of Wild Flora and Fauna).
j) In 1972, the Indian National Man and the Biosphere Committee was established, for Biosphere Reserves.
k) Several projects to conserve individual endangered species like crocodiles (1954), lions (1972) and tigers (1973).
l) The National Wildlife Action Plan, 1982 produced by the IBWL.
m) The Forest (Conservation) Act, 1980, amended in 1988.
n) The Environment (Protection) Act, 1986.
o) The Biodiversity Act, 2002, implemented in 2004.
p) The National Environmental Policy, 2006.

Wildlife disease

Table 17.1: Showing details of various wildlife diseases

Diseases	**Description**	**Animals susceptible**
Anthrax	A highly contagious and lethal cattle illnesscaused by Bacillus anthracis, a rather big spore-forming rectangle bacteria.	This illness mostly affects hoofed animals such as deer, cattle, goats, and sheep.
Black quarter (black-leg)	It is a bacterial illness that is extremcly contagious and lethal to cattle. Clostridium chauvoei causes this bacterial illness. It is a soil-borne disease that usually appears during the rainy season. The sickness is sporadic (1-2 animals) in India.	Buffaloes, sheep, and goats are also vulnerable. Young cattle in good bodily condition, aged 6-24 months, are most impacted.
Foot-and-mouth disease	Foot-and-mouth disease is a highly contagious illness that affects cloven-footed animals. It causes fever, vesicles and blisters in the mouth, udder, teats, and on the area between the toes and above the hoofs. Animals that have recovered from the illness have a particularly rough coat and foot deformity.	The illness is common in India and plays an important role in the cattle business. The illness spreads either directly or indirectly through contaminated water, manure, hay, and pastures. Cattle attendants also transport it. It has been observed in recovered animals, including rats, porcupines, and birds.
Rabies (Mad dog disease)	If an animal with rabies bites another animal or a person, the illness spreads. Rabies germs are found in the saliva of a sick (rabid) animal. This is a lethal illness, although not every dog that bites is sick.	Rabies is a disease that affects dogs, foxes, wolves, hyaenas, and in some areas, bats that feed on blood.

Blue tongue	Bluetongue, a disease transmitted by midges, impacts domestic and wild ruminants as well as camelids. Virus transmission between animals occurs by Cullicoides species midges.	Sheep, on the other hand, are particularly vulnerable. Cattle, while infected more commonly than sheep, do not always exhibit illness symptoms.
Pox	It is an extremely contagious illness. It kills 20 to 50 percent of animals under the age of six months and damages the wool and skin of adults.	Sheep are most probably affected.
Johne's disease	The illness causes gradual emaciation and persistent diarrhoea and intestinal thickening in cattle and buffaloes.	Johne's disease is a kind of chronic infectious enteritis that affects cattle, sheep, goats, buffaloes, and, on rare occasions, pigs.
Mastitis	Microorganisms enter the teat canal and proliferate in the mammary gland, causing infections.	Dairy Cattle
Footrot	Foot rot is a major cause of lameness in cattle, and it most commonly develops when cattle on pasture are forced to travel through mud to get to water and feed.	Cattle
Bovine rhinotracheitis	Bovine herpesvirus-1 (bhv-1) is the culprit behind the extremely contagious, infectious respiratory illness known as infectious bovine rhinotracheitis (ibr).	Cattle
Theileriosis	Theileriases are tick-borne illnesses caused by theileria spp. Both theileria and babesia are members of the piroplasmorina suborder.	Cattle

East coast fever	East coast fever, a cattle acute illness, is characterised by high fever, enlargement of the lymph nodes, dyspnea, and a high death rate. It is a severe disease in eastern and southern Africa, caused by theileria parva and transmitted by the tick vector rhipicephalusappendiculatus.	Cattle
Ringworm	This is the most prevalent infectious skin disease in beef cattle. It is caused by a fungus and is contagious to humans. The condition typically manifests as crusty grey spots on the head and neck, particularly around the eyes.	
Milk fever	Milk fever, also known as parturient hypocalcaemia and parturient paresis, is a condition that has grown in prominence with the introduction of heavy milking cows.	
Calf scour	Scours can develop in calves as a result of bacterial or viral illnesses. Scours is often referred to as "calf scours" or "neonatal calf diarrhoea." Scours is caused by Rotavirus, Coronavirus, Cryptosporidium parvum, Salmonella, and Escherichia coli.	
PPR (goat plague)	Initially, the causal virus was assumed to be an abnormal strain of rinderpest virus that had lost its capacity to infect cattle. Later molecular research revealed that it was separate from rinderpest virus but closely connected to it.	Cattle
Trpanosomia	Trpanosomia Virus	Tiger, Elephant, Sambar, macaque

National parks and Sanctuaries

In 1981, India had 19 National Parks (NP) and 202 Sanctuaries covering 74,763 square kilometres, or roughly 2.3 percent of the entire geographical area. By 1983, there were 44 National Parks and 207 Sanctuaries totalling 87,735 square kilometres, accounting for 2.7 percent of the country's land area. The 207 Sanctuaries comprised an area of 69,698 square kilometres, whereas the 44 National Parks covered an area of 18,037 square kilometres. There have been more additions, and as of June 1988, there were 66 National Parks and 368 Sanctuaries with a total size of approximately 1,41,298 square kilometres, representing around 4 percent of the country's geographic area. There were 73 National Parks and 416 Sanctuaries as of June 1992. India has 88 National Parks and 490 sanctuaries in 2000, covering an area of 1,53,000 km^2. As of 2009, India's various biogeographic zones have 99 national parks and 513 sanctuaries. In 2012, several Indian states have 102 National Parks and 515 wildlife sanctuaries. There were 106 national parks with an area of 40,501.13 km^2 as of June 2021, accounting for approximately 1.33 percent of India's total surface area of under-protected territories of India category II. The country had 564 wildlife sanctuaries as of December 2021, accounting for 3.73 percent of its geographical area.

It's amazing to be in India, but it's even more exciting to be within an Indian National Park. Away from the concrete jungles that surround most of the major cities, the greenery and pure air supplied by Indian wildlife parks are a long cry from the norm. India's wildlife legacy is as diverse as, if not more so than, its cultural heritage. India, one of the world's megadiverse countries, is home to a plethora of animal species found nowhere else on the planet. India is a natural destination for wildlife enthusiasts from all over the world. The national parks have the largest number of tigers in the world, the largest number of one-horned rhinos in the world, they are home to the practically extinct Asiatic lions, and a substantial percentage of the world's elephants. These wildlife creatures are just a few of the more than 500 kinds of mammals that call India home. Aside from mammals, India is home to around 2000 bird species, 500 reptile and amphibian species, and about 30000 insect species, including spectacular butterflies.

The following are some of the most notable National Parks and their famous species in various states around the country:

Table 17.2: Showing National Parks and their distinct species

Animal/Bird	Predominant in Sanctuary/Reserve	State
Wild Ass	Rann of Kutch Wild Ass Sanctuary	Gujarat
One horned Rhinoceros	Kaziranga National Park	Assam
Elephants	Periyar National Park	Kerala
Asiatic Lions	Gir National Park	Gujarat
Great Indian Bustard	Ghatigaon Wildlife Sanctuary & Desert National Park	Madhya Pradesh & Rajasthan
Royal Bengal Tiger	Sunderbans National Park	West Bengal
Olive Ridley Turtle	Gahirmatha Turtle Sanctuary	Orissa
Dolphins	Vikramshila Gangetic Dolphin Sanctuary	Bhagalpur, Bihar
Vulture	Ramdevarabetta Vulture Sanctuary	Karnataka
Bear	Daroji Bear Sanctuary	Hampi, Karnataka
Sangai	KeibulLamjao National Park	Loktak Lake (Bishnupur), Manipur
Barasingha (Swamp deer)	Kanha National Park	Madhya Pradesh
Siberian cranes	Keoladeo national park	Rajasthan (Bharatpur)

Jerdon's Courser	Sri Lankamaleswara Wildlife Sanctuary	Andhra Pradesh
Gharial	National Chambal Sanctuary	U.P., Rajasthan, and M.P.
Grizzled Giant Squirrel	Srivilliputhur Wildlife Sanctuary	Virudhinagar& Madurai, Tamil Nadu
Clouded Leopard	Clouded Leopard National Park	Tripura
Snow Leopard	Hemis National Park	Ladakh
Hoolock Gibbon	Hoollongapar Gibbon Sanctuary	Assam
Golden langur	Chakrashila Wildlife Sanctuary &Raimona National Park	Assam
Hangul (Kashmir Stag)	Dachigam Sanctuary	Jammu and Kashmir
NilgiriTahr	Eravikulam National Park	Kerala
Red Panda	Khangchendzonga National Park	Sikkim

Conservation efforts

a) Project Tiger: This project, a Centrally Sponsored Mission, was launched in 1973 in Jim Corbett National Park with the following objectives:
1) To safeguard the survival of the tiger population in India for scientific, economic, aesthetic, cultural, and ecological reasons.

2) To maintain places of such biological value as a national heritage for the benefit, education, and pleasure of people at all times.

The goal of Project Tiger was to conserve endangered animals while also ensuring the rights of tribal people living in and around tiger reserves. Tiger Reserves are established using a 'core-buffer' method under Project Tiger. On the advice of the National Tiger Conservation Authority (NTCA), state governments must declare an area a tiger reserve. Bangladesh, Bhutan, Nepal, India, China, Lao PDR, Cambodia, Myanmar, Malaysia, Indonesia, Thailand, Vietnam, and the Russian Federation are among the 13 tiger range nations that are part of the Global Tiger Recovery Programme. The Tiger Reserve consists of:

1) Core Zone: The core region is protected from biotic disturbances and forestry operations. Activities such as collecting minor forest produce, human disturbances, grazing, and so forth are prohibited. For the sake of tiger conservation, these areas must be protected without compromising the rights of Scheduled Tribes or other forest dwellers. The State Governments notify these regions in cooperation with an Expert Committee.
2) Buffer Zone: The Act defines the buffer zone as the region around the core area. It provides more habitat for spreading tigers while still allowing for human activities. The boundaries of such regions are established by consultation with the relevant Gram Sabha and an Expert Committee formed for that purpose. Changes to the borders of a Tiger Reserve can be made only on the suggestion of the NTCA and with the approval of the National Board of Wildlife (NBWL).

b) Project Elephant: To assist states with free-roaming populations of wild elephants and to ensure the long-term survival of recognised viable elephant populations in their natural habitats, the Centrally Sponsored Scheme known as Project Elephant was launched in February 1992. States are receiving financial and technical assistance to help them achieve the following project objectives:
   1) Elephants, their habitats, and corridors must be protected.
   2) To address the problem of animal-human conflict.
   3) Welfare of domesticated elephants.

The Monitoring of Illicit Killing of Elephants (MIKE) Program: was launched by a CITES COP decision as a worldwide partnership to follow trends in information linked to the illegal killing of elephants from Africa to Asia and to assess the success of field conservation activities.

Hathi Mere Saathi: is a campaign launched by the MoEFCC in collaboration with the Wildlife Trust of India (WTI) to boost elephant conservation and welfare prospects.

E-8 Countries: India, Botswana, the Republic of Congo, Indonesia, Kenya, Sri Lanka, Tanzania, and Thailand are among the E-8 nations having considerable elephant populations.

E-50:50 forum: The E-8 nations hold the E-50:50 meeting to create a unified global vision for elephant conservation, management, and wellbeing across all range countries.

c) Vulture Conservation: There are nine species of vultures in India. However, their population has been steadily declining, bringing them dangerously close to extinction. Diclofenac Sodium, a nonsteroidal anti-inflammatory medicine used to relieve inflammation and discomfort in certain circumstances, is a likely explanation of the drop. Diclofenac Sodium has been linked to renal failure in vultures, which are unable to break down the chemical diclofenac when they consume the carcass of animals given the medicine. Meloxicam is being utilised as a substitute for Diclofenac in light of this.
   Saving Asia's Vultures from Extinction (SAVE): A coalition of like-minded regional and international organisations called Saving Asia's Vultures from Extinction was established to manage and coordinate conservation, advocacy, and fundraising initiatives to improve the status of vultures in South Asia. This consortium's partners include Bird Conservation Nepal, Bombay Natural History Society, International Centre for Birds of Prey (UK), Nepal National Trust for Nature Conservation, and the Royal Society for the Protection of Birds (UK).
d) Sea Turtle Project: In November 1999, the Sea Turtle Conservation Project was begun in conjunction with UNDP. The project is being carried out by the Wildlife Institute of India (WII). While the initiative is being executed in ten coastal states throughout the country, the state of Odisha receives special attention.
e) Project Hangul: The Kashmir stag, sometimes known as the Hangul, is a Central Asian Red Deer subspecies. It may be found in the Dachigam National Park in Kashmir. The government of

Jammu and Kashmir, in collaboration with the IUCN and WWF, created Project Hangul to combat the state's dwindling population.

f) Captive Breeding: Captive breeding, often known as "captive propagation," is the technique of keeping plants or animals in confined spaces such as wildlife reserves, zoos, botanical gardens, and other conservation institutions. Captive breeding of lion-tailed macaques, for example, in Arignar Anna Zoological Park in Chennai.
g) Crocodile Breeding Project: With UNDP help, the programme began in Orissa and was expanded to six additional states in April 1975. The primary goal was to safeguard three endangered crocodile species: Gavialis gangeticus, Crocodylus palustris, and the saltwater crocodile, Crocodylusporosus.
h) Lesser Cat Project: The project was initiated in 1976 with the aid of WWF in India for the protection of four species of smaller cats found in Sikkim and the northern portion of West Bengal, namely Felis bengalensis Kerr, Felis marmorta Martin, Felis lemruinki Vigors Horsfield, and Felis viverrina Bennet.
i) Project Rhino: Poachers target one-horned rhinos for their horns. Assam's Department of Environment and Forests is implementing the Indian rhino vision 2020. WWF-India, the International Rhino Foundation (IRF), and a number of local non-governmental organisations (NGOs) will assist the effort. The IRV 2020 programme relies heavily on translocations. The motive was to inhabit the designated possible rhino habitat regions, namely Manas NP, DibruSaikhowa WLS, and Laokhowa-Bura Chapori WLS, with a healthy rhino population by translocating from Kaziranga NP and Pobitora WLS. Manas National Park was chosen as the initial destination for rhino relocation. A total of 10 rhinos have been released into Manas National Park since 2008. The program's objective was to expand the total number of rhinos in Assam from around 2000 to 3000 by 2020, with these rhinos scattered among at least 7 protected areas (PA) to guarantee the long-term survival of the one-horned rhino population.
j) Himalayan Musk deer Project: This was started in 1981 to conserve the endangered musk deer, which is on the edge of extinction. Captive breeding has shown positive effects.
k) Project Snow Leopard: It was launched in 2009 with the goal of promoting an inclusive and participatory approach to the conservation of snow leopards and their environment. On October 23, 2013, the world celebrated International Snow Leopard Day.

The day began with the acceptance of the Bishkek Declaration on Snow Leopard Conservation by 12 countries. On this day, the Indian government began the "HimalSanrakshak" community volunteer initiative to conserve snow leopards. Through the Project Snow Leopard, India has been working to protect the snow leopard and its habitat (PSL). Since 2013, India has also been a member of the Global Snow Leopard and Ecosystem Protection Programme (GSLEP). India has chosen three big landscapes for conservation: Hemis-Spiti in Ladakh and Himachal Pradesh, Nanda Devi – Gangotri in Uttarakhand, and Khangchendzonga – Tawang in Sikkim and Arunachal Pradesh. In 2009, Project Snow Leopard (PSL) was established to foster an open and participatory approach to the conservation of snow leopards and their habitat. The Snow Leopard is one of 21 severely endangered species on the Ministry of Environment, Forests, and Climate Change's recovery list.

l) The Manipur Brow-antlered Deer Project: This was started in Manipur in 1981 to conserve the brow-antlered deer (Cerevuseldieldi), which is on the verge of extinction. The ecosystem consists of 35 square kilometres of park and sanctuary. The deer population has expanded from 18 to 27.

m) Ganges Dolphin: The Ministry of Environment and Forests has named the Ganges River Dolphin the country's official aquatic animal. The River Dolphin lives in the Ganges-Brahmaputra-Meghna and Karnaphuli-Sangu river systems of Nepal, India, and Bangladesh. They are listed on Schedule I of the Wildlife Protection Act (1972) and their estimated total population is 2,000. The Ganges Dolphin is one of just four "obligate" freshwater dolphins in the world, the others being the "baiji" of the Yangtze River (China), the "bhulan" of the Indus River (Pakistan), and the "boto" of the Amazon River. These four species only live in rivers and lakes, unlike the majority of marine dolphin species, whose ranges include some freshwater areas. In 2006, a group of international specialists declared the Chinese River Dolphin to be virtually extinct. India's Ganges River Dolphin is in danger from oil poaching, incidental entanglement in fishing nets, river water pollution, and siltation.

CHAPTER 18

MARINE ENVIRONMENT AND ECOSYSTEM

Topography of Marine Environment

All major oceans and seas differ in size, shape, depth, areal extent, and other characteristics such as temperature, density, salinity, and chemical composition. Marine geological, chemical, and biological information and data are huge and extremely important for researching and evaluating the seas. Ocean Relief Features vary from continental characteristics in that the oceanic crust is just 60-70 million years old, whereas continental features are Proterozoic in age (Over 1 billion years old). While there is just one global ocean, it is physically split into various designated zones that span 71 percent of the Earth. Diversification of historical, cultural, geographical, and scientific reasons, the borders between these areas have altered over time. Previously, echo-sounding techniques were utilised; presently, radar soundings and electrical measurements are used. Echo devices are used to determine the clear-cut depths of sea floor and to map ocean relief. The seas, unlike the continents, mix so naturally into one another that it is difficult to distinguish between them. Geographers have classified the earth's oceans into five groups: the Pacific, the Atlantic, the Indian, the Southern, and the Arctic. A large portion of the sea floor is between 3-6 kilometres below sea level. The ocean bottoms are rough, featuring the world's greatest mountain ranges, deepest trenches, and largest plains. These characteristics, like those of continents, are generated by tectonic, volcanic, and depositional processes.

Major relief features

The ocean floors can be categorized into four distinct divisions (Fig 18.1):

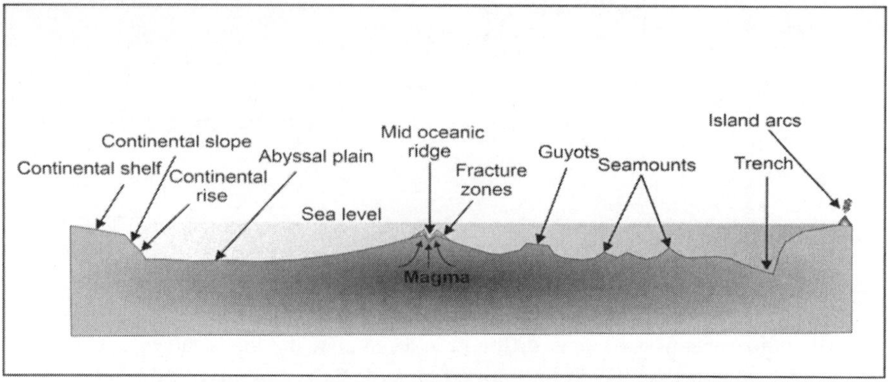

Fig. 18.1: Diagram showing major and minor relief features of the ocean floor

a) Continental Shelf: The continental shelf is the seaward extension of the continental plate that is gently sloping. Each continent's long edges are populated by relatively shallow oceans and gulfs. The continental shelves of all seas cover 7.5 percent of the total ocean area. The continental gradient is 1° or less. The shelf often terminates at a very steep slope known as the shelf break. The continental shelves are covered with varying thicknesses of sediments deposited by rivers, glaciers, and other bodies of water. Massive sedimentary deposits deposited by the continental shelves throughout time provide a source of fossil fuels. The shelf is developed primarily as a result of the submergence of a portion of a continent, a relative rise in sea level, and sedimentary deposits carried down by rivers.
b) Continental Slope: A continental slope is the slope that runs from the continental shelf's outer border to the deep ocean floor. Submarine canyons cut the continental slope in various places. The continental slope denotes the continental shelf's seaward edge. The gradient of the slope region ranges from 2 to 5 degrees. It ranges in depth from 180 to 3600 metres. Continental slopes have extremely few sediment deposits on them, owing to their steepness and increasing distance from the land. A sediment deposit runs at the base of the continental slope. The continental rise is formed by this strip of sedimentary deposits. With depth, the continental slope progressively lessens its steepness. The continental rise occurs when the slope reaches a level of between 0.5° and 1°. With increasing depth, the rise flattens down and merges with the abyssal plain.

c) Abyssal plain: Abyssal Plains are gently sloping ocean basins. Because of terrigenous (marine sediment eroded off land) and shallow sea sediments that bury the uneven terrain, they are the world's flattest and smoothest regions. It occupies about 40 percent of the ocean floor. The depths range from 3,000 to 6,000 metres. Fine-grained sediments such as clay and silt cover these plains. It is characterised by enormous submerged plateaus, ridges, trenches, beams, and oceanic islands that rise above sea level in the middle of seas.
d) Trenches: The trenches are narrow, steep-sided basins (Depressions). These arc the deepest places of the seas. They are generated during ocean-ocean and ocean-continent convergence and are of tectonic origin. They are 3-5 kilometres deeper than the surrounding ocean bottom. Trenches can be found on the outskirts of the deep-sea plain, near the bases of continental slopes, and around island arcs. The trenches run parallel to the adjacent-fold mountains or island chains. Trenches are relatively prevalent in the Pacific Ocean and form a nearly continuous ring throughout the Pacific's western and eastern edges. The Mariana Trench, located off the coast of the Guam Islands in the Pacific Ocean, is the deepest trench, reaching a depth of more than 11 kilometres. They are linked to active volcanoes and strong earthquakes (Deep Focus Earthquakes like in Japan). As a result, they are extremely important in the study of plate movements.
e) Mid Oceanic ridge:

Minor Relief features

Aside from the major relief characteristics of the ocean floor listed above, other smaller but significant features prevail in different sections of the seas (Fig. 18.1) and they are explained below:

a) Seamounts: A seamount is a volcanically produced underwater mountain. The seamount does not reach the ocean's surface. These can reach heights of 3,000–4,500 metres. Emperor seamount, a Pacific Ocean extension of the Hawaiian Islands, is an excellent example.
b) Guyots: Guyot, commonly known as a tablemount, is an isolated underwater volcanic mountain with a flat top located more than 200 metres beneath the sea's surface. They show signs of gradual subsidence through stages, eventually becoming flat-topped

submerged mountains. More than 10,000 seamounts and guyots are thought to occur in the Pacific Ocean alone.

c) Fracture zones: The scars left in plate interiors by transform faults that offset mid-ocean ridge segments are known as fracture zones. This transform faulting occurs at the endpoints of spreading segments where melt supply can be limited or increased, resulting in crustal thickness variations across transform faults.

d) Island arcs: Together with oceanic ridges, island arcs are key structural features of ocean basins. Island arcs, as the name indicates, are often a curving chain of volcanic islands found along the margins of ocean basins.

e) Canyons: A submarine canyon is an abrupt or steep-sided valley carved into the seabed of a continental slope, often extending deep onto the continental shelf (Fig. 18.2), with almost vertical walls and canyon wall heights of up to 5 km from canyon floor to canyon rim, such as the Great Bahama Canyon. The Hudson Canyon is the world's most well-known submarine canyon. Canyons can also be formed by a current known as Turbidity current that has the capacity to change the physical shape of ocean floor by eroding vast areas. Additionally, these currents carry huge amounts of sediment with them everywhere they go, generally in the form of gradients or fans with the largest particles at the bottom and the smallest at the top. Turbidity currents may develop when earthquakes, collapsing slopes, and other geological disturbances loosen mud and sand on the continental shelf. The turbid water then rushes downward, taking up particles and increasing in speed as it flows.

f) Aseismic ridge: Many guyots and seamounts on the seafloor are linked together in chains. These volcanic chains, along with a few other seafloor ridges, are known as aseismic ridges; that is, they are undersea ridges that are not connected with earthquakes. Aseismic features are distinguished from the much broader mid-oceanic ridge, where earthquakes occur along the rift valley.

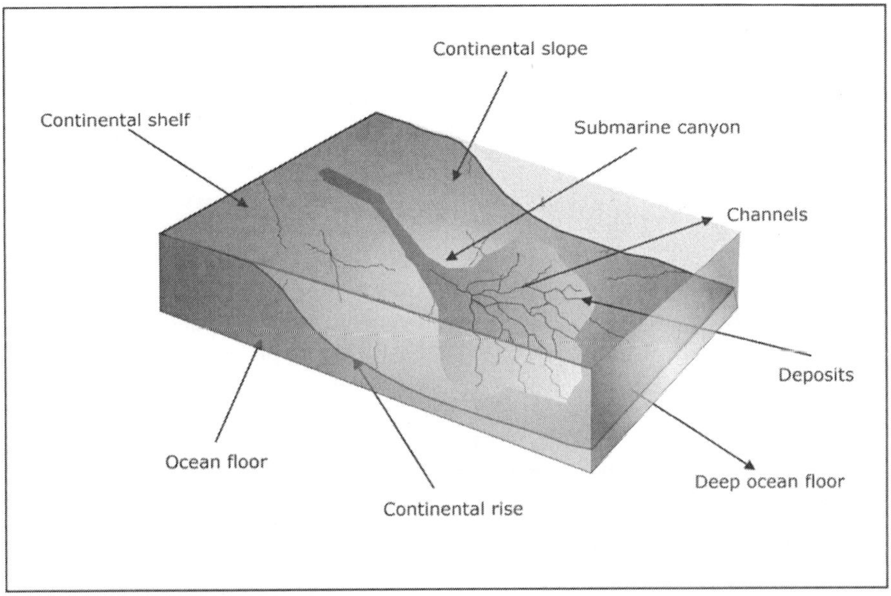

Fig. 18.2: Submarine canyon

Sea floor Spreading and Plate Tectonics

H. Harry Hess proposed the theory of seafloor spreading in 1960. Hess was able to mapped the ocean floor and discover the mid-Atlantic ridge by using sonar. He also discovered that the temperature near the mid-Atlantic ridge was warmer than the surface farther away. He suspected that the high temperature was caused by magma leaking from the ridge. This idea on the shifting location of the earth's surface supports Alfred Wegener's 1912 Continental Drift Theory.

Convection Current theory

At the heart of the Seafloor Spreading Theory is the Convection Current Theory. In the 1930s, Arthur Holmes hypothesised the potential of mantle convection currents. These currents are caused by radioactive materials that cause heat variations in the mantle. According to this idea, the extreme heat created by radioactive elements in the mantle (100-2900 km below the earth's surface) seeks a way to escape and causes convection currents to emerge in the mantle. Oceanic ridges emerge on the seafloor when rising limbs of these currents meet owing to the divergence of the lithospheric plates (tectonic plates), and trenches form where failing limbs meet due to the convergence of the lithospheric plates. Movement of the lithospheric plates is produced by magma movement in the mantle. The process of sea-floor spreading advancements new ocean floor.

Paleomagnetism

Paleomagnetism is the study of magnetic fields that have been retained in rocks, sediment, or archaeological artefacts as a record of the earth's magnetic field. Thus, by examining rocks of various ages, one may identify the polarity of the Earth's magnetic field and magnetic field reversals. The bulk of the rocks that make up the ocean bottom are basaltic ones, which are low in silica and high in iron. Basalt contains magnetic minerals, which when the rock hardens align themselves in the magnetic field's direction. This records the direction in which the magnetic field was at the time. The direction of the earth's magnetic field has regularly cycled (geomagnetic reversal) during geologic time, according to paleomagnetic examinations of rocks. It was essential in reviving the continental drift idea and transforming it into theories of Sea Floor Spreading and Plate Tectonics. The places with the most complete record of the Earth's magnetic field are found near mid-ocean ridges where the seafloor is expanding. When the paleomagnetic rocks on each side of the oceanic ridges were studied, it was discovered that alternating magnetic rock stripes were flipped such that one stripe had normal polarity and the next had reversed polarity. As a result, paleomagnetic rocks (paleo: rock) on either side of mid-ocean or underwater ridges provide the most crucial evidence for the notion of Sea Floor Spreading. Magnetic field records can also reveal the past position of tectonic plates. These oceanic ridges are the points at which tectonic plates diverge (moving apart). The fissure or vent (between the ridge) allowed magma to rise and form into a long narrow strip of rock on each side of the vent. Rising magma takes on the polarity of the Earth's geomagnetic field before solidifying on the oceanic crust. The solidified strip of rock moves away from the vent (or ridge) when the oceanic plates separate, and a new strip of rock takes its place a few million years later when the magnetic field is reversed. As a result, the neighbouring rock strata have opposing polarity, resulting in magnetic striping. This process continues indefinitely, resulting in a succession of thin parallel rock bands on either side of the ridge and an alternating pattern of magnetic striping on the seafloor.

Evidence for Sea floor spreading

a) Evidence from Molten Material: Rocks shaped like pillows (rock pillows) demonstrate that molten material erupted and cooled swiftly from fissures along the mid-ocean ridge.
b) Evidence from Magnetic Stripes: The ocean floor's rocks form a pattern of magnetising stripes that record reversals in the Earth's magnetic field.
c) Evidence from Drilling Samples: Core samples from the ocean bottom demonstrate that older rocks are found farther away from the ridge, while the youngest rocks are located near the ridge's core.
d) Subduction: The process through which the ocean floor dips beneath a deep-ocean trench and back into the mantle; allows for the sinking of a portion of the ocean floor back into the mantle.
e) Deep-Ocean Trench: This happens in subduction zones. Where the Oceanic crust bends downward due to its higher density than the continental crust, deep underwater canyons emerge.

Problems solved with the help of sea floor spreading

a) It answered the dilemma of younger crust being found near the mid-oceanic ridges and older rocks being found further away from the ridges.
b) It also explains why the sediments at the centre of the oceanic ridges are so thin.
c) The spreading of the seafloor also supported Alfred Wegener's hypothesis of continental drift and aided in the development of plate tectonics.

Plate tectonics emerged in the late 1960s by integrating two previously existent notions, continental drift and seafloor spreading. The concept of continental drift is that continents move freely over the Earth's surface, altering their locations relative to one another. According to the idea of seafloor spreading, the sea floor forms at the crest of the mid-oceanic ridge and then moves horizontally away from the ridge crest into an oceanic trench. The ridge's two sides move in opposing directions like slow conveyor belts.

According to plate tectonics, the Earth's solid outer crust, the lithosphere, is divided into plates that move across the asthenosphere, the molten top layer of the mantle. Plate tectonics (given by Alfred Wegener) is a scientific theory that describes the large-scale motion of seven big plates and a greater number of smaller plates in the Earth's lithosphere over hundreds of millions of years. It is a comprehensive theory that attempts to explain the majority of the dynamics of the earth crust as well as the characteristics of endogenetic forces.

Plate Tectonic theory postulates

According to mechanical rigidity, the earth's interiors can be divided into several layers:

a) Lithosphere: The lithosphere is the Earth's rocky outer layer. The brittle crust and the top section of the upper mantle make up this layer. The lithosphere is the Earth's coldest and most rigid layer. Forests, grasslands for grazing, agriculture, and human settlements are all provided by the lithosphere, which is also a rich supply of minerals. It also helps to provide plants with nutrients since the lithosphere is made up of several kinds of rocks, including igneous, sedimentary, and metamorphic rocks.
b) Asthenosphere: The asthenosphere is the plastic layer underneath the lithosphere (The plates of the lithosphere float on the asthenosphere). The lithosphere is considered to have been broken into fragments that are floating on the asthenosphere, a ductile layer (upper part of the mantle). Convention currents generated in the upper mantle are thought to be responsible for the movement of these plates. As rigid units, plates move horizontally over the asthenosphere.
c) Mesosphere: Underneath the asthenosphere is the mesosphere. It includes the lower mantle, where material flows at a slower rate than the asthenosphere. Somewhere under the mesosphere, there is a layer of liquid iron and nickel (along with other elements).

Lithospheric Plates

A plate is a broad lithosphere section that floats on the asthenosphere underneath it and moves independently of the other plates. Continental plates and oceanic plates are the two types of plates. The earth is split into seven major plates and nine smaller plates by La Pichon and they are mentioned below:

a) Major tectonic plates:
   1) Antarctica and the surrounding oceanic plate (Surrounded by divergent boundaries).

2) North American plate (shifting westwards at a velocity of 4–5 cm/year, it is half oceanic–half continental).
3) South American plate (shifting westwards at a velocity of 3–4cm/year, it is half continental–half oceanic).
4) Pacific plate (oceanic plate shifting north-westwards at a velocity of 2–3 cm/year).
5) India-Australia-New Zealand plate.
6) Africa with the eastern Atlantic floor plate.
7) Eurasia and the adjacent oceanic plate (mostly continental, shifting eastwards at a velocity of 2–3 cm/year).

b) Minor tectonic plates:
 1) Arabian plate: mostly covers the Saudi Arabian landmass
 2) Bismarck plate (North Bismarck Plate & South Bismarck Plate)
 3) Caribbean plate
 4) Carolina plate (straddles the Equator in the eastern hemisphere located north of New Guinea)
 5) Cocos Plate
 6) Juan de Fuca Plate (between Pacific and North American plates)
 7) Nazca plate
 8) Philippine plate: between the Asiatic and Pacific plate
 9) Persian Plate
 10) Anatolian Plate ([also known as the Turkish Plate] is a continental tectonic plate comprising most of the Anatolia [Asia Minor] peninsula [and the country of Turkey])
 11) China plate
 12) Fiji plate (located between the Pacific Plate and the Indo-Australia Plate).

Plate Motion

a) Divergent: A diverging boundary is formed when two tectonic plates move apart from one other (Fig. 18.3). Long fissures spew lava, and geysers discharge superheated water around their edges. Earthquakes are common along the fissure. Under the fissure, magma rises from the mantle. It oozes into the crack and hardens into solid rock, forming new crust on the split edges of the plates. The lava from the mantle cools and solidifies into basalt, a dark, dense rock that lays under the ocean's surface. As a result, basaltic oceanic crust is created at divergent boundaries. Such outcomes include mid-ocean ridges, rift valleys, fissure volcanoes, and so forth.

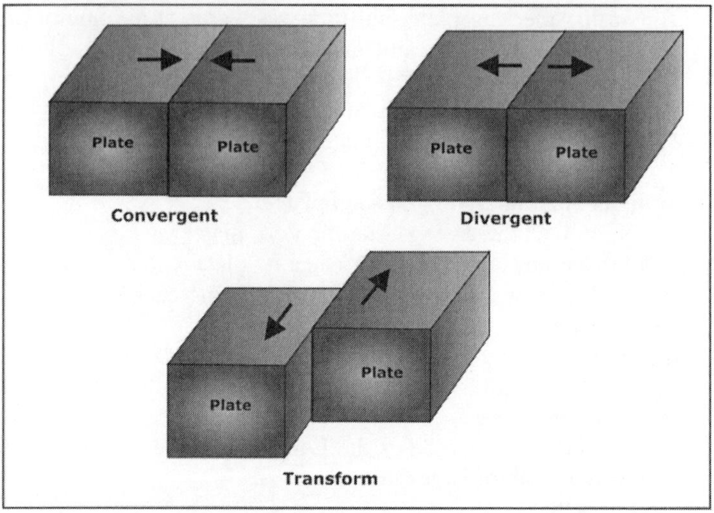

Fig. 18.3: Different types of plate motion

b) Convergent: A convergent boundary is formed when two plates come together. The collision of the two plates buckles one or both plates' edges up into a rugged mountain range, and occasionally bends the other down into a deep undersea trench. Parallel to the boundary, the mountain range, and the trench, a chain of volcanoes typically arises. On both sides of the boundary, powerful earthquakes shake a large region. If the oceanic crust of one of the colliding plates is thick enough, it is driven down into the mantle and begins to melt. Magma rises through and into the opposite plate, forming new crust. The continents are made up of granite, a light-coloured, low-density rock created by magma from melting plates. As a result, continental crust, comprised of granite, is formed near convergent boundaries, whereas oceanic crust is destroyed. The following are the three types of convergent boundaries:

1) Oceanic–continental convergence: A continent may collide with the oceanic crust. Subduction occurs because the oceanic plate is denser. The oceanic plate is sinking beneath the continent as a result of this. In the process of subduction, the thick oceanic plate falls slowly and inexorably into the asthenosphere. The rest of the plate is pulled in by the subducting slab—this "slab

pull" is likely the principal source of most plate movement, drawing the rest of the plate in and after itself.

2) Oceanic–oceanic convergence: The colder, denser oceanic lithosphere descends beneath the warmer, less dense oceanic lithosphere when two oceanic plates collide. As the slab descends further into the mantle, dehydration of hydrous minerals in the oceanic crust releases water. As one of the oceanic plates subducts beneath the other, an oceanic trench forms, shallow and deep-focus earthquakes occur, and volcanic activity begins on the ocean bottom with volcanoes developing.
3) Continental–continental convergence: Between two continental plates, a Continent-Continent convergence occurs. Oceanic sediments are crushed and upthrust between the plates when the plates converge, and these squeezed sediments appear as fold mountains at the plate margins. This form of convergent plate boundary may be seen in the Himalayan Mountains.

c) Transform: A transform fault, also known as a strike-slip boundary, is a fault that runs along a plate boundary and is characterised by horizontal motion. Transform Plate Boundaries are places when two plates slide past one other, with no new landforms being created or destroyed, merely deformation of the current landform. A transform fault is the fracture zone that produces the transform plate boundary. Transform faults are planes of separation in the oceans that are often perpendicular to the mid-oceanic ridges.

Nutrients and Diversity of Organisms

The availability of nutrients required for phytoplankton, such as nitrate, phosphorus, iron, and silica, determines the distribution of life over the ocean's width and depth. Marine populations achieve their peak concentration when physical circumstances provide huge amounts of nutrients. The sources of nutrients must be addressed in order to understand where these places are found. Water erodes the continents as runoff, transporting material to the seas and depositing it as sediment on the continental margins. Nitrates and phosphates, which are the essential nutrients for phytoplankton, are also dissolved and transported by runoff. All garden and agricultural fertilisers contain nitrates and phosphates as fundamental components. Phytoplankton use photosynthesis to mix these nutrients with carbon dioxide and water to make the carbohydrates, proteins, and lipids that the rest of the ocean's biological population eats. Because the continents are the primary providers of these nutrients, the continental margins have the highest densities of marine life. However, as the distance between the continental margins and the open sea increases, the concentration of marine species drops. These discrepancies are due to the huge depths of the world's seas and the large distance between the open ocean and the coastal regions where nutrients are concentrated. The shortage of certain nutrients, notably nitrogen (as nitrates) and phosphorus, can sometimes limit productivity. As a result, they are some of the most investigated chemicals in chemical oceanography. In comparison, nitrogen compounds involved in photosynthesis can be 10 times higher than the overall nitrogen compound concentration measured on an annual basis. This level necessitates the full recycling of soluble nitrogen molecules up to ten times per year. The amount of available phosphates can be changed up to four times each year.

Carbon is also a significant factor in productivity since it is the fundamental component of all organic molecules (including carbohydrates, proteins, and fats). However, many kinds of carbon are abundant in the ocean, thus there is no carbon scarcity for photosynthetic synthesis. As a result, carbon has no effect on production. The ratio of carbon to nitrogen to phosphorus in the tissues of algae when nutrients are not restricting production is 106:16:1 (C:N:P), which is known as the Redfield ratio after American oceanographer Alfred C. Redfield, who originally described it in 1963. This ratio is also found in diatom-feeding zooplankton and in the majority of ocean water samples examined throughout the world. Furthermore, phytoplankton absorb nutrients in the same proportion as they are available in ocean water and pass them on to zooplankton in the same proportion. When these plankton and animals die, the same proportions of carbon, nitrogen, and phosphorus are returned to the water.

Phytoplankton

The word "plankton" refers to any non-motile creatures that cannot swim against the current in both marine and freshwater environments. Water currents carry them along. Planktons range in size from 0.2 millimetres to more than 20 centimetres. From small microbes to large organisms like jellyfish. Their distribution fluctuates depending on the amount of light and nutrients available. Large aquatic species feed on planktons and are dependent on them. Planktons play an important function in the ecology. Photosynthesis produces 50 percent of the total oxygen (O2) generated by phytoplankton. Plankton is distinct from other similar names used to describe other aquatic creatures, such as:

a) Pleuston: refers to creatures that float.
b) Nekton is a term for creatures that are actively swimming.
c) Neuston: creatures that live on the surface of the water
d) Benthos are sessile, burrowing, crawling, and other animals that live on the seabed.

Types of Plankton

Lifecycle, size, and trophic level are all used to classify planktons.

Planktons are classed as follows based on their lifecycle:

a) Holoplankton – Organisms that live their whole lives as planktonic organisms, such as algae and jellyfish.
b) Meroplankton – Organisms that exist as plankton for just part of their life cycle, such as starfish larvae, worms, sea urchins, fish, and so on.

Planktons are classed according to their size as follows:

a) Megaplankton: they are large in size at> 20 cm, e.g., Jellyfish, tunicates pyrosomes, etc.
b) Macroplankto: their size ranges from 2 to 20 cm.
c) Mesoplankton: includes organisms with size of 0.2 to 20 mm.
d) Microplankton: their size varies from 20 to 200 μm, e.g., most of the phytoplankton, protozoans and large protists.
e) Nanoplankton: size ranges from 2 to 20 μm, e.g., protists, diatoms and algae.
f) Picoplankton: size ranges from 0.2 to 2 μm, e.g., bacteria, chrysophytes.
g) Femtoplankton: includes marine viruses with size of< 0.2 μm.

Planktons are categorised into the following categories based on their trophic levels and characteristics:

a) Phytoplankton – They are autotrophs or producers, e.g., cyanobacteria, algae, diatoms, dinoflagellates, etc.
b) Zooplankton – It includes primary consumers, who feed on other plankton, e.g., small protozoans, larvae of fish and other animals.
c) Mycoplankton – This includes fungi.
d) Bacterioplankton – This includes bacteria and is important for nutrient recycling.

e) Virioplankton – This includes viruses.
f) Mixotrophs – They act as both producers and consumers according to environmental conditions. When there is abundant light and nutrients available they perform photosynthesis and in the absence of limited resources, they feed on other plankton to fulfil their nutrient requirements

Factors affecting Phytoplankton Biodiversity

a) Light: Phytoplanktons are only found in the top layers of the ocean. Photosynthesis can take occur in the water if the light intensity is high enough. The rate of photosynthetic activity is affected by the amount of light available.
b) Nutrients: Nitrogen and phosphorus are the two most important inorganic nutrients for phytoplankton growth and reproduction. Silicate (SiO2) is also required in large numbers by diatoms and silicoflagellates. Some phytoplankton species can fix nitrogen and flourish in low-nitrate environments. They also need trace quantities of iron, which restricts phytoplankton development in wide regions of the ocean due to low iron concentrations.
c) Temperature: Temperature, along with other variables, influences how photosynthetic output varies. In general, when the temperature rises, the rate of photosynthesis increases, but then drops abruptly after a certain point is reached. The seasonal fluctuation of phytoplankton production in temperate latitudes is influenced by temperature and light.
d) Salinity: Salinity, in addition to light and temperature, is known to affect primary production.
e) Grazing by Zooplankton: One of the key parameters controlling the size of the standing crop of phytoplankton, and hence the pace of production, is the grazing rate of zooplankton.

Significance of Phytoplankton

a) They create half of the entire quantity of oxygen released during photosynthesis as producers.
b) Plankton is the primary source of fish food for many fisheries.
c) Phytoplanktons are employed in space flight as a source of food as well as for carbon dioxide fixing. Chlorella is a kind of algae that is used as a protein supplement.

d) They play a key role in the recycling of nutrients. Organic matter that has died or decayed can be converted into inorganic nutrients for plants.
e) Some species may convert dissolved air nitrogen to ammonia and nitrates, which are beneficial to phytoplankton development.
f) The equilibrium between carbon dioxide and oxygen is maintained by phytoplankton and zooplankton. Phytoplanktonic growth can boost an ocean's CO2 absorption.

Zooplankton

Zooplankton is an important component of the food chain, nutrient recycling, and organic matter transport from primary producers to secondary consumers such as fish. They are more common in mangrove systems than in nearby coastal seas, and zooplanktivorous juvenile fish make up a major component of the mangrove fish population. The quantity of fish stock is determined by zooplankton. As a result, zooplankton communities are used to measure the ecosystem's productivity, fertility, and health condition based on the quality and variety of their species.

Sea grass

Sea grasses are marine flowering plants (angiosperms) that look like grass. They feature strap-like or oval leaves and a root system, as well as flowers. They thrive in shallow coastal waters with sandy or muddy bottoms and prefer calmer conditions. They are the only higher plants that have evolved to living in saline water. Major sea grass meadows may be found along Tamil Nadu's south coast and in the lagoons of a few Lakshadweep Islands. Around the Andaman and Nicobar Islands, there are a few grass beds as well. The high salinity, transparency of the water, and sandy substratum that characterise the Tamil Nadu coast and the Lakshadweep islands contribute to the abundant growth of seagrasses. Physically, sea grass beds minimise wave and current energy, filter suspended particles from the water, and stabilise bottom sediments to limit erosion. It serves as a habitat for marine invertebrates and fish. Seagrass beds are common in lagoons, and the population of fish and migrating birds is larger in these locations due to the abundance of food and shelter. Sea grasses on reef flats and near estuaries act as nutrient sinks by buffering or filtering nutrient and chemical inputs into the marine environment. The greatest risks to sea grass beds include eutrophication, siltation, trawling, coastal engineering structures, and overexploitation for commercial purposes.

Diatoms

Diatoms are photosynthesizing algae that may be found in practically any aquatic habitat, including fresh and salt water, soils, and almost any wet environment. Diatoms have silica cell walls, and each species has a unique pattern of small holes in the cell wall (frustule) via which they receive nutrients and expel waste. A diatom is a photosynthetic, single-celled creature, which means it produces its own food in the same manner that plants do.

Diatoms are important because they form the foundation of the food chain for both marine and freshwater microorganisms and animal larvae, they are a major source of atmospheric oxygen responsible for 20–30 percent of all carbon fixation on the planet, they can act as environmental indicators of climate change, and they are the basis of some household goods such as pest/mite prevention and mild abrasive cleaners.

Radiolaria

Radiolaria, also known as Radiozoa, are protozoa with diameters of 0.1–0.2 mm that generate complex mineral skeletons, generally with a central capsule separating the cell into endoplasm and ectoplasm. The intricate mineral skeleton is often composed of silica. They are found as zooplankton all over the world. Radiolarians are mostly heterotrophic zooplankton, although many include photosynthetic endosymbionts and are hence classified mixotrophs. As siliceous ooze, the skeletal remnants of certain radiolarians cover a considerable portion of the ocean floor. Radiolarians are an important diagnostic fossil discovered from the Cambrian period forward due to their fast change as species and complicated bones. Radiolarians feature several needle-like pseudopods supported by bundles of microtubules, which help in buoyancy. The cell nucleus and most other organelles are located in the endoplasm, whereas the ectoplasm is loaded with foamy vacuoles and lipid droplets that make the cell buoyant. Symbiotic algae, particularly zooxanthellae, can be found in radiolarians and provide the majority of the cell's energy. Heliozoa have some of similar arrangement, but they lack central capsules and only generate basic scales and spines.

Seaweed

Seaweeds are (thalloid plants) macroscopic algae, which means they lack real tissue differentiation such as roots, stems, and leaves. They have appendages that resemble leaves. Seaweeds, the bigger and more apparent marine plants, are found in the intertidal and shallow subtidal zones of the sea, adhering to rocks, corals, and other submerged strata. Seaweeds occur in shallow coastal waters when large substrata are present. Seaweeds are widely grouped into several classes based on the colour of their pigmentation, such as blue-green, green, brown, red, and so on. Seaweeds are vital as food for people, animal fodder, and plant fertiliser.

Seaweeds are used as a medicine to treat goitre, infertility, and gastrointestinal issues. Seaweeds are used to extract commercially valuable products such as agar-agar and alginates, as well as iodine. Methane and other environmentally significant gases can be created in huge amounts by the biodegradation of seaweeds. Some seaweed extracts have antimicrobial properties. Because of their remarkable propensity to bind and collect metals, seaweeds are also employed as possible indicators of pollution in coastal ecosystems, particularly heavy metal contamination. Rotting seaweed is a rich source of hydrogen sulphide, a very hazardous gas, and has been linked to several cases of apparent hydrogen-sulphide poisoning. It might result in vomiting and diarrhoea.

Algae

This category includes plants with a lack of well-defined body shape. They are frequently referred to as algae. Algae are chlorophyll-containing creatures that are simple, thalloid, autotrophic, and mostly aquatic (both fresh water and sea). They may be found in diverse environments, including damp stones, soils, and wood. Some of them are also found in the presence of fungus (lichen) and animals (e.g., on sloth bear). Algae vary greatly in shape and size. The size ranges from microscopic unicellular forms such as Chlamydomonas to colonial forms such as Volvox and filamentous forms such as Ulothrix and Spirogyra. Some marine species, such as kelps, have huge plant bodies. Algae reproduce by vegetative, asexual, and sexual means. Vegetative reproduction occurs through fragmentation. Each fragment becomes a thallus. Asexual reproduction occurs by the creation of many types of spores, the most common of which are zoospores (capable of swimming through a flagellum). They are flagellated (motile) and produce new plants upon germination. Sexual reproduction occurs by the fusing of two gametes. These gametes can be flagellated and small (as in Chlamydomonas) or non-flagellated (non-motile) but small (as in Spirogyra). This type of reproduction is known as isogamous (this is the fusion of two gametes of comparable size). Anisogamous fusion occurs when two gametes of different sizes fuse, as in several Chlamydomonas species. Oogamous fusion occurs when one big, non-motile (static) female gamete fuses with a smaller, motile male gamete, as in Volvox and Fucus.

Characteristics:

Algae, for example, may photosynthesize like plants and have specialised structures and cell organelles found exclusively in animals, such as centrioles and flagella. Some of the general properties of algae are listed below.

a) Algae are photosynthetic organisms.
b) Algae can be either unicellular or multicellular organisms.
c) Algae lack a well-defined body, so, structures like roots, stems or leaves are absent
d) Algaes are found where there is adequate moisture.
e) Reproduction in algae occurs in both asexual and sexual forms. Asexual reproduction occurs by spore formation.
f) Algae are free-living, although some can form a symbiotic relationship with other organisms.

Types of Algae

Algae come in a variety of forms. However, the following are some of the most common types:

a) Green Algae (Chlorophyceae): Chlorophyceae members are generally referred to as green algae. The body of the plant might be unicellular, colonial, or filamentous. Because of the prevalence of the pigments chlorophyll A and B, they are often grass green. The pigments are concentrated in certain chloroplasts. The chloroplasts of the majority of the members include one or more storage structures termed pyrenoids. Pyrenoids, in addition to carbohydrates, include protein. Food may be stored in the form of oil droplets by certain algae. Green algae have a rigid cell wall that is composed of an inner layer of cellulose and an outside layer of pectose. Vegetative reproduction is often accomplished by fragmentation or the generation of various types of spores. Green algae that are widely seen include Chlamydomonas, Volvox, Ulothrix, Spirogyra, and Chara.

b) Brown Algae (Phaeophyceae): Members of the phaeophyceae, sometimes known as brown algae, are mostly found in marine environments. They vary greatly in size and shape. They range from basic branching, filamentous forms (Ectocarpus) to lavishly branched forms like kelps, which can grow to be 100 metres tall. They include chlorophylls A and C, as well as carotenoids and xanthophylls. They range in colour from olive green to brown, depending on the quantity of the xanthophyll pigment fucoxanthin present. The vegetative cells contain a cellulose wall that is normally coated on the exterior by an algin gelatinous covering. In addition to plastids, the protoplast has a centrally positioned vacuole and nucleus. Vegetative reproduction occurs through fragmentation. Ectocarpus, Dictyota, Laminaria, Sargassum, and Fucus are the most prevalent types.

c) Red Algae (Rhodophyceae): Rhodophyceae members are usually referred to as red algae due to the preponderance of the red pigment r-phycoerythrin in their bodies. The majority of red algae is marine, with higher amounts seen in warmer locations. They may be found in both well-lighted areas near the top of the sea and at considerable depths in oceans where relatively little light penetrates. The red thalli of the majority of red algae are multicellular. Some of them have complicated bodily structures. Polysiphonia, Porphyra, Gracilaria, and Gelidium are the most prevalent members.

Algae Biofuel

Algae may now be used as a fuel source because to recent advances in science and technology. Global demand for petroleum goods, along with worsening environmental health, has pushed the usage of environmentally friendly alternatives such as algae biofuel. As a result, algal fuel is becoming a more realistic alternative to existing fossil fuels. It is utilised in the manufacture of everything from "green" diesel to "green" jet fuel. It is comparable to other biofuels derived from corn and sugar cane.

Corals

Corals are calcareous rocks formed from the skeletons of tiny marine creatures known as polyps. Polyps take calcium salts from saltwater and use them to build rigid skeletons that protect their fragile bodies. Corals grow from these skeletons. Coral colonies are attached to the stony seabed. On the skeletons of dying polyps, new generations emerge. Corals are cemented calcareous stone masses that grow upwards and outwardly as tubular skeletons. A reef is the shallow rock formed by these depositions. These reefs eventually become islands. Corals come in a variety of shapes and colours, depending on the salts or elements they are comprised of. Over time, the steady growth of corals emerges on the water surface in various forms. Small marine plants (algae) deposit calcium carbonate, which aids in coral formation.

Ideal condition for Coral growth

a) Corals grow in tropical waters between the latitudes of 30°N and 30°S.
b) The best depths for coral growth are 45 to 55 metres below sea level, where there is enough of sunshine.
c) The water temperature should be approximately 20 degrees Celsius.
d) Clear salt water is ideal for coral growth, however both fresh and very saline water are detrimental to polyp growth.
e) A sufficient supply of oxygen and tiny marine food, known as plankton, is required for development and survival. Corals develop more quickly on the seaward side because the food source is more plentiful.

Types of Coral Reefs

Coral reefs are categorised based on large-scale reef morphology; the size and shape of a reef, as well as its relationship to neighbouring land. This is typically (but not always) enough to tell one kind from the others. The dominating groups of animals and plants, as well as their ecological interactions, frequently overlap among the various reef types (within a particular biogeographic area). Coral reefs are classified into three types (Fig.18.4):

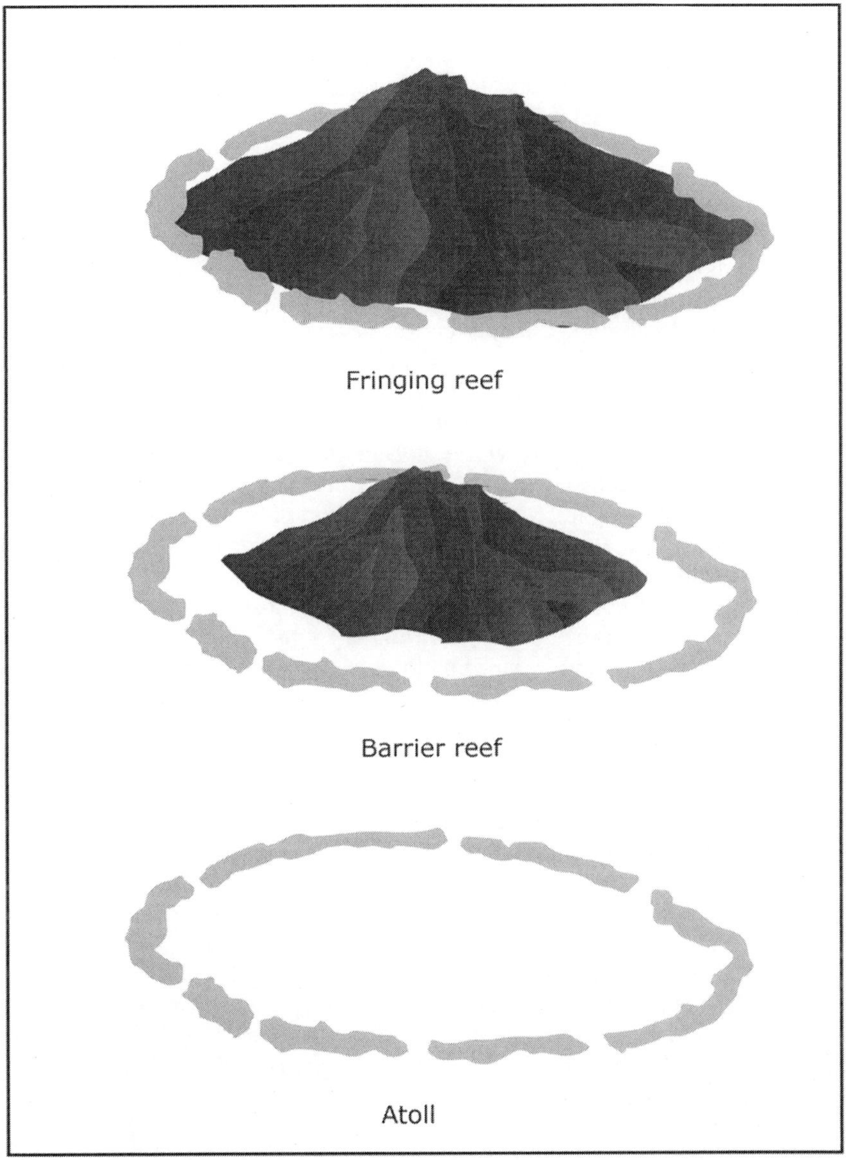

Fig. 18.4: Diagram showing the different types of coral reefs

a) Fringing Reef: It is by far the most prevalent of the three major coral reef types. It is a coral platform connected to a continental shore or an island by a short, shallow lagoon or channel. A bordering reef is a thin strip that ranges in width from 0.5 km to 2.5 km. This reef forms from the deep sea floor, with the seaward side falling sharply into the deep sea. Coral polyps do not spread due to abrupt and substantial changes in depth. A bordering reef's surface is rough because it is covered with coral remnants that form a boulder zone or reef flat.
b) Barrier Reef: The biggest of the three reefs, it stretches over hundreds of kilometres and is several kilometres wide. It wraps around the shore or an island in a broken, uneven ring, running practically parallel to it. A barrier reef is distinguished by its distance from the shore and a larger and deeper lagoon, which is occasionally united with saltwater by one or more channels cutting across the barrier reef. A barrier reef is extremely thick, reaching depths of up to 180 metres below the surface, with the seaward side sloping sharply into the deep sea. A barrier reef's surface is covered with coral debris, stones, and sand. The Great Barrier Reef, off the coast of north-eastern Australia, is the most renowned example of this type of reef, measuring 1900 km long and 160 km wide.
c) Atoll: It is a ring-shaped reef that partially or completely encircles a lagoon. The lagoon may have a flat appearance, but the reef's seaward side drops sharply into the deep sea. The lagoon generally has a depth of about 80–150 metres and may be connected to saltwater via a variety of canals that cut over the reef. Atolls are located at large distances from deep-sea platforms, where underwater features, such as a submerged island or a volcanic cone that may reach a level suitable for coral development, may aid in the construction of atolls. The Pacific has much more atolls than any other ocean. Atolls are well-known examples, such the Fiji atoll and the Funafuti atoll near Ellice Island. The Lakshadweep islands also include a considerable number of atolls. There are various atoll forms in the Indian Ocean. Examples may be found in the Maldives and Chagos island groups, as well as in the Seychelles and Cocos Island groups.

Distribution of Coral Reefs

a) Tropical and subtropical environments are home to the majority of reef-building corals. These are most common between 30° north and 30° south latitudes.

b) The Indonesian/Philippine archipelago boasts the highest concentration of reefs and the most diverse coral species in the world.
c) Other places of reef concentration include Australia's Great Barrier Reef, the Red Sea, and the Caribbean, the latter of which has significantly lesser diversity than any major Indo-Pacific regions.
d) World's major coral reef regions:
1) Caribbean / western Atlantic
2) Eastern Pacific
3) Central and western Pacific
4) Indian Ocean
5) Arabian Gulf
6) Red Sea

Coral Reefs Bleaching

a) Coral reef habitats throughout the world have suffered significant deterioration in recent decades. Coral reef disturbances encompass both manmade and natural phenomena.
b) Recent rapid coral reef deterioration appears to be mostly due to anthropogenic factors (overfishing, overexploitation, increasing sedimentation, and nutrient overloading).
c) Natural disturbances that cause coral reef degradation include intense storms, floods, very high and low temperatures, El niño Southern Oscillation (ENSO) occurrences, and so on.
d) Coral bleaching happens when the link between the coral host and the marine algae that gives coral its colour breaks broken. Without the sea algae, the coral animal's tissue looks translucent, revealing the coral's dazzling white skeleton.
e) When corals bleach, they begin to starve. While some corals can feed themselves, most corals cannot survive without algae. Corals may reclaim their algae, revert to their original colour, and live if conditions return to normal. However, this stress is likely to result in reduced coral growth and reproduction, as well as increased disease susceptibility.
f) Bleached corals frequently die if the stress is not relieved. Coral reefs that have had a high incidence of coral mortality as a result of bleaching might take years or decades to recover.

Causes of Coral Bleaching

Because coral reef bleaching is a general reaction to stress, it can be caused by a range of circumstances, either alone or in combination. As a result, determining the reasons of bleaching incidents is challenging. The following causes have all been connected to outbreaks of coral reef bleaching. The following causes have all been connected to outbreaks of coral reef bleaching:

a) Temperature: Because coral species exist within a very small temperature range, both low and high water temperatures can cause coral bleaching. Bleaching events occur as a result of abrupt temperature reductions associated with significant upwelling occurrences, seasonal cold-air outbreaks, and so on.
b) Solar Irradiance: Bleaching occurs disproportionately in shallow-living corals and on exposed tops of colonies during the summer months, when seasonal temperature and irradiation maxima occur.
c) Subaerial Exposure: Bleaching can occur when reef flat corals are suddenly exposed to the atmosphere during occurrences like as very low tides, ENSO-related sea level dips, or geological uplift.
d) Fresh water dilution: Coral reef bleaching has been linked to the rapid dilution of reef waters caused by storm-generated precipitation and runoff.
e) Other: Other factors include a rise in inorganic nutrient concentrations, sedimentation, oxygen deprivation induced by an increase in zooplankton levels as a result of overfishing, ocean acidification, salinity changes, sea level change owing to global warming, cyanide fishing, and so on.

CHAPTER 19

GLOBAL WARMING AND THE GREENHOUSE EFFECT

Concept

The climate of a region is the long-term average of its weather events. The climate of the Earth is not constant. It has altered several times over billions of years due to natural reasons such as sun spots, ice age glaciations, and so on. "Climate change" refers to a change in climate caused by human activity that modifies the composition of the global atmosphere, in addition to natural climate variability observed over comparable time periods. Worldwide climate change and global warming are, without a doubt, a heated topic. Despite the fact that the component of global warming may not be breaking news, the expected consequences often cause a major catastrophe. The idea of global warming was pondered until 1938, when Guy S. Callender proposed that the warming trend discovered in the nineteenth century was driven by a 10 percent increase in atmospheric carbon dioxide from fossil fuel use. Researchers were unconcerned at this time because they knew that most of the carbon dioxide emitted by humans had broken down safely in the seas and oceans. Hans Suess and Roger Revelle disseminated this concept in 1957, when they discovered a sophisticated chemical buffering system that prevents ocean water from retaining too much carbon dioxide from the atmosphere. Thomas Chrowder Chamberlin proposed the concept that variations in climate might be caused by changes in atmospheric carbon dioxide concentrations in 1899. Now that we are in the midst of a pandemic, we have all realized how connected and essential our well-being is with our environment, ecosystem, and biodiversity, and we consider the climate issue as a true crisis.

Climate change refers to the observable consequences of the ongoing warming trend. Climate change is often defined as significant variations in temperature, rainfall, snow, and wind patterns that persist decades or longer. Humans contribute to climate change by using vast amounts of fossil fuels (coal, oil, and natural gas), as well as deforestation (when forests are cut down or burned, they can no longer store carbon, and the carbon is released to the atmosphere). James Hansen is known as the father of climate change.

Factors responsible for Climate Change

The amount of energy entering and exiting the planet's system has an impact on its temperature. The Earth gets warmed when incoming solar energy is absorbed by the Earth system. The Earth cools when the sun's energy is reflected back into space. Changes in the Earth's energy balance can be caused by both natural and manmade processes. The major factors are explained below:

a) Natural factors: A variety of natural causes can influence the Earth's climate. Continental drift, volcanoes, ocean currents, the earth's tilt, and comets and meteorites are among the most notable. Natural forces have a long-term impact on climate change, lasting thousands to millions of years in terms of geologic time scale.

1) Continental drift: Before 200 million years, the continents that we see now were not the same. It was formed millions of years ago when the continent began progressively drifting apart owing to Plate movement. The physical characteristics of the landmass, their position, and the position of water bodies, as well as the passage of ocean currents and winds, all had an influence on the climate. The drift process continues today, with the Himalayan range increasing by around 1 mm (millimetre) every year as the Indian landmass moves slowly but gradually towards the Asian landmass.
2) Variation in the Earth's orbit: The seasonal distribution of sunlight reaching the Earth's surface is intimately tied to Earth's orbit, and a little change in Earth's orbit causes a change in distribution around the world. The yearly averaged sunlight does not fluctuate much, but the regional and seasonal distribution can shift dramatically. The three components or variations of the earth's orbit are eccentricity, obliquity, and precession, and these variations are known as Milankovitch cycles (Fig. 19.1).

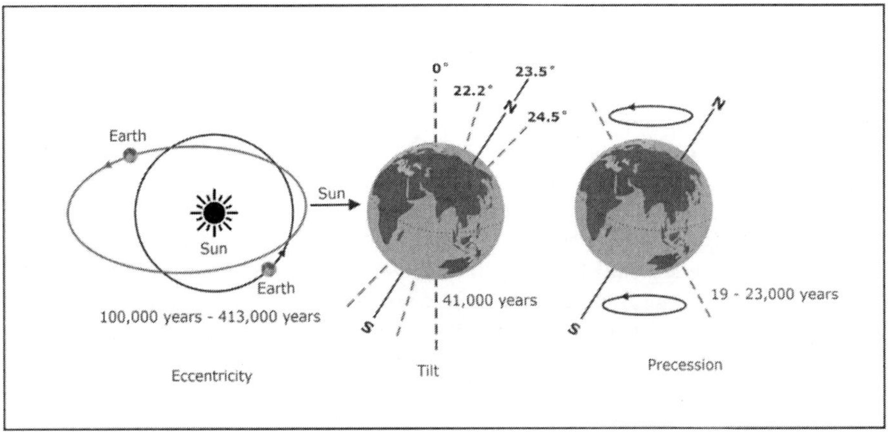

Fig. 19.1: Milankovitch cycles

i. Eccentricity: Eccentricity has two main frequencies: a cycle of 100,000 years and a longer cycle averaging 413,000 years.
ii. Obliquity: The tilting of the earth's axis to its orbital plane during a period of approximately 41.000 years, from 22.1 to 24.5 degrees, is known as obliquity.
iii. Precession: Precession is the movement of the earth's rotation axis, which does not always point in the same direction toward the sky. Rather, the rotation axis indicates a space-side clockwise cycle with a length of 19,000–23,000 years, similar to the spinning of a wobble top.

3) Plate tectonics: Because of temperature variations in the Earth's core, mantle plumes and convection currents drive the Earth's Plates to adjust, resulting in plate rearrangement. This can have an impact on both global and local climate and atmospheric patterns. The geometry of the oceans is influenced by the location of the continents, which governs the patterns of ocean circulation. The positions of the oceans have a vital role in managing the passage of heat and moisture throughout the world, and hence in defining global climate. The formation of the Isthmus of Panama around 5 million years ago, which cut off direct mixing between the Atlantic and Pacific seas, is a recent example of tectonic influence on ocean circulation.
4) Volcanic activity: When a volcano erupts, the gases and dust particles released partially cover the Sun's rays, causing the

weather to cool. Sulphur dioxide combines with water to form small sulphuric acid droplets that can float for years. Even if volcanic activity only lasts a few days, the massive volumes of gases and ash released can have a long-term influence on climate patterns. Dinosaurs became extinct as a result of dust particles from a meteorite impact blocking the sun's light, preventing photosynthesis.

5) Ocean currents and circulation: Ocean currents are a fundamental component of the climatic system that is driven by horizontal wind forces that cause seawater displacement against the sea surface. The climate of the region is greatly impacted by temperature variations in the ocean. On longer time periods, changes to ocean processes such as thermohaline circulation play an important part in spreading heat by moving water at a very slow and extremely deep rate, and in the long-term redistribution of heat throughout the world's ocean. Much of the heat escaping from the seas comes in the form of water vapour, which is one of the most prevalent greenhouse gases on the planet. Water vapour, on the other hand, leads to the production of clouds, which shade the surfaces and have a net cooling effect.

b) Anthropogenic factors: Short-term climate changes are caused by anthropogenic or man-made influences. Variations in the energy balance of the Earth-atmosphere system cause changes in weather and climate. Since the beginning of the twentieth century, scientists have been noticing a shift in the climate that cannot be attributable to any of the "natural' forces of the past. Global warming has happened at a faster rate than any previous human-caused climate change, and it is thus of enormous interest and relevance to the human population. Anthropogenic (human-induced) climate change is caused by greenhouse gases, aerosols, and changes in land use patterns.

1) Greenhouse gases: The Earth has a natural greenhouse effect, in which certain gases (known as greenhouse gases) in the atmosphere allow sunlight to enter but absorb heat radiation. They keep the Earth's average surface temperature about 15 degrees Celsius. Without the natural greenhouse effect, the average surface temperature of the Earth would be roughly -19 degrees Celsius. As human activity has expanded since the beginning of the industrial revolution, so has the production of massive amounts of greenhouse gases, resulting in more heat

absorption in the atmosphere and an increase in global temperature. While greenhouse gases are mostly transparent to incoming solar radiation, they absorb the majority of the infrared released by the earth's surface. The primary greenhouse gases are:

i. Water Vapour: It is the most prevalent greenhouse gas (GHG), although it only stays in the atmosphere for a brief period of time. The amount of water vapour varies greatly depending on time, geography, and altitude. It is not regarded as the most significant greenhouse gas.
ii. Carbon dioxide (CO2): Carbon dioxide (CO2) is the most significant GHG, and it is generated both naturally and as a result of human activity. CO2 is naturally emitted into the atmosphere by volcanic eruptions and animal respiration. It is also emitted as a result of human activities such as deforestation and the use of fossil fuels for energy. CO2 remains in the atmosphere for a long period, increasing its influence. Humans have raised the concentration of CO2 in the atmosphere by 30 percent since the industrial revolution. Deforestation contributes to global warming by reducing the number of plants available to absorb carbon dioxide from the atmosphere.
iii. Methane: Methane is mostly produced through the decomposition of organic materials, such as in landfills and agriculture, as well as by the digestion of ruminants (cows, goats, etc). It is a more potent greenhouse gas than CO2 because it can absorb more heat. It is, however, significantly less plentiful in the atmosphere.
iv. Nitrous oxide: Nitrous oxide is a very powerful greenhouse gas that is abundantly generated in agriculture, particularly in the manufacture and application of organic fertilisers. It is also created when fossil fuels are burned.
v. Fluorinated gases: They are released by a range of industrial processes, including aluminium and semiconductor manufacturing and substitution for Ozone-Depleting Substances. In comparison to other greenhouse gases, several fluorinated gases have extremely high global warming potentials (GWPs). Fluorinated gases are well-mixed in the atmosphere and spread all over the planet when they are emitted. Fluorinated gases, in general, are the most potent and long-lasting form of greenhouse gas released by

human activity. Fluorinated gases includes HFCs (hydrofluorocarbons) which are utilised in the manufacture of refrigerants, aerosol propellants, solvents, and fire retardants. Because they do not damage the stratospheric ozone layer, these chemicals were developed as a substitute for chlorofluorocarbons (CFCs), perfluorocarbons (PFCs), and sulphur hexafluoride (SF6).

2) Aerosols: Atmospheric aerosols have two major effects on climate: they scatter and absorb solar and infrared radiation, and they influence the microphysical and chemical characteristics of clouds, as well as their duration and extent. Solar radiation scattering cools the Earth, but solar radiation absorption by aerosols directly warms the air rather than enabling sunlight to be absorbed by the Earth's surface. Aerosols can impact climate directly by absorbing or reflecting incoming solar radiation, but they can also have an indirect effect on climate by altering cloud formation or cloud features. Winds and upper-level circulation in the atmosphere can carry aerosols thousands of kilometres from their origins. Aerosols are classified into two types:

   i. Natural Aerosols: Windblown mineral dust formed in arid and semi-arid regions is one type of aerosol. Volcanic eruptions, which create sulphate aerosols, and biogenic sources (e.g., phytoplankton), which produce dimethyl sulphide, are two other natural producers of aerosols. Other major biogenic aerosols, such as tarpenes, are naturally produced by particular types of trees or other plants.

   ii. Anthropogenic Aerosols: Human activity has increased the amount of aerosols in the atmosphere in a variety of ways. Ammonia, which is commonly employed in fertilisers or is generated during the combustion of plants and other organic materials, is a major source of nitrate aerosols. Sulphur dioxide is produced when coal and oil are burned, which is another primary source of sulphate aerosols. Dust is a common by-product of agricultural procedures. Burning biomass produces a mixture of organic droplets and soot particles. A diverse range of aerosols are produced by industrial activities. Exhaust emissions from transportation create a rich mix of pollutants that are either aerosols from the start or are transformed to aerosols by chemical reactions in the atmosphere.

3) LULC (Land Use Land Cover): Cutting down forests to make way for farming changed the quantity of sunlight reflected back into space from the ground (the surface albedo). It is believed that half of all land use changes happened during the industrial age, with much of it attributable to the replacement of forests with agricultural farming and grazing pastures across Eurasia and North America. The albedo of previously forested snow-covered land has risen owing to deforestation in the High Latitudes region. This is due to the fact that snow on trees only reflects roughly half of the sunlight that falls on it, but snow on open ground reflects around two-thirds. Overall, the increasing albedo ovcr thc agricultural regions of Eurasian and North American has had a cooling impact. Tropical deforestation, which alters evapotranspiration rates (the amount of water vapour emitted into the atmosphere by evaporation and transpiration from trees), desertification, which increases surface albedo, and the general effects of agriculture on soil moisture characteristics are all examples of climate change. Analysing changing land use surface records with satellite measurements of plant cover attributes reveals that forest clearing for agriculture and irrigated farming in arid and semi-arid regions are two primary causes of climatically significant land cover changes.

Greenhouse gases and global climate change

CO2 levels at Mauna Loa Observatory in Hawaii range from 315 ppm in 1958 to 419.53 ppm in 2021, with a small seasonal variation per year. The two most important human activities that contribute to the gradual rise of CO_2 levels in the atmosphere are the increased burning of fossil fuels and the loss of CO_2 sinks, i.e., the gradual destruction of forest cover. Coal and oil consumption have increased many fold over the years as the primary energy sources. The increase in CO2 levels has an immediate impact on the rise in global temperature. There are several cases where the increase in CO2 levels in the atmosphere has a positive link with the rise in temperature on Earth's surface in the prehistoric period or even in the post-industrialisation phase. Other GHGs, such as NOx, CH4, and CFCs, have been added to the atmosphere at an alarming rate throughout the years. According to NASA, CFCs are increasing at a rate of around 5 percent per year, whereas CH4 is increasing at a rate of about 1 percent per year. Total, assuming the above GHG growth continues at its current rate, the overall rise of each polluting gas that contributes to the greenhouse effect will be on the higher side, with a doubling point somewhere in 2030. As a result, limiting the present pace of GHG emissions is essential. Though the immediate influence of GHGs on climate is slow and invisible, the long-term impact on climate change is worrisome and irreversible.

During the Industrial Revolution, temperatures rose by 1.1 degrees Celsius, and we are now feeling the consequences. Many climate scientists think that world temperatures should not increase beyond 1.5 degrees Celsius to guarantee a sustainable future, and the Conference of Parties (COP 26) agrees on this. It is self-evident that a half-degree difference is significant for ecosystem sustainability. We are only able to go up to a 2 degree rise before massive and potentially irreversible consequences are felt. Carbon emissions must decrease by 2030 if we are to remain below 1.5 degrees Celsius, which means we only have 9 years or 3081 days to reduce world emissions by at least 50 percent. To get to this point, virtually every element of civilization must undergo fast and scientific changes. A climate emergency is required. There is no question that we are in the midst of a crisis, but it also presents a chance to rethink all areas of society with a scientific mentality in order to rebuild our economy and energy sectors, resulting in the creation of millions of employment. This is a once-in-a-lifetime opportunity to make the world a greener, cleaner, and more equitable place for future generations; otherwise, we may find ourselves in the last and most recent period of the quaternary in the Cenozoic era, with no hope for the future, and on the path to extinction with many other species on our planet.

Consequences of Global warming and climate change

Rising global temperatures are accompanied by changes in weather and climate, such as increased floods, droughts, or strong rain, as well as more frequent and severe heat waves. The planet's oceans and glaciers have also seen significant changes, with oceans warming and growing more acidic, ice caps melting, and sea levels increasing. As these changes grow more obvious over the next few decades, they will represent a significant challenge to our society and environment.

a) High Temperature: Greenhouse gases emitted by power plants, vehicles, deforestation, and other causes are warming the Earth. In fact, the five hottest years on record have all occurred since 1997, and the 10 hottest years on record have all occurred since 1990, including the warmest years on record – 2005 and 2010.

Temperature increases are to blame for an increase in heat-related mortality and sickness, rising seas, increasing storm strength, and many other harmful consequences. The Earth's average temperature rose by 1 degree Fahrenheit over the twentieth century to its highest level in four centuries — estimated to be the quickest rise in a thousand years. According to research, if heat-trapping carbon emissions are not controlled, average surface temperatures might rise by 3 to 5 degrees Celsius.

b) Change in Landscape: Increasing temperatures and shifting rain and snow patterns have caused trees and plants to migrate to the poles and up mountain slopes across the planet. These vegetation shifts will have an impact on much of the conservation community's previous work, with the potential to permanently alter the appearance of Conservancy preserves, local land trusts, and even our national parks. As plant populations strive to adapt to climate change by shifting to cooler locations, the animals that rely on them will be pushed to do the same. Some species and populations, such as polar bears and alpine meadows, may be left with no suitable habitat, placing much of our valued biodiversity in danger.

c) Ecosystem risk: Climate change is altering weather and vegetation patterns across the world, pushing animal species to relocate to new, cooler places in order to live. The fast pace of climate change is expected to outstrip many species' abilities to move or adjust. Experts anticipate that if the present pace of warming continues, one-fourth of Earth's species would become extinct by 2050.

d) Ocean acidification and sea level rise: Sea levels increase when the Earth warms because warmer water takes up more space than cooler water, a phenomenon known as thermal expansion. Melting glaciers exacerbate the situation by sending even more fresh water into the seas. Rising seas endanger populous low-lying regions, islands, and dense coastal communities, eroding shorelines, causing property damage, and destroying ecosystems such as mangroves and wetlands that shelter coasts from storms. In the last century, sea levels have increased by 4 to 8 inches. Current forecasts indicate that sea levels will continue to increase by 4 to 36 inches over the next 100 years. Carbon dioxide (CO2) concentrations in the atmosphere have increased CO2 absorption in the ocean, lowering pH and making the oceans more acidic. Many marine organisms, including plankton, mollusks, and other shellfish, would suffer if the water became more acidic. Corals, in particular, can be extremely susceptible to increased acidity since it is difficult for them to form and maintain the skeletal structures required for support and protection.

e) Drought, fire and floods: Higher temperatures cause more moisture to evaporate from land and water, resulting in drought in many locations. Drought-affected areas are more prone to floods when rain comes. Droughts will grow more common and severe as global temperatures increase, possibly wreaking havoc on agriculture, water supplies, and human health. This effect has already been noticed in portions of Asia and Africa where droughts have grown longer and more extreme. Forest fires are also more likely when temperatures are high and conditions are dry.

f) Intensified storms: According to scientific evidence, climate change will intensify hurricanes and tropical storms, leading them to stay longer, unleash greater winds, and do more damage to coastal ecosystems and towns. The biggest issue is rising ocean temperatures, because hurricanes and tropical storms acquire their energy from warm water. Storms will contain more energy when sea surface temperatures increase. Simultaneously, other causes such as rising sea levels, vanishing wetlands, and growing coastal development threaten to exacerbate the damage inflicted by hurricanes and tropical storms.

g) Disease: The dangers of heat-related illness and possibly mortality for the most vulnerable human groups rise as temperatures rise. Scientists have connected the deadly heat waves to climate change and predict that more will follow. Extreme heat waves, for example, killed over 20,000 people in Europe and over 1,500 people in India in 2003. Climate change accelerates the spread of infectious diseases, mostly because warmer temperatures allow disease-carrying insects, animals, and bacteria to live in previously inaccessible locations. Diseases and pests that were historically exclusive to the tropics, such as mosquitos that spread malaria, may find suitable circumstances in previously inhospitable places. Climate change, according to the World Health Organization (WHO), may have caused more than 150,000 fatalities in the year 2000 alone, with a rise in mortality predicted in the future.

h) Economic losses: According to research, if no action is made to reduce global carbon emissions, climate change might cost between 5 and 20 percent of the yearly global GDP. In comparison, it would require 1 percent of GDP to mitigate the most harmful impacts of climate change. Climate change may result in drastically lower lake levels, changing coastal ecosystems and costing millions of dollars in port and shore infrastructure relocation.

i) Food security: Because solar radiation, temperature, and precipitation are the primary drivers of crop development, agriculture has always been strongly dependent on climatic patterns and fluctuations.Humans have been modifying the global climate since the industrial revolution by generating massive volumes of greenhouse gases into the atmosphere, resulting in greater global temperatures, disrupting hydrological systems, and increasing climatic unpredictability.Climate change will have a profound influence on agricultural conditions, food supply, and food security.

Greenhouse effect

A greenhouse is a glass structure that may be used to cultivate plants. The sun's rays warm the plants as well as the air within the greenhouse. The heat trapped within cannot leave and hence heats the greenhouse, which is necessary for plant development. The same is true of the earth's atmosphere. The sun warms the earth's atmosphere during the day. When the earth cools down at night, the heat is reflected back into the atmosphere. The heat is absorbed by the greenhouse gases in the earth's atmosphere throughout this process (Fig. 19.2). This is what keeps the earth's surface warm, allowing living things to survive. However, due to rising amounts of greenhouse gases, the earth's temperature has risen significantly. This has had a number of severe consequences.

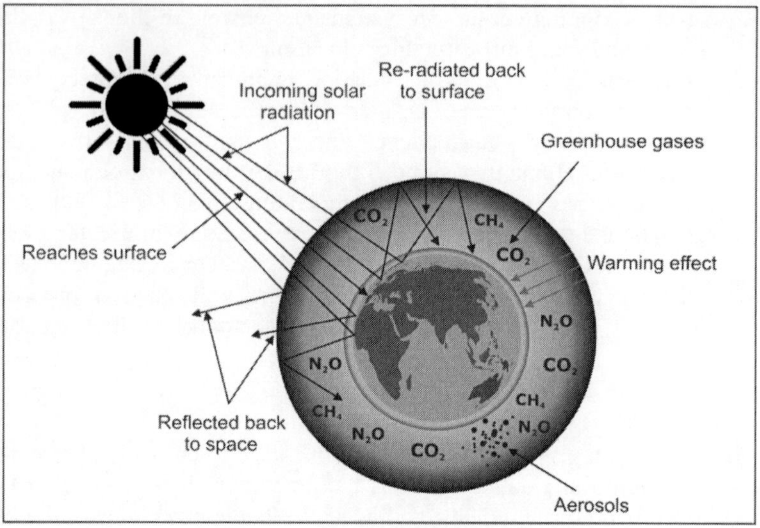

Fig. 19.2: Greenhouse effect

Effects of Greenhouse effect

a) Global warming: It refers to the progressive rise in the average temperature of the Earth's atmosphere. The major source of this environmental problem is the increased production of greenhouse gases such as carbon dioxide and methane from the combustion of fossil fuels, as well as emissions from automobiles, factories, and other human activities.

b) Depletion of Ozone layer: The Ozone Layer shields the Earth from the sun's dangerous UV radiation. It can be found in the stratosphere's higher reaches. The depletion of the ozone layer allows dangerous UV radiation to enter the earth's surface, which can cause skin cancer and radically alter the climate. The build-up of natural greenhouse gases such as chlorofluorocarbons, carbon dioxide, methane, and others is the primary source of this phenomenon.

c) Smog and Air pollution: Smog is produced when smoke and fog combine. It can be caused by both natural and man-made factors. Smog is caused by the build-up of additional greenhouse gases, such as nitrogen and sulphur oxides. Automobile and industrial emissions, agricultural fires, natural forest fires, and the reactivity of these chemicals among themselves are the key contributors to smog production.

d) Acidification of Water bodies: The increase in the overall amount of greenhouse gases in the atmosphere has acidified the majority of the world's water bodies. The greenhouse gases combine with rainwater to form acid rain. This causes water bodies to become acidic. Furthermore, precipitation takes impurities with it and deposits them in rivers, streams, and lakes, producing acidification.

Global Warming potential

The global warming potential of each gas reflects its influence on global warming. In terms of climatic influence, the two most crucial features of a GHG are how well the gas absorbs energy (preventing it from quickly escaping to space) and how long the gas remains in the atmosphere. The Global Warming Potential (GWP) of a gas is a measure of how much energy a gas absorbs over a specific time period (usually 100 years) when compared to carbon dioxide. Gases with a greater GWP absorb more energy per pound than gases with a lower GWP, contributing more to global warming. The following Table 19.1 shows the GWP and lifetime of greenhouse gases:

Table 19.1: Showing global warming potential of greenhouse gases

Gas	GWP (100 years)	Lifetime (years)
Carbon dioxide	1	100
Methane	21	12
Nitrous oxide	310	120
Hydrofluorocarbons (HFCs)	140–11,700	1–270
Perfluorocarbons (PFCs)	6,500–9,200	800–50,000
Sulphur hexafluoride (SF6)	23,900	3,200

Geological Records of Climate Change

Around 55 million years ago, near the end of the Paleocene, temperatures soared by around 6degreesCelsius worldwide and by 10-20 degree Celsius at the poles. Carbon isotope measurements reveal that this warming episode (named the Paleocene-Eocene Thermal Maximum, or PETM) was accompanied by a massive release of carbon into the ocean and atmosphere of 1500-2000 billion tonnes or more. This carbon influx might have been caused primarily by the breakup of methane hydrates beneath the deep sea floor, which could have been initiated by volcanic activity on top of an underlying progressive global warming trend that peaked 50 million years ago in the early Eocene. CO2 levels were already high at the time, but the additional CO2 pumped into the atmosphere and ocean made the water significantly warmer, less oxygenated, and more acidic, resulting in the demise of numerous deep sea floor organisms. Similar abrupt warming occurrences have occurred in the distant past, for example, between 120 and 183 million years ago. All of these episodes required the Earth's temperature to recover for 100,000 years or more, demonstrating that a CO2 emission of this scale may alter the Earth's climate for that length of time. Air bubbles stored in Antarctic ice cores can be used to sample the atmosphere over the last 800,000 years. CO2 and other gas concentrations in these bubbles closely mirror the trend of rising and dropping temperature during glacial and interglacial periods. CO2 levels, for example, ranged from roughly 180 ppm (parts per million) in glacial maxima to around 280 ppm during interglacials. Temperature and CO2 climbed simultaneously over many thousand years during glacial to interglacial warmings, while the best estimate from the end of the last glacial is that temperature began to rise a few decades before CO2 exhibited any reaction. Palaeoclimatologists believe that early warming induced by changes in the Earth's orbit and axial tilt eventually led CO2 to be released from the rising ocean, reinforcing the temperature rise already in train. Additional positive feedback reinforcing the temperature rise would have come from increased water vapour evaporated from the warmer ocean, which is another greenhouse gas, as well as a decrease in sea ice and, eventually, the size of the northern hemisphere ice sheets, resulting in less reflection of solar energy back into space. The current amount of CO2 in the atmosphere is 419 ppm. In the previous 200 years, it has expanded by one-third. Half of the rise occurred in the previous 30 years. When compared to the spectrum of CO2 in air bubbles trapped in ice cores, this quantity and pace of growth are unique (170–300 ppm). There is some evidence that the rate of growth in CO2 in the atmosphere during the sudden global warming 183 million years ago (Early Jurassic) and maybe 55 million years ago (the PETM) was roughly comparable to today's pace. The fossil and sedimentary record demonstrates that the Earth has experienced significant temperature variations during the last 200 million years, from warmer than the current climate to far colder, on many different timelines. While earlier climatic changes may be linked to geological occurrences, it is impossible to link the Earth's warming since 1970 to anything identifiable as having a geological origin (such as volcanic activity, continental displacement, or changes in solar energy received). This recent warming is accompanied by a rise in CO2 and a reduction in Arctic sea ice, both of which would be predicted to warm the climate based on physical theory and geological analogues.

The Intergovernmental Panel on Climate Change analysed several lines of data that clearly suggest that a substantial part of the contemporary rise in CO2 is the consequence of burning fossil fuels, with some contribution from cement manufacturing and some from deforestation.

Major Events of the Debate on Climate Change

a) 1992 Rio de Janeiro The Earth Summit: The United Nations Conference on Environment and Development (UNCED), often known as the Rio de Janeiro Earth Summit, Rio Summit, Rio Conference, and Earth Summit (Portuguese: ECO92), was a significant United Nations conference held from June 3 to 14, 1992, in Rio de Janeiro (Brazil). There were 172 governments in attendance, with 116 sending their heads of state or government. The first document, known as the Rio Declaration, comprised 27 principles intended to assist countries in future sustainable development. Agenda 21 is a non-binding action plan promoting sustainable development. The Forest Principles are technically known as the "Non-Legally Binding Authoritative Statement of Principles for a Global Consensus on the Management, Conservation, and Sustainable Development of All Types of Forests." It is non-binding and includes several recommendations for forestry conservation and sustainable development.
b) Kyoto protocol: The Kyoto Protocol was an international convention that extended the 1992 United Nations Framework Convention on Climate Change by committing state parties to limit greenhouse gas emissions, based on scientific consensus that global warming is occurring and that human-caused CO2 emissions are to blame. The Kyoto Protocol was approved on December 11, 1997, in Kyoto, Japan, and went into effect on February 16, 2005. In 2020, there were 192 parties to the Protocol (Canada withdrew from it in December 2012). After the first commitment period of the Protocol concluded in December 2012, participants in the Kyoto Protocol convened in Doha, Qatar, to accept an addendum to the original Kyoto accord. This was known as the Doha Amendment, and it included additional emission-reduction objectives for these nations

for the second commitment period, 2012–2020. All UNFCCC parties ratified another deal, the Paris Climate Agreement, during the 2015 Sustainable Development Summit in Paris, thus replacing the Kyoto Protocol. This international convention aims to carry out the goals of the United Nations Framework Convention on Climate Change (UNFCCC) by reducing greenhouse gas concentrations in the atmosphere. The following important greenhouse gases are addressed:

1) Carbon dioxide (CO2)
2) Methane (CH4)
3) Nitrous oxide (N2O)
4) Hydrofluorocarbons (HFCs)
5) Perfluorocarbons (PFCs)
6) Sulphur hexafluoride (SF6)

c) World summit on Sustainable development (WSSD): The United Nations World Summit on Sustainable Development, often known as the Earth Summit 2002, was an international convention on the environment and sustainable development that took place from August 26 to September 6, 2002, in Johannesburg, South Africa. The Johannesburg Declaration on Sustainable Development, an international agreement on the environment and sustainable development, was issued during the Earth Summit in 2002. The Johannesburg Declaration reaffirms the majority of the Rio Declaration on Environment and Development and Agenda 21, international accords from the Earth Summit in 1992. The Johannesburg Declaration includes targets and deadlines for accomplishing Agenda 21 goals. Many environmental organisations have criticised the Earth Summit 2002 for failing to produce any new, significant international accords. Development that satisfies the requirements of the present without jeopardising future generations' ability to satisfy their own needs' is called sustainable development (SD). The Brundtland Commission provided the most commonly acknowledged concept of SD in their report Our Common Future (1987).SD advocates for collaborative efforts to create an inclusive, sustainable, and resilient future for people and the earth. Growth in the economy, social inclusion, and environmental conservation are all important factors. It is The Energy and Resources Institute's yearly flagship event (TERI).It is the only global summit taking place in the developing world.It provides a forum for world leaders and practitioners to debate and reflect on global climate challenges.By bringing together the world's most enlightened leaders and intellectuals on a single platform, it aims to propose long-term solutions for the benefit of the global community.It is carrying on the tradition of the Delhi Sustainable Development Summit (DSDS), which began in 2001 with the intention of making'sustainable development' a worldwide shared goal. The Energy and Resources Institute's (TERI) current edition of the 21st World Sustainable Development Summit (WSDS) was inaugurated by Prime Minister Narendra Modi, who discussed planetary resilience in the aftermath of one of the many catastrophic weather disasters throughout the world in recent years. TERI is a pioneer in these areas, and WSDS 2022 will be critical in emphasising our role in delivering technological, policy, and knowledge solutions as India seeks to meet the pledges set at COP26.

d) Copenhagen Scenario: The Copenhagen Accord was reached during the UNFCCC COP15 in Copenhagen in 2009. The agreement called for clear emission promises from all major economies, including China and other big emerging countries for the first time. It is not, however, a legally enforceable agreement. The discussions for the Accord began with the Bali Action Plan in 2007 and lasted two years, culminating in the historic Accord. The US and the BASIC (Brazil, South Africa, India and China) nations wrote the Accord. It agreed to keep the Kyoto Protocol in place. According to the Accord, India committed to reducing its carbon intensity by 20–25 percent by 2020 compared to 2005 levels. However, all of the Accord's pledges are voluntary and not legally obligatory.

El-Nino Climate cycles and Southern oscillation

The El Niño–Southern Oscillation is a cyclical change in winds and sea surface temperatures over the tropical eastern Pacific Ocean. The surface waters of the tropical Pacific Ocean warm or cool by 1 degree Celsius to 3 degrees Celsius every three to seven years, compared to usual. El Niño is the warming phase of the sea temperature, while La Niña is the cooling phase. Thus, El Niño and La Niña are total opposites in the El Niño-Southern Oscillation (ENSO) cycle. The extreme stages of the ENSO cycle are El Niño and La Niña; between these two phases lies a third phase known as ENSO-neutral. These variations from typical surface temperatures can have large-scale effects on ocean processes as well as worldwide weather and climate (Fig. 19.3).

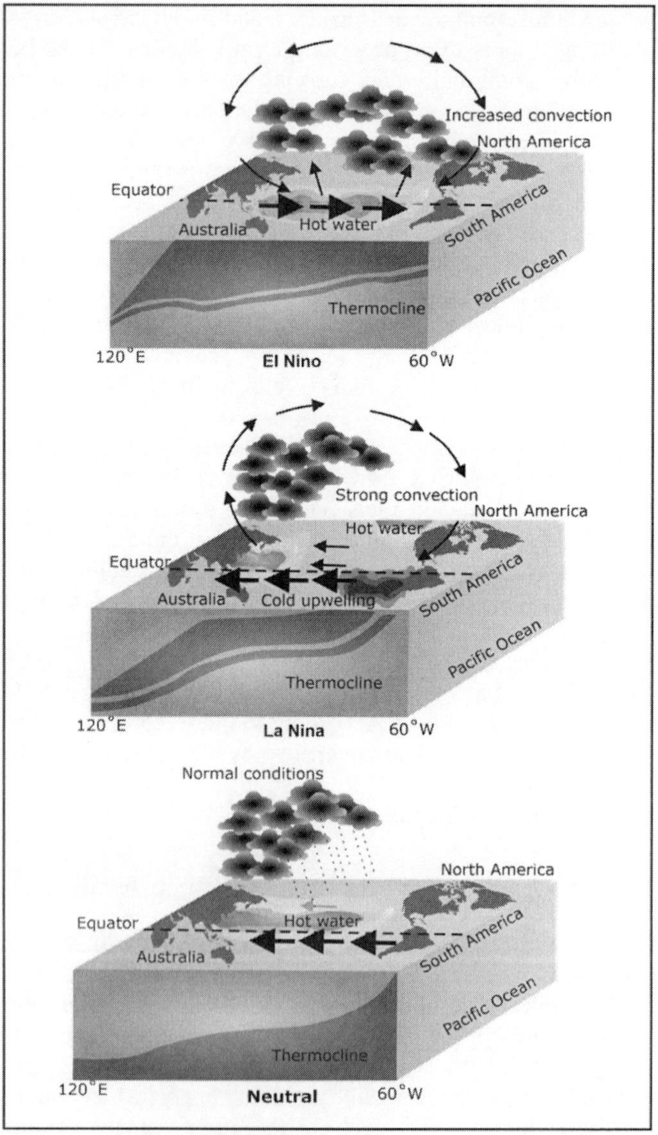

Fig. 19.3: Diagram showing stages of ENSO

ENSO Mechanism

a) The Neutral Phase:
   1) We have a state known as a Neutral phase under "normal" conditions.
   2) The warm surface water is initially pushed away from South America by Easterly Trade Winds (winds flowing from east to west). The Walker Circulation is the pattern of air rising in the west and descending in the east, with westward flowing air near the surface.
   3) In contrast, the depth of the warm equatorial top layer of the water has decreased off the west coast of South America, and the thermocline is closer to the ocean surface. This implies that cold, nutrient-rich water rises to the surface of the ocean.
   4) Warm surface water evaporates around tropical South-East Asia.
   5) This generates higher rainfall in Indonesia and neighbouring countries like as Australia's top end.
   6) Dry air sinks with cloud dispersion across the eastern tropical Pacific and off the coast of Peru, resulting in much drier weather there.
b) El Niño Phase: In Spanish, El niño means "small boy" or "Christ child." The phenomenon was so named because it was observed by South American fisherman in the early seventeenth century. The occurrences, i.e., warm Pacific Ocean waters, tended to occur in December, hence the name.
   1) During an El Niño phase, the Trade Winds lose much of their intensity and may even reverse into a westerly wind (west to east).
   2) As a result, warm surface water from tropical South-East Asia rushes towards South America's shore.
   3) The thermocline dives off the coast of South America. Cold deep water no longer rises to the ocean's top layer.
   4) The source of nutrients is cut off, affecting fishing stocks off the coast of South America.
   5) Warm surface water evaporates nearby, over the eastern tropical Pacific. This causes more rain than normal in the area.
   6) On the other side of the Pacific, during an El Niño, Indonesia and neighbouring nations are drier than typical.
c) La Niña: La Niña, which means as "small girl," is also known as El Viejo, or "cold event." The water temperature in the Eastern Pacific drops below average. As a result, a powerful high pressure system

has formed over the eastern equatorial Pacific. There is now low pressure in the Western Pacific and off the coast of Asia. La Niña causes drought in Peru and Ecuador, major flooding in Australia, high temperatures in the Western Pacific, Indian Ocean, and off the coast of Somalia, and abundant monsoon rains in India. La Niña conditions are really advantageous to the Indian monsoon.

1) Trade winds blow warm water at the ocean's surface from South America to Indonesia in this pattern.
2) As warm water travels west, cold water from the deep comes to the surface near South America's coast.
3) Since La Niña pushes all of this warm water to the western Pacific, regions like Indonesia and Australia have experienced far more convectionalrain than is typical.
4) The cool water in the eastern Pacific, on the other hand, causes fewer rain clouds to develop there. As a result, areas such as the southern United States are drier than is typical.
5) La Niña may also cause an increase in lightning activity in the Gulf of Mexico and along the Gulf Coast.
6) And more tropical cyclones, including hurricanes, are emerging in the deep tropics (near Caribbean islands, for example).
7) La Niña causes drought in Peru and Ecuador, large floods in Australia, high temperatures in the Western Pacific, Indian Ocean, and off the Somalian coast, and considerably better monsoon rains in India.

El Niño Effects worldwide

a) El Niño causes an increase in sea surface temperatures.
b) It also weakens the affected region's trade winds.
c) It can cause drought conditions in India and Australia. This has a significant impact on agricultural productivity. It has also been noticed that El Niño may not induce drought but rather significant rains at times. It causes significant harm in both circumstances.
d) In certain countries, however, it may cause a total reverse, i.e., excessive rainfall.

Effects on India

a) Since 1950, 10 of India's 13 droughts have occurred during El Niño years, with one occurring during a La Niña year. This is because, in general, El Niño brings less rain in India than usual. Because Indian agriculture is primarily reliant on the monsoons, less rainfall during the monsoons often correlates to lower-than-average crop production.

b) The pressure along Peru's coast in South America is higher than in the region bordering northern Australia and South East Asia.
c) Because the Indian Ocean is warmer than the surrounding oceans, it has lower pressure. As a result, moisture-laden winds blow from the western Pacific to the Indian Ocean.
d) Because the pressure on the Indian subcontinent is lower than on the Indian Ocean, moisture-laden winds go farther from the ocean to the lands.

India's Initiative to Mitigate Climate Change

The Indian government has signed and ratified the UNFCCC. The Ministry of Environment and Forests is India's focal agency on climate change problems, including working groups on the UN Framework Convention on Climate Change and the Kyoto Protocol. Current initiatives in India to reduce greenhouse gas emissions include estimating sectoral greenhouse gas emissions, creating a national inventory of greenhouse gas sources and sinks, transferring environmentally sound technology, conserving forests and biodiversity, and managing coastal zones, in addition to the participation of several governmental and non-governmental organisations in climate change research. To balance growth objectives with economic and environmental concerns, India has implemented a number of measures in reaction. To reduce greenhouse gas emissions, several measures are being implemented in the coal, oil and gas, renewable energy, transportation and residential, electricity, and industry sectors. Steel, cement, caustic soda, aluminium, and electric power generation are the primary energy consumers in India. Energy efficient methods and equipment are being promoted in various industries, as well as in the electricity sectors, through technological upgrades and energy conservation initiatives.

Clean development mechanisms projects in India

The Kyoto Protocol's Clean Development Mechanism (CDM) permits developed countries to invest in initiatives that decrease greenhouse gas emissions in developing countries. Through the World Bank Bio-Carbon Fund, countries such as Spain, Sweden, Japan, and Canada have invested in the projects. The money is meant to support initiatives that store carbon in plants and soil. China is now the only country with an active CDM reforestation project. In India, similar CDM reforestation and afforestation initiatives are in the early phases in the states of Himachal Pradesh and Haryana. The Himachal Pradesh Bio-carbon Reforestation Project, initiated by the State Government under CDM, intends to clean the air of developed nations (Spain, Canada, Sweden, and Japan) via forest trees. Produced on deteriorated ground in rural areas. The trees operate as CO2 sinks in the atmosphere. The aforementioned project has begun at Ranhog village of Solan District, Himachal Pradesh. Villagers seek to obtain water and wood from deteriorated land. The project will include chir pine, walnut, reetha, and bamboo trees. Payments for land are dependent on the number of certified emission reductions (CER) generated.

It will take time to fully assess the benefits of such initiatives, but states such as Karnataka, Uttarakhand, Jammu and Kashmir want to join the league because of the perceived benefits of bringing about development by incorporating local people. They have approached the World Bank about doing similar initiatives. However, CDM afforestation and reforestation are not the only ways to sequester carbon. Stopping deforestation is also being emphasised. Standing forests, according to experts, are the most important territorial carbon dioxide reservoirs.

CHAPTER 20

OZONE LAYER

The Ozone Layer and Ozone Layer Depletion

Ozone is a natural gas that is an allotrope of oxygen made up of three oxygen atoms bonded together in a non-linear form. O3 is the chemical symbol for ozone. Ozone gas is significantly blue in colour and has a distinct pungent odour. The ozone layer is mostly present in the earth's lower atmosphere. It has the ability to absorb around 97–99 percent of the damaging UV radiations emitted by the sun, which can affect life on Earth. Millions of individuals would acquire skin disorders and have weaker immune systems if the ozone layer disappeared. Scientists, however, have identified a hole in the ozone layer above Antarctica. This has focused their attention on numerous environmental challenges and solutions. Chlorofluorocarbons, carbon tetrachloride, methyl bromide, and hydrochlorofluorocarbons are the primary causes of the ozone hole (Fig. 20.1).

Ozone layer depletion is the thinning of the upper atmosphere's ozone layer. This occurs when chlorine and bromine atoms in the atmosphere come into contact with ozone and destroy it. One chlorine atom may destroy 100,000 ozone molecules. It is destroyed faster than it is formed. When certain chemicals are exposed to intense UV radiation, they emit chlorine and bromine, which leads to ozone layer depletion. These substances are known as Ozone-Depleting Substances (ODS). Chlorofluorocarbon, carbon tetrachloride, hydrochlorofluorocarbons, and methyl chloroform are all ozone-depleting compounds containing chlorine. Halons, methyl bromide, and hydro bromofluorocarbons are ozone-depleting bromine-containing compounds. Chlorofluorocarbons are the most abundant ozone-depleting substance. It is only when the chlorine atom reacts with some other molecule, it does not react with ozone. The Montreal Protocol was proposed in 1987 to prohibit the use, manufacturing, and import of ODS and to reduce their concentration in the atmosphere in order to conserve the earth's ozone layer.

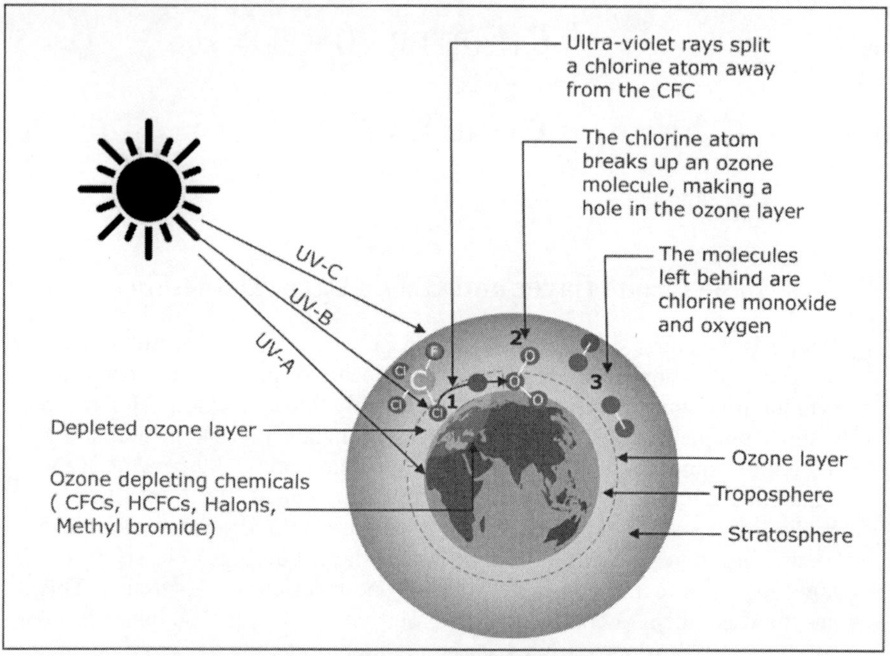

Fig. 20.1: Diagram showing how the ozone layer depletes

Causes of Ozone Depletion

Ozone layer depletion is a key problem that is linked to a variety of reasons. The following are the primary reasons of ozone layer depletion:

a) Chlorofluorocarbons: CFCs, or chlorofluorocarbons, are the primary cause of ozone layer depletion. These are released by solvents, spray aerosols, refrigerators, air-conditioners, etc. Ultraviolet radiations break down chlorofluorocarbon molecules in the stratosphere, releasing chlorine atoms. These atoms react with and destroy ozone.
b) Uncontrolled Rocket launches: According to studies, unchecked rocket launches cause far more ozone layer depletion than CFCs do. If not regulated, this might lead to a significant depletion of the ozone layer by the year 2050.
c) Nitrogenous compounds: Nitrogenous chemicals such as NO2, NO, and N2O are major contributors to ozone depletion. Nitrous oxide (N2O) is produced from solids by denitrification of nitrates in

anaerobic settings and nitrification of ammonia in aerobic situations. This N2O can eventually reach the stratosphere's middle layer, where it is photolytically oxidised to produce nitric oxide, which destroys ozone.

d) Natural causes: Certain natural processes, such as sunspots and stratospheric winds, have been discovered to degrade the ozone layer. However, it only accounts for 1–2 percent of ozone layer loss. Volcanic eruptions are also to blame for the ozone layer's depletion.

Effects of Ozone depletion

The most apparent reason for concern about ozone depletion is its role as a filter of UV light from the Sun. The UV-C band (2.0-2.9 x 10^7 nm) is almost completely removed by the atmosphere. This is also true since UV-C is deadly to microorganisms (thus its usage in germicidal lamps) and may damage both nucleic acids and proteins: in the range of 2.4–2.9 x 10^7 nm, UV-C protection is entirely attributed to ozone absorption. However, the band between 2.9 x 10^{-7} nm and 3.2 x 10^{-7} nm known as "biologically active UV" or UV-B is more essential in terms of ozone depletion, percentage of UV-B penetrates all the way to the ground. As a result, the proportions of UV-B reaching the surface should be particularly sensitive to changes in the ozone column. UV-B has a wide range of impacts on humans, animals, plants, and materials. The majority of these impacts are negative.

a) Human health: The relationship between UV-B and the occurrence of skin cancer is especially emotional: there are two major lines of evidence here. To begin with, skin cancer is more common in people with white skin, as the dark pigment known as melanin is known to be an efficient UV-B filter. The second strand is derived from epidemiology, which is the study of the variables that determine illness incidence in human populations. Increased UV-B exposure can also have other directly damaging consequences on the human body, the two most significant of which being a tendency to inhibit the body's immunological responses and damage to the eyes, particularly in the formation of cataracts.
b) Terrestrial plants: Plants are mostly adapted to present levels of visible radiation, but little is known about how they react to increased levels of UV-B. Sensitive plants often have slower development and smaller leaves; since they cannot photosynthesise

as efficiently as other plants, they produce fewer seeds or fruit. In some situations, these plants' chemical makeup changes, which might impair food quality. More importantly, small changes in plant development caused by UV-B might break the delicate balance in natural ecosystems, affecting the distribution and abundance of plants.

c) Aquatic ecosystems: UV radiation is also harmful to marine life. Although not as significant as visible light or nutrient level temperature, there is evidence that ambient solar UV-B radiation is a limiting factor in marine ecosystems. The potential impact of any increase in UV-B will be significantly dependent on the depth to which it penetrates—more than 20 m in clear water, but only 5 m or so in turbid water. Furthermore, increased UV-B has been proven to harm a variety of tiny aquatic creatures, including zooplanktons, larval crabs and shrimp, and juvenile fish, as well as reducing photosynthesis in phytoplankton.
d) Climate: Another source of concern is ozone's other significant role in the atmosphere. As ozone cycles through its synthesis and destruction, it absorbs solar energy, which is eventually deposited as heat into the stratosphere. This heats the stratosphere and causes the temperature inversion at the tropopause without the ozone layer, there would be no stratosphere. As a result, any loss of stratospheric ozone is expected to cool this region and, as a result, modify the temperature structure of the atmosphere to some extent.
e) Materials: Solar UV radiation has a negative impact on synthetic polymers, naturally occurring biopolymers, and several other commercially important materials. The use of certain materials, particularly plastics, in circumstances requiring frequent exposure to sunshine is only conceivable with the addition of light-stabilizers or surface treatment to protect them from sunlight. Any increase in solar UV-B concentration as a result of partial ozone depletion may thus accelerate photogradation rates of these materials, restricting their outdoor life.

Ozone depletion process

Ozone depletion in the stratosphere causes an increase in UV-B on the earth, which has negative impacts on health, ecosystems, aquatic systems, materials, and so on. Approximately 3–5.5 percent of O3 was depleted in the northern hemisphere between 1969 and 1988, according to estimates. In general, there are three main methods of destroying O3.

a) Hydrogen system (OH system): It destroys only 10 percent of O3 and this reaction can be seen above 40 km over the earth's surface. The reaction is given below:

$H_2O + O\,(ID) \rightarrow 2OH$
$OH + O_3 \rightarrow H_2O + O_2$
$H_2O + O \rightarrow OH + O_2$

Net $O + O_3 \rightarrow 2O_2$

OH can also be formed from oxidation of CH_4 (CH_4 + O(ID) - CH_3 +OH). OH then reacts as above.

b) Nitrogen system (N_2O): 60 percent of ozone depletion is caused by the N_2O system. N_2O generated by the bacterial activity of microorganisms in the ocean and soil rises from the troposphere to the stratosphere, where it combines with 'O' in the presence of light to produce NO, which eventually destroys O3. The detailed reaction is as follows:

$N_2O + O\,(ID) \rightarrow 2NO$
$N_2O + hv \rightarrow NO + O$
$NO + O_3 \rightarrow NO_2 + O_2$
$N_2O + O \rightarrow NO + O_2$

$O_3\,O + O \rightarrow 2O_2$

c) Chlorine system ($CFCl_3$ or CF_2Cl_2 system): Neutral chlorine contributes relatively little to the breakdown of O_3. However, $CFCl_3$ and CF_2Cl_2 are the primary O_3 destroyers. They are inactive in the troposphere but become dissociated once they reach the stratosphere.

$$CFCl_3, CF_2Cl_2 \rightarrow \xrightarrow[180\text{ - }220\text{ nm}]{hv} Cl_2$$

$Cl + O_3 \rightarrow ClO + O_2$
$ClO + O \rightarrow Cl + O_2$

Net $O + O_3 \rightarrow 2O_2$

Ozone Hole and Worldwide Ozone Trends

It refers to a stratospheric area where ozone concentrations are exceptionally low during specific months. Ozone (a molecule of three oxygen atoms) is found mostly in the high atmosphere, or stratosphere, between 10 and 50 kilometres above the earth's surface. Ozone absorbs the sun's damaging Ultraviolet (UV) radiations, eliminating a significant threat to life on Earth. UV radiation have been linked to skin cancer as well as other illnesses and deformities in plants and animals. World Meteorological Organization (WMO), World Weather Watch (WWW), Integrated Global Ocean Services Systems (IGOSS), and Global Climate Observing System are some organisations that assist in monitoring the atmosphere and forming a network of information communication about the atmosphere, including ozone layer monitoring.

Arctic hole

The ozone depletion over the Arctic was substantially greater in 2020. The latest significant Arctic ozone hole was observed in 2011. The hole in the North Pole's ozone layer, discovered in February 2020, had grown to a maximum size of roughly 1 million square kilometres. Scientists think the cause was unique atmospheric conditions, including freezing temperatures in the stratosphere. According to a European Space Agency assessment, the loss of the Arctic ozone layer was caused by frigid temperatures (below -80 degrees Celsius), sunshine, wind fields, and compounds such as chlorofluorocarbons (CFCs). Although Arctic temperatures do not often plunge as low as in Antarctica, intense winds swirling around the North Pole this year confined frigid air within the polar vortex—a rotating whirlpool of stratospheric winds. However, the size of the hole was still minor in comparison to what is often seen in the southern hemisphere. According to 2018 data from the Scientific Assessment of Ozone Depletion, the ozone layer in portions of the stratosphere has recovered at a pace of 1–3 percent every decade since 2000. At these expected rates, the Northern Hemisphere and mid-latitude ozone will recover by 2030, the Southern Hemisphere by 2050, and the polar areas by 2060.

Ozone hole over Antarctica

The ozone holes are most commonly associated with depletions over Antarctica, which form each year in the months of September, October, and November as a result of a unique set of meteorological and chemical conditions that arise at the South Pole, and can reach sizes of up to 25 million square kilometres. Polar vortices are the most responsible for stratospheric ozone loss. A polar vortex is a huge region of low pressure and cold air that surrounds both the North and South Poles of the Earth.As winter approaches, a wind vortex forms around the pole, isolating the polar stratosphere. Thin clouds create ice, nitric acid, and sulphuric acid mixes when temperatures fall below -78 degrees Celsius (-109 degrees Fahrenheit). Chemical interactions on the surfaces of ice crystals in clouds produce active forms of CFCs, resulting in ozone depletion. Temperatures continue to increase in the spring, ice melts, and the ozone layer begins to heal.

In 1986, NASA, in partnership with WMO and UNEP, established the 'Ozone Trends Team,' an international panel of experts entrusted with a thorough re-evaluation of all available ozone data from both ground-based and satellite equipment. In March 1988, their findings were made public. Their main conclusion was that a complete reanalysis of the ground-based data—this time over specific latitude zones and seasons, rather than worldwide and yearly as before—revealed statistically significant long-term declining trends. The results from Antarctica likely inspired this shift in data analysis methodology. In summary, the research revealed a drop in the 2.3 range to 6.2 percent between 30°N and 60°N for the winter months (December to March) from 1969 to 1986, with the highest reductions (about 7 percent) in the upper latitude zone (53-64°N), but declines of around 5 percent as far south as 40°N. There were no notable indications indicating that the loss was accelerated. Furthermore, the observed losses are two or three times greater than the projected declines, while being broadly consistent with the pattern of seasonal and latitudinal variations revealed by the best 2-D model calculations.

The Montreal Protocol

The protocol was agreed on September 16, 1987, and went into effect in January 1989 after being approved by the requisite 11 nations representing two-thirds of world consumption. The initial agreement called for the management of five CFCs and three halons to be phased in. Overall, consumption of halons would be frozen at 1986 levels in 1992, while consumption of CFCs would be reduced to 50 percent of 1986 levels by the year 2000. The protocol's main highlights were:

a) CFCs should be phased out by the year 2000, with 50 percent reductions by 1995 and 85 percent reductions by 1997.

b) Halons should be phased out by 2000, with a 50 percent reduction by 1995.
c) Carbon tetrachloride should be phased out by 2000, with an 85 percent reduction by 1995.
d) Methyl chloroform must be phased out by 2005, with a reduction of 30 percent from 1990 levels by 2000.

India and the Montreal protocol

In 1992, India became a signatory to the Montreal Protocol.

a) As an Article 5 country, India is eligible for Multilateral Fund support in its efforts to phase out ODSs and transition to non-ODS technology.
b) India primarily manufactured and used seven of the Protocol's twenty banned drugs. CFC-11, CFC113, CFC-12, Halon-1301, Halon-1211, Carbon tetrachloride, Methyl Bromide, and Methyl Chloroform are among them.
c) The Ministry of Environment, Forests, and Climate Change is in charge of implementing the Montreal Protocol in India.
d) To carry out the Protocol, the Ministry has established an Ozone Cell.

Ozone depleting substance rule

a) The rules are written under the authority of the Environment (Protection) Act.
b) These Rules provide dates for the phase-out of certain ODSs, in addition to regulating the manufacture, trading, import, and export of ODSs and products containing ODS.
c) These Rules restrict the use of CFCs in the manufacture of different items after January 1, 2003, with the exception of metered dosage inhalers and other medical devices.
d) Similarly, except for critical purposes, the use of halons was restricted after January 1, 2001.
e) Other ODSs, such as carbon tetrachloride, methylchoroform, and CFC, can be used in metered dosage inhalers until January 1, 2010.
f) Furthermore, the usage of methyl bromide is permitted until January 1, 2015.
g) Because HCFCs are employed as a temporary alternative for CFCs, they are permitted until January 1st, 2040.

CHAPTER 21

OCEAN ACIDIFICATION

Ocean Acidification

The oceans are a major CO2 reservoir, absorbing a large amount of it (one-third) created by human activities and successfully mitigating climate change. Ocean acidification is a shift in ocean chemistry characterised by a decrease in ocean pH (i.e., an increase in the concentration of hydrogen ions) caused by the ocean's absorption of carbon molecules from the atmosphere (Fig. 21.1). Some of it interacts with water to generate carbonic acid (H2CO3) in order to achieve chemical equilibrium. Some of these additional carbonic acid molecules combine with water molecules to form bicarbonate ions and hydronium ions, increasing ocean acidity (H+ ion concentration). CO2 interacts with H2O molecules to generate the weak acid H2CO3 (carbonic acid). The majority of this acid degrades into hydrogen ions (H+) and bicarbonate ions (HCO3-). The increase in H+ ions lowers pH (a measure of acidity) and causes the oceans to acidify, or become more acidic or less alkaline. This is known as ocean acidification. Ocean acidification has been referred to as the "evil twin of global warming" and the "other CO2 crisis." An estimated 30–40 percent of the carbon dioxide generated into the atmosphere by human activities dissolves in seas, rivers, and lakes. Eutrophication can also cause huge plankton blooms, and as these blooms collapse and fall to the sea floor, the following respiration of bacteria decomposing the algae reduces seawater oxygen and increases CO2 (a decline in pH). The only way to minimise ocean acidification is to reduce CO and CO2 emissions and pollution. The typical pH of the ocean has fallen over the last century, and there are indications that current levels are beginning to affect organisms that construct their shells out of the minerals aragonite and calcite (both composed of CaCO3). Aragonite is more prone to disintegration in acidic waters than calcite, because it is more soluble. Coral reefs composed of the mineral aragonite are especially sensitive to ocean acidification. According to new research, the amount of coral covering Australia's Great Barrier Reef has been reduced by half since 1985. However, coccolithophores and foraminifera, which are made of calcite and play an important role at the bottom of the marine food chain, are becoming increasingly vulnerable. Furthermore, the future seems to be much more dark; according to certain CO2 forecasts, the ocean's acidity will have increased by 150 percent by 2100 compared to pre-industrial times.

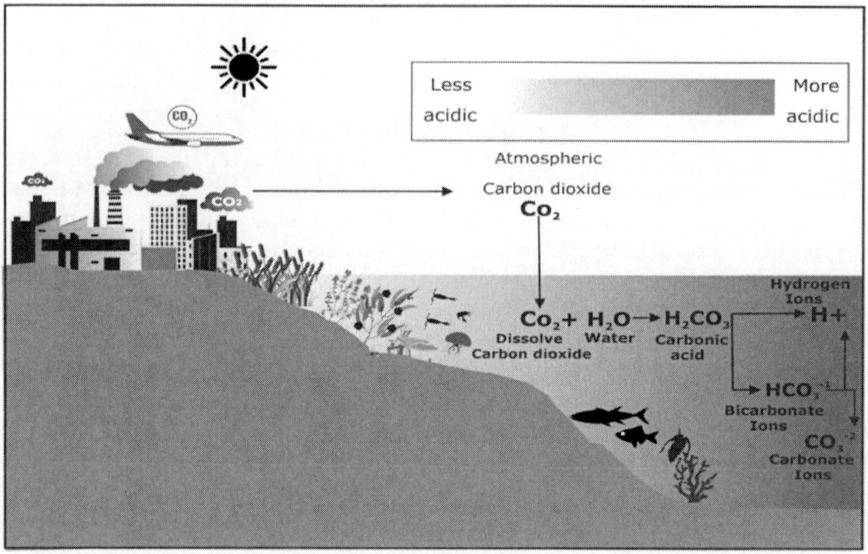

Fig. 21.1: Diagram showing how the ocean acidifies

Changes in the pH of Water

Solutions with low pH values are termed acidic, whereas those with high pH values are called basic. Seven is a neutral number. Since the industrial revolution, ocean pH has fallen from 8.2 to 8.1 and is anticipated to decline another 0.3 to 0.4 pH units by the end of the century. A pH reduction of 0.1 may not appear to be significant, but the pH scale, like the Richter scale for measuring earthquakes, is logarithmic. pH 4 is ten times more acidic than pH 5 and 100 times more acidic than pH 6. Many chemical processes, including those required for life, are sensitive to minor pH fluctuations. Normal blood pH in humans, for example, ranges between 7.35 to 7.45. A 0.2–0.3 decrease in blood pH can result in seizures, comas, and even death. Similarly, even little changes in seawater pH can be hazardous to marine life, affecting chemical communication, reproduction, and growth. Many scientists first focused on the advantages of the ocean absorbing this greenhouse gas from the atmosphere. Scientists first assumed this was a positive thing since it leaves less carbon dioxide in the atmosphere to warm the planet. However, they've recognised in the last decade that the slowed warming has come at the expense of altering the chemistry of the ocean. Worryingly, as the oceans continue to absorb more CO_2, their ability as a carbon sequestration may shrink. This implies that more of the carbon dioxide we produce will remain in the atmosphere, exacerbating global climate change. Ocean acidification reduces the ocean's ability to absorb human CO_2, exacerbating climate change.

Consequences of Ocean Acidification

a) Oceans are becoming increasingly inhospitable to marine life as a result of plastic pollution, overfishing, global warming, and rising acidity caused by the use of fossil fuels.
b) Corals will suffer as a result of ocean acidification. This will have an impact on one million species that have made coral their home.
c) Coral reefs will deteriorate quicker than they can regenerate. When shelled creatures are threatened, the entire food web may be jeopardised.
d) Higher CO_2 concentrations may benefit certain algae and seagrass by increasing photosynthetic and growth rates.
e) Most marine species appear to be more sensitive in their early phases of development.
f) Climate change, pollution, coastal development, overfishing, and agricultural fertilisers will exacerbate acidity changes.
g) These changes will have an impact on the myriad services provided by the ocean.

Ocean Acidification and Carbon's Short and Long Term Destiny in the System

Over long periods (>100,000 years), there is a natural balance between the uptake and release of CO_2 on Earth; the CO_2 released by volcanoes, the largest natural source of CO_2, is taken up by plant synthesis of organic matter and rock weathering on land. However, because rock weathering takes tens of thousands of years, the current anthropogenic CO_2 intake to the atmosphere and ocean will not be removed quickly enough. Over shorter time spans (>1,000 years), the ocean contains an inherent stabilising feedback known as carbonate compensation that connects the ocean carbon cycle to the underlying carbonate rich sediment. Because the top layers of the ocean are supersaturated with $CaCO_3$, little dissolution occurs, but the deep ocean is undersaturated and carbonate dissolves quickly. The lysocline, or depth at which dissolution dramatically increases in the deep ocean, is the initial boundary between these two phases. $CaCO_3$ in the form of decaying shells sinks to the ocean floor. When it sinks in shallow water, the bulk of the $CaCO_3$ is buried in the sediment and retained for a long time, but when it sinks in deep water, virtually all of the $CaCO_3$ dissolves, prohibiting the carbon from being locked away for millions of years. The present increasing rate of atmospheric CO_2 dissolution into the ocean causes an imbalance in the carbonate compensation depth (CCD), which is the depth at which all carbonate is dissolved. As the pH of the ocean decreases, the lysocline and the CCD shallow, exposing more shells trapped in the sediments to suboptimal conditions and forcing them to dissolve, which will help buffer ocean acidification over a thousand years.

Effects of Ocean acidification on Coral reefs

a) The oceans constitute a large CO_2 reservoir, absorbing a substantial amount of it (one-third) created by anthropogenic activities and successfully mitigating climate change.
b) Carbon dioxide is being absorbed from the atmosphere at a pace that exceeds the oceans' natural buffering capability.
c) In certain species, increasing acidity reduces metabolic rates and immunological responses.
d) Carbonic acid, bicarbonate, and carbonate ions are formed when seawater absorbs CO_2.
e) However, increased CO_2 levels in the atmosphere cause a drop in pH, a rise in the concentration of carbonic acid and bicarbonate ions, and a decrease in the concentration of carbonate ions.
f) The decreased availability of carbonate ions makes it more difficult for marine calcifying animals like coral (calcareous corals) and some plankton (calcareous plankton) to generate biogenic calcium carbonate.
g) Acidification threatens commercial fisheries by harming calcifying organisms that are the foundation of Arctic food webs.
h) Coral bleaching is exacerbated by rising acidity because corals are extremely sensitive to changes in water composition.

Impact of Ocean Acidification on Cloud Formation

The majority of sulphur in the atmosphere is exhaled by the ocean, most commonly in the form of dimethylsulfide (DMS) generated by phytoplankton. Some of the DMS generated by phytoplankton penetrates the atmosphere and interacts to form sulphuric acid, which aggregates into aerosols, or minute airborne particles. Aerosols contribute to the formation of clouds, which serve to cool the Earth by reflecting sunlight. However, in acidified ocean water, phytoplankton creates less DMS. This drop in sulphur may result in less cloud formation, rising global temperatures and this will lead us to severe climate change effects. One such effect is cloud seeding. Cloud seeding is the technique of introducing dry ice or, more typically, silver iodide aerosols into the higher layers of clouds in order to induce precipitation and generate rain. Because most rainfall is caused by the formation of ice crystals from super-cooled cloud droplets in the top sections of clouds, the silver iodide particles are intended to promote the formation of new ice particles.

Socio-Economic Impacts

Some socio-economic aspects are given below:

a) Food: Food security may be jeopardised as a result of ocean acidification. The worldwide yearly expenses of mollusk loss due to ocean acidification might exceed \$100 billion by 2100.
b) Coastal protection: Coral reefs, which protect coastlines from the catastrophic effects of storm surges and storms, will be impacted. The coastal estuaries and rivers are being impacted by ocean acidification.
c) Tourism: The effects of ocean acidification on marine habitats might have a significant influence on this business. Acidification might harm the Arctic tourism industry and indigenous peoples' way of life.
d) Carbon storage and climate regulation: As ocean acidification worsens, the ocean's ability to absorb CO2 reduces. Acidifying waters are less efficient in mitigating climate change.

Impacts on the Indian Ocean

a) The Arabian Sea's surface waters are becoming acidic as a result of abundant carbon dioxide in the atmosphere.

b) The northern Bay of Bengal's ocean acidification is mostly caused by contaminants combining with sea water from the Indo-Gangetic lowlands.
c) During the winter, air travelling from land to sea brings all contaminants with it and deposits them in the ocean.
d) A study found that the presence of marine phytoplankton in the western Indian Ocean is quickly declining.
e) Given the levels of ocean warming, research warns that the Indian Ocean may be turned to an ecological desert.
f) The Indian Ocean, which is one of the most pristine and bountiful ecosystems on the planet, will be devastated by ocean acidification in the Arabian Sea and Bay of Bengal.

Mitigation

a) The ultimate remedy to ocean acidification is to reduce global greenhouse gas emissions.
b) Water quality enhancements: Monitoring and managing localised acidity sources such as runoff and contaminants such as fertilisers.
c) Sustainable fisheries management techniques development: Catch regulation to reduce overfishing
d) Habitat management that is sustainable includes increasing coastal protection, lowering sediment loading, and implementing marine spatial planning.
e) Significant study on climate engineering is being conducted to determine the viability and impact of its methodologies.
f) Educate or alert ordinary folks to the dangers of climate change and ocean acidification.
g) Reducing the use of carbon-based energy sources.

CHAPTER 22

SOIL EXTINCTION: THE BIGGEST THREAT TO HUMANITY

Introduction

No species can exist on Earth eternally. Understanding when and why species become extinct is critical for gaining a comprehensive understanding of the repercussions of anthropogenic activity and its influence on ecosystem functioning. Many concerns regarding the fate of belowground organisms remain a mystery, demanding the collaboration of philosophers and applied ecologists in the continued development of soil extinction ecology. Extinction ecology is the study of the factors that cause species to become extinct, either locally or worldwide. So far, the majority of research has concentrated on the extinction dynamics of macroscopic aboveground species. Most species that have ever lived on Earth are now extinct, with extinction events sometimes occurring concurrently in what are known as mass extinctions. There have been five mass extinctions so far, with a sixth mass extinction believed to be occurring. While extinctions can occasionally be traced back to environmental variability alone, the two most major documented drivers of extinction today are habitat loss and invasive species. The global biodiversity is concentrated in the belowground compartment of ecosystems. While our knowledge of belowground species is limited, it has become clear that these organisms may be less functionally redundant than previously imagined. Many justifications for maintaining biodiversity align with our moral need to protect nature, but it is becoming increasingly clear that extinctions may jeopardise our capacity to obtain ecosystem services, which may become the basis for escalating conservation. Because the aesthetic and moral reasons for belowground organisms are less clear, the necessity to avoid species extinction to ensure ecosystem functioning is more important.

Soil is the pinnacle of a miracle, not only on Earth, but throughout the known cosmos. Over 87 percent of everything depends on it, including people. We originate from the soil, eat on the soil, and return to the soil. Global warming, water scarcity, deforestation, and pollution are among heated concerns in today's global environmental debates.

Soil is a substantial carbon sink and the foundation of all life. On average, 36-39 inches of topsoil cover the globe, providing the foundation for 87 percent of life. Our own bodies are made of soil. This body and all other life cannot be healthy unless the soil is rich and healthy. Our soils have been rich in organic content and full with life for thousands of years. However, it has recently turned to sand. Organic content is the live component of soil derived from humus. Over time, anything that is living degrades into humus. It comes from plant litter, leaves, and everything else, as well as animal droppings, in agricultural fields. Plants, trees, bushes, and other types of vegetation must cover the soil. However, most agricultural sites have essentially no vegetation in today's agricultural techniques. We remove soil and organic substance out of the soil every time we harvest. But nothing is replenishing the soil. Even after harvest, the land is mostly left naked and exposed. This causes soil extinction, which implies that the soil loses its organic content and turns into sand. To be alive, soil need at least 3–6 percent organic content. In India, the organic content of the soil is less than 0.5 percent in 62 percent of the country's soil. However, in most parts of the world, the situation is serious. However, humanity is today dealing with a massive, urgent crisis: the extinction of soil.

Extinction Risk Factors for Soil Organisms

Soil organisms supply a wide range of ecosystem services on which all life on Earth is dependent. However, the soil biota is under growing stress as a result of anthropogenic pressures such as soil loss due to erosion, pollution, and climate change. There is a lack of information on present levels of below-ground biodiversity, making it impossible to evaluate the impact of human activities on soil biota. However, there is evidence that soil biodiversity is already declining in many locations. Over the last 20 years, the number of mushroom species in the Netherlands has decreased by 65 percent, while an invasive nematode is already diminishing the richness of earthworm populations in the United Kingdom. Some extinction risk factors for soil organisms are given below:

a) Habitat loss and global change: The extinction ecology of the bulk of belowground species is thus anticipated to differ significantly from that of the macroorganisms studied as far in the literature.

Habitat loss and fragmentation are the most well-documented causes of extinction. For aboveground species, habitat loss is related to the elimination of the autotrophs that dominate a region in some way. In many circumstances, belowground organisms demand a totally different approach on habitat loss. In some cases, the viewpoint is quite similar; for example, one of the most obvious instances of belowground habitat loss is urbanisation. Scientists have discovered that the quantity of predatory nematodes decreases after urbanisation. Extreme drought occurrences have been proven to have a significant influence on belowground habitats, with consequences that are usually equivalent to the aboveground segment of ecosystems. However, climate change can have indirect, far more complex, and less apparent effects on soil qualities, which must be evaluated in the context of habitat loss. Increased atmospheric CO2 indirectly lowered soil aggregate and pore diameters, which likely resulted in the extinction of large-diameter worms and a decline in organic content due to habitat loss. Furthermore, a variety of anthropogenic activities can have long-term consequences on soil biota, perhaps leading to local extinctions. Soil tillage and fertiliser are two prominent examples. Irrigation is another example, which has been connected to local extinctions of earthworm species. It is difficult to quantify the amount to which any of these causes might contribute to the loss of belowground diversity.

b) Specialization and competition: Microbial invasions might be another danger factor for soil biota. While microorganisms have gotten little attention in the invasion field, rising global human movement is projected to increase unintended incursions. Because of the tiny size of the organisms, invasions of belowground microorganisms would be far more difficult to manage than those of macroscopic life forms. The extent to which ideas outlined for aboveground organisms may be applicable to soil biota is uncertain. For macroorganisms, for example, it has been proposed that invading predators might significantly raise the probability of extinction of native biota more than invasion of organisms with remarkable competitive abilities. If this approach applies to belowground species, predatory microarthropods and protist invasions, rather than bacterial or fungal invasions, would demand attention.

Soil Organic Matter

Soil organic matter feeds soil fauna and helps to soil biodiversity by functioning as a reservoir for soil nutrients such as nitrogen, phosphorus, and sulphur; it is the primary contributor to soil fertility. Soil organic carbon strengthens the soil's structure, creating a better physical environment for roots to penetrate. Organic matter absorbs water and may hold up to six times its weight in water, providing a lifeline for flora in naturally dry and sandy soils. Organic matter-rich soils have a better structure, which enhances water infiltration and minimises susceptibility to compaction, erosion, desertification, and landslides. Soils contain almost double the amount of carbon contained in the atmosphere and three times the amount found in plants on a global scale. Carbon dioxide (CO2) is released into the atmosphere as soil organic matter decays; yet, CO2 is removed from the atmosphere when it is generated.

Human activity reduces soil organic matter concentration and biological activity in several ways. However, increasing or simply sustaining soil organic matter content needs a persistent effort that involves returning organic materials to soils and rotations with high-residue crops and deep- or dense-rooting crops. Raising the organic matter content in well-aerated soils, such as coarse sands, and soils in warm-hot and dry climates is extremely difficult because the additional materials decay quickly. In fine textured soils in cold temperate and moist-wet climates with limited aeration, soil organic matter levels can be maintained with less organic residue.

Causes

Climate, texture, hydrology, land use, and vegetation all have an impact on soil organic carbon level.

a) Climate: Because organic matter decays faster at higher temperatures, soils in warmer climes contain less organic matter than soils in colder climates.
b) Soil Texture: Fine-textured soils have more organic matter than coarse soils; they store nutrients and water better, allowing for greater plant development. Coarse soils are better aerated, and the presence of oxygen causes organic matter to degrade more rapidly.
c) Hydrology: The more moist a soil becomes, the less oxygen is available for organic matter to degrade, causing it to collect.

d) Land Use (Tillage): Tillage introduces oxygen into the soil and elevates its average temperature, accelerating the decomposition of organic materials. Organic matter is also lost as a result of erosion, which washes away topsoil and humus. Cropping, on average, delivers less organic matter to the soil than natural plants.
e) Vegetation: Roots contribute much to soil organic matter. Grassland develops deep roots that decompose deep in the soil. Forest soils, on the other hand, rely mostly on surface litter breakdown for organic matter intake. Crops generate more biomass above ground than roots. The amount of organic matter applied to crops is determined by land management methods, such as whether crop leftovers are removed or left behind.

Soil organic carbon content loss can impair the soil's capacity to offer nutrients for long-term plant development. This may result in decreased yields and have an impact on food security. Fewer organic carbon also implies less food for the soil's living organisms, diminishing soil biodiversity. Soil organic matter loss affects a soil's ability to absorb water, resulting in increased run-off and erosion. In turn, erosion diminishes organic matter content by washing away rich topsoil. Under semiarid conditions, this may possibly lead to desertification.

Relationship between Soil organic matter and Micro-organisms

Organic matter in the soil and soil organisms are intricately linked. Microorganisms are the catalysts for most nutrient-releasing activities, and microbial biomass is the live component of soil organic matter. They allow crops to thrive and soils to be productive. Microbial development and activity, on the other hand, are dependent on the availability of carbon and other mineral nutrients, as well as a favourable physical and chemical environment. The way soil is maintained through tillage and cropping patterns has a significant influence on life under the surface. In soil, microorganisms breakdown organic materials and debris in a variety of ways. Microorganisms release nutrients from organic materials during decomposition, while others breakdown minerals to release nutrients, and still more bring underlying soil and nutrients to the surface for plants to consume. The Key points are given below:

a) In soil, microorganisms breakdown organic materials and detritus in a variety of ways.

b) The size and content of microorganisms vary.
c) Soil structure is improved by microorganisms.
d) Microorganisms release nutrients from organic materials during decomposition, while others breakdown minerals to release nutrients, and still more bring underlying soil and nutrients to the surface for plants to consume.
e) Some microbes have a symbiotic connection with plants, in which they obtain sugars from the plants and assist the plants in obtaining nutrients from organic debris.
f) Some protect plants from infections that are harmful to them.
g) Some microorganisms in soil may break down poisons and pesticides, preventing toxins and pesticides from entering groundwater.
h) Without microbes, the earth's surface would be stacked high with dead organic matter, eventually producing lifeless soil with little output for food production.

Mitigation to Combat Soil Extinction

By placing the ground under cover from vegetation and replenishing the soil with plant litter and animal dung, at least 3–6 percent organic content in the soil may be restored. All that is required is for the soil to be covered by vegetation of any sort, including plants, trees, bushes, grasses, and so on. Experts estimate that if we act now, we will be able to reverse the soil condition within the next 15–20 years. If we wait another 20 years to intervene, however, it will take 150–200 years to regenerate. To address soil extinction, governments might employ the following strategies:

a) Reducing emissions.
b) Afforestation.
c) Ploughing on a contour (cultivation against the direction of the wind).
d) cultivation in strips (cultivation in strips).
e) Government measures to mitigate flooding and the actions taken by policymakers to restore at least 3–6 percent organic content in soil.
f) Reclamation of contaminated land.
g) To prevent illicit deforestation, security measures and strong rules must be implemented.
h) Increasing the number and range of government-protected forests.
i) Infrastructure building (roads, dams, etc.) should be carefully planned to limit the loss of forest land.

j) Investing in novel agricultural technology (such as hydroponics) and assisting farmers in implementing environmentally friendly farming practises (such as cyclic agriculture).
k) Wind breaches in the terrain, much like trees at the edge of a fence.
l) Agriculture that is organic.
m) Controlling or restricting shifting cultivation is a good idea.
n) Drainage system construction.
o) Gullies, ravines, and other natural features are levelled.
p) Market regulation of chemical fertilisers and insecticides.
q) A proper understanding of the importance of conservation.

CHAPTER 23

MITIGATION STRATEGIES FOR CLIMATE CHANGE

Carbon Sequestration and Types

Carbon capture and storage, commonly known as CCS or carbon sequestration, refers to methods that are supposed to combat global warming by absorbing CO2 from power plants, industrial locations, or even straight from the air and storing it permanently underground. Carbon sequestration refers to the long-term storage of carbon dioxide or other types of carbon in order to mitigate or avoid global warming. It has been proposed as a method of slowing the concentration of greenhouse gases in the atmosphere and oceans caused by the combustion of fossil fuels.

Steps to Carbon capture and storage

Carbon capture and storage consists of three major steps:

a) CO2 capture and separation from other gases
b) moving collected CO2 to a storage site
c) keeping that CO2 out of the atmosphere (underground or deep in the ocean).

A plethora of approaches for sequestering atmospheric carbon are being investigated. These can be divided into three major categories (Fig.23.1):

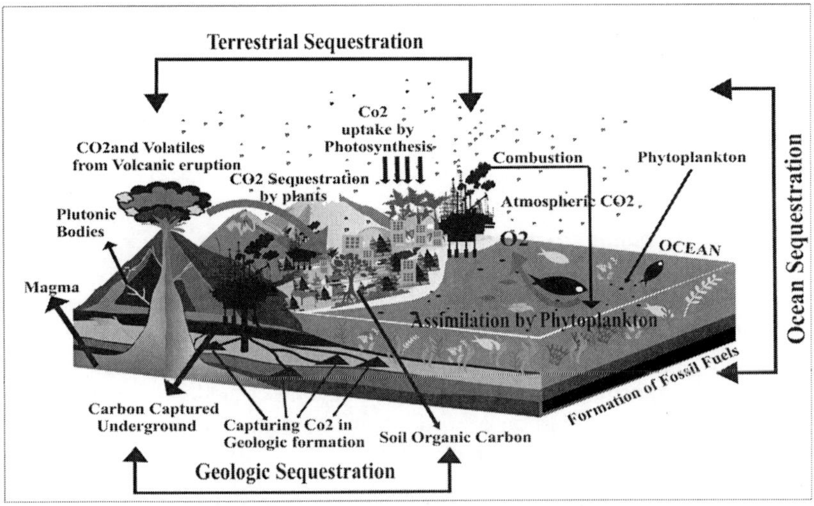

Fig. 23.1: Carbon sequestration and types

a. Ocean Sequestration: Carbon sequestration in the ocean implies the plankton (floating tiny organisms) at the ocean surface will transform the carbon dioxide, via photosynthesis in the photic zone, into oxygen, just as the trees and land plants do on land. The oceans absorb approximately 25 percent of the greenhouse gas carbon dioxide produced by humans each year. When the ocean releases carbon dioxide, it creates a positive atmospheric flow. However, if the oceans become more acidic owing to the emission of carbon dioxide, they may potentially alter their chemistry in the future.
b. Geological Sequestration: This type of carbon storage works by storing the carbon dioxide of industrial sources underground. This method involves using porous rocks to store the carbon dioxide. Injection into deep-sea basalt formations offers distinct and substantial benefits over other possible geological storage methods. Climate scientists have long suspected that highly reactive basaltic rocks might provide a solution to the carbon storage issue. Basalts are found all over the globe and contain significant quantities of calcium and magnesium ions, which chemically combine with CO_2 to form calcite, dolomite, and magnesite. This method allows the use of fossil fuels while reducing greenhouse gas emissions. Geologic sequestration can be done by the following methods:

i. Hydrodynamic Trapping: Under low-permeability cap rock, carbon dioxide can be trapped as a gas (much like natural gas is stored in gas reservoirs).
ii. Solubility Trapping: Carbon dioxide can be dissolved into a liquid, such as water or oil.
iii. Mineral Carbonation: Carbon dioxide has the ability to react with minerals, fluids, and organic matter in a geologic formation to generate stable compounds/minerals, primarily calcium, iron, and magnesium carbonates.

c. Terrestrial Sequestration: Soils and plants, which serve as natural carbon sinks, store a considerable amount of carbon. Increased carbon fixing via photosynthesis, slowing down or decreasing decomposition of organic matter, and altering land use practices may increase carbon absorption in these natural sinks. Terrestrial sequestration can be done by the following methods:

i. By Forests: grasslands and forests absorb about 25 percent of the global greenhouse gas emissions. When the trees die, their leaves and branches fall to the ground, releasing the carbon into the soil.
ii. In Soils: The carbon in soil can be sequestered by plants through photosynthesis. Unfortunately, some agro ecosystems can deplete the organic carbon levels in the soil. Fortunately, soil can also store carbon through carbonates.
iii. By Grasslands: Grasslands and rangelands are more resilient than forests when it comes to storing carbon. They can also store more carbon underground due to their porous soil.

Carbon Sink

Carbon sinks are environmental processes or any natural or man-made reservoir that absorbs and deposits certain carbon-containing chemical molecule for a boundless amount of time, lowering the concentration of CO2 in the atmosphere. Plants, oceans, and soil are the primary natural carbon sinks. Since the 1997 Kyoto Protocol, which encourages the use of CO2 sinks as a form of carbon offset, public understanding of their importance has expanded. Plants absorb carbon dioxide from the atmosphere and utilize it in photosynthesis; part of this carbon is transferred to soil when plants die and disintegrate. The oceans are a key carbon dioxide storage system. Carbon dioxide is also taken up by marine creatures for photosynthesis, whereas some carbon dioxide just dissolves in saltwater. The world's greatest carbon sinks are the ocean, atmosphere, soil, and forests. Protecting these key ecosystems is critical for combating climate change and maintaining a stable climate. They are, however, under growing threat.

Green and blue carbon sequester greenhouse gases from the atmosphere, unlike black and brown carbon, which contribute to them.

Green Carbon

Green carbon is carbon that is removed by photosynthesis and stored in the plants and soil of natural ecosystems, and it is an important component of the global carbon cycle. Many plants and most crops have short lifetimes and lose a significant amount of carbon at the end of each season, but forest biomass accumulates carbon over decades and millennia. Furthermore, forests may store significant amounts of CO2 in very short periods of time, generally many decades. Afforestation and reforestation are two strategies for increasing biological carbon sequestration.

Blue Carbon

Blue Carbon refers to the coastal, aquatic, and marine carbon sinks stored by the respective flora, marine organisms, and sediments. Coastal habitats, in particular, such as tidal marshes, mangroves, and seagrasses, collect carbon from the atmosphere and ocean via natural mechanisms, storing it in plants and depositing it in the sediment underneath them. These coastal ecosystems are extremely effective at sequestering and storing carbon; each square mile of these systems may extract carbon from the atmosphere and seas at rates greater than established tropical forests. Furthermore, coastal ecosystems have been shown to store massive amounts of carbon in organic rich sediments—up to five times more carbon than many temperate and tropical forests.

Importance of such Ecosystems

a) Preventing deterioration and destruction of coastal habitats and encouraging restoration are important tools for mitigating climate change.
b) Natural systems such as coastal habitats such as mangroves, tidal marshes, and sea grasses are among the most at risk of disappearing.
c) When they are gone, they not only stop sequestering carbon, but they also release their carbon stores, becoming new sources of

climate change-causing carbon emissions that can persist for decades.

Initiative and International Cooperation

a) The Blue Carbon Initiative is the first integrated initiative with a global objective focused on mitigating climate change via the conservation and restoration of coastal marine ecosystems.
b) Conservation International (CI), the International Union for Conservation of Nature (IUCN), and the United Nations Educational, Scientific, and Cultural Organization (UNESCO) are working with governments, research institutions, non-governmental and international organisations, and communities around the world to develop management approaches, financial incentives, and policy mechanisms to ensure the conservation and restoration of coastal Blue Carbon ecosystems.
c) Participate with local, national, and international governments to ensure that policies and laws promote coastal Blue Carbon conservation, management, and finance.
d) Implement better systems for counting coastal carbon;
e) Create incentive mechanisms for Blue Carbon initiatives, such as carbon payment programmes.
f) Implement initiatives across the world to show the viability of coastal Blue Carbon accounting, management, and incentive agreements.
g) Support scientific study on the role and value of coastal Blue Carbon ecosystems in the mitigation of climate change.

Carbon Credit

A carbon credit is a tradable certificate or permit that represents the right to emit one tonne of carbon dioxide or its equivalent (tCO_{2e}). One carbon credit is equal to one tonne of carbon dioxide, or carbon dioxide equivalent gases in some markets.

How to Earn

A carbon credit is awarded to an organisation that generates one tonne less of carbon or carbon dioxide equivalent than the standard level of carbon emission permitted for their outfit or activity.

How it Works

Countries that have signed the Kyoto Protocol under the UNFCCC have established gas emission standards for their businesses that must be reached. In such instances, a corporation has two options for lowering emissions.

a) It can reduce greenhouse gas emissions by implementing new technology or improving on current equipment to meet the new emission standards.
b) It can collaborate with developing countries to assist them establish new eco-friendly technologies, allowing the developing country or its enterprises to 'earn' credits. This credit is converted into a licence for the company to release greenhouse gases in its home nation. However, only a fraction of a firm's carbon credits in a developing nation can be transferred to a corporation in a developed country.

Issues regarding carbon credits

a) Developing nations, notably India, China, and Brazil, benefited significantly from the Kyoto Protocol's CDM carbon market.
b) India has registered 1,703 CDM projects, the second highest number in the world.
c) The approval of the Paris Agreement would alter the carbon market landscape.
d) Targets for developing nations: Unlike in the case of the Kyoto Protocol, emerging countries must now set mitigation goals.
e) CDM transition: If the CDM transition judgement is not favourable, India may lose money because it has around 4 billion unsold certified emission reductions (CERs).
f) Several nations, including India, are requesting that existing carbon credits be carried forward in order to fulfil new climate commitments.
g) Accounting rules: According to Article 6.4 of the Paris Agreement, a government can acquire CER credits from public and private companies to satisfy its NDC objectives, but these reductions must represent extra efforts on the part of such entities.
h) Share of Proceeds (SOP) to the Adaptation Support: While poor nations want SOP to be applied equitably to Articles 6.2 and 6.4 to fund adaptation, rich countries prefer to limit its use to Article 6.4.

Issues discussed in COP 26

a) Pledges to reduce emissions: In 2015, governments pledged to limit greenhouse gas emissions to 2 degrees Celsius, ideally 1.5 degrees Celsius. The deadline for nations to make stronger promises to reduce emissions is this year, and large emitters such as China and India have yet to submit enhanced NDCs.
b) Fossil Fuels: The UN has called on OECD (Organisation for Economic Co-operation and Development) countries to phase out coal by 2030, but G20 environment ministers have not agreed on a timetable.
c) Finance: Developed nations promised to contribute $100 billion per year until 2020 to help developing countries deal with the impacts of climate change, but this goal has yet to be met. Climate justice requires emerging nations to have equal access to global carbon space. A carbon market facilitation mechanism that adheres to the UNFCCC principles would aid in the acceleration of the transition to low-carbon development.

Carbon Offsetting

Carbon offsets are credits for greenhouse gas emissions reductions generated in another site, such as wind farms, which generate renewable energy and reduce the demand for fossil-fuel driven electricity. Carbon offsets are measured and sold in metric tonnes of CO2 equivalent (CO2e). Purchasing one tonne of carbon offsets results in one tonne less carbon dioxide in the atmosphere than would otherwise exist. Carbon offsetting is sometimes the quickest approach to achieve the deepest reductions inside enterprises, and it also provides additional advantages on-site, such as employment possibilities, community development programmes, and training and education. To be credible, a carbon offset must meet essential quality criteria, such as proof that it is additional (the reduction in emissions would not have occurred without the carbon finance), that it will be retired from the carbon market so that it cannot be double counted, and that it addresses issues such as permanence (it delivers the reductions stated) and leakage (the reduction in emissions in one area does not cause an increase in emissions in another).

Example

In the medium term, Business R is unable to decrease 100 tonnes of CO2 emissions. However, there is a project somewhere in the world that can easily save 100 tonnes upon a capital injection. Take, for example, a community in Gujarat who want to switch from carbon-intensive kerosene to solar panels as an energy source—but they can't afford the solar panels. By purchasing carbon offsets, you give financial support to subsidise the cost of installing solar panels on houses, resulting in a reduction of 100 tonnes of CO2. As a result, Business R has decreased worldwide net CO2 emissions by 100 tonnes.

Carbon Tax

Carbon tax is a potential alternative to the protocol's present "cap and trade" system. This tax is dependent on the quantity of carbon in a fuel such as coal, for example. The goal of this tax is to reduce the usage of fossil fuels while also providing an incentive to adopt alternative energy sources. If a carbon tax were imposed, it would be phased in, beginning with a modest amount and gradually increasing to allow for the development of better industries and technologies. There are five key reasons why a carbon price may be preferable than the "cap and trade" system.

a) Predictability: The tax might aid in the prediction of energy costs, which could benefit investments in energy efficiency and alternative fuels.
b) Implementation: A carbon tax may be implemented considerably faster than the legalities associated with the "cap and trade" technique.
c) Understandable: The carbon price is simpler to understand and may thus be accepted more widely by ordinary people.
d) Simple: Because of its simplicity, a carbon tax is less susceptible to manipulation by special interest organisations.
e) Refund: The carbon tax, like other types of taxes, may be subject to public rebates.

How does it work

Carbon will be priced per tonne by the government. It translates to a tax on the following:

a) electricity
b) natural gas or oil

Carbon Tax India

a) In 2010, India introduced a carbon price of Rs 50 per tonne of coal produced and imported.
b) It was raised to Rs 100 in 2014.
c) It was raised to Rs 200 in 2015.
d) The carbon tax is now Rs 400 per tonne.

Geoengineering

Geoengineering is a type of climate engineering or human climate intervention that aims to change long-term climate patterns on Earth. Many of today's geoengineering plans centre on two main goals: reversing global warming and removing carbon dioxide from the atmosphere. Large-scale geoengineering programmes, on the other hand, might aim to modify any part of the Earth's ecosystems or climate system, perhaps addressing the consequences of global warming, ocean acidification, melting arctic ice, or volcanic eruptions that break through the Earth's surface. Solar geoengineering was addressed in President Lyndon Johnson's first report on climate change in 1965, and it was considered in studies by the US National Research Council (NRC) in 1977, 1983, and 1992. The approach is based on two arguments: to minimise climate hazards, net emissions must be zero; and reducing emissions does not eliminate dangers from gases already in the atmosphere. One type of carbon dioxide removal technology comprises bio-energy with carbon capture and storage, direct air capture, and ocean fertilisation (adding nutrients to stimulate marine food production to absorb CO2). The other category is solar radiation management, sometimes known as solar geo-engineering, which involves reflecting sunlight to limit the amount of infrared radiation trapped by greenhouse gases.

The primary goal of geoengineering is to modify and cool the Earth's environment, so mitigating environmental harm and resulting climatic changes and making the world more livable. Geo engineering is merely a theoretical concept at this time. Some of the options include hoisting parasols, deploying mirrors in space, whitening the stratosphere with sulphate aerosols, whitewashing building tops to reflect sunlight, and dumping iron filings into the ocean to stimulate carbon-consuming algae.

Some big plans to stop global warming

a) Volcano Replica: Many million tonnes of sulphurdioxide gas can be emitted by a volcanic eruption, forming a cloud that blocks some of the sun's energy. Some scientists believe that by injecting sulphur into the atmosphere, they can block solar radiation and perhaps cool the earth. Those droplets are very effective in scattering the sun's light back into space. Furthermore, because sulphur does not heat the stratosphere as much as other aerosols, it would not interfere with the cooling effect. Sulphur dioxide is a better choice for atmospheric seeding than hydrogen sulphide.
b) Shoot mirrors into space: Geoengineers intend to launch a mirror the size of Greenland and strategically put it between the earth and the sun in order to deflect enough sunlight to restore the Earth's temperature to pre-industrial levels.
c) Seeding the Sea with iron: Iron, according to scientists, may be the key to turning things around. Iron is preferred by phytoplankton, which live near the surface. They are also capable of extracting carbon from the atmosphere during photosynthesis. When the organisms die after approximately 60 days, the carbon they have eaten sinks to the bottom of the ocean. Scientists believe that by injecting iron into the water and promoting phytoplankton growth, global warming may be reversed.
d) Whitening the clouds with wind powered ships: Scientists believe that, like volcanic eruptions, cloud tops will reflect solar energy. Spraying a large amount of seawater into the sky by wind-powered remotely triggered ships in order to whiten the clouds and deflect solar energy.
e) Fake trees: "artificial tree," a scaled-down version of an earlier prototype capable of capturing one tonne of carbon each day in the atmosphere. Panels covering the tree's surface (about 50 square metres) will be constructed of an absorbent resin that interacts with carbon dioxide in the air to generate a solid. It is similar to a furnace filter in that it can remove pollutants from the air. The panels, or "boxes," may be removed and subjected to 113 F steam, which cleans the filter effectively. The solid releases the carbon it has trapped as a result of the chemical interaction with the steam, which can then seize as liquid CO_2. Klaus Lackner, an engineering professor at Arizona State University, developed this approach after studying trees for two decades. He created a mechanical tree that might one day absorb up to 32 tonnes of CO_2 in a single year, compared to plants that absorb 1 tonne per year. The 'leaves' of

such a tree may absorb 1,000 times more carbon dioxide than real leaves. To put this into perspective, a hundred million of these mechanical trees could remove 3.2 billion tonnes of CO2 in a single year. That equates to 10 percent of worldwide carbon dioxide emissions every year. However, removing carbon dioxide from the atmosphere is only half the battle; it must next be sequestered, or permanently stored.

Drawbacks of geoengineering

a) Scientists are unsure if they will be able to halt some of these geoengineering programmes once they begin. Geoengineering treats the symptoms of global warming while potentially undermining attempts to address the main cause.
b) People may believe they do not need to limit their own carbon emissions, and businesses may continue to operate as usual, expecting researchers to clear up the mess.
c) It may introduce unknown climatic dangers. Geoengineering efforts may have unforeseen consequences on Earth systems. Because the negative effects of iron seeding or aerosol injections cannot be completely understood until they are implemented, these activities pose moral risks to scientists.
d) It might be ineffective. Unproven technologies are used in geoengineering efforts. These geoengineering methods must be viewed with caution until they are conclusively proved to perform without harmful side effects.
e) It might be financially impossible. Some kinds of geoengineering, like as reforestation, are quite inexpensive and simple to undertake. Others need massive government investment and public support.

CHAPTER 24

HEALTH AND ENVIRONMENT

Introduction

Human health is affected by the environment in two ways. The first is directly—via toxins released into the air and water by factories and vehicles. The present illness pattern, particularly in metropolitan areas, is intimately connected to degradation in air and water quality. The quality of the working environment has an impact on human health. Authorities have long recognised that employees in specific sectors, such as textile mills or stone quarries, suffer from occupational diseases. Although the owners of these firms and quarries are required by law to provide workers with devices to reduce their exposure to contaminants, in most situations, such efforts are not implemented. As a result, a rising number of employees, particularly those in the informal sector, suffer from diseases for years while receiving little or no compensation. The World Health Organization recognises the following environmental factors as having an impact on human health (Table 24.1):

Table 24.1: Environmental factors that affect human health, WHO

Diseases	Polluted air	Poor sanitation and Waste	Polluted Water	Polluted food	Unhealthy housing	Climate change
Acute respiratory infections	◙				◙	
Diarrheal diseases		◙	◙	◙		◙
Other infections		◙	◙	◙	◙	
Malaria and other vector borne diseases		◙	◙		◙	◙
Injuries and poisonings	◙		◙	◙	◙	◙
Mental health conditions					◙	
Cardiovascular diseases	◙					◙
Cancer	◙			◙		◙
Chronic respiratory diseases	◙					

The degradation of the environment, such as the presence of stagnant pools in villages and cities, provides a breeding ground for vectors. Many of these disease-carrying vectors have evolved resistance to insecticides. As a result, malaria, once considered to be eradicated, has reached epidemic proportions in several sections of the country. Worse, the malarial parasite has developed resistance to an increasing number of insecticides and medications. For example, P. falciparum, which causes aggressive cerebral Udlaria, is resistant to chloroquine, the most frequently used malaria drug. Similarly, kala-azar is difficult to eradicate because the organisms that disseminate it find an increasing number of breeding sites.

Overcrowding and unsanitary living conditions in cities, particularly slums, have also aided the development of other contagious illnesses such as TB and cholera. So long as such conditions exist, health treatments have become practically worthless. It has become clearly obvious that improving environmental circumstances should be a major component of healthcare, as curative care alone will not enough as more people suffer to chronic diseases.

Human Diseases

A disease is a harmful alteration in the body's state caused by external variables that might be dietary, chemical, biological, or psychological in nature. Infectious agents, toxic substances, physical conditions, and psychological stress all play roles in the beginning or progression of human illnesses. To further appreciate how these elements influence us, consider the following broad types of environmental health concerns (Table 24.2):

Table 24.2: Some major environmental health problems of the world

Diseases	Description
Diarrhea	Diarrhea is primarily caused by a virus, although it can also be caused by contaminated food. It might also be a symptom of another condition, such as inflammatory bowel disease or irritable bowel syndrome.
Malaria	Malaria is a potentially fatal disease caused by parasites transferred to humans through the bites of infected female Anopheles mosquitos. It is both preventable and treatable. Malaria has affected 241 million people globally in 2020, according to estimates.
Parasitic worms	Parasitic worms, commonly known as helminths, are huge macroparasites that may be seen with the naked eye in adults. Many are soil-transmitted intestinal worms that infect the gastrointestinal system. Climate change causes temperature rises and changes in weather patterns, which can influence disease vectors and human populations indirectly. Temperature rises promote the growth of arthropod vectors, which transmit many parasitic species, as well as the parasites themselves.
Anemia	Anaemia is caused by a shortage of or malfunction of red blood cells in the body. This results in decreased oxygen supply to the body's organs. Fatigue, skin pallor, shortness of breath, light-headedness, dizziness, or a fast heartbeat are some of the symptoms.

Respiratory diseases	Chronic obstructive pulmonary disease (COPD), asthma, occupational lung disorders, and pulmonary hypertension are among the most frequent. Other risk factors, in addition to tobacco smoking, include air pollution, occupational toxins and dusts, and repeated lower respiratory illnesses throughout infancy.
Trachoma	Trachoma is the most common avoidable cause of blindness globally. It's quite infectious. Trachoma usually affects both eyes. Mild itching and irritation of the eyes and eyelids are the first symptoms. They may escalate to impaired vision and discomfort in the eyes.
Goiter and Cretinism	Iodine deficiency is widely recognised as a cause of cretinism in foetuses and newborns, which is characterised by low birth weight, small height, intellectual disability, deafness, and mutism. Iodine insufficiency in later life produces goitre, which is frequent in areas with poor iodine consumption. Iodine shortage occurs when iodine levels in the soil are inadequate, resulting in low concentrations in food products and insufficient iodine consumption in the population. When iodine needs are not satisfied, the thyroid may be unable to manufacture enough thyroid hormone.
Tetanus	Tetanus is a bacterial illness caused by Clostridium tetani. When bacteria enter the body, they release a toxin that causes painful muscular contractions. Lockjaw is the other name of it.
Polio	Polio, also known as poliomyelitis, is a crippling and potentially fatal disease caused by the poliovirus. The virus transmits from person to person and has the potential to infect a person's spinal cord, resulting in paralysis.
Corona Virus	Several studies have found that air pollutants like as PM2.5, PM10, CO, NO2, SO2, and O3 increase COVID-19 mortality rates. However, some studies indicates that COVID-19's indirect action may aid in the reduction of air pollution. The presence of particulate matter has been connected to an increase in COVID-19 occurrences.

Infectious diseases

Pathogenic (disease-causing) organisms continue to constitute the main environmental health danger for the majority of the world's population. Although we have focused most of our emphasis on harmful synthetic chemicals, we should also be cognizant of the biological threats to which we are exposed. In less developed nations, where almost 80 percent of the world's population lives, infectious agents, parasites, and nutritional deficiencies continue to be the leading causes of sickness and mortality (death). Gastrointestinal illnesses (diarrhoea, dysentery, and cholera) are likely to be the leading cause of mortality globally. Malnutrition and diarrhoea, once again, create a vicious cycle. People who are malnourished are more prone to infection, and diseases make it more difficult to obtain, absorb, and retain food. Most, if not all, of these diseases might be avoided with improved cleanliness and nutrition. Malaria, parasitic worm infections, TB and related respiratory risks, tetanus, and other infectious illnesses have claimed the lives of millions of people, mainly in developing countries.

Chemicals

Poisonous chemicals in the environment are becoming a growing cause of concern for individuals in developed countries, even though humans are frequently exposed to toxic compounds in nature. Chemical agents are classified into two types: those that are hazardous and those that are toxic. Toxic substances comprise poisonous compounds that are cytotoxic in nature. Hazardous substances include combustible chemicals, explosives, irritants, sensitizers, acids and alkalis, and so on. Furthermore, allergies, mutagens, teratogens, and carcinogens are fairly abundant in nature, posing serious health risks in communities. Physical agents, such as radiation, can potentially pose major health risks to people. Cells contain a bioaccumulation process that allows them to selectively absorb and store a wide range of chemicals. This permits them to gather nutrients and critical minerals, but it also allows them to absorb and store toxic chemicals. As a result, the impacts of pollutants are amplified in the environment via food chains. Biomagnification happens when a predator at a higher trophic level accumulates and concentrates the toxic load of a large number of species at a lower trophic level. Some of these dangerous chemicals/compounds are highly unstable and disintegrate quickly in most environments. DDT and other chlorinated pesticides, for example, represent a major danger to biotic populations due to their lengthy persistence.

Diet

Diet also has a significant impact on health. Malnutrition, in particular, poses a major danger to the protection against infectious illnesses as well as hazardous chemical exposures. In addition to calories, we require particular nutrients in our diet such as proteins, vitamins, and minerals. The following are some specific types of health problems:

a) Water-borne pathogens: A vast range of human infections may be identified in the faeces of humans and other animals. The majority of human pathogens are viruses, protozoans, helminths, or bacteria. Pathogenic organisms can be found in both raw sanitary sewage and land runoff. Bacterial pathogens from the genera Salmonella and Shigella were responsible for approximately 20 percent of the reported outbreaks and illnesses related with water-borne infections. Another deadly and acute intestinal sickness produced by the bacteria is cholera. Vibrio cholera outbreaks occur seldom as epidemic illnesses. A large number of enteric viruses, which infect numerous human organs, were discovered in water. These include Poliovirus, Hepatitis A, Reovirus, Adenovirus, and Rotavirus, among others. A number of helminth illnesses were discovered to exist in humans, with cysts detected in waste water contaminated water bodies. Ascariasis and Taenia infections were the most common helminth infections.
b) Respiratory hazards due to air pollution in urban centres: Respiratory hazards are becoming more prevalent as a result of the rapid growth in air pollution concerns. Over the last few decades, urban air pollution issues have exacerbated a number of respiratory diseases ranging from bronchitis to pneumonia.
c) Climate changes and human health effects: Global climate has altered dramatically throughout time, notably in terms of temperature. Such a shift has major health consequences, notably in terms of allergies, vector-borne infections, and infectious diseases.
d) Vector-borne diseases: In the trophics, there are several vector-borne diseases. Malaria, Kala-azar, Dengue, Encephalitis, and Yellow fever are a few examples. According to research, climate change might increase the African population at risk of malaria by 90 million by 2030, and the worldwide population at risk of dengue by 2 billion by the 2080s.
e) Metal toxicity and diseases (Minamata disease): A large number of heavy metals exhibit poisoning signs. The toxicity induced by lead, mercury, and arsenic is notable among them. One such metal

poisoning condition is Minamata disease. Fishermen and their families in the Japanese city of Minamata began to exhibit signs of what became known as the Minamata sickness in the early 1950s. The first symptom was loss of sensation, which was followed by difficulty in walking, slurred speech, decreased eyesight, and hearing loss. Unfortunately, many people died as a result of paralysis and comas.

f) Radiation damage: It is generally understood that radiation causes both immediate and long-term harm to several human organs. Both cosmic rays and UV-radiation are harmful to human health. Cancer development in various organs is one of the biggest health dangers caused by low level chronic radiation exposure. The frequency of risk for many forms of cancer varies greatly depending on exposure and skin type.

Health Impact Assessment

In general, health hazards are classified into five categories: communicable diseases, non-communicable diseases, malnutrition, injury, and mental disorders. The pattern of health impact assessment in any of these is determined by the types of health hazards, their causes, and the size of the population under consideration. Most environmental management programmes are now limited to sanitation and hygiene. The network of primary health care clinics is only designed to treat common illnesses. Health morbidity statistics owing to pollution-related concerns are extremely scarce, current health rules exclusively address workplace safety and chemical safety. The current environmental impact assessment methodologies do not thoroughly examine the health and social consequences of development initiatives. As a result, health remains an extrinsic component in all environmental analyses.

CHAPTER 25

WORLDWIDE CLIMATE CHANGE ORGANIZATION

Milestones in the History of Climate Change

- Svante Arrhenius, a Swedish scientist, describes the greenhouse effect for the first time in the early twentieth century. From the early twentieth century through 2021, the following events occurred in the history of climate change:
- 1957–58: Roger Revelle and Hans Suess wrote a formal study on climate change, which they later submitted to the international political committee in the run-up to the 1972 Stockholm Conference. As part of the International Geophysical Year, continuous observations of atmospheric CO begin.
- 12–23 February 1979: The World Meteorological Organization (WMO) hosted the first World Climate Conference in Geneva, determining that manmade CO2 emissions might have a long-term influence on climate.
- The World Climate Programme held a meeting in Bellagio, Italy, from November 9 to 13, 1987, where it was pointed out that the implications of rising sea levels will overwhelm any direct temperature effects of climate change in coastal places throughout the world. The meeting took place in the aftermath of the discovery of an ozone hole and the establishment of the "Montreal Protocol on Substances that Deplete the Ozone Layer" in 1987. The World Commission on Environment and Development (WCED), often known as the Brundtland Commission, called on UNEP and WMO to take more action in November 1987.
- 27–30 June 1988 (Toronto Targets): Following the Brundtland Commission's findings, a "Conference on the Changing Atmosphere: Implications for Global Security" was held in Toronto, Canada. It advocates for the creation of a World Atmosphere Fund, which would be supported in part by a tax on fossil fuel usage in developed nations.

- The UNEP and WMO formed the Inter-governmental Panel on Climate Change (IPCC) in November 1988, with three working groups to collect and review information on climate change
- November 1989: The Maldives hosted a small states conference on sea level rise, which was quickly followed by the founding of the Alliance of Small Island States (AOSIS).
- April 1990 (Early Dissent): Disagreements between developing and industrialised nations surfaced often throughout discussions. Selected Developing Countries Conference on Global Environmental Issues in New Delhi. Climate change responsibilities are placed on developed countries. Developing nations agree not to accept any obligations.
- May 16, 1990 (On the UN Agenda): The Bergen (Sweden) Declaration of the United Nations Economic Commission for Europe (UNECE) highlights certain nations' willingness for CO) Stabilisation. The UNEP Governing Council is preparing for a climate convention.
- 27–30 August 1990: Climate change differences between poor and developed countries became clear. The Fourth Plenary Session of the Inter-governmental Panel on Climate Change (IPCC) was held in Sundsvall, Sweden, and adopted legally the IPCC First Assessment Report on global temperature, which is expected to rise by 3 degrees Celsius by the end of the twenty-first century.
- November 1–7, 1990: Second World Climate Conference in Geneva (Switzerland), where 137 countries agreed to develop a global climate treaty.
- 4–14 February 1991: Some (developing countries) are more equal than others. The first meeting of the Inter-governmental Negotiating Committee (INC-1) in Washington, DC, elects a bureau and adopts particular measures to persuade poor nations to participate. The paper attempted to estimate that emerging countries were responsible for a significant portion of global methane emissions.
- 19–28 June 1991 Political positions are formed. Divisions within G-77 as well as divisions within the North. At the INC-2 meeting in Geneva, India will present a non-paper based on the concepts of per capital equity, as a result of the CSE's research on Global Warming in an Unequal World.
- 9–20 September 1991: At the INC-3 meeting in Nairobi, certain poor nations, like India, expressed willingness to pursue particular

initiatives to reduce greenhouse gas emissions if the Global North covered the whole additional costs of these projects.

- Ministers from OECD countries accepted responsibility for climate change on December 2–3, 1991.
- December 9–20, 1991: A wider agreement among developing nations begins to emerge during the INC-4 meeting in Geneva. The G-77 nations emphasise that the primary responsibility for reducing the effects of climate change lies with the industrialised countries, who have continued to reject such a plan.
- February 18–20, 1992: Global North and South disagreements remain, but efforts to finalise a draught are underway. Industrialised nations suggested modest responsibilities for greenhouse gas emission stabilisation at the INC-5 meeting in New York.
- 9 May 1992: At the United Nations Headquarters in New York, the United Nations Framework Convention on Climate Change (UNFCCC) was formally approved.
- June 5, 1992: The United Nations Framework Convention on Climate Change (UNFCCC) was opened for signing at the United Nations Conference on Environment and Development (UNCED) in Rio de Janeiro, Brazil; 154 nations, including India, sign it.

UNFCCC

The United Nations Summit Conference on Environment and Development (UNCED) in Rio de Janeiro on June 5, 1992, established the first global legal instrument on climate change, the UN Framework Convention on Climate Change (UNFCCC). Countries joined the UNFCCC in 1992 to consider what they might do collaboratively to prevent average global temperature rises and the accompanying climate change, as well as to cope with whatever effects which were, by then, unavoidable. The Convention presently has 197 signatories. The Convention's ultimate goal is to keep greenhouse gas concentrations at a level that would avoid hazardous anthropogenic (human-caused) interference with the climate system. It argues that such a level should be attained within a time frame sufficient to allow ecosystems to naturally adjust to climate change, to guarantee that food supply is not jeopardised, and to allow economic development to proceed in a sustainable way. Following its implementation, the following important COP are listed below:

Conference of Parties

Table 25.1: Showing key points of COP

Conferences	Timeline of Important events
COP1 (March 26–April 7, 1995)	The first Conference of Parties (COP1) was convened in Berlin following the Rio summit, with 117 delegates in attendance. India broke the gridlock in emissions-cutting discussions by putting on hold a Green Paper (a revised AOSIS proposal) pushing industrialised nations to cut CO2 emissions by 20 percent by 2000. In Berlin Mandate, a timeframe for discussions by 1997 was agreed, as well as the formation of an Ad hoc Group on Berlin Mandate (AGBM). The restructured GEF serves as an interim financial mechanism.
COP2 (July 8–19, 1996)	Following the release of the Second Assessment Report of the IPCC's Working Group II, the COP2 meeting in Geneva adopted the Geneva Declaration. Action by developed nations to cut emissions was emphasised, explaining that the AGBM process is aimed at reaching agreement on legally enforceable targets.
COP3 (December 1–10, 1997)	The COP3 meeting in Kyoto drew 10,000 attendees, observers, and media attention. The Kyoto Protocol to the UN Framework Convention on Climate Change was approved. All industrialised nations agreed to cut greenhouse gas emissions by 5.2 percent below 1990 levels between 2008 and 2012. Japan agreed to cut 6 percent, the US 7 percent, and the EU 8 percent below 1990 levels. The Kyoto Protocol was open for signing in New York on March 16, 1998.
COP4 (November 2–12, 1998)	The COP4 meeting in Buenos Aires (Argentina) established an Action Plan with year 2000 deadline for sectors crucial to the implementation of the Kyoto Protocol.
CoP5 (October/November, 1999)	Parties continued their efforts to fulfil the Buneos Aires Action Plan at the COP5 meeting in Bonn.
COP6 (November, 2000)	Climate talks fell down altogether during the COP6 meeting in Hague, The Netherlands, as the EU refused to cave in to US demands.

COP6.5 (July, 2001)	COP6.5 was conducted in Bonn to break the impasse of COP6. Agreement was obtained on 'crunch' concerns during the resumed session of COP6. The United States withdrew from the process, undermining the Kyoto Protocol.
COP7 (October–November, 2001)	COP7, conducted in Marrakech, Morocco, established a framework for meaningful and effective action to improve the implementation of Convention Article 4.5. The framework addressed five core topics and action areas: technology requirements and needs assortments; technology information; enabling environments; capacity building; and technology transfer methods.
CoP8 (October 23–November 1, 2002)	COP8 was held in New Delhi, with 1456 delegates from 167 nations in attendance. Though Prime Minister Atal Bihari Vajpayee talked of India's pledges to sustainable development and reaffirmed poor countries' solid promises to emission reductions, the Delhi Declaration failed to negotiate or organise like-minded nations to strive for a bold, unambiguous vision. What he said could not become a reality in the Declaration. The United States remained steadfast in its refusal to make any required obligations, instead attempting to lay the groundwork for bilateral climate change agreements with other countries. As a result, the United Statescontinued its efforts to trade greenhouse gases outside of the Kyoto Protocol. At this summit, India has just gone through a momentous movement.
COP9 (2003)	The COP9 meeting was held in Milan, Italy. On 21st of June, 2004, The Indian government issued its first "National Communications on Climate Change," a document that detailedgreenhouse gas emissions and their current and future implications.
COP10 (December 6–17, 2004)	COP10 took place in Buenos Aires, Argentina. There has been little progress in unifying world leaders behind a global crisis. The summit resulted in a significant reduction in political will to address global warming and climate change. Not only the United States, but also Russia and Italy, have urged that future promises to reduce greenhouse gas emissions be made on a voluntary basis.

COP12 (November 6–17, 2006)	The UNFCCC COP12 meeting was place in Nairobi, Kenya. The most crucial issue, requiring promises to reduce carbon emissions, was again blocked, with the United States refusing to budge from its previous position. The EU advocated for overall emission reductions. The emerging countries, particularly China, India, and Brazil, refused to adopt any plan that would jeopardise their economic growth push via draconian cuts. The only significant progress made was in establishing a fund to assist underdeveloped nations in adapting to climate change. The first problem to be discussed and agreed upon was the principles and modalities of a five-year programme to gather knowledge on climate change and its consequences.
COP13 (December, 2007)	The Bali Action Plan (BAP) was adopted during COP13 in Bali, Indonesia. The nations agreed on a plan of action known as the BAP—a thorough mechanism to allow the full and effective implementation of the agreement. Nothing could be accomplished. The strategy strained relations between developed and poorer countries.
COP14 (December, 2008)	COP14 took place in Poznan, Poland. Climate negotiations failed on every level.
COP15 (December, 2009)	The 15th Conference of Parties was held in Copenhagen, Denmark. All of the core ideas of the Kyoto Protocol looked to be dead in the Copenhagen Accord. The developed world was not committed to legally reducing emissions. The Accord was isolated by the BASIC countries (Brazil, South Africa, India, and China).
COP16 (December, 2010)	COP16 was convened in Cancun, Mexico, and all nations except Bolivia agreed on a document that was substantially similar to the Copenhagen Accord, which COP15 had declined to accept. Developing nations agreed to wipe off developed countries' previous debt, have their own domestic emission targets/actions, and enable third-party verifications of these objectives, making it enforceable. Developed nations pledged to create $100 billion over time, promote knowledge transfer and finance for REDD, and address self-deforestation.

COP17 (December, 2011)	COP17 took place in Durban, South Africa. The meeting resulted in an agreement on a management structure for a Green Climate Fund in the future The fund will distribute $100 billion per year to assist poor nations in adapting to climate change. The design of the new Green Climate Fund for developing countries was finished with a decision on its governance and other practical measures. This paves the door for the Fund to become live in 2012. The conference resolved that the second commitment period will begin on January 1, 2013 and last until 2017 or 2020. The end date and emission goals for industrialised nations participating will be set at the UN climate summit in Qatar at the end of 2012.
COP18 (November–December, 2012)	COP18 took place in Doha, Qatar. Doha agreed on a work plan for discussions on the new global accord and enhancing ambition under the Durban Platform. The Technology Mechanism established in Cancun was operationalized by agreeing on governance arrangements for the Technology Executive Committee and a Climate Technology Centre and Network; a work programme was launched to develop modalities and procedures for the new market mechanism established in Durban.
COP19 (November 1–23, 2013)	COP19 took place in Warsaw, Poland. The Warsaw conference established a timetable for nations to submit their contributions to lowering or capping greenhouse gas emissions under the new global climate pact by 2015. It also agreed on methods to speed efforts to deepen emission reduction throughout the rest of this decade, as well as to establish a mechanism to redress climate change losses and damage in vulnerable developing nations. The conference reached resolutions that would improve the implementation of previously agreed-upon measures such as climate financing, REDD+, and openness in emissions reporting.
COP20 (December 1–12, 2014)	Lima, Peru, hosted COP19. Nations closed by defining the contents of the new agreement, which is anticipated to be approved in Paris in late 2015, as well as the basic conditions for how all nations can submit contributions to the new accord during the first quarter of next year. These Intended Nationally Determined Contributions (INDCs) will serve as the framework for climate action from 2020, when the new agreement takes effect. Transparency and confidence-building reached new heights as numerous developed countries agreed to be questioned about their

	emissions targets as part of a new procedure known as a Multilateral Assessment. The Lima Ministerial Declaration on Education and Awareness-Raising encourages countries to integrate climate change into school curriculum and national development goals.
COP21 (30 November–12 December, 2015)	COP21, also known as CMP11, was held in Paris, France. The Paris Agreement was signed. It aspires to keep global temperatures "far below" 2.0 degrees Celsius (3.6 degrees Fahrenheit) above pre-industrial levels and "endeavour to reduce" them even further, to 1.5 degrees Celsius. Rich countries should assist poorer countries by giving "climate financing" to help them adapt to climate change and transition to renewable energy. The pact expects wealthy countries to maintain a $100 billion annual financing guarantee beyond 2020.
COP22 (7–18 November, 2016)	It took place in Marrakech, Morocco. Moving forward on creating the Paris Agreement's rule book was a critical consequence of the Marrakech climate summit. The Marrakech Partnership for Climate Action was established.
COP23 (6–17 November, 2017)	The Conference of the Parties (COP23) to the UN Framework Convention on Climate Change (UNFCCC) was held in Bonn, Germany, at the UNFCCC Secretariat's headquarters. Countries continued to negotiate the finer points of how the deal will operate beginning in 2020. This was the first round of talks since the United States, under Donald Trump's presidency, announced its decision to withdraw from the Paris Agreement earlier that year. Even though it was held in Bonn, it was the first COP to be hosted by a small-island developing country, with Fiji holding the presidency.
COP24 (2–15 December, 2018)	The 2018 United Nations Climate Change Conference, also known as the Katowice Climate Change Conference or COP24, was the 24th Conference of the Parties to the UN Framework Convention on Climate Change. It was place in Katowice, Poland. It sought to bring together professionals and influencers from across the globe who are driving the implementation of the Sustainable Development Goals for a series of high-level conversations on the interconnections between climate change and the 2030 Agenda.
COP25 (2–13 December, 2019)	COP25 took place in Madrid, Spain. COP 25 did accomplish something: the adoption of a worldwide Gender Action Plan, which focused on communicating and methodically incorporating gender considerations

	into climate policies. Over 1,000 towns and local governments have joined the Cities Race to Zero to increase climate action in order to limit global warming to 1.5 degrees Celsius. The following unsolved concerns from COP 24 were addressed during COP 25: Carbon Market Greenhouse gas emission reduction targets Individual targets of the member nations Other forms of international cooperation under "Article 6" of the Paris Agreement Functioning of international emissions trading systems Compensation for poor countries to deal with adverse impacts of climate change
COP26 (31 October–12 November, 2021)	The international climate conference COP 26 was held in Glasgow. The major objective was to achieve worldwide net zero emissions by mid-century and maintain warming to a maximum of 1.5 degrees Celsius. The Intergovernmental Panel on Climate Change's Sixth Assessment Report (AR6), released in August 2021, issued a warning to the world's states about global temperature rise and associated concerns. The report discusses how human activity has warmed the atmosphere, ocean, and land, causing extensive and fast changes in the atmosphere, ocean, cryosphere, and biosphere. This contributes to the intensification and frequency of extreme climatic events such as droughts, heavy rains, rising sea levels, and heatwaves.

REDD & REDD+

REDD (Reducing Emissions from Deforestation and Forest Degradation) is a global effort to incentivize poor nations to conserve, better manage, and save their forest resources, therefore contributing to the global battle against climate change. REDD+ goes beyond simply preventing deforestation and forest degradation by including incentives for positive conservation, sustainable forest management, and increased forest carbon stores. REDD+ envisions a flow of positive incentives for demonstrable reductions in deforestation or for improving the quality and extent of forest cover. It operates on the premise of establishing a monetary value for the carbon stored and improved in standing forest biomass and soil. Countries that cut emissions and practise sustainable forest management will be eligible for funding and resources as incentives. The REDD+ method encompasses significant advantages such as improved livelihoods, biodiversity protection, and food security services.

REDD+ benefits for India

India's ongoing efforts to conserve and grow its forest and tree resources may be rewarded for offering carbon services to the world community in addition to conventional commodities and services to local people. The REDD+ incentives will be passed on to the local communities active in forest preservation and management. This will secure the long-term protection of our forests against destruction. A REDD+ programme for India is anticipated to capture more than 1 billion tonnes of extra CO2 over the next three decades and deliver more than USD 3 billion in carbon service incentives under REDD+.

India's position and initiatives related to REDD and REDD+

a) India argues that REDD should be seen in the context of REDD+, rather than in isolation or in a shortened form, because reduction of deforestation and protection and enhancement of forests are two sides of the same coin and should be regarded as such.
b) India's position was finally accepted at the 13th Meeting of the Conference of the Parties (COP13) in Bali, when elements of conservation, sustainable forest management, and forest carbon stock enhancement were added to the then-existing text of reducing deforestation and forest degradation as part of the Bali Action Plan.
c) Under its National Action Plan on Climate Change, it has launched an ambitious Green India Mission initiative.
d) In December 2008, India filed a response to the UNFCCC on "REDD, Sustainable Management of Forests (SMF), and Afforestation and Reforestation (A&R)."
e) A Technical Group has been formed to create methods and processes for assessing and monitoring the contribution of REDD+ efforts.
f) A National REDD+ Coordinating Agency is being formed.
g) A National Forest Carbon Accounting Program is being established.
h) In 2012, India hosted the Conference of Parties (COP11) of the Convention on Biological Diversity (CBD) to commemorate the twentieth anniversary of the Rio Convention.
i) The Indian Network for Climate Change Assessment (INCCA) issued a study on the impact of climate change on India's forests in November 2010.
j) The increase in Net Primary Productivity (NPP) is expected to be between 20 and 57 percent.

The GEF

Article 11 of the UNFCCC establishes a financial mechanism for convention implementation that will operate under the supervision and accountability of the UNFCCC COP. According to Article 11(1), the COP is responsible for determining the financing mechanism's policies, programme priorities, and eligibility requirements related to the convention. Article 21 appoints the GEF to operate as the temporary financial mechanism. The World Bank created the Global Environment Facility (GEF) in 1991 in collaboration with the United Nations Development Programme (UNDP) and the United Nations Environment Programme (UNEP) to provide funds to safeguard the global environment. The Instrument for the Establishment of the Restructured Global Environment Facility, which was enacted in 1994 and later updated in 2002, outlines the GEF's governance, operational, financial, and administrative supervision procedures (GEF Instrument). The GEF presently has six areas of focus:

a) biological diversity
b) climate change
c) international waters
d) land degradation, primarily desertification and deforestation
e) ozone layer depletion
f) persistent organic pollutants.

Climate Smart Agriculture

While agriculture is the sector most sensitive to climate change, it is also a large contributor, accounting for around 14 percent of greenhouse gas emissions (IPCC 2007). Nonetheless, agriculture may contribute to the answer by assisting people in feeding themselves and adapting to changing conditions while reducing climate change. Agriculture has the potential to sequester or absorb carbon into the soil rather than emit it. This is possible without sacrificing production or yields. Higher yields, more carbon in the soil, and improved resistance to drought and heat are all conceivable. This is known as the "triple win": measures that boost yields (poverty reduction and food security), make yields more robust in the face of extremes (adaptation), and make the farm a solution to the climate change issue rather than a source of it (mitigation). These triple successes are likely to need a bundle of actions that are country and location specific. Climate Smart Agriculture is the name given to this kind of farming.

Climate-smart agriculture employs tried-and-true practises

For example:

a) Conservation tillage increases the organic content of the soil, which increases its water holding capacity, making yields more robust and decreasing erosion.
b) Promoting soil carbon capture also aids in climate change mitigation. Another example is integrated soil fertility management, which may save fertiliser expenditures while increasing soil carbon and yields.
c) Climate smart agriculture emphasises landscape techniques, such as integrated planning of land, crops, forestry, fisheries, and water, to exploit synergies.
d) These may be enhanced further by incorporating better weather forecasting, more robust food crops, and risk insurance to cover losses caused by weather fluctuations.
e) If yields grow and become more stable as a consequence of such techniques, farm revenue will improve. Farmers' adaptive capacity is enhanced by a steadier income.
f) A growing number of countries are demonstrating that it is possible.
g) China has been a pioneer in this field, with initiatives like the Loess Plateau becoming renowned worldwide.
h) Brazil has also invested in high-quality research and extension, resulting in these three outcomes.
i) In Kenya, small-holder farmers are already getting cash rewards as part of a trial programme for novel farming practises that will trap more carbon in the soil while enhancing soil fertility.

Green Economy

The term "green economy" does not have a common meaning or universal standards. The Rio+20 result document places green economy in the context of sustainable development and poverty eradication, and it states that each country's approach would differ depending on its own circumstances and goals. As a result, India's green economy is viewed in the context of sustainable development and inclusive economic growth, including poverty eradication. The government makes an effort to balance the three components of sustainable development (social, economic, and environmental). The green economy focuses on the fundamental adjustments that are necessary to make economic systems more sustainable. Green

Economy focuses on solutions to the underlying reasons of unsustainable economic development. A Green Economy is one in which income and employment development are fuelled by public and private investments that decrease carbon emissions and pollution, improve energy and resource efficiency, and avoid biodiversity and ecosystem loss.

Sustainable energy, green jobs, low-carbon economies, green policies, green buildings, agriculture, fisheries, forestry, industry, energy efficiency, sustainable tourism, sustainable transportation, waste management, water efficiency, and all other resource efficiency are all part of the Green Economy. Decarbonizing the economy, committing the environmental community to justice and equity, and conserving the biosphere are three priority in the transition to a green economy.

Adapting the Green Economy

a) An energy audit can help you minimize your building's carbon footprint and save money on energy expenditures. Overfishing threatens to damage future fish populations in many places of the world. We can avoid this by promoting sustainable fishing methods.
b) Deforestation is responsible for over 20 percent of global greenhouse gas emissions.
c) Forests that are sustainably maintained can continue to support populations and ecosystems without negatively impacting the environment or climate.
d) Reduce your reliance on paper products by using electronic files.
e) Supporting certified sustainable forest products helps to ensure a healthy environment and long-term livelihoods.
f) Carpooling or utilising public transportation decreases environmental and economic consequences while improving the community.
g) Short-distance walking or biking is excellent for your health and the environment.
h) Small measures toward sensible water use can help protect this valuable resource.
i) Water is one of our most vital resources, hence resource efficiency is critical to a Green Economy.
j) The growth of clean, renewable energy sources such as solar, wind, and tidal will help the green economy.
k) The pressure on our natural resources is reduced by recycling acceptable products and composting food waste.

l) Moving to a green economy has the potential to deliver sustainable development and poverty eradication on an unprecedented scale, with unparalleled speed and efficacy.

Ecological Footprint

The "ecological footprint" quantifies human demand on Earth's ecosystems. It is a standardised measure of demand for natural capital that can be compared to the planet's ecological capacity to regenerate, and it represents the amount of biologically productive land and sea area required to supply the resources consumed by a human population and to assimilate associated waste. Currently, humanity's total ecological footprint is estimated at 1.5 planet Earths—in other words, humanity uses ecological services 1.5 times as fast as Earth can renew them.

The "carbon footprint" is the amount of carbon being emitted by an activity or organisation. The carbon component of the ecological footprint converts the amount of carbon dioxide released into the amount of productive land and sea area required to sequester it and tells the demand on the Earth caused by the use of fossil fuels. The carbon footprint accounts for 54 percent of the ecological footprint and is the fastest-growing component, increasing 11-fold since 1961.

Every year, the Global Footprint Network (GFN) publishes an ecological footprint report that maps consumption and the need for natural resources to support it. The lifestyle embraced by industrialized countries is unsustainable, and it will take five Earths to meet their needs. On the other hand, where one planet is sufficient, the Indian way of life is sustainable. According to the Earth Overshoot Report, the Ecological Footprint of industrialized nations ranges from 8 to 4, while India stands at 0.9.

IntergovernmentalPanel on Climate Change (IPCC)

In December 1988, the United Nations General Assembly passed a resolution on the issue, endorsing the UNEP/WMO proposal to establish the Inter-Governmental Panel on Climate Change (IPCC). It was founded in 1988 by the United Nations Environment Programme (UNEP) and the World Meteorological Organization (WMO) to give the world's governments with a clear scientific understanding of what is happening to the world's climate. The Secretariat coordinates all IPCC activities and communicates with governments. WMO and UNEP assist the secretariat, which is based at WMO headquarters in Geneva. It is available to all United Nations (UN) and World Meteorological Organization (WMO) members. The IPCC now has 195 member nations.

The IPCC's initial task, as outlined in UN General Assembly Resolution 1988, was to prepare a comprehensive review and recommendations on the state of knowledge in climate science, the social and economic impact of climate change, and possible response strategies and elements for inclusion in a possible future international climate convention. The IPCC is a scientific organisation. It examines and evaluates the most recent scientific, technological, and socioeconomic information produced throughout the world that is crucial to understanding climate change. It neither conducts research nor monitors climate-related data or characteristics. Thousands of scientists from across the world contribute to the IPCC's work on a volunteer basis. To ensure an impartial and full assessment of current knowledge, review is an integral aspect of the IPCC process. Governments take part in the review process as well as the plenary sessions, where major decisions about the IPCC work programme are made and reports are accepted, adopted, and approved. Governments acknowledge the authority of their scientific substance by approving the IPCC reports. As a result, the organization's work is both policy-relevant and policy-neutral, never policy-prescriptive. The IPCC has consistently provided the most up-to-date information. The Assessment Studies, the most comprehensive scientific reports on climate change produced worldwide, have been presented on a regular basis by the IPCC. It has also answered to the UNFCCC's request for scientific and technical information through Special Reports, Technical Papers, and Methodology Reports. It has also developed methodology and recommendations to assist UNFCCC Parties in preparing their national greenhouse gas inventories. In accordance with its mandate and as reaffirmed in various Panel decisions, the IPCC prepares at regular intervals comprehensive Assessment Reports of scientific, technical, and socioeconomic information relevant to understanding human-induced climate change, potential impacts of climate change, and mitigation and adaptation options. Assessment Reports are often released in many volumes, one for each of the IPCC Working Groups and, subject to Panel approval, a Synthesis Report. They are written in a non-technical way that is appropriate for policymakers. They are divided into two parts: a lengthy report and a Summary for Policymakers.

The Economics of Ecosystem and Biodiversity (TEEB)

In 2007, Germany and the European Commission created the Economics of Ecosystems and Biodiversity (TEEB). Pavan Sukhdev is in charge of research. It is a global campaign to raise awareness of the global economic advantages of biodiversity. Highlighting the rising cost of biodiversity loss and environmental degradation, as well as bringing together experts from science, economics, and policy to allow effective solutions Through its five deliverables, it will analyse, explain, and mainstream the urgency of activities. These include:

a) Science and economic underpinnings, policy costs, and the consequences of inactivity
b) Opportunities for policymakers at the national and international levels
c) Decision-making assistance for local administrations
d) Risks, opportunities, and metrics in business
e) Ownership by citizens and consumers

UN Environment Programme

The United Nations Environment Programme (UNEP) has been the global authority that sets the environmental agenda, promotes the coherent implementation of the environmental dimension of sustainable development within the United Nations system, and serves as an authoritative advocate for the global environment since its inception in 1972. UNEP's objective is to inspire, enlighten, and enable nations and peoples to improve their quality of life without sacrificing that of future generations by providing leadership and encouraging cooperation in environmental care. UNEP seeks to achieve transformative change for people and nature by addressing the core causes of the three global crises of climate change, biodiversity loss, and pollution and waste. Climate Action, Chemicals and Pollution Action, Nature Action, Science Policy, Environmental Governance, Finance and Economic Transformations, and Digital Transformations are the seven interconnected action subprogrammes used by UNEP. UNEP increases awareness and advocates for effective environmental action through various programmes, including World Environment Day. UNEP is headquartered in Nairobi, Kenya, and operates through divisions, regional, liaison, and out-post offices, as well as an expanding network of collaborating centres of excellence. Through the UN Environment Assembly, the world's highest-level environmental decision-making body,

UNEP works closely with its 193 Member States and representatives from civil society, business, and other important groups and stakeholders to solve environmental concerns.

National Oceanic and Atmospheric Administration

The National Oceanic and Atmospheric Administration (NOAA) is a scientific and regulatory agency of the United States Department of Commerce that forecasts weather, monitors oceanic and atmospheric conditions, charts the seas, conducts deep sea exploration, manages fishing, and protects marine mammals and endangered species in the United States' exclusive economic zone.

Purpose and function

NOAA's Specific role includes:

a) Environmental Information Products are supplied. Through the National Weather Service, NOAA provides information on the state of the seas and the atmosphere to its customers and partners, such as weather alerts and forecasts. Climate, ecosystems, and commerce are all covered by NOAA's information services.
b) Environmental stewardship services are provided. NOAA is a guardian of the United States' coastal and marine habitats. NOAA supervises the use of these habitats in collaboration with federal, state, local, tribal, and international agencies, regulating fisheries and marine sanctuaries and safeguarding vulnerable and endangered marine animals.
c) Performing Applied Scientific Research NOAA's mission is to provide accurate and impartial scientific information in four areas of national and worldwide importance: ecosystems, climate, weather and water, and commerce.

Five Fundamentals of NOAA's

a) Instruments and data collecting networks are used to monitor and observe Earth processes.
b) Understanding and explaining Earth systems through data study and analysis.
c) assessing and forecasting the evolution of these systems across time.

d) Providing important information to the public and partner organisations through engaging, advising, and enlightening them.
e) Environmental resource stewardship.

World Meteorological Organization (WMO)

The World Meteorological Organization (WMO) is a United Nations specialised organisation charged with encouraging worldwide collaboration in atmospheric research, climatology, hydrology, and geophysics. The World Meteorological Organization evolved from the International Meteorological Organization, a voluntary organisation created in 1873 to facilitate the exchange of meteorological data and research. Proposals to change the IMO's status and organisation culminated in the 1947 World Meteorological Convention, which formed the World Meteorological Organization. The Convention went into effect on March 23, 1950, and the WMO commenced functioning as an intergovernmental agency inside the UN system the following year. The World Meteorological Organization (WMO) is made up of 193 nations and territories and allows the "open and unlimited" exchange of data, information, and research between its members' meteorological and hydrological organisations. It also works with nonprofit groups and other international organisations on environmental, climate change, resource management, and socioeconomic development issues. The World Meteorological Organization is managed by the World Meteorological Congress, which meets every four years to define policies and goals and is headquartered in Geneva, Switzerland. The Congress is managed by an Executive Council lead by the President, who is presently Germany's Gerhard Adrian.

Strategy plan:

a) Disaster risk reduction
b) The Global Framework for Climate Services (GFCS)
c) The WMO Integrated Global Observing System (WIGOS)
d) Aviation meteorological services
e) Polar and high mountain regions
f) Capacity development
g) Governance

World Wide Fund for Nature (WWF)

The World Wildlife Fund Inc. (WWF) is a non-governmental worldwide organisation created in 1961 that strives to preserve wildness and reduce human effect on the environment. It was previously known as the World Wildlife Fund, and this is still its official name in Canada and the United States. WWF is the world's biggest conservation group, with over five million supporters worldwide, operating in over 100 countries, and funding around 3,000 conservation and environmental initiatives. Since 1995, they have funded over $1 billion in over 12,000 conservation activities. In 2014, the WWF received 55 percent of its financing from individuals and bequests, 19 percent from government sources (such as the World Bank, DFID, and USAID), and 8 percent from companies. The WWF's mission is to halt the destruction of the planet's natural environment and to create a future in which humans live in harmony with nature. Since 1998, WWF has issued the Living Planet Report every two years; it is based on a Living Planet Index and an ecological footprint estimate. Furthermore, WWF has launched numerous major global initiatives, such as Earth Hour and the Debt-for-nature exchange, and its present work is centred around six areas: food, climate, freshwater, wildlife, forests, and oceans.

Panda symbol

The WWF's huge panda emblem was inspired by Chi Chi, a panda that was brought from Beijing Zoo to London Zoo in 1958, three years before WWF was founded. Her distinctive physical traits and position as an endangered species, as well as her fame as the only panda dwelling in the Western hemisphere at the time, were deemed perfect to suit the organization's requirement for a powerful identifiable emblem that would transcend all language boundaries. The organisation also needed an animal that would stand out in black and white print. Sir Peter Scott then developed the logo based on early designs by Gerald Watterson, a Scottish naturalist.

Earth Justice

Earthjustice (Originally known as the Sierra Club Legal Defense Fund) is a non-profit public interest group established in the United States that focuses on environmental litigation. It is headquartered in San Francisco and includes 14 regional offices around the United States, as well as an international programme, a communications team, and a policy and legislative team in Washington, DC. The Sierra Club Legal Defense Fund was established in 1971, although it was completely independent of the Sierra Club. In 1997, it renamed itself Earthjustice to better reflect its function as a legal advocate for hundreds of regional, national, and worldwide organisations. As of September 2018, the organisation had offered free legal assistance to over 1,000 clients ranging from the Sierra Club, World Wildlife Fund, and American Lung Association to smaller state and neighbourhood organisations like the Maine Lobstermen's Association and Friends of the Everglades.

Three key goal areas:

a) The Wild: The Wild focuses on cases in the Arctic to protect the ecosystem from climate change and fossil fuel drilling; cases in the ocean ecosystem to protect from overfishing, pollution, and habitat loss; cases in the wild to preserve wild places and support biodiversity; and cases related to saving wildlife from extinction that are threatened by clear cutting, oil drilling, dams, dewatering streams, and climate change.
b) Healthy Communities: Healthy Communities focuses on instances that safeguard against pollution and harmful substances at the regional and national levels. Earthjustice use lawsuits to implement and enforce the Clean Air and Clean Water Acts, as well as to reform the toxics regulatory system.
c) Clean energy and stable climate: Clean Energy and a Stable Climate focuses on situations that limit the use of fossil fuels, remove obstacles to and provide incentives for the use of renewable energy sources (for example, through the Right to Zero campaign), and encourages ecological resiliency to endure rising global temperatures. Earthjustice uses lawsuits to develop and enforce national regulations for coal ash waste, retire outdated coal-fired power stations, restrict coal exports, and halt coal mining. Cases in this critical area also focus on blocking fracking, oil and gas extraction on public lands, enhancing environmental and health safeguards, and restricting investments in fossil fuel infrastructure.

Greenpeace

Greenpeace is a non-profit global advocacy organisation. The network is made up of 26 autonomous national and regional organisations spread across Europe, the Americas, Africa, Asia, and the Pacific, as well as a coordinating body, Greenpeace International, located in Amsterdam, the

Netherlands. Greenpeace was formed in Canada in 1971 by Irving and Dorothy Stowe, two American immigrant environmental campaigners. Greenpeace's stated objective is to guarantee the Earth's ability to nourish life in all its richness, and the organisation campaigns on global concerns such as climate change, deforestation, overfishing, commercial whaling, genetic engineering, and nuclear issues.

Aims and Objectives:

a) Stop the earth from rising above 1.5 degrees Celsius to avoid the most disastrous consequences of climate collapse.
b) Conserve biodiversity in all of its manifestations.
c) Slow the rate of hyper-consumption and teach ourselves to live within our means.
d) Promote renewable energy as a solution capable of powering the entire planet.
e) Encourage global disarmament and nonviolence.

Greenpeace India is the Indian section of the worldwide environmental NGO Greenpeace, which has a presence in 55 countries spanning Europe, America, and Asia. Greenpeace India is a legally established society with headquarters in Bengaluru and branches in Delhi, Chennai, and Patna.

Environment Defense Fund

The Environmental Defense Fund, sometimes known as Environmental Defense, is a non-profit environmental advocacy organisation established in the United States. The organisation is well-known for its work on topics such as global warming, ecosystem restoration, oceans, and human health, and it advocates for combining strong research, economics, and legislation to create effective environmental solutions. It is a neutral organisation whose work frequently argues for market-based solutions to environmental concerns. The group's headquarters are in New York City, with offices around the United States and scientists and policy professionals operating globally. Austin, Texas; Boston, Massachusetts; Boulder, Colorado; San Francisco and Los Angeles, California; Raleigh, North Carolina; and Washington, D.C. are among the US regional offices. With offices in London, Brussels, Mumbai, and Beijing, the group's international footprint is expanding.

Centre for International forestry research, Bogor, Indonesia

The Centre for International Forestry Research (CIFOR) is a non-profit scientific study organisation that conducts research on forest use and management in poor nations, with a concentration on tropical forests. CIFOR is the forestry and agroforestry research centre of the Consultative Group on International Agricultural Research (CGIAR), a network of 15 research centres around the world that focus on agricultural research for sustainable development, working closely with governments and other partners to help develop evidence-based solutions to problems related to sustainable agriculture and forestry.

The goal of CIFOR-ICRAF research is to provide practical information and solutions to revolutionise how land is utilised and food is produced: protecting and repairing ecosystems, responding to global climate, hunger, biodiversity, and desertification problems. It also contributes to the UN Sustainable Development Goals, Rio Convention obligations, including the Paris Agreement, and the UN Framework Convention on Climate Change.

Research Themes:

a) Improving smallholder management and production systems (food security and nutrition).
b) Increasing smallholder income generation and market integration.
c) Improving laws and structures to increase social assets and safeguard rights in forest- and tree-dependent communities Understanding tree species dangers and developing genetic conservation methods.
d) Conserving and characterising high-quality tree crop and wild related germplasm.
e) Improved Silviculture, monitoring, and management approaches for numerous uses.
f) Creating tools and strategies for resolving disputes over the allocation of benefits and resource rights.
g) Understanding the forces driving forest transition.
h) Understanding the effects of forest transformation on ecosystem services and livelihoods.
i) Using forests, trees, and agro forestry to mitigate climate change.
j) Improving adaptation to climate change.

k) Understanding the synergies between mitigation and adaptation to climate change.

Chartered Institute of Ecology and Environmental Management

The Chartered Institute of Ecology and Environmental Management (CIEEM) is the professional organisation that represents and supports ecologists and environmental managers, mostly in the United Kingdom but also in Ireland, mainland Europe, and across the world. CIEEM envisions a healthy natural environment for present and future generations. CIEEM was founded in 1991 and presently has over 6,000 members representing local governments, government agencies, industry, environmental consulting, teaching/research, and non-governmental organisations. CIEEM, formerly known as IEEM, received a royal charter in 2013, recognising its high degree of expertise.

Activities of CIEEM

a) CIEEM offers a number of services (including conferences, training, events, counselling, and advice) to help professional ecologists and environmental managers improve their competency and standards, as well as to promote ecology and environmental management as a profession.
b) Members of the CIEEM can become Chartered Ecologists and/or Chartered Environmentalists.
c) CIEEM is a member of the Society for the Environment as well as the Environmental Policy Forum (EPF). CIEEM is also a member of the IUCN-UK Committee and a Greener UK supporter.
d) CIEEM is a partner in the United Nations Decade of Biodiversity 2011–2020, as well as a signatory to the Countdown 2010 pact to help safeguard biodiversity and a partner in the 2010 International Year of Biodiversity UK collaboration.

Copenhagen Climate Council

The Copenhagen Climate Council (now Sustainia) is a worldwide partnership between international business and science formed by Erik Rasmussen, creator of Monday Morning, Scandinavia's foremost independent think tank situated in Copenhagen. The Copenhagen Climate Council has come together to raise global awareness of the importance of the UN Climate Summit (COP15) in Copenhagen in December 2009, as well as to provide technical and public support and assistance to global decision makers when negotiating a new climate treaty to replace the Kyoto Protocol from 1997.

Purpose

The Copenhagen Climate Council's mission is to raise global awareness of the significance of the UN Climate Summit (COP15) in Copenhagen in December 2009. The Copenhagen Climate Council aims to provide creative yet feasible solutions to climate change in the run-up to this key UN summit, as well as analyse what is needed to make a new global treaty successful. The council will strive to foster productive conversation between government and industry so that when the world's political leaders and negotiators convene in Copenhagen, they will be armed with the most compelling reasons for drafting a treaty that can be backed by global business. Through the promotion and demonstration of creative, constructive, and significant corporate leadership and ideas. The goal of the Copenhagen Climate Conference is to show that securing an effective global climate pact is not only possible, but also imperative. The following concepts underpin the strategy:

a) Increasing international awareness of the significance of the Copenhagen UN Climate Summit and the Kyoto Protocol's successor treaty.
b) Encouragement of healthy conversation between government, industry, and science.
c) Inspiring global business leaders by demonstrating that addressing climate change has the potential to open up enormous opportunities for innovation and economic prosperity.

Klima Action, Malaysia

Klima Action Malaysia (KAMY) is a Malaysian climate activist organisation founded by a group of young people with the goal of empowering women, youth, and marginalised groups. They were established in 2019 and are one of Malaysia's few grassroots climate activism organisations.

Aims:

The grassroots movement calls for climate policy based on peace, fairness, and democracy, and it urges governments to take action on climate change. Since April of 2019, they have been actively participating in activities such as multi-language advocacy, climate protests, capacity development workshops, and community awareness in an effort to empower communities and reach a wider population.

Extinction Rebellion

Extinction Rebellion (abbreviated as XR) is a worldwide environmental movement whose declared goal is to use peaceful civil disobedience to compel government action to avert climate system tipping points, biodiversity loss, and the risk of societal and ecological collapse. In May 2018, Extinction Rebellion was founded in the United Kingdom. Extinction Rebellion's website put forward the following goals at the time of the group's formation in the United Kingdom:

a) The government must speak the truth by declaring a climate and ecological emergency and collaborating with other institutions to highlight the importance of change.
b) The government must act now to reverse biodiversity loss and achieve net-zero greenhouse gas emissions by 2025.
c) The government must establish and be governed by a citizens' assembly on climate and ecological justice.

CHAPTER 26

INDIA AND CLIMATE CHANGE

India's position on Climate Change

In 2010, India released 2,136.84 million tonnes of CO_2 equivalent greenhouse gases. In 2010, the energy sector was the leading source of emissions, accounting for 71 percent of total emissions. Electricity generation, fuel burning in industries, transportation, and fugitive emissions are all part of the energy industry. Industrial operations and product consumption contributed 8 percent to the national GHG inventory, while agriculture and waste sectors contributed 18 percent and 3 percent, respectively. About 12 percent of emissions were offset by carbon sink activity in forests and croplands, bringing the total national GHG emissions to 1,884.31 million tonnes of CO_2 equivalent. India's per capita GHG emissions in 2010 were 1.56 tCO_2 equivalent, which is less than one-third of global emissions and significantly lower than many developed and developing countries. A reduction in GDP emission intensity of roughly 12 percent between 2005 and 2010 was accomplished in comparison to our voluntary promise to lower GDP emission intensity by 20–25 percent by 2020, compared to 2005 levels. India is the fifth-largest GHG emitter, accounting for around 5 percent of world emissions. India's emissions climbed by 65 percent between 1990 and 2005 and are expected to grow by another 70 percent by 2020. As previously stated, India is particularly vulnerable to climate change due to rising natural catastrophes, dwindling natural resources, and a heavy reliance on agriculture and rain. Regardless of resource constraints, India is taking several steps to adapt to and mitigate climate change by boosting energy efficiency, creating a circular economic model, and encouraging the use of renewable energy. According to World Bank report, India will continue to be a low-carbon economy. India's principal focus is on adaptation, with a particular emphasis on mitigation. India has previously released a thorough National Action Plan on Climate Change, the details of which are available to the public. As a global public good, India encourages collaborative research in future low-carbon technologies and access to intellectual property rights (IPRs).

Climate and Weather Changes in India

Some of the observed changes in climatic indicators in India have been aggregated in India's National Communication (NATCOM) to UNFCCC. There is yet to be established a strong relationship between the reported changes indicated below and warmth caused by human climate change.

a) Surface Temperature: Since the late 1800s, the average surface temperature of the Earth has risen by around 1.62 degrees Fahrenheit (0.9 degrees Celsius), owing mostly to rising carbon dioxide and other man-made emissions into the atmosphere. The majority of the warming has happened in the last 35 years, with the five hottest years on record occurring since 2010. Over the last century, surface air temperatures have increased by 0.4° C on a nationwide scale. Warming has been recorded throughout the west coast of India, as well as in central India, the interior peninsula, and north-eastern India. However, cooling patterns have been noted in regions of north-west India and south India. The 1999 super cyclone in Orissa killed over a million people and caused significant property damage in coastal areas. During the Hud-Hud storm in 2014, Andhra Pradesh went through a similar scenario. In 2019, India and its neighbouring nations experienced a severe heat wave from mid-May to mid-June.

b) Precipitation: Monsoon rainfall has already been declining since the 1950s. A 2 degrees Celsius increase in global average temperatures would make India's summer monsoon very unpredictable. At 4 degrees Celsius warming, a highly wet monsoon, which now occurs once every 100 years, is anticipated to occur every 10 years by the end of the century. Dry years are projected to be dryer, while rainy years will be wetter.

c) Droughts: Evidence suggests that portions of South Asia have been drier since the 1970s, with more droughts occurring. Droughts have far-reaching repercussions. Droughts hit more than half of India's crop area in 1987 and 2002–2003, resulting in a significant drop in crop production. Droughts are anticipated to become increasingly common in several parts of India, particularly in north-western India, Jharkhand, Orissa, and Chhattisgarh. Crop yields are anticipated to plummet dramatically by the 2040s due to high heat. Drought-prone states like as Madhya Pradesh, Karnataka, Western Rajasthan, Andhra Pradesh, Southern Gujarat, Haryana, Madhya Pradesh, and southern Bihar are constantly experiencing dryness.

d) Ground water: Even in the absence of climate change, 15 percent of India's groundwater resources are overexploited. Water tables are expected to continue to fall as a result of increased demand for water from a growing population, more affluent lifestyles, as well as the services sector and industry.
e) Sea level rise: Over the previous century, the global sea level risen by around 8 inches. However, the rate in the recent two decades has roughly doubled that of the previous century and is increasing slightly each year. For more than 40 years, coastal tide gauge data in the north Indian Ocean show that sea level increase has been between 1.06–1.75 millimetres each year. These rates are consistent with the IPCC's global sea level rise predictions of 1–2 millimetres per year. Chiru in Rajasthan reported 50.8 degrees Celsius, the hottest temperature in India in 2016. Thermal expansion of a warmer ocean accounts for a significant portion of sea-level rise (as much as 0.3–0.8 metres over the past century).
f) Ocean acidification: The acidity of surface ocean waters has increased by around 30 percent since the beginning of the Industrial Revolution. This rise is due to people generating more carbon dioxide into the atmosphere, which is then absorbed by the seas. The quantity of CO2 absorbed by the ocean's top layer is growing by around 2 billion tonnes every year.
g) Warming ocean: The oceans have absorbed a large portion of the additional heat, with the top 700 metres (about 2,300 feet) of ocean rising by more than 0.4 degrees Fahrenheit since 1969.
h) Indian Agriculture: Because a substantial portion of India's arable land is rain-fed, agricultural output is highly dependent on rainfall and its pattern. Agriculture will be harmed not just by a rise or decrease in overall rainfall volumes, but also by changes in rainfall pattern. Any shift in rainfall patterns poses a significant risk to agriculture, and so to the economy and food security. Summer rainfall provides for over 70 percent of total annual rainfall in India and is critical to Indian agriculture. However, research suggest that summer rainfall will decrease by the 2050s. As temperatures rise, semi-arid parts of western India will receive more rainfall than usual, whereas central India may see a 10 percent to 20 percent decline in winter rainfall by the 2050s. Small climatic changes may have a big impact on water supplies, especially in arid and semi-arid regions like north-west India. Climate change and its influence in the form of extreme weather events is one of the most pressing problems for a country's food security. The forecasted 1–2.5 degree

Celsius temperature rise by 2030 is expected to have a significant impact on agricultural output. High temperatures can shorten crop duration, alter photosynthesis, increase crop respiration rates, and impact pest population. In India, a 1.5 degrees Celsius rise in temperature and a 2 millimetre decrease in precipitation can affect rice output by 3 to 15 percent.

i) Impact on the Himalayas: Glaciers are receding practically everywhere on Earth, including in the Alps, Himalayas, Andes, Rockies, Alaska, and Africa. The Himalayas have some of the most extensive snow and ice resources, and its glaciers provide water to perennial rivers such as the Indus, Ganga, and Brahmaputra. Glacial melt may have an effect on their long-term lean-season flows, affecting the economy in terms of water supply and hydropower generation. While certain glaciers have receded in some Himalayan locations in recent years, the trend is not constant across the whole mountain chain, according to existing monitoring data on Himalayan glaciers. As a result, it is too early to identify long-term patterns or their causality, for which numerous explanations exist.

Current Actions and Adaptations

In the context of climate change, adaptation refers to the actions done to mitigate the negative effects of climate change, such as moving people living near the seashore to cope with increasing sea levels, or switching to crops that can endure higher temperatures. Mitigation includes actions to minimise greenhouse gas emissions that cause climate change in the first place, such as converting to renewable energy sources such as solar or wind energy, or nuclear energy, rather than burning fossil fuel in thermal power plants. Current Indian government investment on climate adaptation surpasses 2.6 percent of GDP, with agriculture, water resources, health and sanitation, forecasts, coastal-zone infrastructure, and extreme weather events being particular areas of concern.

a) Agriculture: Two risk-financing programmes help with climate adaption. The Crop Insurance Scheme helps farmers insure against climate risks, while the Loans Support Mechanism makes it easier to extend credit to farmers, particularly for crop failure owing to climatic unpredictability.
b) Crop improvement: The current programmes cover measures such as arid-land crop development and pest control, as well as capacity

building for extension workers and non-governmental organisations (NGOs) to encourage better vulnerability-reduction techniques.

c) Drought proofing: The current programmes aim to reduce the negative effects of drought on crop and livestock production, as well as land, water, and human resource productivity, in order to eventually lead to drought proofing of the affected areas. They also aim to promote overall economic development and improve the socioeconomic conditions of the resource poor and disadvantaged populations who live in the programme areas.

d) Forestry: India has a robust and fast expanding afforestation programme. The Forest Conservation Act of 1980, which aimed to stop the clearing and degradation of forests through strict, centralised control of the rights to use forest land and mandatory requirements of compensatory afforestation in the case of any diversion of forest land for any non-forestry purpose, accelerated the afforestation process. Furthermore, a vigorous afforestation and sustainable forest management programme resulted in yearly reforestation of 1.78 mha between 1985 and 1997, and is currently 1.1 mha. As a result, carbon stocks in Indian forests have risen to 9-10 gigatonnes of carbon over the previous 20 years (from 1986 to 2005).

e) Water: The National Water Policy emphasises that non-traditional methods of water utilisation, such as inter-basin transfers, artificial recharge of groundwater, and desalination of brackish or sea water, as well as traditional water conservation practises such as rainwater harvesting, including roof-top rainwater harvesting, should be used to increase utilisable water resources. Many states currently have obligatory water collection programmes in place in a number of cities.

f) Coastal regions: In coastal areas, limits have been placed between 200 and 500 metres from the HTL (high tide line), with specific restrictions applied up to 200 metres to safeguard delicate coastal ecosystems and prevent their exploitation.

g) This meets the coastal population's concerns as well as their livelihood. Construction of coastal protective infrastructure and cyclone shelters, as well as plantation of coastal forests and mangroves, are some particular efforts implemented in this respect.

h) Health: The primary goal of current health programmes is to monitor and control vector-borne illnesses such as Malaria, Kala-azar, Japanese Encephalitis, Filaria, and Dengue. Programs also provide for emergency medical relief in the event of natural disasters, as well as training and development of human resources for these responsibilities.

i) Disaster management: The National Disaster Management programme assists victims of weather-related disasters and administers disaster relief efforts. It also supports proactive disaster prevention programmes, such as information distribution and disaster-management personnel training.

National Action Plan on Climate Change

To address the problem of climate change, we must act on several fronts at the same time. The National Action Plan is dependent on the development and use of new technologies. The Plan would be implemented through suitable institutional frameworks tailored to the efficient fulfilment of each Mission's objectives, including public-private partnerships and civil society activity. The emphasis will be on fostering climate change understanding, adaptation and mitigation, energy efficiency, and natural resource conservation. The National Action Plan is built around eight National Missions that reflect multifaceted, long-term, and integrated strategies for attaining critical goals in the context of climate change.

a) National Solar Mission: The NAPCC's mission is to encourage the development and use of solar energy for power production and other applications, with the ultimate goal of making solar competitive with fossil-based energy choices. The plan contains specific targets for growing the use of solar thermal technology in urban areas, industry, and commercial enterprises; boosting photovoltaic output to 1000 MW/year; and installing at least 1000 MW of solar thermal power generation. Other goals include the formation of a solar research centre, enhanced international collaboration on technological development, local manufacturing capacity enhancement, and increased government financing and international assistance.
b) National Mission for Enhanced Energy Efficiency: Initiatives aimed at boosting energy efficiency were projected to save 10,000 MW by 2012. Building on the Energy Conservation Act of 2001, the plan suggests: mandating specific energy consumption reductions in large energy-consuming industries, with a system for companies to trade energy-savings certificates; energy incentives, such as lower taxes on energy-efficient appliances; and financing for public-private partnerships to reduce energy consumption

through demand-side management programmes in municipal, building, and agricultural sectors.

c) National Mission on Sustainable Habitat: To encourage energy efficiency as a key component of urban design, the plan proposes: Extending the current Energy Conservation Building Code with a stronger emphasis on urban waste management and recycling, including waste-to-energy generation; incentives for using public transportation, as well as stronger enforcement of car fuel efficiency regulations and the use of price strategies to encourage the purchase of efficient vehicles.

d) National Water Mission: With water shortages expected to intensify as a result of climate change, the initiative aims to enhance water usage efficiency by 20 percent through price and other methods.

e) National Mission for Sustaining the Himalayan Ecosystem: The strategy aims to protect biodiversity, forest cover, and other ecological resources in the Himalayan area, where glaciers, a vital source of water for India, are expected to decrease due to global warming.

f) National Mission for Green India: Goals include reforesting 6 million hectares of degraded forest land and increasing forest cover from 23to 33 percent of India's area.

g) National Mission for Sustainable Agriculture: The strategy intends to help agriculture adapt to climate change by developing climate-resilient crops, expanding weather insurance arrangements, and changing agricultural practises.

h) National Mission on Strategic Knowledge for Climate Change: The strategy anticipates a new Climate Science Research Fund, enhanced climate modelling, and expanded international collaboration to achieve a better grasp of climate science, impacts, and problems. It also promotes private-sector initiatives, including as venture capital funds, to develop adaptation and mitigation technology.

Indian Network on Climate Change Assessment

The Ministry of Environment and Forests (MoEF) founded the Indian Network on Climate Change Assessment (INCCA) in October 2009 in an effort to promote domestic climate change research and build on the country's climate change expertise. INCCA is a MoEF network-based programme with over 120 universities and over 250 scientists. The goal is to implement more science-based policymaking across the country, based on measurements, monitoring, and modelling. The INCCA has been conceptualised as a Network-based Scientific Programme designed to address and assess the drivers and implications of climate change through scientific research, climate change assessments on various aspects of climate change, associated vulnerabilities, and adaptation, decision support systems, and capacity building for climate change management. The INCCA will conduct research on the consequences of climate change in various Indian regions and sectors, and will provide recommendations for appropriate adaptation and mitigation measures. The goal is to create an independent organisation of Indian scientists who can produce scientific papers at the scientific level concerning the impact of climate change on diverse sectors, which may present a genuine picture and influence the global discussion. The INCCA's reports will be included in India's National Communication (Nat Com) to the United Nations Framework Convention on Climate Change (UNFCCC).

National Initiative on Climate resilient Agriculture

In the period 2010–2011, the ICAR established the National Initiative on Climate Resilient Agriculture (NICRA), with a budget of Rs.350 crores for the XI Plan. This effort will largely improve the resilience of Indian agriculture, which includes crops, livestock, and fisheries. The major goal was to improve Indian agriculture's resilience by adapting crops, livestock, and fisheries to climate variability and change through the development and implementation of improved production and risk management strategies. Strategic research in crops, horticulture, livestock, natural resource management, and fisheries has been planned in a network style at ICAR's premier research institutes. To begin, the programme would prioritise crops such as wheat, rice, maize, pigeonpea, peanuts, tomato, mango, and banana; livestock such as cattle, buffalo, and small ruminants; and economically valuable marine and freshwater fish species.

Project components:

a) Strategic adaptation and mitigation research
b) Farmers' fields are being used for technology demonstrations to deal with present climatic unpredictability.
c) Grants for sponsored and competitive research to solve crucial research shortages.
d) Building the capacity of various stakeholders.

Major research themes

a) Assessment of key manufacturing zones' vulnerability
b) Linking weather-based agro-advisories to contingency planning.
c) Evaluating the consequences of main climatic stressors (drought, heat, frost, flooding, etc.) in major food and horticultural crops.
d) Continuous monitoring of greenhouse emissions in open fields in big manufacturing systems.
e) Improving water and nutrient use efficiency and conservation to develop adaptation and mitigation measures agriculture.
f) Changes in pest dynamics, pest/pathogen-crop linkages, and the introduction of novel pests and diseases in response to climate change.
g) Animal adaptation techniques based on dietary and environmental modifications.
h) Using a better understanding of spawning behaviour to harness the favourable impacts of temperature in inland and marine fisheries.

Indian Solar Loan Programme

The United Nations Environment Programme (UNEP) launched a three-year credit facility in Southern India in April 2003 to assist rural people in financing the purchase of Solar Home Systems. Canara Bank and Syndicate Bank, together with their eight affiliate Regional Rural Banks, collaborated with UNEP to design and manage a Loan Program through their branch offices throughout Karnataka and a portion of neighbouring Kerala State. UNEP offers technical assistance, vendor qualification, and other initiatives to strengthen institutional capacity for this type of finance, in addition to financial help in the form of interest rate subsidies for borrowers.

India's response to Climate Change

a) National Action Plan on Climate Change: The NAPCC describes current and future climate mitigation and adaptation policies and projects. Solar Energy; Enhanced Energy Efficiency; Sustainable Habitat; Water; Sustaining the Himalayan Ecosystem; Green India; Sustainable Agriculture; and Strategic Knowledge for Climate Change are identified as eight main "national missions" running through 2017. The majority of these missions have severe adaption requirements.

b) National Clean Energy Fund: In 2010, the Indian government established the National Renewable Energy Fund (NCEF) to finance and promote clean energy activities, as well as to fund clean energy research in the country. The fund's corpus is developed by levying a cess of INR 50 (subsequently raised to INR 100 in 2014) per tonne of local or imported coal.
c) Paris Agreement: India has made three pledges under the Paris Agreement. By 2030, India's greenhouse gas emission intensity of GDP will be decreased by 33–35 percent from 2005 levels. In addition, non-fossil fuels would account for 40 percent of India's power capacity. Simultaneously, India will develop an extra 'carbon sink' of 2.5 to 3 billion tonnes of CO2 equivalent by 2030 through increased forest and tree cover.
d) International Solar Alliance: ISA was inaugurated on November 30, 2015, during the United Nations Climate Change Conference in Paris by India and France, in the presence of Mr. Ban Ki Moon, former Secretary-General of the United Nations.
e) BS(Bharat Stage) Emission Norms: Vehicle emissions are one of the leading causes of air pollution, prompting the government to implement BS 2000 (Bharat Stage 1) vehicle emission standards in April 2000, followed by BS-II in 2005. In 2010, BS-III was rolled out across the country. However, in 2016, the government opted to follow worldwide best practises and forgo BS V entirely in order to jump to BS-VI standards.

PART 5:

MISCELLANEOUS

CHAPTER 27

ENVIRONMENT AND AGRICULTURE

Introduction and Importance

Over the last 200 years, global food production has typically kept up with the exponential expansion in human population. The fast growth in global food production has resulted from very substantial improvements in both the productivity of existing farmland and the area of land cultivated. Furthermore, increasing transportation and refrigeration availability and efficiency have enabled humans to export food from one area or continent to another, alleviating issues caused by local droughts or epidemics. Despite the fact that overall food production on the planet has expanded greatly in recent years, per capita grain output will continue to drop since population growth has been far higher during the last five decades. Most advances in global food production from the commencement of agriculture until the early 1900s occurred from an increase in the total area of farmland under cultivation. However, mechanisation of farming operations has resulted in a continuous greater output to satisfy the expanding food supply need during the last few decades. High agricultural yields need the use of specialised seed varieties (high-yielding cultivars), an active fertiliser programme, and insect control. In many areas, irrigation is also required for excellent yields.

Agriculture is derived from the Latin terms'ager' or 'agri', which means soil, and 'cultura', which means cultivation. Agriculture is a wide phrase that includes agricultural cultivation, animal husbandry, fishing, forestry, and other related activities. Agriculture, which accounts for 17.2 percent of the country's GDP, provides a living for around two-thirds of the people. The industry employs 56.7 percent of the country's workforce and is the single largest private sector occupation. Agriculture contributes for approximately 14.7 percent of overall export profits and supplies raw materials to a wide range of industries (textiles, silk, sugar, rice, flour mills, milk products). The agriculture industry serves as a bulwark in preserving food security and, by extension, national security. Horticulture, animal husbandry, dairy, and fisheries have a vital role in improving the general economic situation, as well as the health and nutrition of the rural masses.

Some related terms

a) Silviculture is the cultivation of forest trees.
b) Sericulture is the practise of cultivating silkworms for the purpose of producing raw silk.
c) Apiculture is the human-maintained care of honey bee colonies, most typically in hives.
d) Olericulture is the study of cultivating vegetables that deals with the cultivation of non-woody (herbaceous) plants for nourishment.
c) Viticulturc is thc scicncc of growing, producing, and studying grapes.
f) Floriculture is a horticultural specialty that focuses on the production of blooming and attractive plants for gardens.
g) Arboriculture is the study, care, and cultivation of individual trees, shrubs, vines, and other perennial woody plants.
h) Pomology is a field of horticulture concerned with the cultivation, production, harvesting, and preservation of fruit, among other things.
i) Aeroponics is the method of growing plants without the use of soil or aggregate media in an air or mist environment.
j) Hydroponics is a method of growing plants in water without soil that uses several fertiliser solutions.
k) Growing plants in regular soil is referred to as geoponic in farming practise.

Problems of Indian Agriculture

a) Land holdings are being fragmented.
b) Small and marginal farmers exist.
c) Variation by region.
d) Seasonal rainfall is a factor.
e) Land production is low.
f) Uncovering unemployment is on the rise.
g) Agricultural product marketing is in disarray.
h) Land reformation is ineffective.

Evolution of Man and Agriculture

Agriculture's growth is staged in relation to human civilization. They are:

a) Hunting: Hunting used to be the principal source of food. It is a significant vocation that has existed for a very long time.
b) Pastoral: Humans received their nourishment from domesticated animals such as dogs, horses, cows, buffalos, and so on. They resided on the outskirts of the forest and were responsible for feeding his tamed animals. He would have moved from one location to another in quest of food to feed his animals. It was not pleasant, and they may have benefited by remaining in one location along the river bed.
c) Crop culture: By dwelling near the riverbed, he had adequate water for his animals and domesticated crops, so he began cultivating. As a result, he has begun to settle in a location.
d) Trade: When he began producing more than he needed, the excess was exchanged; this is the foundation of trade. Trade developed as agriculture flourished. This resulted in the creation of infrastructure such as roads, routes, and so on.

From the crop cultivation stage, agriculture became civilised. Some significant occurrences from various eras that contributed to the advancement of scientific agriculture are given below in table 27.1:

Table 27.1: Advancement of scientific agriculture

Period	**Events**
Earlier than 10000 BC	Hunting and gathering
7500 BC	Cultivation of wheat and barley
3400 BC	Wheel was invented
3000 BC	Bronze was used for making tools
2900 BC	Irrigated farming started
2300 BC	Cultivation of chickpea, cotton and mustard
2200 BC	Rice cultivation started
1500 BC	Cultivation of sugar cane
1400 BC	Use of Iron
1000 BC	Use of Iron plough
1500 AD	Cultivation of orange, brinjal and pomegranate
1600 AD	Several crops were introduced to India, including potato, tapioca, tomato, chilies, pineapple, groundnut, tobacco, rubber, and American cotton.

Agriculture Revolution in India

a) Green revolution: The primary goal of the green revolution was to boost agricultural production in developing countries via the application of technology and agricultural research. The major emphasis was on changing India into a modern industrial system through the employment of technology, such as high yielding variety (HYV) seeds, mechanised farm equipment, irrigation infrastructure, pesticides, and fertilisers. M.S. Swaminathanis regarded as the father of the GreenRevolution.

b) Round Revolution: Instead of a single yearly increase, the Potato Revolution planned to double or triple potato production.

c) Grey Revolution: This revolution began in response to the green revolution. It was established to remedy the mistakes made during the Green Revolution.

d) Pink Revolution: The Pink Revolution refers to a technical development in the poultry and meat processing industries. The construction of meat testing facilities, cold storage for growth, and other infrastructural facilities is part of the revolution. Durgesh Patel is regarded as the father of Pink Revolution.

e) White Revolution: The revolution was linked to a significant rise in milk output in the country. The White Revolution phase aimed to make India self-sufficient in milk production. VergheseKurien is considered to be the father of the White Revolution.

f) Blue Revolution: The Blue Revolution intends to establish an enabling environment for the integrated exploitation of the country's entire fisheries potential. The Blue Revolution also aspires to significantly enhance the financial situation of fishermen and fish growers while keeping sustainability, bio-security, and environmental issues in mind. Dr Arun Krishnan is regarded as the father of Blue Revolution.

g) Yellow Revolution: As a result of the Yellow Revolution, India went from being a net importer to a net exporter of oilseeds. Annual oilseed harvests reached an all-time high of 25 million tonnes in the early 1990s. Sam Pitroda is regarded as the father of the Yellow Revolution.

h) Protein Revolution: The Protein Revolution is a second Green revolution fuelled by technology. There was an establishment of a Rs.500 crore Price Stabilisation Fund to assist farmers in dealing with volatility. Kisan TV was also launched, which gives real-time information on new techniques, water conservation, and organic farming. Narendra Modi and Arun Jaitely is regarded as the father of Protein revolution.

i) Black revolution: The Indian government intended to increase ethanol output and blend it with gasoline to make biodiesel. Blending ethanol with transportation fuels is projected to improve farmer profits while supplementing limited and environmentally favourable hydrocarbon supplies.
j) Silver Revolution: The Silver Revolution is a strategy that results in a dramatic rise in egg and poultry output. Indira Gandhi was the driving force behind this movement. This was accomplished through the use of hybrid cocks and hens, as well as the application of applied science to improve egg output.
k) Golden Revolution: In India, the era from 1991 to 2003 is known as the Golden Revolution. The Golden Revolution is associated with honey production and horticulture. It is a component of India's significant agricultural revolutions. NirpakTutejis regarded as the father of the Golden Revolution.
l) Red Revolution: Vishal Tewari is regarded as the father of Red Revolution. The red revolution in India is a huge revolution that has boosted the country's tomato and meat output. Following this transformation, India's global ranking in meat and tomato output improved.

Significance of these revolutions

a) These agricultural technologies have lifted millions of people out of poverty by creating rural income prospects for farmers and farm labourers, as well as lowering consumer costs.
b) With the aid of the Green Revolution, India has become self-sufficient in food grain production. The Green Revolution boosted the number of food grains significantly, and as a result, India exited the PL-480 agreements, under which India purchases food grains from the United States.
c) The dairy industry saw a white revolution in the 1970s, thanks to the use of co-operatives. Farmers in the programme boosted milk output collectively, and India became a net exporter of milk. Milk production has increased exponentially, providing nutritional security to the people. Milk availability per capita has reached an all-time high. The farmers' additional income grew, and the people's food and nutritional security improved.

d) The blue revolution enhanced the production of sea food. This fisherman community benefits while also providing food and nutritional security.
e) These revolutions also offered money and jobs for women, resulting in women's empowerment and the emancipation of millions.
f) Milk output increased from 17 million tonnes before independence to 108.5 million tonnes during the White Revolution.
g) During the previous five decades, fish production increased from 0.75 million tonnes to approximately 7.6 million tonnes thanks to the Blue Revolution.
h) Since independence, oil seed output has expanded fivefold (from 5 million tonnes to 25 million tonnes) because to the yellow revolution.
i) Similarly, egg production rose from 2 billion at independence to 28 billion, sugarcane output went from 57 million tonnes to 282 million tonnes, and cotton production increased from 3 million bales to 32 million bales, demonstrating India's success.
j) India is the world's greatest producer of fruits and vegetables. India ranks second in terms of milk and vegetable production.

Crops and their classifications

Agronomy is a Greek word originating from "agros," which means "field," and "nomos," which means "management." It is a subset of agriculture that focuses on crop production and soil management. Crops are plants farmed on a big scale for food, clothing, and other human use.

a) Crop classification based on the type of produce:
   1) Food crops: The term 'food crops' refers to plants that are produced by man and offer food for human consumption. Rice, wheat, maize, millets (jowar, bajra, ragi; pulsesgram, tur (arhar), etc are the examples.
   2) Cash Crops: A cash crop, sometimes known as a profit crop, is an agricultural crop farmed for profit. It is usually acquired by parties other than a farm.
   3) Cotton, jute, sugarcane, tobacco, oilseeds, peanut, linseed, sesamum, castor seed, rapeseed, mustard, and so on are examples.
   4) Plantation Crops: A plantation is a large-scale estate dedicated to cultivation that focuses on income crops. For example, tea,

coffee, coconut, areca nut, rubber, and spices such as cardamom, chilies, ginger, and turmeric.

5) Horticulture: Horticulture is the cultivation of plants for food, materials, comfort, and beauty in decorating. For example, vegetables such as onion and tomato; and fruits such as apple, orange, mango, banana, citrus fruits, and so on.

b) Crop classification based on climate:
   1) Tropical: These crops thrive in warm and hot climates. These crops have a short response time. Examples include rice, sugarcane, jowar, among others.
   2) Temperate: These crops thrive in chilly climates. These crops respond well to extended days. Examples include wheat, oats, gram, potato, apple, among others.

c) Classification based on growing season:
   1) Kharif/Rainy/Monsoon crops: These are crops cultivated during the monsoon season, from June through October–November. Warm, rainy weather is required throughout the key phase of crop growth, as is a short day duration for blooming. Cotton, rice, jowar, and bajara are a few examples of this classification.
   2) Rabi/winter/cold seasons crops: Crops are cultivated throughout the winter season, from October–March. Crops thrive well in cold, dry weather. Flowering requires increased day length. Wheat, gramme, sunflower, and other grains are examples of this type of crop.
   3) Summer/Zaid crops: Crops cultivated throughout the summer months of March– June. Warm, dry weather is required for the main growth stage, and extended day duration is required for flowering. For example, groundnuts, watermelon, pumpkins, and gourds.

d) Classification based on duration of crops:
   1) Seasonal crops: This is where a crop's life cycle is completed in a single season. Examples include: rice, Jowar, wheat, and so on.
   2) Two seasonal crops: These crops have two seasons in which to complete their life cycle. Cotton, turmeric, and ginger are examples.
   3) Annual crops: Annual crops require a full year to complete their life cycle. For example, sugarcane.
   4) Biennial crops: Biennial crops take two years to complete their life cycle. For example, bananas and papayas.

5) Perennial crops: These are crops that live for numerous years are known as perennial crops. Fruit crops, for example, mango, guava, and so on.

e) Classification based on water availability:
   1) Rain fed crops: These crops have poor yields but are less impacted by changes in rainfall. Farmers were protected from the dangerous nature of farming in rainfed and arid places. Rainfed Crops are prone to breakage during the monsoon season owing to water stress. This water stress might be caused by rainfall variability, sowing delay, crop management method variety, and soil type variability. Prolonged pauses might cause e harvests to fail partially or completely. Crop cultivation is mostly determined by the availability of rain water. Jowar, Bajara, Mung, and so forth are examples.
   2) Irrigated crops: These crops require a huge amount of water to be harvested. They need a significant financial commitment as well as the availability of a set of transportation infrastructures. Citrus and flowers, for example, cannot rely on the seasons and require a continual supply of water, as does rice, which requires copious water for development. In other cases, such as the grape or olive grove, there may be a dry or irrigated crop, with the latter producing significantly more. For example, chilli, sugarcane, banana, papaya, and so on.

f) Classification based on root system:
   1) Tap root system: The main root penetrates the soil deeply. Tur, Grape, Cotton, and so forth are examples.
   2) Fiber rooted: Fiber-rooted crops are those whose roots are fibrous and spread into the soil. Cereal crops, such as wheat and rice, are examples.

g) Classification based on the number of cotyledons:
   1) Monocots or monocotyledons: seeds with just one cotyledon. For example, all grains and millets.
   2) Dicots or dicotyledonous: Crops with two cotyledons in the seed are called dicots or dicotyledonous. For example, all legumes and pulses, as well as nearly all trees.

h) Classification based on length of photoperiod: Most plants are impacted by the relative duration of the day and night, especially for floral initiation; this effect is known as photoperiodism, and plants are categorised as follows based on the length of photoperiod necessary for floral ignition:

1) Short-day plants: Flowering occurs when days are fewer than 10 hours long. Rice, Jowar, green gram, black gram, and so on are examples.
2) Long day plants: Long days of more than 10 hours are required for floral initiation. Wheat, barley, and so on are examples.
3) Day neutral plants: Photoperiod has little effect on phase transition in these plants. Cotton, sunflower, and so on are examples.

Types of Crops in India

a) Millets: These are the staple meals of the poor. These are excellent in nutritional content and are also known as coarse grains. Rajasthan, Karnataka, Maharashtra, Madhya Pradesh and Uttar Pradesh are the top millets producing states. The three principal millets are sorghum (jowar), pearl millet (bajra), and finger millet (ragi).
b) Pulses or Legumes: Protein is abundant in pulses. They are grown in a temperature range of about 20–27 degrees Celsius. Sandy-loamy soil is the soil type required for such crops and a rainfall of around 25–60 cm. Tur (arhar), urad, moong, masur, peas, and gram are the most important pulses farmed in India. All of these crops, with the exception of arhar, are leguminous crops that help restore soil fertility by fixing nitrogen from the air. As a result, they are typically planted in rotation with other crops.
c) Oil Seed crops: Various oil seeds are planted, accounting for around 12 percent of the country's total cropped area. Groundnut, mustard coconut, sesamum (til), soyabean, castor seeds, cotton seeds, linseed, and sunflower are the main oil-seeds produced in India. The majority of them are edible and are used as cooking means. Some of these, however, are also utilised as raw materials in the manufacture of soap, cosmetics, and ointments. For example, groundnut or peanut, sunflower, castor, linseed, rapeseed, and mustard.
d) Sugarcane: It is both a tropical and a subtropical crop. It thrives in hot, humid climates with temperatures ranging from 21–27 degrees Celsius and yearly rainfall ranging from 75cm to 100cm. Irrigation is essential in low-rainfall areas. It may be cultivated on various sugarcane and sugar beet varieties.
e) Starch crops or Tuber crops: Tuber crops include root crops like beets and carrots, tuber crops like potatoes and sweet potatoes, and

root crop leaves like beet tops. For example, potato, cassava, sweet potato, radish, and so on.

f) Fiber crops: Field crops produced for their fibres, which are historically used to manufacture paper, fabric, or rope. Fiber crops are distinguished by a high content of cellulose, which gives them their strength. Fibers can be chemically changed, as in viscose (used to make rayon and cellophane). For example, cotton, jute, mesta, sun hemp, sisal hemp, and so forth.
g) Plantation crops: A plantation is a vast farm or estate, typically located in a tropical or subtropical nation, where commodities are cultivated for sale in distant markets rather than for local consumption. For example, tea leaf, coffee seed, rubber, cocoa seed, palm oil, sugarcane, coconut, and so forth.
h) Spices: A spice is a seed, fruit, root, bark, or other plant material that is primarily used to taste, colour, or preserve food. For example, ginger, garlic, chilli, cumin, onion, coriander, cardamom, pepper, turmeric, and so forth.
i) Medicinal plants: Medicinal plants are crops that are used to make medications. For example, tobacco, mint, and so on.
j) Beverages: Crop products used to make drinks. Tea, coffee, and chocolate are examples (plantation crops).

Tillage

Tillage is the mechanical manipulation of soil with tools and implements in order to create perfect circumstances for seed germination, seedling establishment, and crop development. Tillage activities in different types have been conducted since the beginning of plant cultivation. To place the seeds, primitive man utilised instruments to disrupt the earth. Tillage is derived from the 'Anglo-Saxon' terms Tilian and Teolian, which imply 'to plough and prepare soil for seed sowing, cultivate, and raise crops.' Jethrotull, considered the founder of tillage, proposed that rigorous ploughing is required to break up the soil into fine particles.

Tilth is the physical state of the soil as a result of tillage (or) as a result of tillage. A coarse tilth, fine tilth, or moderate tilth is possible.

Objectives

a) To make a good seed bed that will aid in seed germination.
b) To improve crop growth by creating favourable soil conditions.
c) To successfully manage weeds.

d) To improve the soil's ability to absorb rainwater.
e) To evenly distribute manure and fertiliser in the soil.
f) Aeration of the soil
g) To provide sufficient seed-soil contact for water transfer to seed and seedling roots.
h) To remove the hard pan and increase soil depth
i) To accomplish these goals, the earth is disturbed / opened up and turned over.

Types of Tillage

Tillage activities can be classified as follows:

a) On season Tillage: On-season tillage refers to tillage activities performed within the same crop season or at the start of the crop season. They might be both preliminary and post-cultivation.
   1) Preparatory Tillage: This refers to tillage procedures performed to prepare the field for crop production. When the soil is in a workable condition, it comprises of deep opening and loosening of the soil to achieve an acceptable tilth as well as incorporating or uprooting weeds and crop residues. The various types of preparatory tillage are given below:
      i. Primary tillage: Primary tillage or ploughing is the ploughing process performed after crop harvest to prepare the ground for agriculture. Ploughing is the process of breaking up dense soil with various ploughs. Primary tillage implements include the country plough, mould board plough, bose plough, tractor, and power tiller pulled implements.
      ii. Secondary tillage: Secondary tillage refers to the tillage activities conducted on the land following initial tillage in order to improve soil tilth. Secondary tillage is a finer or lighter activity used to clear the soil, break up clods, and integrate manure and fertilisers. Harrowing and planking are done to accomplish these goals. Planking is used to break hard clods, level the soil surface, and mildly compact the soil. Secondary tillage tools include harrows, cultivators, Guntakas, and spades.
      iii. Layout of sea bed: This is also a part of the preparatory tillage. Leveling boards, buck scrapers, and other tools are utilised, and markers are used to lay out the seedbed.

2) After cultivation (Inter Tillage): The tillage activities performed in the standing crop after sowing or planting and before to crop plant harvesting are referred to as after tillage. This is also known as interculture or post-seeding/planting cultivation. It comprises harrowing, hoeing, weeding, earthing up, drilling, and fertiliser side dressing, among other things. Inter cultivation tools include spades, hoes, and weeders.

b) Off season Tillage: Off-season tillage refers to tillage activities used to prepare the soil for the upcoming main season crop. Post-harvest tillage, summer tillage, winter tillage, and fallow tillage are all examples of off-season tillage.

Special Purpose Tillage

Tillage activities meant to suit certain goals are referred to as special purpose tillage. They are:

a) Sub soiling: To decrease compaction, a specific tillage operation (chiselling) is undertaken to break the hard pan underlying the plough layer. Subsoiling is necessary every four to five years, when heavy machinery is used for field operations like as sowing, harvesting, and transportation. The benefits of subsoiling include obtaining a larger volume of soil for crop cultivation, allowing surplus water to seep below to recharge the permanent water table, reducing runoff and soil erosion, and allowing crop plant roots to penetrate deeper to take moisture from the water table.

b) Clean tillage: It refers to the process of cultivating the soil of a complete field in such a way that no living plant is left undisturbed. It is used to manage weeds, pathogens in the soil, and pests.

c) Blind tillage: It refers to tillage performed after seeding or planting the crop (in sterile soil) either during the pre-emergence stage of the crop plants or when they are in the early phases of growth so that crop plants (sugarcane, potato, etc.) are not injured but additional plants and wide leaved weeds are uprooted.

d) Dry tillage: Dry tillage is used for crops that are seeded or planted in dry land conditions with enough moisture for seed germination. This is appropriate for crops such as broadcasted rice, jute, wheat, oilseed crops, pulse crops, potato crops, and vegetable crops. Dry tillage is done in soil that has enough moisture (21–23 percent). Dry tillage makes the soil more permeable and softer. Furthermore, the soil's water holding capacity and aeration are improved. These are more favourable circumstances for soil microorganisms.

e) Wet tillage or puddling: Wet tillage or puddling refers to tillage performed on terrain containing standing water. The puddling technique is repeatedly ploughing in standing water until the soil becomes soft and muddy. Puddling forms an impermeable layer beneath the surface, reducing deep percolation losses of water and providing a soft seed bed for rice planting. To include green manures and weeds, puddling is done in both directions. Wet tillage degrades soil structure, and soil particles separated during puddling settle later. Wet tillage is the sole method of preparing ground for the transplanting of semi-aquatic agricultural plants such as rice. Planking following wet tillage levels and compacts the soil. Puddling speeds up transplanting.

f) No tillage: This method involves planting a fresh crop in the leftovers of the preceding crop without any prior soil tillage or seed bed preparation, and it is feasible when all weeds are controlled using herbicides. Zero tillage is appropriate for soils with a coarse textured surface horizon, strong internal drainage, high biological activity of soil fauna, favourable initial soil structure, and a sufficient amount of crop waste as mulch. These conditions are common in humid and sub-humid tropics Alfisols, Oxisols, and Ultisols.

Advantages:

1) Zero tilled soils have a more homogeneous structure and a greater quantity of earthworms.
2) Because of reduced mineralization, the organic matter content increases.
3) The presence of mulch reduces surface run-off.

Disadvantages:

1) In zero tillage, a greater amount of nitrogen is required for organic matter mineralization.
2) Perennial weeds might be an issue.
3) There are a lot of volunteer plants and a lot of bugs.

Cropping Pattern and Cropping System

Traditionally, greater food production has resulted from cultivating more land. However, in many parts of the world, particularly in Asia, all of the land that can be economically farmed is already in use. In the future, the majority of the additional food demands will have to come from increased output from existing cultivated land. A significant portion of this increase is anticipated to result from increasing the number of crops grown every year on a given plot of land using better crop varieties. Such repeated cropping has the potential to boost both food production and land damage. The notion of cropping systems is as old as agriculture in India. Farmers favoured a mixture. cropping, particularly in dry land circumstances, to reduce the chance of crop failure. Even in There is mention of first and second harvests in the Vedas, showing the presence of sequential cultivation.

A system is defined as a collection of interconnected and interacting components. A cropping system is a collection of crop systems that comprise a farm system's cropping operations. Cropping systems include all of the components necessary for the production of a certain crop as well as the interrelationships between them and the environment. In other words, a cropping system is often a mix of crops in both time and area. Combinations in time happen when crops develop at various times, while combinations in space happen when crops are interplanted. When it comes to annual crops, a cropping system is typically defined as the combination of crops grown in a single year.

Cropping pattern: The annual sequence and spatial arrangement of crops or crops and fallow on a certain region.

Cropping system: Cropping patterns employed on a farm, as well as their interactions with farm resources, other farm operations, and available technology, which define their composition.

Intensive Cropping

Through improved ground preparation, the transition period between crops is reduced. It is feasible when enough resources are available. Consider the cultivation of garden land. In an intense cropping system, cropping intensity is higher. Intercropping, relay cropping, sequential cropping, ratoon cropping, and other crop intensification techniques are examples. All of these approaches are referred to as multiple cropping. Intensive cropping is the practise of growing many crops on the same plot of land at the same time.

Need:

a) Cropping systems must adapt in response to climate, soil, and water availability in order to make the most use of available natural resources.
b) Population growth has put pressure on land to raise productivity per unit area, unit time, and unit resource utilised.
c) This farming strategy should offer enough food for the household, fodder for cattle, and enough monetary flow to cover home and agricultural expenditures.

Cropping Intensity: Cropping intensity is defined as the number of crops planted on a plot of land each year. Cropping intensity is greater than 100 percent (about 140–150 percent) in Punjab and Tamil Nadu. Cropping intensity is lower in Rajasthan.

Multiple cropping: Cropping intensification in time and spatial dimensions is referred to as multiple cropping. In a given year, growing two or more crops on the same land. The various forms of multiple cropping are given below:

a) Inter cropping: Growing two or more crops on the same field at the same time. Crop intensification occurs in both time and place. Intercrop competition exists during all or part of crop development.

Advantages:

1) Improved use of development resources such as light, nutrients, and water.
2) Weed extermination.
3) Yield stability means that even if one crop fails due to unanticipated circumstances, another crop will yield and provide money.
4) Successful intercropping results in increased equivalent yields (base crop yield + intercrop yield), as well as higher cropping intensity.
5) Incidences of pests and diseases have decreased.
6) Soil health and the agro-ecosystem are being improved.

Types:

1) Mixed intercropping: Growing two or more crops at the same time with no specific row organisation is known as mixed intercropping. Also known as mixed cropping. In rainfed circumstances, for example, sorghum, pearl millet, and cowpea are blended and aired.
2) Row intercropping: Row intercropping is the practise of growing two or more crops at the same time, with one or more crops grown in rows. Intercropping is another term for it. Maize + greengram (1:1), Maize+blackgram (1:1), Groundnut+Rredgram (6:1).
3) Strip intercropping: Strip intercropping is the practise of growing two or more crops in strips broad enough to allow autonomous cultivation but narrow enough for the crops to interact agronomically. Strip of groundnut + redgram (6:4).
4) Relay intercropping: Relay intercropping is the practise of growing two or more crops at the same time during the life cycle of each. After the first crop has reached the reproductive stage of development but before it is ready for harvest, a second crop is planted. Relay cropping is another term for it. Rice is a fallow pulse.

b) Sequential cropping: Growing two or more crops in succession on the same field in the same agricultural year is known as sequential cropping. The following crop is sown after the previous crop has been harvested. Crop intensification only occurs in the time dimension. There is no rivalry between crops.

1) Double, triple, and quadruple cropping: Growing two, three, and four crops in succession on the same land in a year. For example, double cropping rice and cotton; triple cropping rice and pulses; and quadruple cropping tomato, ridge gourd, amaranthus greens, and baby corn.
2) Ratoon cropping: The cultivation of crop regrowth after harvest, although not always for grain. Sugarcane, for example, is ratoon; sorghum is ratoon (for fodder).

Farming Systems

Farm: A distinct piece of land on which crop and livestock businesses are carried out under shared administration.

Farming: Farming is the way of transforming solar energy into economically viable plant and animal goods.

System: A system is a collection of interconnected and interacting components.

Wetland farming: Wet land is soil that has been flooded or watered by a lake, pond, or canal, and the land is always submerged. Wetland farming is the process of cultivating crops on soils that are inundated by natural water flow for the majority of the year.

Garden land farming: Growing crops using supplemental irrigation by raising water from subterranean sources is known as garden land farming.

Dry land farming: Dryland farming is the process of growing crops that are entirely dependent on rainfall and soil moisture. This is done in locations with yearly rainfall of less than 750mm. Crops may experience moisture stress on a regular basis as a result of unpredictable monsoon distribution or failure.

Rain fed farming: Crop production in locations when rainfall exceeds 750mm (i.e., assured rainfall areas). Moisture stress will be at its lowest here. The importance of soil protection is emphasised.

Mixed farming: Mixed farming is described as a farming method on a specific farm that incorporates crop production, rearing livestock, poultry, fisheries, beekeeping, and so on in order to support and meet as many of the farmer's demands as feasible. Subsistence farming is a key goal of mixed farming. While increased profitability without compromising ecological balance is critical in the farming sector.

Specialized farming: Specialized farming refers to a farm in which 50 percent or more of total crop output income is produced from a single crop.

Diversified farming: A diversified farm has numerous producing firms or sources of revenue, but no source of income accounts for more than 50 percent of total income. It is sometimes referred to as general farming.

Difference between Rain fed and Irrigated farming:

Table 27.2: Showing major difference between rain fed and irrigated farming

Sl.no.	Rain fed farming	Irrigated farming
1	Crops are planted at specific times of the year where rainfall is abundant.	Throughout the year, depending on the availability of water
2	Crops/crop types that are drought tolerant or require less water are employed.	Crops or kinds of crops are chosen based on the requirement.
3	Crop length is determined by rainfall duration/growing season, which is usually brief (LGP)	Depending on the situation
4	Mixed cropping is advantageous.	In most cases, pure cropping is used.
5	Due to moisture restrictions, only one or two harvests each year are possible.	Depending on water availability, more than two crops are planted every year.
6	The field is ploughed deeply to maximise rainwater infiltration.	There is no need for extensive ploughing to conserve soil moisture.
7	Land is prepared shortly following rains.	The land is prepared in accordance with the best time to plant.
8	Crop failure is projected as a result of inadequate soil moisture or drought.	There is no danger of crop failure.

Crop Rotation

Growing several crops on a plot of land is a planned succession. The objective of crop rotation is to maximise the use of available resources in order to harvest the most from a given plot of land while minimising the impact on soil health. For example, rice-red gram-banana.

Principles:

a) Because legumes fix atmospheric nitrogen into the soil and provide organic matter to the soil, they should be planted before non-leguminous crops.

b) Crops with tap roots (deep rooted crops like cotton) should be followed by crops with fibrous roots (shallow rooted crops like sorghum or maize). This allows for the correct and homogeneous utilisation of soil nutrients.
c) More intensive crops should be followed by less intensive crops since crops such as potato, sugarcane, maize, and others require more inputs such as better tillage, more fertilisers, more irrigation, and so on.
d) Crop selection should be based on necessity or demand.
e) Crops from the same family should not be planted consecutively since they serve as alternate hosts for insect pests and illnesses.
f) Crop selection should take into account farmers' financial situation.
g) The crop chosen should also be appropriate for the land and climate.

Sustainable Agriculture

It is a type of agriculture that aims to fulfil the demands of the current generation without jeopardising the future generations' resource base. It is defined as a farming method that uses manure, crop rotation, and little tillage, as well as there being a low reliance on synthetic fertilisers, pesticides, and antibiotics. It is a balanced management system of renewable resources such as soil, wildlife, forests, crops, fish, cattle, plant genetic resources, and eco-systems that provides food and livelihood for current and future generations while preserving or enhancing productivity and ecosystem services. Land degradation and soil erosion must be avoided in order for agriculture to be sustainable. It must replace nutrients while also controlling weeds, pests, and diseases through biological and cultural approaches. The primary challenge to agriculture's long-term growth is the degradation of natural resources.

Major components of Sustainable agricultural system:

a) Soil and water conservation to prevent soil productivity deterioration
b) Use of limited irrigation water efficiently without causing soil salinity, alkalinity, or a high ground water table.
c) Crop rotations that reduce weed, disease, and pest issues while increasing soil productivity and reducing soil erosion
d) By combining organics, inorganics, and bio-fertilizers, integrated nutrient management enhances soil health and lowers environmental contamination.

e) Crop rotation, weather monitoring, the adoption of resistant cultivars, planting timing, and biological pest control are all examples of integrated pest management.
f) Weed management strategy based on preventative measures, tillage, timely intercropping, and crop rotation to increase plant health.

Organic Farming

Organic farming is a production method in which all types of agricultural goods, including grains, meat, dairy, eggs, fibres such as cotton, flowers, and processed food products, are produced organically. Synthetic fertilisers, pesticides, growth regulators, and livestock feed additives are avoided or limited in organic farming.

Need and Scope:

a) Increased health consciousness and awareness
b) Global customers are increasingly seeking organic food that is deemed safe and devoid of hazards.

c) Organic food is more profitable and remunerative on a worldwide scale.
d) The fact that the agricultural sector contains numerous organic nutrient resources such as livestock, water, crop residue, aquatic weeds, forest litter, urban, rural solid wastes and agro industries, bio-products demonstrates the potential of organic farming.
e) India has enormous potential for organic farming since it has a local market for organic products.

Principles:

a) To generate adequate quantities of high-quality food.
b) To interact constructively and positively with natural systems and cycles.
c) Consider the overall social and environmental effect of organic production and processing systems.
d) To promote and improve biological cycles in the farming system, including microorganisms, soil flora and fauna, plants, and animals.
e) To maintain and improve soil fertility over time.
f) To preserve the genetic variety of the production system and its surroundings, including wildlife habitat conservation.

g) To encourage the responsible use and care of water, water resources, and all aquatic life.
h) To employ renewable resources as much as feasible in locally coordinated production systems.
i) To provide adequate living conditions for all livestock while taking into account the fundamental aspects of their innate behaviour.
j) Reduce all types of pollution.
k) To provide a quality of life for everyone involved in organic production and processing that fulfils their fundamental requirements and provides an appropriate return and enjoyment from their labour, including a safe working environment.
l) To work toward a socially equitable and environmentally responsible manufacturing, processing, and distribution chain.

Advantages:

a) Nutrition - Food with improved soil health has a significantly higher mineral content.
b) Poison-free - Free of harmful chemicals such as pesticides, fungicides, and herbicides that can harm your health.
c) Food has a better flavour.
d) Food lasts longer and may be preserved for extended periods of time.
e) Because of healthy plants, disease and insect resistance has increased.
f) Weed competitiveness - Healthier crops that can compete.
g) Low input costs - No expensive chemicals are needed, and nutrients are generated in-situ (in the farm)
h) Drought resistance.
i) More lucrative - Organic fruit has a higher food value, thus people are ready to pay a premium for it.

Disadvantages:

a) Productivity - Low productivity is sometimes described as the quantum nutrition used being substantially lower.
b) Cultivation necessitates greater labour, particularly for weed management.
c) Farming organically takes tremendous skill. For example, consider pest control solutions.

d) Inconvenience in management compared to easier management methods such as fertiliser application in traditional ways.

Some common terms of Organic farming

a) Eco-farming: It is farming that promotes mutually reinforcing ecological approaches to food production. It tries to maintain soil chemically, physiologically, and physically in the same way as nature would if left alone. Soil would then take care of the plants that grew on it. The mantra and slogan of ecological farming is "feed the soil, not the plant."
b) Biological farming: Farming in relation to biological diversity.
c) Bio-dynamic farming: Farming that is biologically organic, environmentally sound, and sustainable.
d) Macrobiotic agriculture: The macrobiotic movement promotes consuming seasonal, locally sourced whole foods cultivated utilising natural farming practises in order to limit and eliminate food production through industrial monoculture methods.

Permaculture

In 1978, Bill Mollison, an Australian ecologist, and one of his pupils, David Holmgren, coined the term "permaculture." It's an abbreviation for "permanent agriculture" or "permanent culture." It is characterised as a design system for constructing human settings that are sustainable. It designs integrated systems of food production, housing, suitable technology, and community development on the basis of ecological. Permaculture is based on the idea of caring for the planet and connecting with nature in mutually beneficial ways. The design of ecological landscapes that provide food is a major concept in Permaculture. The utilisation of multi-purpose plants, cultural methods such as sheet mulching and trellising, and the use of animals to recycle nutrients and graze weeds are all highlighted.

Characteristics:

a) It is one of the world's most comprehensive, integrated systems analysis and design approaches.
b) It can be used to generate productive ecosystems for human use or to assist damaged ecosystems restore their health and wildness.
c) It may be used in any environment, no matter how degraded it is.

d) It recognises and acknowledges traditional knowledge and experience.
e) Incorporates global sustainable agriculture methods, land management techniques, and strategies.
f) It serves as a link between traditional and emerging earth-tuned cultures.
g) It advocates for pesticide-free organic agriculture.
h) It seeks to enhance the symbiotic and synergistic relationships among site components.
i) It is site-specific, client-specific, and culture-specific in design.

Integrated farming system

Integration of agricultural operations such as cropping systems, animal husbandry, fisheries, forestry, and so on for optimal resource utilisation and farmer success. The components of a farming system should be chosen and implemented based on the availability of land, the kind of land, water, money, resources, the farmer's technical expertise, market facilities, and so on.

Benefits:

a) Other than revenue from regular farming, stable income.
b) In the case of unexpected crop failures, risk coverage is provided via subsidiary allocation.
c) Possibility of employment.
d) Increased output.
e) Increased organics returns and recycling.
f) Simple to implement for marginal and submarginal farmers.
g) Farm operations are generally improving.
h) Improved use of farm land, labour, time, and accessible manures.

Elements Required in Plant Growth

Plants require 16 components to survive. They are classified according to their relative abundance in plants. Plants contain several non-essential elements, accounting for about 60 components. When the soil solution has a high concentration of Al ions, Al is absorbed. When plant components are burnt, the resulting plant ash includes all necessary and non-essential elements with the exception of C, H, O, N, and S. Micronutrients are absorbed 30–100 times more quickly than macronutrients.

Macro nutrients: These are the nutrients that plants require in relatively substantial quantities for growth and development. N, P, and K are three examples.

a) Nitrogen (N) is an important component of proteins and is found in a variety of other chemicals with significant physiological importance in plant metabolism. It is a component of chlorophyll, which is the principal observer of light energy required for photosynthesis. It also gives plants robust vegetative growth and a rich green colour.
b) Phosphorus (P) is a key component of the enzymes that assist the crop in storing light energy. It is an integral component of nucleic acids, which are the transmitters of genetic information, and it is crucial in encouraging root development.
c) Potassium (K) is engaged in mechanisms that guarantee carbon absorption and photosynthesis transfer throughout the plant for growth and sugar and protein storage. Potassium ions are also necessary for water control and absorption. Furthermore, the presence of adequate potassium provides resistance to cold, drought, and some illnesses.

Secondary nutrients: Secondary nutrients are those nutrients that plants require in modest levels and are assigned secondary significance in their supply and control. Ca, Mg, and S are three examples.

a) Calcium: Plant development, cell division, and expansion all require calcium. Calcium influences the formation of root and shoot tips as well as storage organs since it is a component of cell membranes. Calcium is also necessary for pollen growth and leaf fall prevention.
b) Magnesium: Magnesium is found in chlorophyll and acts as an enzyme activator.
c) Sulphur: Sulphur is a component of two important amino acids, which are among the numerous protein building components. It can also be present in vitamin B1 and a number of essential enzymes.

Minor nutrients: Micronutrients are nutrients that plants use in lesser quantities for growth and development, but are just as significant and necessary to plants as macronutrients. Fe, Mn, Zn, Cu, B, Mo, Cl, and Ni are a few examples.

a) Iron: Fe deficiency is more common in crops grown on calcareous or alkaline soils. Citrus and deciduous fruit crops, as well as sorghum cultivated on neutral to alkaline soils, display Fe chlorosis. Soybeans, beans, corn, strawberries, avocado, vegetable crops, and various ornamentals are among the other crops. Fe deficiency first manifests itself in young leaves, and as a result, growth ends. Young leaves acquire interveinal chlorosis, which spreads swiftly throughout the entire leaf. In extreme circumstances, the leaves become completely white.
b) Manganese: Interveinal chlorosis occurs in broad-leaved plants. Mn shortage in numerous crops is described by terms such as grey speck of oats, marsh spot of peas, and speckled yellows of sugar beets. Wheat plants with low Mn levels are frequently more sensitive to root rot pathogens.
c) Zinc: Corn, beans, citrus, flax, grapes, onions, rice, and soybeans are all susceptible to Zn deficiency. Alfalfa, clovers, sorghum, sudan grass, and tomatoes are somewhat sensitive crops. Zn deficiency symptoms include white bud in maize and sorghum, small leaf in cotton, mottle leaf or frenching in citrus crops, and fern leaf in potatoes.
d) Copper: Cu deficiency has been recorded in a variety of plants, however it is more common in crops grown in peat and muck soils. Alfalfa, wheat, onions, carrots, clover, corn, and fruit trees are among the crops most vulnerable to Cu deficiency.
e) Boron: Symptoms of B deficiency include thickened, wilted, or curled leaves; thickened, cracked, or water-soaked petioles and stems; and discolouration, cracking, or rotting of fruit, tubers, or roots. Apple internal cork is caused. Citrus fruits have inconsistent skin thickness, lumpy fruit, and sticky deposits as a result. Internal tissue degradation in root crops appears as darkening regions and is referred to as brown heart or black heart.

Manures

Plant and animal wastes are employed as a source of plant nutrients in manures. After breakdown, they release nutrients. Manures are classified as bulky organic manures or concentrated organic manures.

Role of Manures:

a) Organic manures bind the sandy soil and increase its capacity to store water.
b) They aerate the clayey soil and encourage root development.
c) They add a small percentage of plant nutrients as well as micronutrients that are essential for plant growth. Microbial activity is increased, which aids in the release of plant nutrients, for example, bulky organic manures—farmyard manure, compost from organic waste, night soil, sludge, sewage, sheep folding, green manures, concentrated organic manures- oilcakes (edible, non-edible), blood meal, fish meal, bone meal.
d) Becausc of the gradual release of nutrients, organic manures should be applied before sowing or planting.

Classification of organic manures

a) Farmyard manure: This is conventional manure that is widely supplied to farmers. Farm yard manure is a decomposed mixture of bovine faeces and urine, as well as straw and litter used as bedding material and remnants from the cattle's feed.
b) Compost Manure: Compost is well-rotted organic manure produced by organic matter decomposition. Composting is primarily a biological process in which aerobic (requires oxygen for deep growth) and anaerobic (works in the absence of air or free oxygen) microorganisms degrade organic materials and reduce the C:N ratio of garbage. Compost is a well-rotted manure that is the end result of composting.
c) Sheep and goat droppings: This is also an excellent source of organic manure. It is composed of 0.5 to 0.7 percent N, 0.4 to 0.6 percent P2O5, and 0.3 to 1.0 percent K2O. It works on all sorts of crops.
d) Concentrated organic manures:
    1) Oil cakes: There are several types of oil cakes that include not just nitrogen but also some P and K as well as a high proportion of organic content. There are two kinds of oil cakes. Edible oil cakes are ideal for cow nutrition. Non-edible oil cakes are not suited for cow feeding. Oil cakes are organic manure that acts quickly. Despite the fact that they are insoluble in water, their nitrogen became accessible to plants in approximately a week or 10 days after application.

2) Bone meal: Bone meal, the earliest phosphatic fertiliser utilised, is made up of bones from slaughterhouses, corpses of all animals, and meat industries. It also has some Nitrogen in it.
3) Fish meal: Fish dung or meal is produced by drying non-edible fish, fish corpses, and fish industry wastes. It includes 4.0–10.0 percent nitrogen, 3.0–9.0 percent phosphorus, and 0.3–1.5 percent potassium. Fishmeal is a fast-acting organic manure that may be applied to any crop on any soil.
4) Green manuring: It is a method of strengthening the physical structure and fertility of the soil by ploughing in green plants produced in the field or bringing green plants from outside and incorporating them into the soil. For instance, Sannhemp, Dhaicha, Pillipesara, Shervi, Urd, Mung, Cowpea, Berseem, Senji, and so on.
5) Green leaf manuring: Green leaf manuring is the process of putting green leaves and green twigs from legume plants or trees to a field and then ploughing them into the soil. Glyricidia, wild Dhaicha, Karanj (pongamia), and leucaena are a few examples.
6) Bio-fertilizer: Bio-fertilizers are preparations containing live or latent cells of efficient strains of nitrogen fixing, phosphorus solubilizing, or cellulolyotic microorganisms that are applied to seed or composting areas with the goal of increasing the number of such microorganisms and accelerating those microbial processes that augment the availability of nutrients that are easily assimilated by plants. Bio-fertilizers use specialised microorganisms to capture atmospheric nitrogen, which can be free-living in soil or symbiotic with plants.

Watershed Management

A watershed is a region of land and water bordered by a drainage divide where surface runoff gathers and flows out through a single outlet into a larger river (or) lake. Watershed management is a method of optimising the use of land, water, and vegetation in a given region in order to alleviate drought, moderate floods, minimise soil erosion, enhance water availability, and boost fuel, feed, and agricultural output on a long-term basis.

Micro Irrigation: Micro irrigation is described as the application of a small volume of water at a low pressure and frequency. The system has a large network of pipes that run at low pressure. Outlets for emission water, also known as emitters, are provided at predetermined intervals.

Sprinkler Irrigation: Water is sprayed into the air and allowed to fall on the ground surface in a manner similar to rainfall in the sprinkler technique of irrigation. Water under pressure is forced through tiny orifices or nozzles to produce the spray.

Drip Irrigation: Drip irrigation, often known as trickling irrigation, involves dripping water into the soil at extremely low rates using a series of tiny diameter plastic pipes equipped with emitters. Water is administered near to plants, soaking just the soil in which the roots develop, as opposed to surface and sprinkler irrigation, which wets the whole soil profile.

Terracing: A terrace is an earthen embankment or ridge built over a slope to manage runoff and decrease soil erosion. It shortens the length of the hillside slope, minimising sheet and rill erosion and preventing gully development.

Agro-Climatic Zones of India

The Planning Commission has divided India's geographical area into 15 agro-climatic zones. They are as follows:

a) Western Himalayan Region: The Western Himalayan Region includes Jammu and Kashmir, Himachal Pradesh, Uttar Pradesh, and Uttaranchal. The region is made up of cold region skeletal soils, podsolic soil, mountainous soil, and hilly brown soil. In undulating terrain, there are sharp slopes.
b) Eastern Himalayan Region: Assam, Sikkim, West Bengal, and other states in the north-east. This area receives a lot of rain and has a lot of forest cover. Shifting cultivation is performed across about one-third of the farmed area, causing soil deterioration through excessive runoff, soil erosion, and flooding.
c) Lower Gangetic Plains Region: West Bengal, where the soils are primarily alluvial and prone to flooding.
d) Middle GangaticPlains Region: Irrigation covers around 39 percent of this region's gross cultivated area.
e) Upper GangaticPlains Region: Irrigation is accomplished using canals and tube wells. A high potential for ground water extraction.

f) Trans-GangaticPlains Region: These places have the most seeded land, the most irrigated land, the highest agricultural intensity, and the highest ground water use. Punjab, Haryana, Delhi, and Rajasthan make up this region.
g) Eastern Plateau and Hills Region: This region is made up of Maharashtra, Uttar Pradesh, Orissa, and West Bengal. Canals and tanks are used for irrigation. The soil is shallow to medium depth.
h) Central Plateau and Hills Region: Madhya Pradesh, Rajasthan, and Uttar Pradesh are the three states which make up this region.
i) Western Plateau and Hills Region: This zone's average rainfall is 904 mm and is made up of Rajasthan and Madhya Pradesh.
j) Southern Plateau and Hills Region: Andhra Pradesh, Karnataka, and Tamil Nadu are the three states which make up this region. Dry farming is used there, with a cropping intensity of 111 percent.
k) East Coast Plains and Hills Region: Orissa, Andhra Pradesh, Tamil Nadu, and Pondicherry make up the region. Canals and tanks are used there for irrigation.
l) West Coast Plains and Ghats Region: Tamil Nadu, Kerala, Goa, Karnataka, and Maharashtra are among the states which make up this region. Cropping patterns, rainfall, and soil types vary.
m) Gujarat Plains and Hilly Region: Gujarat. Most of this zone is desert, with little rainfall. Irrigation is provided through tube wells and wells.
n) Western Dry Region: Rajasthan. Hot sandy desert with sporadic rainfall, rapid evaporation, and little flora. The ground water is frequently deep and brackish. Famine and drought are typical occurrences in this area.
o) The Island Region: Lakshadweep, Andaman and Nicobar Islands. These areas are typical tropical, with 3000 mm of rainfall distributed over eight to nine months. The area is mostly forested and undulated.

CHAPTER 28

REMOTE SENSING AND GIS

Introduction

The practice of acquiring information about objects or regions using electromagnetic radiation (light) without establishing direct touch with the object or area is known as remote sensing. As a result, remote sensing is becoming more popular among individuals. The human eye performs distant sensing activities such as reading the newspaper, seeing cars pass in front of you, and looking at a speaker in class. Human eyes record the sun light reflected by these things, and your brain understands the colours, grey tones, and intensity fluctuations. These data are then transformed into useful information. The human eye, on the other hand, is confined to a narrow portion of the overall electromagnetic spectrum, precisely 400 to 700 nm. Remote sensing employs a variety of technologies and equipment to make electromagnetic radiation beyond this range visible to the human eye, including near infrared, middle infrared, thermal infrared, and microwaves.

Remote sensing and geographic information systems (GIS) have become interesting and attractive fields with rising potential. Many organisations invest significantly in these domains. The question here is why these disciplines have become so relevant in recent years. There are two major causes behind this. Scientists, researchers, students, and even ordinary individuals are increasingly interested in better understanding our surroundings. By environment, we mean the geographical space of their research area as well as the events that occur there. In other words, we've realised that physical space, as well as the data that describes it, is a part of our everyday lives; practically every decision we make is impacted or driven by some aspect of geography. Advances in sophisticated space technology (which can provide large volumes of spatial data), combined with declining costs of computer hardware and software (which can handle these data), have made Remote Sensing and GIS not only affordable to complex environmental / spatial situations, but also to a wider audience. Remote sensing is currently used extensively in many environmental fields, including geography, geology, zoology, agriculture, forestry, botany, meteorology, oceanography, and civil engineering.

Electromagnetic Energy

Remote sensing is based on measuring electromagnetic (EM) radiation. EM energy may manifest itself in a variety of ways. The Sun is the most significant source of EM energy at the Earth's surface, since it provides us with (visible) light, heat, and UV-light, which may be detrimental to our skin. Many remote sensing sensors measure reflected sunlight. Some sensors, on the other hand, detect energy released by the Earth or generate their own. Understanding the remote sensor principle requires a fundamental understanding of EM energy, its properties, and its interactions by the following concept explain below:

a) Waves and Photons: Electromagnetic (EM) energy may be represented in two ways: as waves or as energy-carrying particles known as photons. Electromagnetic energy is assumed to move over space in the form of sine waves in the wave model. Electrical (E) and magnetic (M) fields that are perpendicular to each other describe these waves. As a result, the term electromagnetic energy is employed. The vibration of both fields is perpendicular to the wave's travel direction (Fig. 28.1). Both fields travel across space at the speed of light c, which is around 299,790,000 m/s and can be rounded up to $3*10^8$ m/s. One feature of electromagnetic waves is very crucial for understanding remote sensing. This is the wavelength, λ, which is defined as the distance between successive wave crests. The length of a wave is measured in metres (m), nanometres (nm = 10^{-9} m), or micrometres (μm = 10^{-6} m).

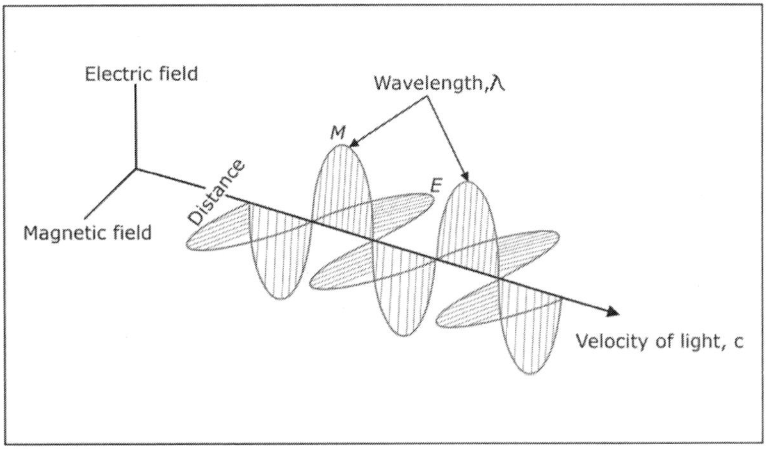

Fig. 28.1: Diagram showing propagation of Electromagnetic energy

The frequency, v, is the number of times a wave passes a fixed location in a given amount of time. Frequency is generally expressed in hertz (Hz), which equals one cycle every second. Because the speed of light is constant, wavelength and frequency are inversely related:

$$C = \lambda * \nu$$

In this equation, C represents the speed of light ($3 * 10^8$ m/s), λ is the wavelength (m), and ν is the frequency (cycles per second, Hz).

The higher the frequency, the shorter the wavelength. In contrast, the lower the frequency, the longer the wavelength (Fig. 28.2).

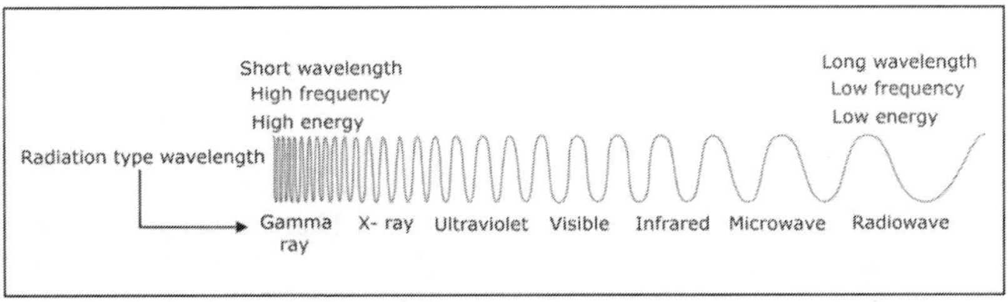

Fig. 28.2: Diagram showing the relationship between frequency and wavelength

The 'wave' model, as stated above, can represent the majority of the properties of EM radiation. However, for certain applications, EM energy is more readily described by particle theory, in which EM energy is constituted of discrete units called 'photons.' When measuring the quantity of energy recorded by a multispectral sensor, this technique is used. The amount of energy carried by a photon of a certain wavelength is thus given by:

$$Q = h * \nu = h * \frac{c}{\lambda}$$

where Q is photon energy (in J), h is Planck's constant (6.6262* 10^{34} J s), and ν is frequency (Hz). As a result, the longer the wave length, the lower the energy content. Gamma rays (about 10^{9} m) are the most energetic, whereas radio waves (> 1 m) are the least.

b) Sources of EM energy: Remote sensing is limited to the use of electromagnetic radiation as a characteristic of a variety of physical processes. All materials with a temperature greater than absolute zero (0K) have the ability to emit electromagnetic radiation. Objects on or near the earth's surface can reflect or scatter incoming electromagnetic radiation released by a source, which can be artificial, such as a flash light, laser, or microwave radiation, or natural, like the sun. This means that both the Sun and the Earth emit energy in the form of waves. A blackbody is matter that is capable of absorbing and re-emitting all EM radiation that it absorbs. Both the emissivity and absorptivity of blackbodies are equal to one. The amount of energy emitted by an object is a function of wavelength and relies on its absolute temperature and emissivity. Stefan-Law Boltzmann's defines this principle in physics.
c) Electromagnetic spectrum: All matter with a temperature greater than absolute zero (K) emits electromagnetic waves of varying wavelengths. The electromagnetic spectrum is widely used to describe the complete range of wavelengths (Fig.28.3). It encompasses gamma rays to radio waves. The visible part (0.4 - 0.7 µm) of the spectrum is frequently referred to as 'light.' It takes up a minor fraction of the electromagnetic spectrum. It is worth noting that this is the only part of the spectrum that we can identify with the idea of colour. The primary colours or wavelengths of the visible spectrum are blue, green, and red. Red has the longest visible wavelength while violet has the shortest.

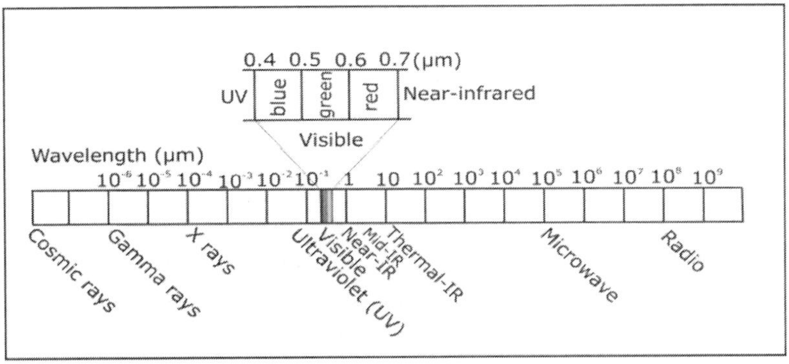

Fig. 28.3: Electromagnetic spectrum

Principles of Remote sensing

The process of most remote sensing requires an interaction between incoming radiation and the targets of interest. This is demonstrated through the usage of imaging systems, which incorporate the seven features listed below. However, remote sensing includes the detection of radiated energy as well as the usage of non-imaging sensors. The overall process of remote sensing is shown in (Fig. 28.4).

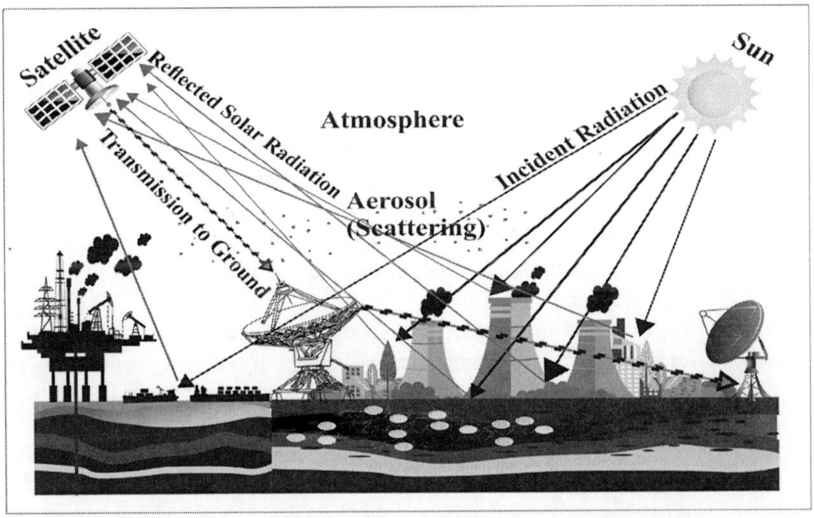

Fig. 28.4: Diagram showing remote sensing process

a) Energy source (A): The initial prerequisite for remote sensing is an energy source capable of illuminating or supplying electromagnetic energy to the subject of interest.
b) Radiation and the Atmosphere (B): As the energy moves from its source to its destination, it will come into contact with and interact with the atmosphere. This contact may occur again when the energy travels from the target to the sensor. The incoming light and radiation can be affected by particles and gases in the atmosphere. These effects are created by scattering and absorption processes. Scattering happens when particles or large gas molecules in the atmosphere contact with electromagnetic radiation, causing it to be deflected from its intended course. The amount of scattering depends on various parameters, including the wavelength of the radiation, the number of particles or gases in the atmosphere, and the distance the radiation travels through the atmosphere. There are three forms of scattering that occur.
    1) Rayleigh scattering: When particles are exceedingly tiny in comparison to the wavelength of the radiation, Rayleigh scattering occurs. These particles might be dust specks or nitrogen and oxygen molecules. Shorter wavelengths of energy are dispersed substantially more than longer wavelengths due to Rayleigh scattering. In the high atmosphere, Rayleigh scattering is the dominating scattering process. This effect is responsible for the sky seeming "blue" during the day. The shorter wavelengths (blue) of the visible spectrum are dispersed more than the other (longer) visible wavelengths as sunlight travels through the atmosphere. At sunrise and sunset, light has to travel longer through the atmosphere than at midday, and the scattering of shorter wavelengths is more thorough, resulting in a higher amount of light.
    2) Mie scattering: When the particles are nearly the same size as the wavelength of the light, Mie scattering occurs. Mie scattering is commonly caused by dust, pollen, smoke, and water vapour, and it affects longer wavelengths than Rayleigh scattering. Mie scattering occurs mostly in the lower atmosphere, where larger particles are more plentiful, and predominates when clouds are present.
    3) Non selective scattering: The final important scattering mechanism is known as nonselective scattering. When the particles are much larger than the wavelength of the radiation, this happens. This form of scattering can be caused by water

droplets and large dust particles. The term "nonselective scattering" refers to the fact that all wavelengths are dispersed equally. Because blue, green, and red light are scattered in roughly equal amounts by this sort of scattering, fog and clouds seem white to human eyes.

Absorption: When electromagnetic radiation interacts with the atmosphere, the other major process at action is absorption. Unlike scattering, this mechanism causes molecules in the atmosphere to absorb radiation at different wavelengths. The three primary atmospheric elements that absorb radiation are ozone, carbon dioxide, and water vapour.

c) Interaction with the target: Once the energy arrives at the target via the atmosphere, it interacts with it based on the qualities of both the target and the radiation.
d) Recording of energy by the sensor: We need a sensor (remote - not in contact with the target) to receive and record the electromagnetic radiation after it has been scattered by or emitted by the target.
e) Transmission, Reception and Processing: The sensor's energy must be transmitted, frequently in electronic form, to a receiving and processing unit, where the data is converted into an image
f) Interpretation and Analysis: The processed photograph is evaluated visually, digitally, or electronically to derive information about the target.
g) Application: The last stage of the remote sensing process occurs when we utilise the information extracted from the picture about the target to better comprehend it, reveal new information, or aid in addressing a specific problem.

Types of Remote Sensing

Depending on the source of energy which illuminate the object under study, the remote sensing techniques are classified in two types:

a) Passive remote sensing: Passive remote sensing techniques make use of natural energy sources such as the Sun or artificial light. Passive sensor systems based on the reflection of the Sun's radiation can only function during the day. Passive sensor devices that monitor longer wavelengths linked to Earth's temperature do not rely on the Sun for light and may be used at any time. Passive sensor systems, for example, must contend with the Sun's fluctuating lighting conditions, which are heavily impacted by atmospheric circumstances. Photographic, thermal, electric field sensing, chemical, infrared, and seismic technologies are examples of passive sensor-based technologies.

b) Active remote sensing: Active remote sensing systems, such as radar and laser, have their own energy source. Active sensors send out a regulated beam of energy to the surface and detect how much energy is reflected back to the sensor. The fundamental advantage of active sensor systems is that they may be used at any time of day or night, have a regulated illumination signal, and are usually not impacted by the environment. Microwaves are used in the majority of devices because they are generally resistant to weather conditions. The difference between active remote sensing systems lies in what they transmit (light or waves) and what they determine (e.g., distance, height, atmospheric conditions, etc.).

Aerial Photography

As early as the mid-nineteenth century, aerial photos were made from balloons and kites. Gasper Felix Tournachon captured the first aerial image from a captive balloon at an altitude of 1,200 feet over Paris in 1858. Any camera may be used to take aerial images. Aerial photography from light planes using handled 35mm cameras has been used in many successful applications. Aerial photographs have been used in India for aerial surveys and interpretation in certain sectors such as geology since 1920. Around 1899, attempts were made to exploit terrestrial pictures generated from photo-theodolites for survey purposes. With the introduction of the multiplex stereo plotting device in 1948, photogrammetric mapping methods were established. Later, from 1954 to 1956, similar equipment was supplemented with the purchase of new stereo plotting devices. Since then, the Survey of India (India's national mapping agency) has kept up with technical advances in photogrammetric mapping and aerial photography.

Aerial photography is described as the science of photographing the earth's surface from the air utilising various platforms, usually aeroplanes. The sun provides the energy (electromagnetic radiation or EMR), and the photosensitive film serves as a sensor to capture the pictures. Variations in the grey tones of a photograph's numerous pictures represent differing levels of energy reflected off the objects as captured on the film.

Characteristics of Aerial Photography

a) Synoptic view: Aerial photography or recording across a vast region provides a birds-eye view from above. These technologies enable the discrimination and detection of tiny scale characteristics as well as spatial interactions between them.
b) Time freezing ability: They are characterised as nearly permanent records of the current circumstances on the Earth's surface at one moment in time, and may also be utilised as a historical document.
c) Capability to stop action: They give a stop-action picture of the dynamic state and are used to examine variable/dynamic phenomena such as flooding, moving wildlife, traffic, oil spills, forest fires, changing dynamics in natural phenomena, and so on.
d) Three-dimensional perspective: Aerial pictures give a stereoscopic perspective of the Earth's surface from which horizontal and vertical measurements may be taken.
e) Spectral and Spatial resolution: Aerial films are sensitive to electromagnetic waves with wavelengths (0.3 μm to 0.9 μm) higher than the human eye's spectral sensitivity (0.4 μm to 0.7μm).
f) Availability: Airborne images may be taken at any time and scale to create a lasting record of any region.

Advantages

a) It gives you a bird's-eye view of everything. A bird's-eye perspective of the terrain, for example.
b) This method of photography is both time-saving and cost-effective.
c) It has the capacity to freeze time. It, for example, retains the current ground conditions at a given moment indefinitely.
d) It can give you a stereoscopic perspective of the landscape. A three-dimensional perspective is an example.
e) Aerial photography gives up-to-date data.
f) It's less expensive than traditional surveying.
g) It is less dangerous than traditional surveying.
h) It can reach even the most distant and challenging locations, such as steep mountains and deep forests.
i) At various sizes, global coverage is easily accessible.
j) No map can compare to the present photographic perspective of the ground it gives.
k) Update the objective record with the area on a daily basis.

Disadvantages

a) It is more expensive and takes more skill to understand than a map.
b) It is deficient in marginal data.
c) Without symbols, ground features are difficult to detect or understand, and other ground characteristics, such as buildings in wooded regions, might cover them.
d) Aerial photography does not have a consistent scale.
e) Many distortions are present, including relief displacement and vertical exaggeration. As a result, without correcting the distortion, distances, directions, and areas cannot be determined directly from aerial pictures.

Operating Earth resource satellites and sensors system

Perhaps no two technologies have sparked more interest and application across a broader variety of disciplines than the marriage of remote sensing and space exploration. Although many components of the process are still in the early stages, studying the Earth from space has progressed from pure research to global, day-to-day use. We currently rely on spaceborne sensors to help with jobs ranging from weather prediction, crop forecasts, and mineral exploration to pollution detection, rangeland monitoring, and commercial fishing. All of this has happened in a relatively short time, and the status of remote sensing from space is changing as new and/or improved spacecraft are sent into Earth orbit.

Earth resources satellites are classified into three types. The first set of satellites captures visible and near-visible wavelengths. The five Landsat series spacecraft, which are first generation earth resource satellites, are a famous example of this category. The four IRS spacecraft, as well as the enhanced SPOT series, may be regarded second generation earth resources satellites of the same group. The Heat Capacity Mapping Mission spacecraft, notably the Explorer series, carry sensors that record thermal infrared wavelengths. Group three satellites are outfitted with microwavelength sensors. This category includes the seasat series and the ERS.

LandsatSatellite Program Overview

The Landsat series is an Earth Observation project led by the USGS and NASA that represents the world's longest operating system of satellites for moderate-resolution optical remote sensing for land, coastal regions, and shallow waterways. Landsat was the first Earth Observation satellite to be launched in July 23,1972, with the purpose of monitoring the world's land. It was quickly followed by successors, however the Landsat-6 spacecraft failed to reach orbit and communication with the satellite was never established. The series is still running today, making Landsat the longest-running continuous Earth imaging programme in history. Agriculture, botany, cartography, civil engineering, environmental monitoring, forestry, geography, geology, land resources analysis, landuse planning, oceanography, and water quality analysis are just a few of the uses for Landsat imagery. So far, nine Landsat satellites have been deployed, transforming this experimental project into an operational global resource monitoring programme, as listed below:

a) Landsat 1: The Landsat 1 satellite was launched in 1972. It was the first satellite launched as part of the United States' Landsat programme. It was an improved version of the Nimbus 4. It was the first satellite to be equipped with a multispectral scanner. It covered mineral resources, geology, agricultural and forestry resources, water resources, environmental pollution, marine resources, meteorological phenomena, and so on.
b) Landsat 2: It was first introduced in 1975. The primary goal of Landsat 2 was to collect medium-resolution earth data. Landsat 2 was decommissioned in 1982.
c) Landsat 3: The Landsat 3 satellite was launched in 1978. Previously, satellites were operated by commercial firms. Unlike previous satellites, Landsat 3 was entirely operated by NASA. In 1983, it was deactivated. It was utilised in over 31 nations. These nations employed Landsat 3 to assess the natural resources of their own areas.
d) Landsat 4: The Landsat 4 satellite was launched in 1982. The satellite's primary mission was to deliver scientific data. The mission's goal was to photograph the Earth. In 1983, the satellite lost 50 percent of its solar power. As a result, it was unable to communicate data to Earth. This resulted in the early release of Landsat 5.
e) Landsat 5: It was first introduced in 1984. Landsat 5's primary goal was to acquire photographs of the Earth. In 2013, it was deactivated. Guinness World Records named Landsat 5 the longest-running Earth Observatory satellite. The spacecraft has completed almost 150,000 orbits around the Earth. It has also sent 2.5 million photographs.

f) Landsat 6: The Landsat 6 satellite was launched in 1993. It was launched aboard the Titan II spacecraft. However, it was unable to enter orbit. This was due to the explosion that happened in the satellite's liquid fuel system.
g) Landsat 7: The Landsat 7 satellite was launched in 1999. The satellite's primary purpose was to deliver cloud-free photographs of the Earth. It updated the worldwide satellite picture archive. The US Geological Survey supervised and operated it. Landsat 7 was accompanied by Earth Observing-1. It was one minute behind Landsat 7. However, due to fuel depletion, it was deactivated in 2011. NASA stated in 2020 that Landsat 7 would be refuelled as part of the OSAM-1 mission. This was NASA's first announcement regarding recharging a live spacecraft. However, it was not possible, and Landsat 7 will be deactivated in 2021.
h) Landsat 8: The Landsat 8 spacecraft is an Earth Observation satellite. It was first introduced in 2013. It houses a thermal infrared sensor and an operating land imager for studying the earth's surface temperature. These instruments are also used to research global warming. The three primary goals of Landsat 8 are as follows: For a period of five years, collect medium resolution multispectral data of the world's landmasses. To guarantee that the data from previous Landsat missions are sufficiently consistent in terms of spectral properties, geometry, and coverage characteristics. To make data products available to the broader public
i) Landsat 9: The Landsat 9 satellite is scheduled to launch in September 27, 2021. The Landsat 9 is a spacecraft for Earth observation. It will be released in September 2021. The United States Geological Survey runs it. NASA planned and built the Landsat 9 satellite. The tasks of Landsat 9 are basically identical to those of Landsat 8. It is equipped with a thermal infrared sensor, an operational land imager, as well as optical and thermal sensors. The various characteristics of Landsat series is given below in Table 28.1:

Table 28.1: Landsat Satellite Characteristics

System	Sensors	Radiometric Resolution	Temporal Resolution	Orbit Altitude	Launch Date	Decommission or Expiration Date
Landsat 1 (ERTS-A)	RBV and MSS	6 bits	18 days	907-915 km	23-Jul-1972	6-Jan-78
Landsat 2 (ERTS-B)	RBV and MSS	6 bits	18 days	900 km	22-Jan-1975	5-Feb-82
Landsat 3	RBV and MSS	6 bits	18 days	917 km	5-Mar-1978	31-Mar-83
Landsat 4	MSS and TM	8 bits	16 days	705 km	16-Jul-1982	15-Jun-01
Landsat 5	MSS and TM	8 bits	16 days	705	1-Mar-1984	MSS: January 6, 2013
Landsat 6	ETM	8 bits	16 days	705 km	Launch failed October 5, 1993	
Landsat 7	ETM+	8 bits	16 days	705 km	15-Apr-1999	April 6, 2022
Landsat 8 (LDCM)	OLI and TIRS	12 bits	16 days	705 km	11-Feb-2013	Operational
Landsat 9	OLI 2 and TIRS 2	14 bits	8 days	705 km	27-Sept-2021	Operational

SPOT Satellite Programme

France, Sweden, and Belgium collaborated to construct the System Pourl'observation de la Terre (SPOT), an earth observation satellite programme. SPOT-1, the first satellite in the series, was launched on February 21, 1986 from Kourou Launch Range in French Guiana atop an Ariance Launch vehicle (AIV). This is the first earth resource satellite system to integrate a linear array sensor with push broom scanning. This allows for side-to-side oft-nadir viewing and full-scene stereoscopic imaging from two separate viewing positions in the same region. Thematic Mapper (TM) and High Resolution Visible (HRV) data produced by SPOT sensors have been widely utilised for urban planning, urban growth assessment, transportation planning, and other traditional natural resource applications.

SPOT -1, 2 and 3

SPOT-I, 2, and 3, like the Landsat satellites, have a circular, near-polar, sun-synchronous orbit. SPOT-I, 2, and 3 have a nominal orbit with an altitude of 832 km and an inclination of 98.7°. They cross the equator about 10:30 a.m. local solar time, crossing somewhat later in northern latitudes and slightly sooner in southern latitudes. SPOT, for example, passes areas at a latitude of 40° N about 11:00 a.m. and areas at a latitude of 40° S around 10:00 a.m. SPOT-I, -2, and -3 orbit patterns repeat every 26 days. This means that at this frequency, any location on Earth may be photographed using the same viewing angle. SPOT-I, -2, and -3 sensor payloads consist of two identical high resolution visible (HRV) image systems and auxiliary magnetic tape recorders. Each HRV is intended to work in one of two sensing modes: a 10-m-resolution 'panchromatic' (black and white) mode over the 0.51 to 0.73 μm range, or a 20-m-resolution multispectral (colour IR) mode over the 0.50 to 0.59 μm, 0.61 to 0.68 μm, and 0.79 to 0.89 μm ranges.

SPOT-4

The SPOT programme, like its Landsat cousin, has been designed to offer long-term data collecting continuity while improving the technological capabilities and performance of the sensing devices involved. SPOT-4, which was released on March 23, 1998, exemplifies this continuity with an enhanced performance design philosophy. The high resolution visible and infrared (HRVIR) sensors and the Vegetation instrument are the principal imaging systems on board this spacecraft.

SPOT-5

With the successful launch of SPOT-5 on May 3, 2002, the SPOT programme entered a new phase. The satellite is equipped with two high resolution geometric (HRG) sensors, one high resolution stereoscopic (HRS) instrument, and a vegetation instrument comparable to SPOT-4. HRG systems are designed to give high spatial resolution, with either 2.5or 5mresolution panchromatic images, 10m resolution in the green, red, and near-IR multi-spectral bands, and 20-m resolution in the mid-IR band. Instead of the high resolution red band used for the SPOT-4 HRVIR, the panchromatic band has a spectral range identical to that of the HRV on SPOT-1, 2, and 3 (0.48 to 0.71 μm). The HRS instrument includes stereo panchromatic imagery collection to aid in the creation of digital elevation models (DEMs) at a resolution of 10 m.

Indian Remote Sensing (IRS) Satellite

Beginning with IRS-IA in 1988 and continuing with three subsequent IRS series satellites in the 1990s, the Republic of India successfully built, launched, and operated multiple intermediate resolution Indian Remote Sensing (IRS) satellite systems. IRS-IA and -B incorporated the Linear Imaging Self-Scanning Sensor (LISS)-I and -II, multispectral sensors similar to the Landsat MSS and TM systems. IRS-IC and IRS-ID are equipped with three sensors: LISS-III, a panchromatic sensor, and a Wide Field Sensor (WiFS). The IRS P6 satellite's LISS IV sensor has the same spatial resolution (5.8 m) as IRS 1D PAN data and the same spectral resolution as IRS 1D LISS III data with a spatial resolution of 23.5 m. The LISS IV sensor has three spectral bands in the electromagnetic field's green, red, and near infrared regions. It can be inclined up to (plus/minus) 26° in the across-track direction, allowing for a 5 day revisit time and 70 *70 km stereo pairs. This opens up a whole new realm of micro-level applications. The various characteristics of IRS are given below in the Table 28.2:

Table 28.2: Showing characteristics of IRS

Platform	Lifetime	Altitude	Equator crossing	Coverage	Sensors
IRS-1A,1B	1A: 1988-1995 1B: 1991-present (3 years	904 km	10:30 a.m.	22 Days	LISS I LISS II
IRS-1C,1D	1995-present 1997-present (5 years)	817 km	10:30 a.m.	24 Days 5 Days 5Days	LISS III PAN WiFS
Resourcesat-1	2003-present (5 years)	817 km	10:30 a.m.	5-24 Days	LISS IV LISS III AWiFS

Sensors								
Characteristics	**LISS-I**	**LISS-II**	**LISS-III**			**LISS-IV**	**WiFS**	**AWiFS**
Resolution (m)	72	36	23	50	6	5.8	188	56-70
Swath width (km)	142	74	142	148	70	24-70	774	370-740
Sensor channel	04	04	03	01	PAN	03	02	04
Spectral Bands (µm)	0.45-0.52 (blue) 0.52-0.59 (green) 0.62-0.68 (red) 0.77-0.86 (near IR)	0.45-0.52 (blue) 0.52-0.59 (green) 0.62-0.68 (red) 0.77-0.86 (near IR)	0.52-0.59 (green) 0.62-0.68 (red) 0.77-0.86 (near IR)	1.55-1.70 (mid-IR)	0.5-0.75	0.52-0.59 (green) 0.62-0.68 (red) 0.77-0.86 (near IR)	0.62-0.68 (red) 0.77-0.86 (near IR)	0.52-0.59 (green) 0.62-0.68 (red) 0.77-0.86 (near IR) 1.55-1.70 (mid-IR

Application of Remote sensing

The use of remote sensing can be divided into three phases: the first, before 1950, represents the time of the first applications of aerial photography to ecology; the second, 1950 to 1972, represents the transition of applications from photography to unconventional imagery systems (for example, thermal infrared scenery and side-looking air-borne radars) and from low altitude aircraft to satellite platforms; and the third, 1972 to the present, emphasises the use of satellite platforms. Currently, remotely sensed data may be utilised for a wide range of practical applications, many of which are related to 'Earth Resources' in general. They are divided into nine primary groups for general discussion:

a) Atmospheric sciences: Because weather observations rapidly become out of date, weather forecasters require access to information from broad regions as fast and frequently as feasible. Satellites are ideally suited to providing this type of data that meets the requirements for both coverage and timeliness. A detailed description of the current meteorological state is critical for accurate short-term weather forecasting, especially for projecting precipitation motions and development during the next six hours. Satellite visual data are very important for short-term weather forecasting since they give accuracy and detail. As a result, meteorologists are increasingly relying on weather satellite data to help in the understanding of synoptic and small-scale weather systems. The NIMBUS-7 satellite's stratospheric aerosol measurement (SAM II) sensor measures the optical depth of the stratosphere and may thus offer information on air quality. Weather satellite data in graphical form may also disclose exceptional resolution in the position, extent, and intensity of cloud systems, as well as troposphere temperature variations with sufficient accuracy. These data can help anticipate windspeed, cloud movement, tornadoes, and other weather events. Overall, remotely sensed data, as well as a broad-based data base containing inventories of the following parameters, may be obtained:
   1) The energy and radiation exchange between the earth and the atmosphere, especially the net radiation balance at the top of the atmosphere, which is the major driving factor behind the planet's atmospheric circulations.
   2) The global distribution of moisture in the atmosphere is linked to the distribution of energy in the atmosphere.

3) The absorption and radiation of heat are affected by the global temperature distribution over land and water.
4) Cloud cover distribution has a significant impact on the albedo of the earth/atmosphere system and its constituent pieces, and it shows horizontal latent heat transfer patterns.
5) The temperature of the sea surface, which is directly related to the evaporation of latent heat.
6) The energy transmission within the earth/atmospheric system is governed by wind movement and air circulation.
7) Synoptic weather system climatology, encompassing frequency and regional dispersion over long periods of time.

b) Geology: LANDSAT imagery, in particular, has played a significant role in the investigation of geological structures, the identification of rock types and their mapping, and the determination of mineralized deposits on the earth's surface throughout the last several decades. TM data, in addition to LANDSAT, is quite beneficial for these types of research. Geothermal mapping with scanner data is mostly used in geothermal energy exploration studies as part of alternative energy resource investigations or for monitoring active volcanoes, notably in eruption prediction research. It is also feasible to determine the degree of fire in coal mining enterprises using an infrared linescanner. Lineament characteristics can also be determined with the use of radar imaging. SONAR, a significant sensor, has been employed in sea floor mapping since the 1990s. Remote sensing is becoming a significant and valuable technique for mapping vast, remote areas, and it has several applications in geosciences, including geologic and geo-structural mapping, mineral and water exploration, hydrocarbon exploration, natural hazards assessments, and geomorphology. Recent developments in remote-sensing imagery collection and picture availability can assist geoscientists in efficiently exploring and preparing maps, as well as evaluating the geo-potential of any given place on the world. Remote-sensing data processing methods have enhanced the ability to identify geological formations and regional features, which can aid in mineral exploration in complicated and poorly known places.

c) Biospheric survey: Even in the most technologically advanced countries, up-to-date and accurate assessments of total acreage of different crops in production, anticipated yields, stages of growth and condition (health and vigour) are frequently incomplete or late in relation to the information needed by agricultural managers.

These managers are constantly faced with decisions on planting, fertilising, watering, pest and disease control, harvesting, storage, and crop quality evaluation. Remote sensing data is used to forecast marketing aspects, analyse crop failure consequences, assess damage from natural catastrophes, and assist farmers in selecting when to plough, water, spray, or harvest. Colour-infrared film is sensitive to the green, red, and near infrared (500–900 mm) sections of the spectrum and is commonly employed in aerial and space photography surveys for land use and vegetation studies. The green region of the visible spectrum, to which the human eye is sensitive, is reflected by living plants. Furthermore, it reflects 10 times as much in the near infrared (700–1,100 mm) spectrum, which is just outside the range of human eyesight. When photosynthesis decreases, whether due to natural maturation or stress, there is a corresponding drop in near infrared reflectance. In an infrared film, living or healthy vegetation appears as varied shades of red. Because of the reduction in near infrared reflectance when sick or stressed, the colour response will shift to brown or yellow. Because blue light is blocked through filtering, this film is also useful for haze penetration. IRS- IA LISS-II, LISS -III, LISS -IV scenes are often used for forest research and survey works in India.

d) Hydrospheric survey: Many of the hydrological properties of significance for enhanced water resource management are, fortunately, easily detectable and evaluated using remote sensing technologies. Although water occasionally reflects visible light in a similar way to other surface characteristics, it greatly absorbs near infrared radiation. As a result, standing water appears very black in the near infrared, in contrast to soil and plants, which seem brilliant in this region of the spectrum. Surface water may therefore be easily detected and monitored in the optical wavebands in the absence of cloud. In this regard, measurements of the physical, chemical, geometrical, and optical properties of the ocean's coastal and open zones are necessary. Sea, ice, temperature, current suspended sediments, sea condition bathymetry, and water and bottom colour would all be measured. There are several remote sensing capabilities available for providing the relevant information using one of the measuring technique combinations. Images of sea-surface temperature and chlorophyll pigments from NOAA and NASA satellites show the occurrence of wind-driven coastal upwelling and the associated high biological production.

e) Cryospheric survey: Floating ice exists on many of the world's navigation routes for part of the year, while it is present all year in the high Arctic areas. It impedes or inhibits a wide range of maritime operations, including ship passage, offshore resource exploitation and transfer, and offshore commercial fishing. Furthermore, it may be a significant cause of damage, leading in the destruction of boats and equipment, the loss of life, and serious ecological disasters. Remote sensing techniques, both from aircraft and satellite platforms, are particularly valuable for acquiring this ice information. The planes are outfitted with glass domes for visual observation. SLAR (Side Looking Airborne Radar) is utilised for obtaining all weather information concerning capabilities and fewer profilometers for precise surface roughness assessment.
f) Environment: Remote sensing offers information on the quality, protection, and enhancement of land and water resources. Simultaneously, information on man's impact on the environment is collected, allowing monitoring to take place when necessary: of course, not all harmful impacts on the environment are caused by man. Remote sensing can also help with the prompt reaction to natural disasters such volcanic eruptions, earthquakes, hurricanes, tornadoes, and forest fires.
g) Geobotanical survey: Remote sensing plays an important role in mineral prospecting on forested terrain. The application of remote sensing in this context is a relatively new contribution to the geobotany and biochemistry disciplines. Geobotany is concerned with the affinities of certain plants, plant associations, or plant morphological anomalies to soils having greater than usual levels of heavy metals such as copper, lead, zinc, or tin. Plant tissue may be analysed using biochemistry to determine its heavy metal concentration. Geobotanical indicator species, plant association alterations, and biogeochemical anomalies may all be identified via the remote sensing platform's synoptic view. As a result, remote sensing has grown in popularity as an exploration tool in recent years.
h) Oceanography: Satellites, with their ability to cover large regions of land and water on a regular basis, are suitable for coastal and oceanic research. Satellite data is also less expensive than other data sources. For a country like India, which has a long coastline and a large exclusive economic zone, satellite-based remote sensing has the potential to bring significant economic advantages. Satellite

data may now be used to estimate the location of potential fishing grounds, river meandering patterns, and coastal shelf upwelling zones. Many aspects of coastal water and land mass have been found using NOAA, IRS, and LISS-I data since the 1990s. Remote sensing data might be used to examine pelagic fish migration patterns, create geographic distribution charts of chlorophyll-a, predict fisheries, and model ecosystems.

i) Forestry: Forests are an important resource because they provide food, shelter, wildlife habitat, fuel, and daily necessities such as medicinal components and paper. Forests serve a significant role in regulating $CO2$ supply and exchange on Earth, operating as a vital connection between the atmosphere, geosphere, and hydrosphere. Tropical rainforests, in particular, are home to an enormous diversity of species that are better adapted to, and hence survive, changing environmental circumstances more so than monoculture forests. This diversity also serves as a home for a variety of animal species and is a major source of therapeutic compounds. The primary concerns in forest management include depletion caused by natural causes (fires and pests) or human action (clear-cutting, burning, and land conversion), as well as monitoring of health and growth for successful economic exploitation and conservation. Remote sensing applications in forestry include the following:
    1) Reconnaissance mapping: National forest/environmental agencies must meet objectives such as updating forest cover, monitoring depletion, and assessing biophysical features of forest stands.
    2) Commercial forestry: Inventory and mapping applications are important to commercial forestry enterprises and resource management agencies because they collect harvest information and update inventory information for timber supply, broad forest type, vegetation density, and biomass measurements.
    3) Environmental monitoring: Conservation authorities are concerned with the amount, health, and variety of the Earth's forests, which includes deforestation, species inventory, watershed protection, coastal protection (mangrove forests), and so on.

Geographical Information Systems

Demand for data on topography and specific topics of the earth's surface, such as natural resources, increased dramatically in the late twentieth century. Photogrammetrists were able to map enormous regions with excellent precision thanks to stereo aerial photography and remotely sensed data. The same technique provided great advantages for reconnaissance and semi-detailed mapping to earth resource experts, geologists, soil scientists, ecologists, and land use specialists. The thematic maps that resulted were a source of important information for resource exploitation and management. The study of land assessment emerged from the necessity to fit the requirements for producing food and maintaining populations to the available climate, soil, water, and technological resources. Surveying and topographical mapping, thematic cartography, civil engineering, geology, geography, hydrology, spatial statistics, soil science, surveying and photogrammetry, rural and urban planning, utility networks, and remote sensing and image analysis have all seen parallel developments in automated data capture, analysis, and presentation. Many of these specialised industries have been integrated and even taken over by military applications. As a result, there has been a lot of effort duplication and jargon for different applications in different nations. The fusion of activities in formerly diverse but closely related sectors has resulted in the emergence of the general purpose GIS.

Geographical Information System (GIS) is a technology that allows for the collection and utilisation of geographic data to aid in agricultural growth and other sectors. A digital map is often more valuable than a paper map since it may be coupled with other sources of data for evaluating information with a graphical display. GIS software allows you to synthesis vast volumes of heterogeneous data by merging distinct layers of information to better organise and retrieve the data. GIS is a strong tool for agricultural scientists to give better service to farmers and the farming community by addressing their questions and assisting in improved decision making to conduct planning activities for agricultural development.

A Geographical Information System (GIS) is a system for recording, storing, analysing, and managing geographically related data and attributes. The geographical information system is also known as a geospatial information system or a geographic information system. It is an information system that can integrate, store, modify, analyse, share, and display spatially related data. In a broader sense, GIS is a software tool that allows users to create interactive searches, analyse geographical data, change data, generate maps, and display the results of all of these actions. GIS technology is quickly becoming an indispensable tool for combining diverse maps and remote sensing data to create models that can be utilised in real time. Geographical information system (GIS) is the study of using geographic ideas, applications, and systems.

Components of GIS

GIS allows users to utilise a computerised system to enter, manage, alter, analyse, and display spatially related data. GIS components such as software, hardware, data, user, and methodologies are required to execute various tasks with GIS.

a) Software: GIS software includes the capabilities and tools required to store, analyse, and display geographic data. A database management system (DBMS), tools for the entry and modification of geographic information, tools that facilitate geographic query, analysis, and visualisation, and a graphical user interface (GUI) for quick access to tools are key software components. GIS software is either commercial software or free software produced in the Open Source sector. Commercial software, on the other hand, is copyright protected, can be costly, and has a limited number of licensees. Arc Map, Intergraph, Global Mapper, QGIS, and other commercial GIS tools are now available. Arc Map is the most common software package among these. AMS/MARS, for reference, are examples of open source software.
b) Hardware: The computer on which a GIS runs is referred to as its hardware. GIS now runs on a variety of hardware platforms, ranging from centralised computer servers to desktop PCs used in standalone or networked setups.
c) Data: The data is the most crucial aspect of a GIS. Geographic or spatial data, as well as associated tabular data, might be gathered in-house or purchased from a commercial data source. Spatial data might take the shape of a map or remotely sensed data like satellite imaging and aerial photography. These data formats must be georeferenced (latitude and longitude). Tabular data can take the form of attribute data that is associated with geographical data in some way. Most GIS software has built-in Database Management Systems (DBMS) that allow you to establish and maintain a database to assist organise and manage data.
d) User: Without users to administer the system and establish strategies for its application, GIS technology has limited utility. GIS users range from technical professionals who create and manage the system to individuals who utilise it in their daily activities. These consumers are primarily concerned with the findings of the analyses and may have little interest in or understanding of the techniques of analysis. The GIS program's user-friendly interface enables nontechnical users to easily use GIS

analytical capabilities without needing to know intricate software instructions. A basic User Interface (UI) may consist of menus and pull-down visual displays, allowing the user to do needed analyses with a few key clicks rather than learning individual instructions in detail.

e) Methodology: A effective GIS follows a well-designed strategy and business rules, which are the models and operational methods that are specific to each corporation.

Applications of GIS

At a variety of scales, maps have grown in importance as legal documents that communicate property ownership and jurisdictional borders, as well as decision-making aids (for example, in urban planning). The GIS allows for the entire modelling of utility networks, such as those that serve water, power, and telecommunications to a large number of users. A system like this may work at several sizes, simulating service connections to customers, service districts, and full facility inventory and layouts including transformers, valves, conduits, and schematic designs. The following are some GIS applications:

a) Navigation: Web-based navigation maps promote safe waterway navigation. For improved routing, ferry routes and shipping routes are established. ArcGIS facilitates safe navigation by providing precise topographic and hydrographic data. The DNR's Coastal Resources Division just begun using GIS to locate, record, and categorise these no historic wrecks. This section provides public information that alerts residents to the locations of these vessels through a web map. The web map will be updated on a regular basis to keep the boating public informed of these coastal dangers and to reduce the risk of accident and harm.

b) Surveying: Surveying is the process of determining the position of things on the earth's surface. Land surveying is the process of measuring the distances and angles between various sites on the earth's surface. GNSS measurements are being used by an expanding number of national and regional governments and organisations. GNSS is used for topographic surveys with centimetre-level precision. These data can be used in the GIS system. GIS technologies may be used to estimate area and create digital maps.

c) Geology: GIS is used by geologists in a variety of applications. The Geographic Information System (GIS) is used to examine geologic characteristics, analyse soils and strata, evaluate seismic data, and construct three-dimensional (3D) representations of geographic features. GIS may also be used to evaluate rock information and determine the ideal dam site location.

d) Planning and Community development: GIS enables us to better comprehend our world and respond to global concerns. Today, GIS technology is continuously expanding, bringing many new features and improvements to planning. By using known parts of science and GIS to tackle unknown parts, we may improve our quality of life and attain a brighter future. Developing and implementing GIS tools and knowledge enables us to incorporate spatial intelligence into our thinking and behaviour.
e) Earthquakes: The occurrence of an earthquake is one of nature's most terrifying and devastating events. There is a requirement for understanding of global earthquake incidence patterns. A GIS-based user interface system for searching earthquake catalogues would be extremely beneficial to earthquake engineers and seismologists in identifying earthquake behaviour patterns in the spatial and temporal domains.
f) River crossing site selection for bridges: The stability of the slope heading down to and up from the river crossing is a significant geotechnical factor. It is recommended that historical data on erosion and sedimentation be collected. Assessing the amount of river channel contraction, degree of curvature of river bend, type of bed and bank materials, including flood flow and flow depth, may all be done in GIS within expected time and accuracy. This data has frequently been used to choose river crossing sites for bridges.
g) Emissions: Modelling and visualisation of air toxics diffusion and danger in areas around industrial operations.
h) Hazard analysis: Linking drawings and databases in order to do hazardous operations assessments for chemical processes.
i) Forestry management: Imaging and digital elevation modelling are used to assess forest damage caused by fire, logging, pesticides, and acid rain, as well as to identify changes in forest resources.

Benefits

Because the data is linked to a common referencing system, it is easy to use the same data to various applications as well as to associate diverse data sets previously unavailable for joint analysis. Topology allows new questions to be asked and encourages a new style of analysis that is in many cases fundamentally better than those used traditionally. GIS has the following advantages:

a) Better decision making: This usually has to do with making better location judgments. Real estate site selection, route/corridor selection, zoning, planning, conservation, natural resource exploitation, and so on are common examples. People are starting to recognise that choosing the right site is critical to an organization's success.
b) Improved communication: GIS-based maps and visualisations substantially aid in situational knowledge and storytelling. They are a new language that facilitates communication among many teams, departments, disciplines, professional sectors, companies, and the general public.
c) Better information record: Many organisations are primarily responsible for keeping authoritative records regarding the state and change of geography (geographic accounting). Zoning, population census, property ownership, and administrative borders are all instances of cultural geography. Forest inventories, biological inventories, environmental measures, river flows, and a variety of other geographic accountings are examples of physical geography. With comprehensive transaction support and reporting facilities, GIS provides a robust framework for administering these sorts of systems.
d) Managing geographically: GIS is becoming increasingly important in government and many huge organisations for understanding what is going on. Senior administrators and executives at the highest levels of government interact using GIS information products. These products offer a graphical framework for thinking, comprehending, and prescribing action. Briefings on diverse spatial patterns and linkages, such as land use, crime, the environment, and defense/security concerns, are examples.

Geographical Positioning System

The satellite navigation system, sometimes known as the global positioning system (GPS), is a modern tool that geographers utilise to locate spatial occurrences. This approach is based on trilateration concepts, which have been utilised by surveyors for decades. GPS has traditionally been utilised in field mapping, georeferencing, and digital elevation model assessment. To comprehend how a GPS works, it is necessary to recognise the system's five components. They are as follows:

a) The satellite kind (American NAVSTAR or Russian GLONASS systems)
b) The satellite control and monitoring stations.
c) The National and regional GPS base station networks
d) The GPS receivers.
e) Users

Applications

Geographers have traditionally utilised a variety of methodologies to identify the relative and actual location of spatial phenomena. These have ranged from highly exact and precise surveyor techniques to the more generic way of the social scientist. However, a novel approach for finding spatial phenomena has recently gained popularity among geographers. This is the satellite navigation system, sometimes known as the global positioning system (GPS). These portable locational devices can be installed on a vehicle, carried in a backpack, or held in the hand to record location at nearly any point on the earth's surface. GPS is a network of satellites and central systems that enable a specially constructed receiver to establish its location anywhere on the planet 24 hours a day, seven days a week. There are two major systems: the American (NAVSTAR) and the Russian (GLONASS). GPS technology is generally practicable and successful in a wide range of study areas where traditional surveying and mapping approaches would be impossible. GPS technology is now routinely employed for field mapping and georeferencing resource. Application of GPS in some important fields are given below:

a) Aviation: Most contemporary aircraft employ GPS receivers to offer real-time aircraft position to pilots and passengers. They also offer a map of possible destinations based on the aircraft's location. This tool is also used by airline operators to determine which route between locations is the quickest, safest, and most fuel-efficient. They also use the app to follow the aircraft and direct the pilot in the event of a change in weather conditions or any other difficulty.
b) Marine: Boat captains require a very precise navigation tool to navigate across waterways to their destinations. These programmes verify that the channels are clear and that there are no impediments to their navigation. They are also used in marine departments to

map and place dredging activities in rivers, sandbars, and wharves to ensure that other vessels know how deep they should go.

c) Farming: Farmers have a season for planting, weeding, and harvesting, and because the seasons recur, they install GPS receivers on their tractors and other farming equipment. This allows them to chart their plantations and return to the exact same period for planting or weeding the next season. This method is successful, especially in foggy conditions with limited sight, because the machine will continue to run because it is guided by GPS rather than visual reference. Furthermore, its great precision makes it appropriate for mapping soil sample sites, allowing farmers to discover regions with acceptable farming soils.

d) Science: This is one field that intensively uses the GPS app especially since there are numerous branches and departments in the science field and this include physics, biology and earth science to mention a few. Prior to the development of this software, biologists would tag animals with metal and plastic bands and follow them to various locations while monitoring them. However, with the creation and improvement of the navigation app, scientists have been able to attach the animals with collars, and the software can automatically record the animal's movement, which is then relayed to the researcher via satellite. This means they can track the animals' movements without having to physically relocate them. Most of the sciences taught in schools will have more practical proof, especially now that GPS data is so reliable. The programme is also used by earth scientists to investigate how landscapes change over time. Or any geographic place in which they are interested.

e) Surveying: Surveying is one of the critical uses of GPS, since it is utilised in mapping and measuring various measures on the earth's surface and underwater. It is used to determine land boundaries, map sea bottoms, and identify changes in structure form. The nicest part about a high precision GPS software is that surveyors may use it to set up a reference marker over a single location. They may also use it to map the boundaries of certain features in a moving configuration. They may quickly enter in the facts from the application into a programme that will assist them provide their consumers with a thorough chart using the data collected from the application.

f) Military: The GPS app system was developed originally by the US Department of Defense, and it has subsequently been adopted by various armed units across the world. Other governments have even opted to build satellite navigation networks as a defensive measure in times of conflict. The programme is being utilised for a variety of purposes, including mapping the position of vehicles and other hardware such as missiles during a battle. This is a tactic used to protect soldiers while simultaneously managing resources.

g) Telecommunications: Telecommunications, particularly mobile phones, employ this software to give its customers with precision, dependability, and stability in their operations. Although other clocks can do this, the GPS facilitates the calculation of synchronised time zones via satellite signals.
h) Heavy vehicle guidance: This technique is also employed in heavy tack machines used in mining and construction. In highway construction, for example, improved in-cabin vehicle navigation and control systems have replaced marking pegs and surveyors. This simplifies the job because the driver merely needs to adhere to the surveyor's pre-programmed site plan. As a result of this programme, builders may obtain near tolerance for location, gradient, and level.

CHAPTER 29

HYDROGEOLOGY

Introduction

Liquid water can be found in three different forms on our planet: as very large, medium, and small bodies of standing water, such as oceans, seas, and numerous lakes; as bodies of flowing water, such as major rivers, streams, rivulets, and springs; and as subsurface water, in the form of films around grains, droplets in pore spaces, and cavities in rocks, filling them partially or completely over variable areas and creating underground reservoirs. The subsurface water is further classified into two types:

a) Vadose water: Vadose water occurs from the surface below to a varied depth and moves downhill under the effect of gravity. Its motion is usually referred to as Infiltration. The zone of aeration is the thickness of soil and rock through which vadose water infiltrates. Therefore, the soil and rocks in the aeration zone are not saturated with water.
b) Groundwater: Groundwater includes all subterranean water that reaches a depth below which all pore spaces, openings, and other cavities of the soil and rock are totally filled with water is included. The groundwater reservoir in a particular location is defined by the thickness, length, and breadth of the saturated strata, or aquifer. The flow of water in this zone (of saturation) is primarily influenced by a hydrostatic head. Percolation is a word that is widely used to describe this phenomenon, which is characterised by laterality. The top surface of the saturation zone is known as the water table, and it is crucial in the study of groundwater reservoirs.

Age of Groundwater

Water's subterranean residence period has historically been a source of much debate. However, with the development of radioisotopes, it is now feasible to determine the age of groundwater. The isotopes hydrogen-3 (tritium) and carbon-14 have shown to be the most beneficial. Tritium, which has a half-life of 12.33 years, is created by cosmic radiation in the upper atmosphere and then brought to earth by rains and hence underground, where it begins to decay as a function of time. Carbon-14 has a half-life of 5,730 years and is created in the atmosphere at a known constant amount. This isotope may be found in groundwater as dissolved bicarbonate, which comes from the biologically active layers of the soil, where CO is produced by root respiration and the decomposition of organic matter. Tritium may be used to estimate groundwater residence durations of up to 50 years, but carbon-14 can be used to estimate ages ranging from a few hundred to 50,000 years. Ground waters with ages ranging from a few years to thousands of years have been discovered using age-dating methods. Water samples retrieved from deep wells in the Egypt and Saudi Arabia have been dated to be between 20,000 and 30,000 years old.

Hydrologists use a number of methods to determine the age of groundwater. Chlorofluorocarbons (CFCs) are frequently used to treat very fresh groundwater. CFCs are man-made chemicals that are environmentally stable. CFC concentrations in the atmosphere grew from the 1930s to the 1990s, and hydrologists now have a better understanding of how CFC concentrations have evolved over time. A pocket of water 2.6 billion years old has been discovered in a mine 2 miles below the Earth's surface in Ontario, Canada. It is the most ancient pocket of water known so far, dating back to the start of multicellular life.

Hydrological Cycle

Within the Earth's hydrosphere, water flows in various phases (liquid, solid, and gaseous). The hydrological cycle is the cyclic movement of water between the seas, atmosphere, lithosphere, and organisms in its different forms. The hydrologic cycle, which is made up of the cyclic processes is indicated in (Fig. 29.1) is the centre of hydrology. Water vapour is transported over the world by air circulation as it evaporates from the seas and land surfaces. On land and in the oceans, water vapour condenses and precipitates. The precipitated water may be intercepted by plants, move across the ground surface as streamflow, infiltrate into the earth as subsurface flow, or discharge as surface runoff. Direct evaporation from soil and vegetation surfaces, as well as transpiration through plant leaves, make up evaporation from the land surface.

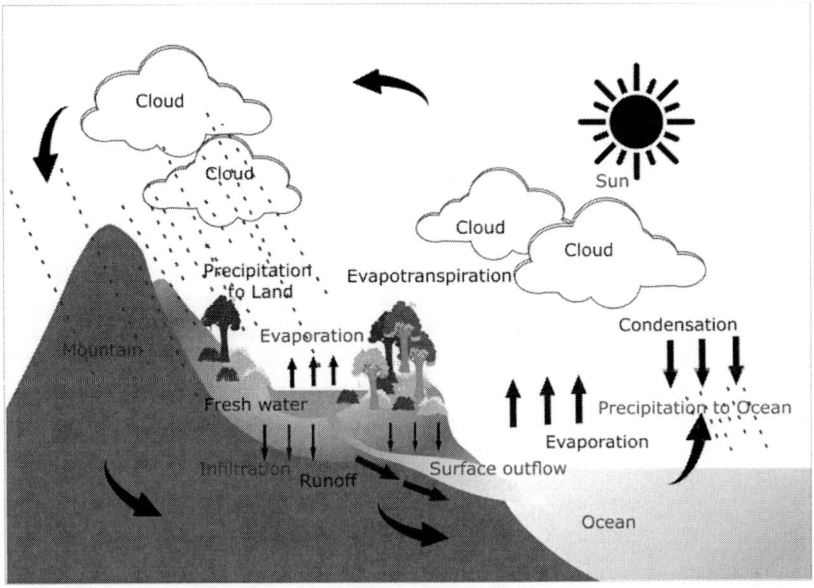

Fig. 29.1: Hydrological cycle

Evapotranspiration is the term for all of these processes together. Infiltrated water may percolate deeper into the earth, recharging groundwater and eventually becoming springflow or seepage into streams, resulting in stream flow. A hydrologic system is a structure or volume in space that absorbs water and other inputs, processes them internally, and generates them as outputs. The atmospheric water system, the surface water system, and the subsurface water system make up the global hydrologic cycle, which may be described as a system with three subsystems. Drainage basins, catchments, and watersheds all refer to the topographic region that gathers and releases surface stream flow through a single outlet. Small drainage basins are commonly referred to as catchments, but no exact area boundaries have been defined. The drainage basin divide, also known as the watershed divide or catchment divide, is the line that separates territory that drains toward a stream from land that drains away from that stream.

Groundwater system

Groundwater is a component of the hydrologic cycle, which is the earth's water circulation system. Water-bearing rocks in the earth's crust serve as water transmission conduits and storage reservoirs. Water enters these formations from the ground surface or from sources of surface water, then flows slowly over varying distances until natural flow, plants, or humans bring it back to the surface. Surface water is the source of almost all groundwater. Precipitation, stream flow, lakes, and reservoirs are the main sources of natural recharge. Excess irrigation, canal seepage, and water used specifically to supplement groundwater sources all contribute to artificial recharge. Along coasts where hydraulic gradients slope downward in an inland direction, even seawater can enter underground and cause salt water intrusion. Water in the ground goes downhill through the unsaturated zone due to gravity, whereas it moves in a direction controlled by the surrounding hydraulic condition in the saturated zone. When water emerges from the earth, it is referred to as groundwater discharge. Flow into surface water bodies, such as streams, lakes, and seas, accounts for the majority of natural discharge; flow to the surface appears as a spring. Evaporation from beneath the soil and transpiration from plants can both restore groundwater near the surface to the atmosphere. Pumping from wells is the most common artificial groundwater discharge.

Hydrological Budget

A water budget is a hydrological technique for calculating the amount of water flowing into and out of a system. To put it another way, it's a tally of all water stored and transferred on the land surface (rivers, lakes), subsurface (aquifers, groundwater), and in the atmosphere (precipitation, evaporation). The premise behind a water budget is that the amount and rate at which water flows into and out of a region balances the rate of change of water stored in that area. Hydrological scientists employ this notion as the foundation for successful water-resource sustainability, management, and environmental planning. There are numerous methods to describe the water budget. While there are several versions of the water budget calculation, they all follow the same basic principle. The water budget for a small watershed is defined by the United States Geological Survey (USGS) is as follows:

$$P + Qin = ET + \Delta S + Qout$$

Where:

P = precipitation (rain, snow, etc.)
Qin = water flow into the watershed
ET = quantity of evapotranspiration from soils, surface-water, plants, etc.

ΔS = change in water storage
Qout = sum of water flowing out of the watershed

Genetic classification of Water

The hydrologic cycle encompasses almost all groundwater, as well as surface and atmospheric (meteoric) waters. Groundwater from other sources may enter this cycle in small volumes. Connate water is defined as water that has been out of touch with the atmosphere for at least a significant portion of a geologic time; it mostly consists of fossil interstitial water that has migrated from its initial burial place. This water may have come from oceanic or freshwater sources and is typically highly mineralized. Magmatic water is water derived from magma; plutonic water is used when the separation is deep, while volcanic water is used when the separation is shallow (perhaps 3 to 5 km). Juvenile water is new magmatic or cosmic water that has not previously been a part of the hydrosphere. Finally, metamorphic water refers to water that has been or is linked with rocks during metamorphism. The interrelationships of different genetic kinds of groundwater are depicted in (Fig. 29.2).

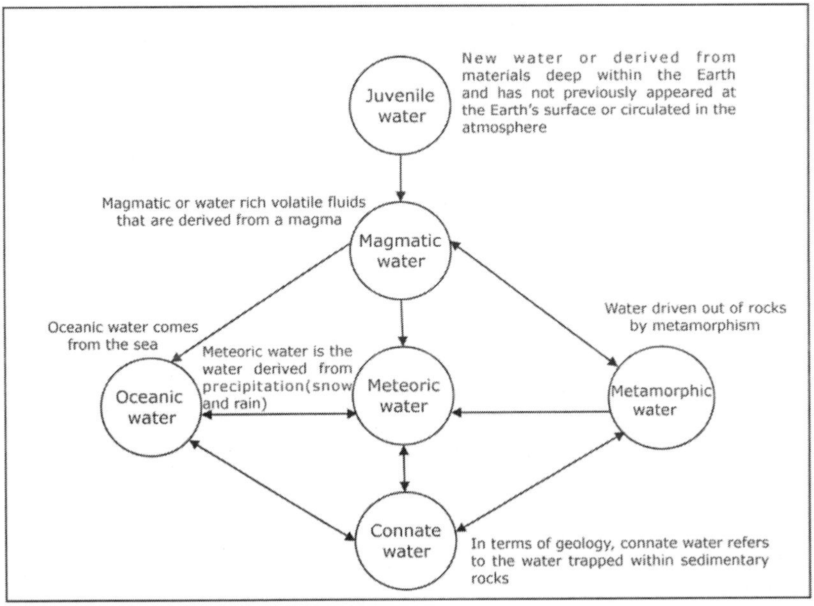

Fig. 29.2: Genetic classification of water

Zonal Distribution of Sub surface Water

Water that flows beneath the surface of the earth may be found in two primary zones or environments: vadose water and phreatic water. There are three main sorts of settings in the vadose water zone: soil water, intermediate vadose water, and capillary water (Fig.29.3). The soil water condenses into a thin layer at the land's near-surface depth. It may be found at depths ranging from 1 to 10 metres, and is kept in place by the root zone of the vegetation cover as well as soil chemicals. This water is critical for the survival and expansion of the world's vegetable cover. By transpiration and evaporation, it is lost to the atmosphere. The intermediate vadose zone is located directly under the soil water zone. It is, in reality, a zone of non-saturation, meaning that the water in this zone is going downhill due to gravity. It's usually only a few millimetres thick, and in other cases it's completely gone. The top two zones are sometimes referred to as the aeration zone. The capillary edge, or zone of capillary water, is only found under the vadose zone in fine-grained soils and rocks. The coarse sediments are devoid of it. Groundwater is forced upwards by capillary action in the fine particle size zone, occasionally to heights of 2–3 metres above the saturated zone underneath. The existence of the capillary fringe is often required for the growth of flora in various deserts. The well-known force of surface tension causes water to ascend (rather than travel downward) in the capillaries of tiny particles.

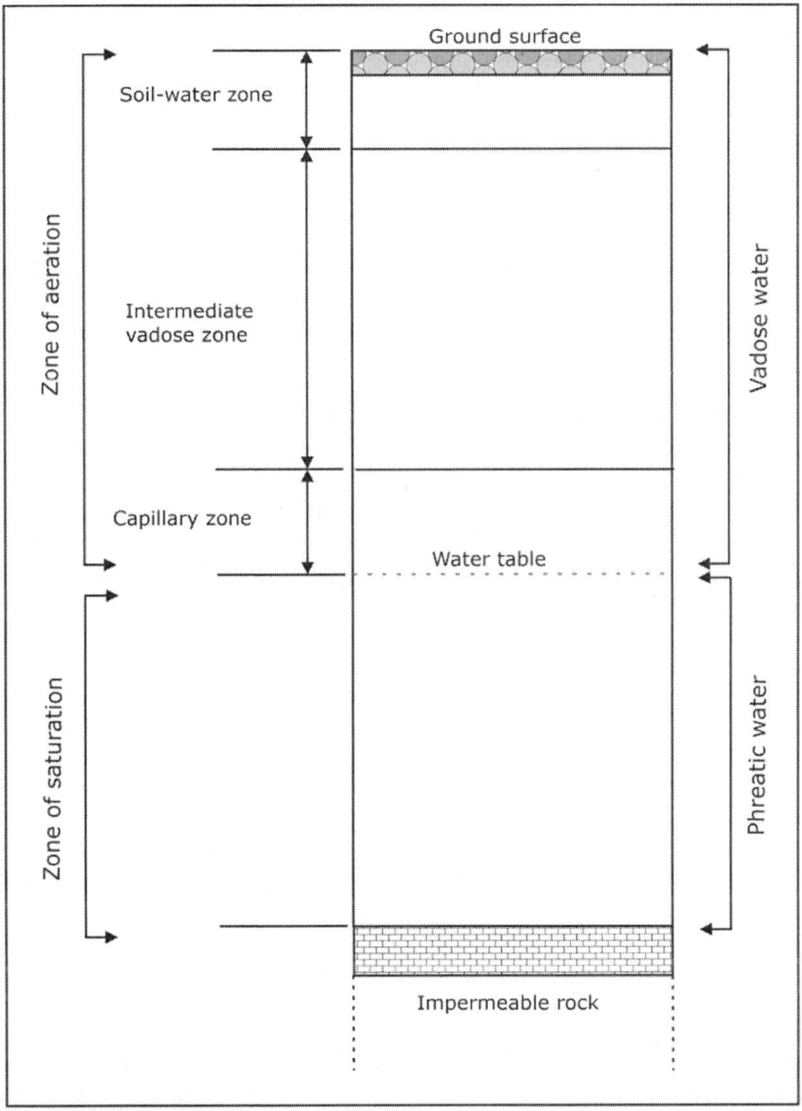

Fig. 29.3: Diagram showing zonal distribution of sub surface water

The Phreatic Water Zone, also known as the saturation zone, is located below the capillary fringe, and the water contained in this zone is referred to be groundwater in the true sense. The water table in the zone is marked by the upper surface of the water. The porous and permeable layers or masses of rocks in this zone have all of their open spaces, such as pores, cavities, and fractures, entirely filled with groundwater. All of these openings are intricately linked, allowing a well bored into this zone to draw water from the whole saturated rock mass known as the aquifer. There is no or very little downward movement of groundwater in this zone because all of the openings are entirely filled with water. The most common movement is a sort of lateral flow, which is regulated by the water's head (or level in various locations) moving from a higher head to a lower head to produce a level surface everywhere. The primary goal of all groundwater exploration programmes is to discover this zone and characterise its extent, geometry, and nature.

In many circumstances, depending on rainfall and recharge, a large rise and decrease in the level of zone of saturation is noted as a recurring phenomenon throughout the year. The depths between the zone of vadose water (aeration) and the zone of permanent saturation are marked by a third zone, the zone of intermittent saturation. This zone will only give water during and soon after rainy seasons, then become nonyielding for the remainder of the year.

Groundwater Movement

Groundwater is found in pore spaces between particles in gravel, silt, and sand aquifers, whereas groundwater is located in fractures within the rock in consolidated aquifers. The quantity of water that a specific type of rock can store is determined by its porosity, which is the volume of pore space between the sediment grains or cracks in the rock. The pore spaces must be linked for water to travel through rock. The speed of groundwater movement through the aquifer is determined by the size of the spaces inside the soil or rock, how effectively these spaces are linked, and the slope (or pressure gradient) of the water table.

Water bearing Characteristics of Geologic unit

All rocks are unable and unsuitable for retaining and transporting groundwater. The various characteristics of Geologic unit are given below:

a) Aquifer: An aquifer is any subsurface layer of water-bearing rock or geological structure that provides enough groundwater for wells and springs. An Aquifer, in geological terminology, is a body of saturated rock or geological formation through which water may easily pass (permeability) into wells and streams. The water table is the top of the water level in an aquifer. An aquifer is filled with water that flows into the ground from rain or melting snow. Water may penetrate through the aquifer's soil in some regions, while in others it may enter through joints and fissures in rocks and move lower until it hits less permeable materials. Aquifers are known to function as reservoirs and may dry up if people drain them faster than nature can replenish them. When found under favourable geological and geographic circumstances, gravels, limestones, and sandstones form good aquifers.

Types of Aquifers

1) Confined aquifer: Confined aquifers are reservoirs of water that accumulate in permeable rock and are surrounded by two impermeable rock layers or rock bodies. Confined Aquifers are aquifers that have been discovered to be overlain by a restricting rock layer or rock bodies, commonly comprised of clay, which may provide some protection against surface pollution. The non-permeable geological barriers found between the aquifers lead the water within them to be under pressure that is higher than air pressure. Groundwater flows vertically or horizontally across aquifers at rates that are typically controlled by gravity and geological formations in these places. Confined aquifers are sometimes known as "Artesian aquifers" since they are located primarily above the base of confined rock strata. Water levels in punctured wells derived from artesian aquifers fluctuate owing to pressure changes rather than quantity of stored water. The punctured wells function primarily as conduits for water transfer from replenishment sites to natural or manmade end locations.
2) Unconfined aquifer: Unconfined aquifers, in contrast to confined aquifers, are found near the ground surface and contain no layers of clay (or other impermeable geologic material) above the water table, while resting substantially above impervious clay rock strata. The water table is the topmost barrier of groundwater inside an unconfined aquifer. Groundwater in an unconfined aquifer is more sensitive to contamination by surface pollution than groundwater in confined aquifers due to simple groundwater penetration by terrestrial contaminants. The level of groundwater fluctuates depending on the amount of groundwater stored in the aquifer, which impacts the rise or fall of water levels in wells that get their water from aquifers. Perched aquifers (Fig. 29.4) are unconfined

aquifers that form when groundwater bodies are isolated from their primary groundwater source by relatively impermeable rock layers with modest areal extents and aeration zones above the main body of groundwater. The amount of water found in this sort of aquifer is generally minute and only available for brief periods of time.

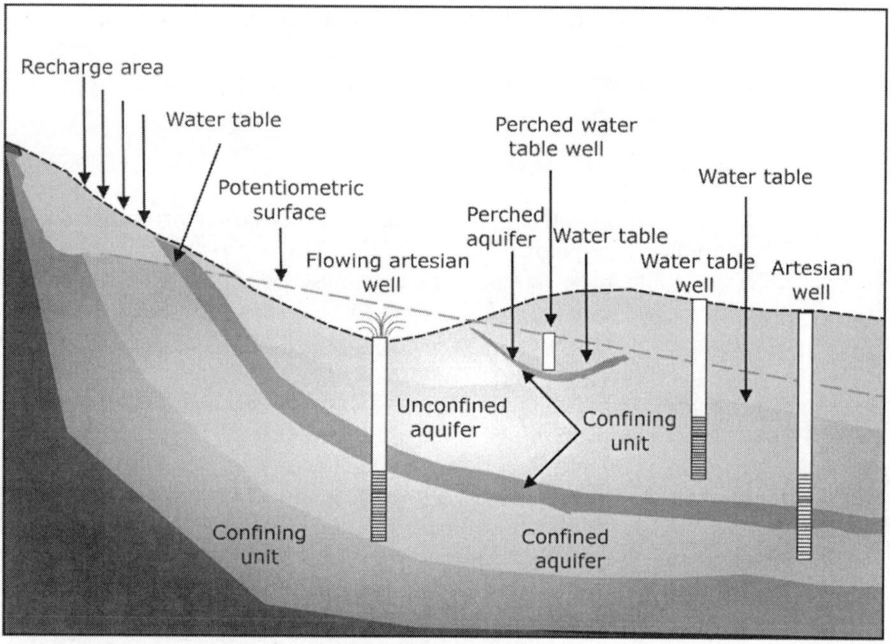

Fig. 29.4: Diagram showing different types of aquifers

Petro-Physical properties of Aquifers

Aquifer petro physical features are characteristics that aid in identifying and describing aquifers. Among the properties under consideration are:

1) Hydraulic conductivity: Hydraulic conductivity can be defined as the ease with which a fluid (groundwater) flows through a medium (in this case a geological formation or rock), which differs from intrinsic permeability in that, while it describes the medium's water-transmitting property, it is unaffected by temperature, pressure, or the fluid passing through the geological formation. The hydraulic conductivity of a soil, rock, or geological formation is determined by a number of physical parameters, including porosity,

particle size and distribution, particle arrangement, and other factors. In general, hydraulic conductivity decreases with particle size in unconsolidated porous media, hence clayey materials have low hydraulic conductivity compared to sands and gravels, which have high hydraulic conductivity (150 m/day for coarse gravels, 45 m/day for coarse sand, and 0.08 m/day for clay). This is due to the fact that small particle size arrangements (fine grained) in geological formations containing primarily clayey materials, while porous, are not permeable enough to allow groundwater flow within them, whereas sands and gravels (medium to coarse grained) have medium to coarse particle size arrangements, resulting in porous and permeable geological formations or rocks that allow for greater ease of groundwater flow. Mathematically hydraulic conductivity could be defined by the formula below:

$$K = k\frac{\rho g}{\mu}$$

where K is the hydraulic conductivity (cm/s or m/s), k denotes intrinsic permeability, ρ denotes fluid density, and μ denotes fluid dynamic viscosity.

Darcy's Law

In 1856, Henri Darcy, a French engineer, established a link between the rate of water flow through a porous medium and other characteristics such as hydraulic gradient and the length of the medium's column. Darcy's Law was named after his experimental findings, and it is a key equation in groundwater flow (Fig. 29.5). The rate of groundwater flow down a column of saturated sand is directly proportional to the difference in hydraulic head (The difference in hydraulic head at two points divided by the length is known as hydraulic gradient) at the column's ends, and inversely proportional to the column's length, according to this law. Mathematically Darcy's law could be defined by the formula below:

$$Q = -KA\frac{\mathrm{dh}}{dl}$$

where Q is the rate of water flow, K is the Hydraulic conductivity, (dh/dl) is the Hydraulic head and A is the column cross section area.

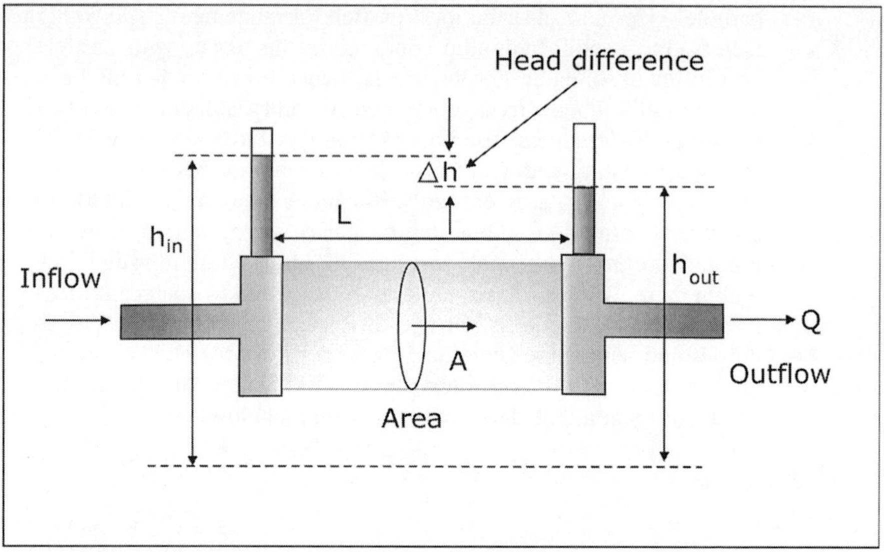

Fig. 29.5: Darcy's law

Limitations

Darcy's law is important for assessing whether hydraulic fluids may flow from a hydraulically fractured to a freshwater zone because it provides a circumstance where fluid movement from one zone to the other decides whether or not hydraulic fluids can reach the freshwater zone. Darcy's law can be applied to a wide range of circumstances, however it does not match these assumptions.

- It is not suited to soils with significant voids, such as gravel or boulders, since water in these soils is instantly drained owing to gravity, making it impossible to sustain laminar flow.
- Unsaturated and saturated flow.
- Flow in fractured rocks and granular media.
- Transient flow and steady-state flow.
- Flow in aquitards and aquifers.
- Homogeneous and heterogeneous systems flow.

2) Porosity: The measure of the enclosed voids or interstices given as a ratio of the volume of voids to the total volume is known as porosity of a geological formation, rock, or soil. It's also known as

the ratio of the volume of pores in a rock or soil sample to the overall volume of the rock matrix. When a rock is formed by cooling from an igneous melt, induration from loose sediment, or soil formation from weathering of rock elements, it has an inherent porosity called primary porosity, which is reduced over time by cementation and compaction. Secondary porosity refers to the formation of joints, fissures, fractures, or solution cavities inside rocks after the must has been inserted. The sum of primary and secondary porosities equals total porosity. The packing arrangement of particle sizes and the homogeneity of a rock's grain-size distribution impact the rock's porosity to a considerable extent. A cubic packing would offer a porosity of 47.65 percent, the highest and most perfect a rock with uniform spherical grains can attain. However, if the rock's packing arrangement were to change to that of a rhombohedral, the porosity would drop to 25.85 percent since the vertices of the rhombus would be formed by the centres of the eight neighbouring spheres. Theoretically, porosity, *n*, is expressed by the relationship:

$$Porosity = \frac{Volume\ of\ intersices}{Total\ volume\ of\ rock} * 100$$

3) Transmissivity: This term was introduced by Theiss in 1935 to express the capability of an entire aquifer to transmit water. Transmissivity (T) is a characteristic of an aquifer that allows it to transmit water. It can also be defined as the amount of water that can be transmitted horizontally through an aquifer unit by its full saturated thickness under a hydraulic gradient of 1 or as the rate at which water of prevailing kinematic viscosity is transmitted through a unit width of aquifer under a unit gradient. Mathematically, it is expressed as:

   $T = Kb\ (m^2/day)$

   where T is the transmissivity, K is the hydraulic conductivity, b is the saturated thickness of the aquifer.
4) Specific yield: When sediments or rocks are drained owing to a drop in the water table, the specific yield is the quantity of water that is really accessible for groundwater pumping. The distribution of pores, their shape, and grain size all influence the yield. Specific yield is a significant component in land surface models for

calculating water table depth, and it may have a big impact on the interaction between soil moisture and groundwater, as well as the water balance between the land surface and the atmosphere.

5) Specific retention: When an aquifer is drained, the volume percentage of water held behind by adhesion and capillary forces is referred to as specific retention.
6) Permeability: The most essential characteristic in groundwater is permeability. Permeability refers to how readily water may pass through rock or unconsolidated silt, as well as how easy it will be to extract the water for human needs. Geoscientists and engineers measure the permeability of a geological material using a variety of various units, the most common of which is hydraulic conductivity.
7) Storage coefficient: Storativity, also known as storage coefficient, is the amount of water extracted from a unit area of an aquifer for a unit reduction in hydraulic head in confined aquifers; in unconfined aquifers, it is equal to the specific storage times the thickness of the aquifer plus the specific yield. Storativity is a dimensionless quantity that is always larger than zero.

   b) Aquiclude: An aquiclude is a rock body or formation that is porous enough to store enough water but does not allow for easy and rapid flow through it due to other qualities. It must be considered as a nearly impervious rock mass. Aquicludes are best represented by compacted clay formations.
   c) Aquifuge: It is a completely impermeable rock formation with no potential of water storage or movement. Pores and other interstices are practically non-existent in such a structure. Compact interlocking granites and quartzites are two examples.
   d) Aquitard: It is a less frequent name for an aquifuge or aquiclude that has become locally leaky due to partial perviousness growth induced by profuse jointing or fissures. Aquitards include shale, clay, and numerous crystalline rocks.

Springs

A spring is a concentrated outflow of groundwater that appears as a flowing stream near the ground surface (Fig.29.6). Seepage regions, which are distinct from springs, imply a slower passage of groundwater to the ground surface. Depending on the amount of the seepage, the climate and the geography, water in seepage zones may pond and evaporate or flow. Springs are categorised according on their source, rock structure, discharge, temperature, and variability. All springs are classified into those caused by non-gravitational forces and those caused by gravitational forces. Volcanic springs, which are connected with volcanic rocks, and fissure springs, are caused by fissures extending to considerable depths in the earth's crust, fall under the first type. Thermal springs are the most common kind. Water flowing under hydrostatic pressure causes gravity springs.

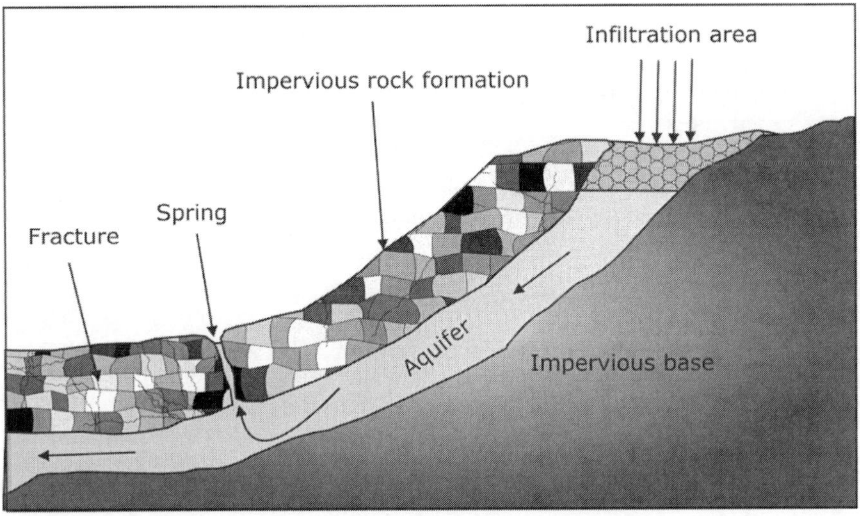

Fig. 29.6: Diagram showing spring and its relation with aquifer

Types

a) Depression springs: When the topography contacts the water table, generally owing to surface stream incision, depression springs emerge in unconfined aquifers. The springs are called depression or gravity springs because they are generated by the earth's gravitational force. These are generally found on cliffs and hilltops.
b) Contact springs: Contact springs form at the junctions of somewhat permeable rocks and low-permeability rocks. At such points, spring water emerges.
c) Artesian springs: It results from the release of water under pressure from confined aquifers.
d) Impervious rock springs: Occurs in tabular channels or fractures of impervious rocks.
e) Tabular or fracture springs: Fracture springs form when low-permeability rocks have permeable fracture zones. Groundwater

movement is mostly facilitated by the cracks that make up the porosity and permeability of aquifers. Where these fissures meet the earth surface, springs occur.

f) Karst spring: The name 'Karst' comes from a Slavic word that implies stony, barren terrain. It's also the name of a Slovenian area near the Italian border that's famed for its sinkholes and springs. All such terrain is referred to as karst by geologists. Cavities arise in carbonates rocks (limestones, dolomites, etc.) as the rock material dissolves owing to a chemical reaction. A spring or a series of springs is formed when water flows through these holes and apertures.
g) Fault springs: Faulting can also provide circumstances where groundwater under hydrostatic pressure (such as in confined aquifers) can move up along fault openings and produce a spring.

Wells

A well is a deep excavation or shaft drilled into the ground for the purpose of gaining access to natural resources. Water wells, oil wells, gas wells, and other types of wells are all available. For almost 8,000 years, wells have been utilised in numerous cultures across the world. All communities rely on wells for survival. Wells offer a steady and adequate source of water for domestic consumption, agriculture, and industry in many regions. People can't exist and grow without groundwater in places where surface water is rare, such as deserts, and they utilise wells to get subsurface water.

Types

a) Gravity well: It's also known as a water table well, and it's a vertical or almost vertical hole that pierces the saturation zone under the ground. The surface level of water in such a well, which is at atmospheric pressure and represents the water-table of the land surrounding the well when at fest, is its most important feature.
b) Tube wells: A tube well is a form of subterranean water well in which a long stainless steel tube or pipe is drilled. A strainer is installed on the bottom end, and a pump raises water for irrigation. The needed well depth is determined by the water table depth. Boreholes and tubewells are key adaptation technology techniques for supplying residential water amid water shortages and droughts. Freshwater is extracted from underground or deeper groundwater aquifers via these devices. The majority of water withdrawal wells

drilled into aquifers are gravity wells. These are the locations of the Tube Wells when water has to be pumped out of these wells.

c) Galleries: To intercept water, horizontal tunnels or open ditches are carved out through a water-bearing structure. These are often excavated perpendicular to the aquifer's water flow direction.

d) Artesian Wells: These are the holes bored into the artesian or restricted aquifers. Water from such wells usually runs to the ground surface, although it may also overflow to a certain height.

Chemical Composition of Groundwater

The chemical composition of groundwater is determined by the geological nature of the soil. Because water is continually in contact with the ground in which it stagnates or circulates, a balance between the soil's and water's composition develops: water that circulates in a sandy or granitic substratum is acidic and contains few minerals. Water in limestone includes bicarbonates and is alkaline. Other important factors to consider include the changing composition of rainfall and atmospheric dry deposition over groundwater recharge areas, evapotranspiration's modification of atmospheric inputs, differential uptake by biological processes in the soil zone, and mixing with seawater in coastal areas. The predominant dissolved components of groundwater in the terrestrial aquatic environment are the six major ions sodium (Na+), calcium (Ca2+), magnesium (Mg^{2+}), chloride (Cl^-), bicarbonate (HCO^-_3, and sulphate (SO_4^{2-}). Regardless of whether the water is dilute rainfall or has a salinity greater than saltwater, these cations and anions often account for more than 90 percent of the total dissolved solids concentration. Minor ions include potassium (K+), dissolved iron (Fe2+), strontium (Sr 2+), and fluoride (F), and aqueous solutions frequently include trace elements and metal species. Human activities can introduce pollutants into groundwater, causing certain ordinarily minor ions to reach quantities equal to major ions. Nitrate is an example, since excessive nitrogenous fertiliser application can cause nitrate concentrations in soil water and groundwater to exceed 50 mg L1. The detail chemical composition of groundwater is shown in table 29.1

Table 29.1: Chemical Composition of Ground water

Major ions (>5 mg L−1)					
Bicarbonate	Sodium	Chloride	Sulphate	Calcium	Magnesium
Minor ions (0.01–10.0 mg L−1)					
Nitrate	Carbonate	Iron	Potassium	Strontium	Phosphate
Fluoride	Boron				
Trace constituents (<0.1 mg L−1)					
Aluminium	Arsenic	Barium	Bromide	Cadmium	Caesium
Cobalt	Gold	Iodide	Chromium	Lead	Lithium
Manganese	Nickel	Phosphate	Radium	Selenium	Silica
Silver	Thorium	Tin	Titanium	Uranium	Vanadium
Zinc					
Dissolved gases (trace to 10 mg L−1)					
Nitrogen	Oxygen	Carbon dioxide	Methane	Hydrogen sulphide	Nitrous oxide

Salt Water Intrusion

Saltwater intrusion refers to the migration of seawater into freshwater aquifers as a result of natural or human-caused causes. Increases in saltwater levels or declines in groundwater levels produce seawater intrusion. Pumping out fresh water quickly lowers the height of the freshwater in the aquifer, resulting in a depression cone. For every 1 foot of freshwater depression, the salt water rises 40 feet, forming a cone of ascension. Take a look at the images below. Water quality can be affected by intrusion not just at pumping well sites, but also at other well sites and underdeveloped aquifer areas. The process by which salty water penetrates a region holding freshwater, such as the groundwater, is known as saltwater intrusion. Saline water ingression is another name for it. It results in land deterioration as a result of an increase in water and, as a result, soil salinity.

Saltwater intrusion is mostly caused by groundwater extraction. In many coastal locations of the United States, groundwater is the primary supply of drinking water, and extraction has risen over time. The inland spread of saltwater is limited by the increased pressure produced by the freshwater column due to its higher elevation under baseline circumstances. Groundwater extraction can lower the freshwater table, lowering the pressure on the freshwater column and enabling the denser saltwater to migrate onshore laterally. On the contrary, Saltwater intrusion can occur as a result of the development of canals and drainage networks. Canals and the deepening of existing waterways for navigation purposes both offer conduits for saltwater to be transferred inland.

Indian Example

Overextraction of groundwater is causing salty water incursion in the Thoothukudi district of Tamil Nadu. As a result, the water pressure in the groundwater table drops, allowing salty water from the sea, which is 3 kilometres distant, to enter. The decline in water pressure has also been attributed to a decrease in rainfall. Other contributing variables include the presence of water-intensive weeds such as Prosopis juliflora and inadequate water management.

Groundwater Exploration

As stated above, earth materials differ greatly in their ability to store water (porosity) and transfer water (hydraulic conductivity). As a result, our capacity to locate subsurface water supplies or predict flow patterns is limited by our understanding of the distribution of porosity and permeability, which is determined by geologic materials. The distribution of geologic materials in a region is usually the result of many separate geologic processes interacting. Tectonic processes, for example, are responsible for the underlying fragmented igneous and metamorphic rocks of India.

Investigating groundwater conditions is difficult and expensive since practically everything we want to know is buried far down and out of mind. It would be wonderful to have an accurate picture of the distribution of geologic materials, hydraulic characteristics, hydraulic heads, and chemical conditions. Many sophisticated ways for investigating and mapping the distribution of groundwater and groundwater-bearing materials have been devised. Most groundwater field projects include some subsurface profiling to collect material samples and to build holes in which to place wells, piezometers, and other monitoring equipment. The most common techniques are given below:

a) Geologic: Geologic investigations enable wide regions to be swiftly and economically analysed for groundwater development potential. The initial phase in a geologic study is the collection, analysis, and hydrogeologic interpretation of existing topographic maps, aerial photographs, geology maps and logs, and other relevant data.

Geologic field reconnaissance and a study of current hydrologic data on streamflow and springs, well yields, groundwater recharge, discharge, levels, and water quality should be supplemented wherever possible. Because no expensive equipment is required, this procedure should be used as the initial step in any subsurface water investigations.

b) Remote sensing: Photographs of the Earth obtained from aeroplanes or satellites at different electromagnetic wavelength ranges can give important information on groundwater conditions. Remote sensing technology has advanced quickly in recent years, but its application to water resources is still being explored. Additionally, the easy availability of data from commercial enterprises and government authorities has increased their use. Stereo vision analysis of black-and-white aerial images has grown in popularity. Differences in geology, soils, soil moisture, vegetation, and land usage can be distinguished by observable patterns, colours, and relief. As a result, remote sensing data can distinguish between rock and soil types, indicating permeability and areal distribution, and so regions of groundwater recharge and discharge.

c) Excavation: Hand digging with shovels or hand augers is an affordable but restricted method of sampling shallow unconsolidated materials. Hand digging is usually limited to a few metres. Hand augers inserted into the ground allow for somewhat deeper sampling, especially in soft sediments. Power excavators can dig deeper, up to four or five metres for bigger machines. A backhoe can dig 10 or more test pits this deep each day. This is a simple and economical method for mapping unconsolidated surficial deposits. It is difficult to excavate considerably below the water table using either hand or backhoe excavation, especially in more porous materials like sands or silts.

d) Direct push probes: Several novel direct-push exploration methods have been developed since the 1980s. Small-diameter probes and sensors are put directly down into unconsolidated materials without first drilling a borehole. Probes are typically made up of a tiny drilling pipe (37–49 mm outer diameter) with a cone-shaped tip and additional instruments integrated close below the tip. Hydraulic or pneumatic jacks attached to a big vehicle use static, impact, or vibrational forces to drive the probe into the earth. Probes, as opposed to drilling, are a reasonably rapid and low-cost approach to explore to intermediate depths, even below the water table. Soil engineers have been using cone penetrometer probes for decades. Probes perform well in a wide range of sands, silts, and clays, but struggle to penetrate thick or cobble-bearing materials.

e) Drilling: Drilling processes may create holes hundreds or thousands of feet deep. Drilling makes a hole in either unconsolidated or solid rock, with samples collected as the hole progresses. Wells, piezometers, or other instruments can be mounted in the drilled hole to gather relatively undisturbed samples. Drilled holes range in diameter from a few centimetres for shallow investigations to a metre or more for large water supply wells. Drilling has a high cost and is commonly utilised for groundwater research due to its flexibility. The following drilling methods are most typically employed for groundwater investigations:
    1) Hollow-stem auger: Hollow-stem augers are commonly employed for shallow environmental and geotechnical investigations. A rotary drilling rig, a heavy vehicle capable of spinning the augers while exerting downward pressure, is used to screw them into unconsolidated deposits. Hollow-stem augers perform better above the water table in softer materials and are confined to unconsolidated deposits. Soils at the bottom of the hole might liquefy and run up into the hollow-stem when the plug bit is pulled below the water table, especially in homogeneous sands or silts. The split-spoon sampler and the thin-walled tube are the two most commonly used techniques for sampling unconsolidated materials. Both are steel cylinders that are typically 24 inches long.
    2) Rotary: The hole is progressed in rotary drilling by spinning a drill bit connected to the bottom of a string of hollow drilling pipe. To keep the hole from caving in, a fluid, which can be mud, water, or air, is cycled down the interior of the drill pipe, out the drill bit, and back up to the surface in the annulus outside the drill pipe. When drilling in rock, rotary drilling methods typically employ an uncased hole unless the rock is extensively fractured or deformable.
    3) Cable tool: Drill bits and samplers are raised and lowered from the surface using a cable tool in cable tool drilling. The hole is progressed by continuously dropping a chopping bit on the bottom of the hole. Cuttings are removed by inserting a hollow bailer into the hole, which fills up but has a flap that prevents cuttings from falling back out when the bailer is lifted.

f) Well construction: The majority of wells are put in holes created by rotary or cable-tool drilling. The well is made up of a permeable part of casing called the well screen that is coupled to a solid casing that reaches all the way to the bottom. Water can easily flow through the screen, such that the head in the well casing equals the head in the formation around the borehole in a nonpumping well. The annulus between the screen and the borehole wall is backfilled with filter pack, which is often sand or gravel. To decrease resistance to flow between the formation and the well, the filter pack should be more permeable than the formation itself. A coarse filter pack can be formed from the surrounding unconsolidated sediments in sand and gravel water-supply wells through a process known as well development. Surging water in and out of the well screen mobilises and suspends small particles from the formation near to the screen during development.
g) Geophysical surveys: Shallow environmental and engineering challenges are addressed using a wide range of geophysical approaches. These approaches may be used to map the subsurface distribution of materials, water, or contaminants in an efficient and non-invasive manner. This part will provide a quick overview of three techniques often used in shallow groundwater studies:
    1) Resistivity: A continuous subsurface electrical field is established in a conventional resistivity survey by pushing direct current through two current electrodes placed at the ground surface or down boreholes. Differences in potential (voltage) are monitored between two additional electrodes situated between the current electrodes. The observed voltages and currents are then simulated using a mathematical model, yielding estimates of the electrical resistance (or conductivity) of underlying materials. Because current is largely transferred by flowing ions in the pore water, the concentration of ions (ionic strength) in the pore water and the volume of pore water present are the key parameters determining resistivity in most geologic materials. The lower the resistivity, the higher the water content and ionic strength. Clay minerals, with their charged surfaces and attracted ion boundary layers, also contribute to low resistivity.
    2) Electromagnetic surveys: Electromagnetic (EM) approaches produce low-frequency (less than 10 MHz) electromagnetic fields and currents in the subsurface by using alternating current in a transmitting coil. Conductive subsurface materials will respond to the transmitting coil's main field with induced

currents and their accompanying secondary electromagnetic field. The total of the main and secondary fields is measured as induced current in a receiver coil that is a fixed distance away from the transmitting coil. The EM characteristics of the intervening subsurface materials may be estimated using the field observed at the receiver coil.

3) Ground penetrating Radar (GPR): GPR is a technology that relies on the reflection of high-frequency electromagnetic waves in the 10–1000 MHz range of the spectrum. A radar wave pulse generator and a receiver are typically combined into a single device that is towed over the ground surface. The down-and-back travel periods of the reflected pulse are measured in the same way as seismic reflection methods are. The approach produces vertical cross-sections that indicate reflecting layers or objects at depth using estimations of radar wave velocities. GPR can penetrate further into materials with limited electrical conductivity. These materials attenuate the signal less than highly conductive materials. The normal penetration limit is around 20 metres, although it can be as much as 50 metres in dry and frozen materials or as low as 2 metres in wet clay or silt.

h) Seismic refraction: The propagation of seismic waves through earth materials is measured using seismic techniques. Seismic waves spread outward from a sound source at the surface, which can be an explosive charge or a mechanical impact in seismic surveys. To detect refracted waves, the refraction approach uses a long array of geophones, whereas the reflection technique employs a condensed array near the source to detect reflected waves. Reflection surveys are commonly employed to map the top crust in preparation for oil and gas development. Shallow mapping using refraction surveys is popular in groundwater supply and pollution investigations.

i) Borehole logging: A geologist is generally present when a hole is dug to explain the samples and drill cuttings. The geologic log of a borehole is a compilation of these descriptions, as well as notes concerning the rate of drilling, drill fluid losses, and so on. Borehole surveys, in addition to geologic logs, can be conducted. The following are brief summaries of the most prevalent borehole survey techniques:

1) Temperature: A thermometer is dragged down the borehole to measure the temperature of the borehole fluid as it varies with depth. Sudden fluctuations in fluid temperature frequently

indicate where water is entering or exiting the borehole. This is particularly effective for detecting transmissive cracks in crystalline rock.

2) Calliper: A calliper pulled through a bedrock borehole correctly measures the borehole diameter as well as the mapping of fractures.
3) Borehole Imaging: Borehole walls in dry boreholes can be imaged using optical video cameras intended to provide a continuous strip shot down the length of the shaft. Similar strip pictures may be created in fluid-filled boreholes utilising acoustic and electrical imaging. Because they function above and below the fluid level, these nonoptical approaches are more adaptable. Borehole photos are very effective for characterising cracks and huge rock holes.
4) Flow meter: A borehole flow metre is a device that measures the velocity of flow up and down a borehole's axis. Some employ impellers, some use heat pulses, while yet others use electromagnetic sensors.
5) Resistivity: Borehole resistivity measures voltage differences in a continuous DC electrical field, just as surface resistivity. One of the two current electrodes is usually towards the top of the borehole on the earth surface, while the other is down the borehole in the drilling fluid. The voltage electrodes are positioned at various depths in the borehole. A borehole survey produces a vertical profile that shows the distribution of electrical resistance in the formations surrounding the borehole.
6) Spontaneous potential: The inherent voltage differences between one electrode at the ground surface and one at deep in the borehole fluid are measured using this approach. There is no induced current in this approach, unlike resistivity methods, simply the ambient electrical field. It can help define the elevation of lithologic contacts.
7) Natural gamma radiation: A probe detects gamma radiation emitted by radioactive atoms in the strata surrounding the borehole. Some sensors distinguish between different energy levels of gamma radiation. This approach is beneficial for distinguishing sedimentary formations since the major source of gamma radiation in them is ^{40}K, which is abundant in clay minerals.
8) Neutron radiation: A radioactive probe produces neutrons, which are collected and dispersed by collisions with atom

nuclei outside the borehole. Some neutrons reflect back and are detected by a probe sensor. Backscattered gamma radiation is observed in several probes. The differences in findings are primarily related to differences in water content. Water molecules' hydrogen nuclei have the same mass as neutrons and are particularly effective in scattering and capturing them.

9) Gamma-gamma radiation: A gamma radiation generator on the probe generates it, and a sensor monitors back-scattered gamma radiation. The bulk density of the surrounding rock has the greatest influence on the outcomes. This approach may be used to determine rock porosity if the density of minerals in the rock is known.

Groundwater Recharge and Discharge

The majority of groundwater comes through recharge in upland areas, when water infiltrates from precipitation on the ground surface. Some water reaches the subsurface via seeping out of the bottom of surface waters, which occurs more frequently in arid regions than in humid ones. In low-lying locations, groundwater flows from the saturated zone back to the earth surface, often at springs or the bottom of surface waterways. Because groundwater constantly travels towards lower head, these exit sites are always lower than the water table when groundwater enters the system as recharge. Under highland locations, recharge enters the groundwater system. It subsequently migrates to discharge sites at low points in the terrain. Wetlands, springs, streams, and lakes are frequently found here. Because of the uneven distribution of hydraulic conductivities in the subsurface, the pathlines that water follows are sometimes quite irregular. Water tends to flow parallel to the layer borders in a high-conductivity layer. Water, on the other hand, prefers to follow the shortest course through a low-conductivity layer, flowing roughly perpendicular to the layer edges. Groundwater's residence duration can range from days to hundreds of thousands of years. Longer residence durations are seen in bigger groundwater basins and deeper regions of the flow system with high flow. Water in aquifers moves much rapidly than water in aquitards.

Concept of Artificial Recharge

People artificially recharge groundwater basins in order to augment the natural supply of groundwater. Artificial recharge is described as enhancing the natural circulation of surface water into subsurface formations by building, water spreading, or intentionally modifying natural circumstances. Artificial recharge initiatives are intended to accomplish one or more of the following goals:

a) Maintain or improve the economic value of natural groundwater.
b) Operate surface and groundwater reservoirs in concert.
c) Combat negative situations such as groundwater depletion, an unfavourable salt balance, and saline water incursion.
d) Provide subterranean storage for surface waters that are either local or imported.
e) Significant land subsidence should be reduced or stopped.
f) Provide established wells with a localised subsurface distribution system.
g) Reclaimed wastewater should be treated and stored for later usage.
h) Use hot or cold water to save or extract energy.

Recharge methods

Artificial recharge initiatives, in most cases, not only function as water conservation mechanisms, but also as a means of addressing overdraft issues. A sufficient volume of water must be acquired in order to deposit water underground for future usage. To artificially recharge groundwater, a number of technologies have been devised. The following are the most commonly used methods:

a) Basin method: Water can be replenished by releasing it into basins constructed by dykes, levees, or excavation. Basin sizes and forms are often adjusted to land surface slope. Silt-free water assists in the prevention of basin closure after submergence. When the bottom surfaces of most basins are dry, they must be scarified, disked, or scraped to increase infiltration rates. A single basin would generally serve to replenish local storm runoff, but when stream flow is redirected for recharging, a sequence of basins, sometimes parallel to the natural stream channel, becomes useful. Basins are the most popular type of recharge due to their general viability, effective use of area, and simplicity of maintenance.
b) Stream channel method: Water spreading in a natural stream channel refers to efforts that extend the time and geographic area over which water is replenished from a naturally losing channel. This includes upstream streamflow management as well as channel modifications to increase infiltration. Upstream reservoirs control unexpected runoff and, ideally, limit streamflow to rates that do not exceed the absorptive capacity of downstream channels. Stream

channels can be upgraded to enhance penetration by enlarging, levelling, scarifying, or ditching.

c) Ditch and furrow method: Water is distributed in a sequence of shallow, flat-bottomed, closely spaced ditches or furrows in the ditch-and-furrow method to maximise water-contact area. The ditch typically follows the ground contour and meanders back and forth over the land via steep switchbacks, tree-shaped, where the main canal progressively branches into smaller canals and ditches, and lateral, where a succession of minor ditches run laterally from the main canal. The widths of the ditches range from 0.3 to 1.8 m. Checks are occasionally installed in ditches on particularly steep slopes to reduce erosion and enhance the wetted area.
d) Flooding method: Water may be channelled to spread evenly across a broad region in relatively level terrain. In practise, canals and earth-distributed gullies are frequently required to discharge water at regular intervals throughout the upper end of the flooded region. To prevent upsetting the mind, it is preferable to produce a thin sheet of water over the ground that travels at a low pace. According to tests, locations with undisturbed plant and soil cover had the highest infiltration rates. Flood spreading has the lowest cost for land preparation when compared to other spreading methods.
e) Irrigation method: Water is often purposely dispersed in irrigated regions by irrigating fields with extra water during dormant, winter, or non-irrigating seasons. Because the distribution system is already in place, there are no extra costs for land preparation. Pumping a considerable amount of the water supply has the advantage of increasing the water table and therefore lowering electricity costs.
f) Pit method: An ideal facility for groundwater recharge is a pit dug into a permeable rock. Because the expense of excavation and material disposal is significant, it is most cost effective to employ abandoned excavations, such as gravel pits. Pits can successfully access materials with greater infiltration rates in places where shallow subsurface strata, such as hardpans and clay layers, hinder the downward movement of water.
g) Recharge well method: A recharge well is one that allows water from the surface to freshwater aquifers. Disposal wells and drain wells are other names for recharge wells. The flow of a recharge well is the opposite of that of a pumping well, however its structure may or may not be the same. Well recharging is viable where deep,

constrained aquifers must be recharged or where space is at a premium, such as in metropolitan areas.

h) Incidental recharge: When water enters the ground as a result of a human activity with no other objective than artificial groundwater replenishment, this is known as incidental or unplanned recharge.
i) Induced recharge: The above-mentioned direct techniques of artificial recharge require transferring surface water to a point where it enters the subsurface. The method of induced recharge is distinguished from these by extracting groundwater at a place beside a river or lake so that reducing the groundwater level induces water to enter the ground from the surface source. Artificial recharge is provided by wells positioned directly close to and mostly fed by surface water. In places supplied by a perennial stream, induced infiltration assures a continuous water supply even if overdraft circumstances prevail in surrounding areas supplied exclusively by natural recharge. The approach has been shown to be successful in unconsolidated formations of permeable sand and gravel that are hydraulically coupled to a stream and an aquifer.

Depletion and Management of Groundwater in India

According to the Central Ground Water Board (CGWB, Ministry of Jal Shakti), several sections of India are facing rapid groundwater depletion, with 230 billion metre cubes of groundwater pumped out each year for irrigating agricultural lands. The overall projected groundwater depletion in India is between 122 and 199 billion metre cubes. The various reasons for groundwater depletion are given below:

a) Increased water demand for household, industrial, and agricultural purposes, as well as restricted surface water supplies
b) Due to the hard rock terrain, storage facilities are limited, with the added drawback of a paucity of rainfall, particularly in central Indian regions.
c) The Green Revolution allowed water-intensive crops to be cultivated in drought-prone/water-stressed areas, resulting in over-extraction of groundwater.
d) Frequent pumping of groundwater without waiting for recharge results in rapid depletion.
e) Electricity subsidies and a high MSP (Minimum Support Price) for water-intensive crops

f) Water contamination, such as that caused by landfills, septic tanks, leaking underground gas tanks, and misuse of fertilisers and pesticides, which causes harm and depletion of groundwater supplies.
g) Inadequate groundwater control fosters the depletion of groundwater resources with no consequences.
h) Deforestation, improper agricultural techniques, chemical effluents from industry, and a lack of sanitation all contribute to groundwater contamination, rendering it useless.

Government Initiative for Groundwater Management

The National Aquifer Mapping and Management program (NAQUIM)

The Central Ground Water Board (CGWB) is implementing the National Aquifer Mapping and Management initiative (NAQUIM).

NAQUIM intends to map aquifers (water-bearing formations), characterise them, and prepare Aquifer Management Plans to aid in the sustainable management of groundwater resources. NAQUIM was established as part of the Ground Water Management and Regulation Scheme to identify and describe aquifers in order to produce ground water management plans.

Atal Bhujal Yojana

The Atal Bhujal Yojana (ABHY) is a Rs.6,000 crore government-funded project for sustainable groundwater management that includes community participation. It proposes public participation through the formation of 'Water User Associations,' water budgeting, and the creation and implementation of Gram-panchayat-level water security plans, among other things. The Ministry of Jal Shakti is in charge of it (earlier known as Ministry of Water Resources, River Development and Ganga Rejuvenation). The programme has the backing of both the Indian government and the World Bank. Gujarat, Haryana, Karnataka, Madhya Pradesh, Maharashtra, Rajasthan, and Uttar Pradesh have been identified as overburdened and water-stressed areas for project execution.

Rain water harvesting

Rain water harvesting is the gathering and storage of rain water that falls from roofs, parks, roadways, and open spaces, among other places. This runoff water can be kept or restored into the ground water. Rainwater harvesting has been used in India for over 4000 years. It is just a simple procedure of collecting and storing rainwater. Rainwater harvesting devices have been used as a source of drinking water, irrigation water, and cattle water since prehistoric times. To combat groundwater depletion, Tamil Nadu was the first state to make rainwater collection mandatory for all buildings. The initiative began in 2001 and has since been spread throughout Tamil Nadu's rural communities. Posters are being distributed throughout Tamil Nadu, particularly rural regions, to raise awareness about rainwater harvesting. A rainwater harvesting system is made up of the following components:

a) Catchments: A water harvesting system's catchment is the surface that directly gathers rainfall and supplies water to the system. It can be a paved area, such as a building's terrace or courtyard, or an unpaved area, such as a lawn or open ground. Water harvesting can also be done with a roof consisting of reinforced cement concrete (RCC), galvanised iron, or corrugated sheets.
b) Coarse mesh: A coarse mesh is installed on the roof to prevent debris from passing through.
c) Gutters: Rainwater is collected and transported to a storage tank by channels that run around the edge of a sloping roof. Gutters can be either semi-circular or rectangular in shape.
d) Conduits: Conduits are tubes or drains that transport rainwater from a catchment area or rooftop to a collecting system. Conduits can be made of any common material, such as polyvinyl chloride (PVC) or galvanised iron (GI).
e) First flushing: A first flush device is a valve that guarantees runoff from the first rainstorm is flushed out and does not enter the system. This is necessary because the initial spell of rain transports a disproportionately bigger amount of contaminants from the air and catchment surface.
f) Filter: The filter removes suspended contaminants from rainwater collected from the roof. A filter unit is a chamber that contains filtering media such as fibre, coarse sand, and gravel layers that remove trash and filth from water before it reaches the storage tank or recharge structure. Charcoal can be used to improve filtering.
g) Storage tanks or various other recharge structures: Used to store the filtered water which is ready to use.

Advantages

a) Less money is spent.
b) Reduces the amount of water used.

c) Reduces the need for water.
d) Reduces the amount of water that must be imported.
e) Encourages the conservation of both water and energy.
f) Groundwater quality and quantity are improved.
g) Landscape irrigation does not necessitate the use of a filtration system.
h) This technology is simple to understand, install, and use.
i) It decreases soil erosion, stormwater runoff, floods, and fertiliser, pesticide, metal, and other sediment contamination of surface water.
j) It's a great supply of water for landscape irrigation because it's devoid of pesticides, dissolved salts, and minerals.

Disadvantages

a) Maintenance is necessary on a regular basis.
b) Installation necessitates some technical knowledge.
c) The supply of rainwater can be limited by both limited and no rainfall.
d) It may attract mosquitoes and other waterborne diseases if not placed properly.
e) Storage limitations are one of the major downsides of rainwater gathering systems.

Common methods of urban rain water harvesting

Some of the most well-known urban rainwater gathering and management methods are:

a) Surface rain water harvesting: In metropolitan areas, surface runoff collection is the best option. Streams of surface runoff rains are channelled and stored in specially constructed reservoirs, either on the surface or beneath, for future use. This ensures a consistent supply of clean, drinkable water as well as water for routine household use.
b) Roof top rainwater harvesting: Rooftop rainwater collection is the most popular urban harvesting alternative. The system may be put in a single house, as well as in schools, universities, and other institutional buildings with more accessible roof area. The core notion of rooftop rainwater gathering is simple and straightforward. A container must be built or positioned beneath the roof level to

hold rainwater captured and steered towards it by pipes made of wood, bamboo, or PVC. Though this is the general idea, industrial-grade rooftop rainwater collection in large buildings is a little more complicated.

c) Recharge pits: Construction of recharge pits to retain rainwater is another typical rainwater gathering procedure. Recharge pits can be any size or shape, depending on the amount of rainfall in a given location. Filling these recharge pits with coarse sand, stones, and gravels acts as a natural filter, removing dirt and contaminants brought by the initial flow of rainfall.

Ground Water Pollution: Geological Consideration

The permeability of the material between the surface and the aquifer, the permeability of the aquifer, the slope of the surface, and the quantity of precipitation all influence the vulnerability of aquifers to pollution. Confined aquifers are substantially less vulnerable than unconfined aquifers, and deeper aquifers are less sensitive than shallow aquifers. Surface water tends to flow off rather than penetrate on steeper slopes (and this can reduce the possibility of contamination). The danger of contamination is much lower in dry locations than in wet places. Ground water is polluted for a variety of causes, including geological factors, industrial pollution, and even man-made cultural activities. Table 29.2 depicts the primary geological characteristics and probable ground water pollution.

Table 29.2: Geological conditions

Rivers and stream valleys	Karst areas	Areas with thin surficial Aquitard	Areas with thick surficial Aquitard
Contamination potential			
High: Alluvial aquifer	High: Bed rock aquifer Low: Deep aquifer	High: Drift aquifer, Bedrock aquifer Low: Deep rock aquifer	High: Shallow drift aquifers Low: Deep drift aquifer, Bedrock aquifer, Deep rock aquifer
Zone of free contamination	Contamination by surface run off and infiltration	Zone of possible contamination by infiltration	

CHAPTER 30

MAJOR ENVIRONMENTAL ACTS AND POLICIES IN INDIA

Introduction

Originally, our constitution did not include any explicit provision for environmental protection. However, following the 1972 United Nations Conference on Human Environment in Stockholm, the Indian constitution was revised to incorporate environmental preservation as a fundamental responsibility. The Wild Life Protection Act 1971 was the first environmental legislation passed in 1972. The 42nd amendment (Fundamental Duties) Clause (g) to Article 51A of the Indian constitution makes protecting and improving the natural environment a fundamental responsibility. According to Article 51A, Clause (g), it shall be the responsibility of every citizen of India to safeguard and develop the natural environment, including forests, lakes, rivers, and animals, and to have compassion for living beings. There is a direction provided to the state as one of the Directive Principles of State Policy about environmental conservation and enhancement. According to Article 48A, the State should work to maintain and develop the environment, as well as to safeguard the country's forests and wildlife. Article 21 of the Indian Constitution guarantees Indian citizens the right to a healthy environment. In India, the Department of Environment was founded in 1980 to maintain the country's environmental health. In 1985, this was renamed as Ministry of Environment and Forests (MoEF).

The Water (Prevention and Control of Pollution) Act of 1974

The Water (Prevention and Control of Pollution) Act was enacted in 1974 to provide for the prevention and control of water pollution, as well as the maintenance or restoration of the country's water's wholesomeness. In 1988, the Act was revised. The Water (Prevention and Control of

Pollution) Cess Act was passed in 1977 to provide for the imposition and collection of a cess on water consumed by people engaged in certain types of industrial operations. The Act gives State Pollution Control Boards regulatory power to set and enforce effluent regulations for factories. The Act gives State Pollution Control Boards regulatory power to set and enforce effluent regulations for industries. A Central Pollution Control Board fulfils the same tasks for Union Territories, as well as developing policies and coordinating the actions of several State Boards. The Act empowers the SPCB and CPCB to test equipment and collect samples for analysis. Prior to its revision in 1988, the Act's enforcement was accomplished through criminal proceedings brought by the Boards. The 1988 amending statute gave the SPCB and CPCB the authority to shut a defaulting industrial unit.

Major water quality issues:

The primary source of river contamination is matter originating from various human activities.

a) Pollution caused by pathogenic bacteria
b) Oxygen Depleting organic pollution
c) Salinity
d) Toxicity (micro-pollutants and other industrial pollutants)

Surface water:

a) Eutrophication
b) Oxygen depletion
c) Ecological health
d) Ground Water

Ground water:

a) Fluoride
b) Nitrate
c) Arsenic
d) Iron
e) Seawater intrusion

The Water Cess Act, 1977

The Water Cess Act was enacted to produce funds to cover the expenditures of the Central and State Pollution Control Boards.

The Act promotes economic incentives for pollution prevention and imposes a cess (tax) on water effluent emission on local governments and some authorised companies. After subtracting collecting charges, the Central Government pays the central and state boards such sums as appear required. To promote capital investment in pollution management, the Act provides a polluter with a 70 percent reduction in the relevant cess if they install effluent treatment equipment.

Air (Prevention and Control of pollution) Act, 1981

The Air (Prevention and Control of Pollution) Act of 1981, often known as the Air Act, was enacted by the Indian Parliament to prevent and control the negative effects of air pollution in the country. This act is regarded as the government of India's first tangible effort toward combating air pollution. Parliament passed the National Air Act in order to implement the conclusions made at the United Nations Conference on the Human Environment in Stockholm in June 1972. The primary goals of this Act are to enhance air quality and to prevent, control, and mitigate air pollution in the country.

Definitions under Air act, 1981

a) Section 2(a) defines an 'air pollutant' as any solid, liquid, or gaseous material that has the potential to injure or damage the environment, persons, plants, animals, or even property.
b) The term "noise" was introduced to the list of dangerous chemicals in a 1987 modification to the statute.
c) The presence of any harmful pollutant that renders the air unbreathable is defined as "air pollution" under the Clean Air Act.
d) Section 2 (g) of the Act also established the Central Pollution Control Board (CPCB), whose authority extended across India. To carry out the CPCB's recommendations, the act also called for the establishment of the State Pollution Control Board (SPCB) for each of India's states.

Important provisions of this Act

a) The framework of the Air Act is similar to that of the 1974 Water Act.
b) The Air Act extended the power of the Water Act's central and state boards to include air pollution management.
c) States that did not have water pollution boards were forced to establish air pollution boards.
d) All industries operating inside specified air pollution control regions must receive a "permission" (permit) from the State Boards under the Air Act.
e) After discussing with the central board and taking into account its ambient air quality guidelines, states are obligated to impose emission regulations for industry and autos.
f) The Act empowers the SPCB to test equipment and collect samples for examination from any chimney, fly ash, dust, or other source.
g) Prior to its revision in 1988, the Act's enforcement was accomplished through criminal proceedings brought by the Boards.
h) The 1988 amending statute gave the SPCB and CPCB the authority to shut a defaulting industrial unit.
i) Notably, the 1987 amendment added a mechanism for citizen suits to the Air Act and expanded the Act to address noise pollution.
j) Failure to comply with Central Pollution Control Board regulations would result in a one-year jail sentence. It can be prolonged to six years with a fine and an extra fee of 5000 Rs each day if the orders are not satisfied.

Environmental protection Act, 1986

A devastating catastrophe shocked the entire nation in 1984. A gas leak catastrophe at the UCIL pesticide facility in Bhopal on a December night killed over 4,000 people. Following the Bhopal disaster, the Indian government established the Environment Protection Act of 1986 under Article 253 of the Constitution. The legislation was approved in March 1986 and went into effect on November 19, 1986. The Act's objective is to carry out the resolutions of the 1972 United Nations Conference on the Human Environment. The choices concern the preservation and enhancement of the human environment, as well as the avoidance of risks to humans, other living animals, plants, and property. The Act serves as an "umbrella" for legislations intended to create a framework for Federal Government, as well as to coordinate the actions of different central and state bodies established by prior Acts, such as the Water Act and the Air Act. The major emphasis in this Act is on "environment," "pollution," "pollutants," and "hazardous substances." This Act grants the Central Government complete authority to safeguard and improve the condition of the environment.

Important Environmental Protection Act of 1986 Provisions

a) The Act comprehensively describes the environment, pollution, contaminants, and hazardous substances. The "environment" includes water, air, and land, as well as the inter-relationship that exists among and between water, air, and land, and human beings, other living creatures, plants, microorganisms, and property; "environmental pollutant" means any solid, liquid, or gaseous substance present in such concentration that may be, or tend to be, injurious to the environment; "environmental pollution" means the presence of any environmental pollutant in the environment. A "hazardous substance" refers to any substance or preparation that is capable of causing harm to humans, other living animals, plants, microorganisms, property, or the environment due to its chemical or physicochemical qualities or handling.
b) The Act is founded on the premise that the polluter pays.
c) The statute grants the Central Government broad authority to develop environmental regulations.
d) The Central Government has the authority to mandate public participation in the environmental protection process.

Important points under this Act

a) The Act expressly forbids the release of environmental contaminants in excess of regulatory criteria.
b) The Act specifies how dangerous chemicals should be handled.
c) The statute makes an exception for locust stands. Any ordinary citizen can now approach the court. Section 19 states that anybody, including authorised government personnel, may submit a complaint with a court alleging a violation of the Act.
d) This "Citizens' Suit" clause requires the individual to notify the Central Government of the suspected environmental crime within 60 days.
e) The minimum penalty for violating any provision of this legislation is five years in prison or a fine of one lakh rupees. The statute

further states that if the violation persists after the date of conviction, the fine is 5000 rupees per day.

f) If a failure or violation continues for more than a year, the offender may face imprisonment of up to seven years.
g) The statute provides immunity to government officers for any conduct performed in accordance with the requirements of the act.
h) The statute takes precedence over other environmental regulations.
i) The statute prohibits civil courts from having jurisdiction to hear any litigation or case arising from a Central Government action, instruction, or order.
j) By virtue of this Act, the Central Government has been given significant powers such as coordinating state action, planning and executing national programmes, establishing environmental quality standards, particularly those governing the emission or discharge of environmental pollutants, imposing restrictions on the location of industries, and so on. Direct orders included the ability to shut, restrict, or control any industry, power of entrance for investigation, testing of equipment, and other reasons, as well as the ability to examine a sample of air, water, soil, or any other substance from any location.

National Forest policy, 1988

The Imperial Forest Department, established in 1864, aimed to impose British authority over forests through several legislations, including the Indian Forest Act of 1865. It gave the British government the authority to declare any area covered with trees a government forest and establish management guidelines for it. The Indian Forest Act of 1927 was heavily influenced by previous Indian Forest Acts enacted by the British. The Indian Forest Act of 1878 was the most well-known. Both the 1878 legislation and the 1927 Act intended to consolidate and conserve areas with forest cover or significant wildlife, as well as to govern the movement and transit of forest produce, as well as the tax levied on timber and other forest produce. The British Administration gained jurisdiction over all wastelands, including woods, by the Forest Act of 1878. This Act also provided the government with the authority to demarcate reserved and protected forests. Local rights were denied in the case of protected woods, although some government-granted advantages to local people can be revoked at any moment. The forests were divided into three categories under this Act: reserved forests, protected forests, and village forests. The Forest Conservation Act of 1980 mandated that central approval be obtained before engaging in sustainable agroforestry in forest regions. Violations or absence of a permission were considered criminal offences. It aimed to reduce deforestation, protect biodiversity, and rescue wildlife. Although this Act gives more optimism for forest protection, it did not achieve its goal.

The 1988 National Forest Policy sought to maintain, conserve, regenerate, and develop forests. The key themes of the National Forest Policy 1988 were the preservation and restoration of ecological equilibrium, as well as the protection of forests as national heritages containing numerous types of flora and animals. The ultimate goal of the National Forest Policy was to ensure environmental stability and ecological balance through preserving forests as a natural treasure. In 1988, the National Forest Policy made a large and categorical move away from economic considerations and toward an emphasis on the ecological importance of forests and participatory management.

Objectives:

a) Conserving the country's natural heritage by providing the surviving natural forests with a diverse range of flora and wildlife, which symbolise the country's remarkable biological richness and genetic resources.
b) Checking soil erosion and denudation in river, lake, and reservoir catchment regions in the interest of soil and water conservation, minimising floods and droughts, and slowing reservoir siltation.
c) Checking the extent of sand dunes in Rajasthan's desert areas and along the shore.
d) Increasing the country's forest/tree cover significantly through extensive afforestation and social forest programmes, particularly on all deforested, degraded, and unproductive regions.
e) Increasing forest production to satisfy critical national demands.
f) Encourage effective use of forest products and maximise wood substitution.

Biological Diversity act, 2002

The Indian parliament approved the Biological Diversity Act in 2002 to safeguard biodiversity and allow the sustainable management of biological resources with local populations. The Act was passed in order to satisfy the standards of the United Nations Convention on Biological Diversity (CBD), to which India is a signatory. The major goal of the Act is to ensure the protection of biological variety, the sustainable use of its components, and the equitable use of its resources in order to avoid misuse or the eventual annihilation of biodiversity. Because India is one of the world's most ecologically varied countries, this legislation is necessary to safeguard its biological legacy.

Salient features:

a) Regulation of access to the country's biological resources.
b) Biological diversity conservation and sustainability.
c) preserving local populations' understanding of biodiversity.
d) Secure benefit sharing among local people as biological resource conservers and bearers of knowledge and information about biological resource utilisation.
e) Threatened species protection and rehabilitation.
f) Involvement of state government entities in the wide scope of Biological Diversity Act implementations through the formation of dedicated committees.
g) Any disputes connected to the National Biodiversity Authority's or a State Biodiversity Board's determination of benefit sharing or order issued under this Act must be filed with the National Green Tribunal (NGT).

National Biodiversity Authority

The Government of India established the National Biodiversity Authority (NBA) within the Ministry of Environments and Forests in 2003 to carry out the terms of the legislation. The NBA is a formal, self-governing organisation based in Chennai. State Biodiversity Boards (SBB) and biological management committees for each local body were also established in the 29 states. The Central Government, in cooperation with the NBA, is required by this legislation to:

a) Alert threatened species and prevent or restrict collection, restoration, and conservation. To learn more about endangered species, see the IUCN Red List.
b) Establish institutes to serve as repository for various types of biological resources.

Functions:

a) Monitoring and prevention of violations of the Act.

b) Advising the government on how to effectively maintain biodiversity in India.
c) Prepare a paper outlining how the government can choose biological heritage locations.
d) Take concrete actions to avoid the award of intellectual property rights over locally exploited biological resources or related traditional knowledge.

Schedule Tribes and other Forest Dwellers Act, 2006

The Forest Rights Act of India, officially known as the Scheduled Tribes and Other Traditional Forest Dwellers (Recognition of Forest Rights) Act, is sometimes known as the Tribal Land Act or the Tribal Rights Act. It addresses the rights of forest people (including Scheduled Tribes) to land and other resources, which have been denied to them throughout the years due to the country's continued application of colonial-era forest regulations.

The Importance of the Forest Rights Act (FRA) of 2006

a) For the first time, community rights and rights over common property resources (CPR) have been acknowledged.
b) Individual rights of tribal and marginal groups, as well as other rights, have been underlined by this legislation.
c) The notion of revenue villages has emerged since the act mentions converting all forest villages, old settlements, un-surveyed villages, and other villages into these.
d) It protects the livelihood and food security of Forest Dwellers, Scheduled Tribes, and Other Forest Dwellers while also strengthening the forest conservation regime.
e) Community Forest Resources are monitored and maintained in a way that maintains the traditional links between marginal communities and these resources. It is well known that these cultures have historically used forest resources for sustainable development.
f) This act protects intellectual property rights as well as traditional knowledge about cultural variety and biodiversity.
g) It broadens the mandate of the Constitution's 5th and 6th Schedules, which safeguard indigenous tribes' claims to land or forests.
h) The Forest Rights Act of 2006 protects the rights of displaced populations. One of the causes of the Naxal movement, which affects states such as Chhattisgarh, Odisha, and Jharkhand, is tribal

alienation. The Act attempts to include tribes by naming IFR and CFR. On the linked website, you may learn more about the Status of Tribal Populations in India.

i) FRA 2006 recognises and protects the rights of marginalised and indigenous populations to participate in development initiatives.
j) Forest rights can also be claimed by any person or group who has lived in forest land for at least three generations (75 years) prior to the 13th of December, 2005 for legitimate subsistence requirements.
k) The act would guarantee that people have the ability to manage their own forests, regulating the exploitation of forest resources by authorities, improving forest governance, and better managing tribal rights.

Bio medical waste management rules, 2016

In June 2015, the proposed Bio-Medical Waste Rules were released, asking public feedback. The suggestions/objections submitted were subsequently reviewed by the Ministry of Environment, Forestry, and Climate Change's Working Group, and the Ministry released the Bio-Medical Waste Management Rules, 2016.

Salient features:

a) To facilitate waste segregation at the source, bio-medical waste has been divided into four categories rather than ten.
b) Within two years, phase out the usage of chlorinated plastic bags, gloves, and blood bags.
c) The rules' scope has been broadened to cover vaccine camps, blood donation camps, surgery camps, and any other type of healthcare activity.
d) On-site disinfection or sterilisation of laboratory waste, microbiological waste, blood samples, and blood bags in accordance with WHO or NACO guidelines.
e) The state government will grant land for the establishment of shared bio-medical waste treatment and disposal facilities.
f) If a service of 'common bio-medical waste treatment facility is accessible within 75 kilometres, no occupier shall install an on-site treatment and disposal facility.
g) The new guidelines establish more strict criteria for incinerators in order to decrease pollutant emissions into the environment.
h) Inclusion of Dioxin and Furan emission limitations.

i) Create a barcode system for disposal bags or containers containing bio-medical waste.
j) All health care professionals should be trained and immunised on a regular basis.
k) The state government has agreed to donate land for the establishment of a common bio-medical waste treatment and disposal facility.
l) Operator of a shared bio-medical waste treatment and disposal facility to enable timely collection of bio-medical waste from HCFs and assistance to HCFs in training.

E-waste management rules, 2016

Discarded electronic or electrical equipment and gadgets are referred to as electronic trash. E-waste is used electronics that are meant for reuse, salvage, resale, disposal, or recycling. Informal or uncontrolled e-waste processing, particularly in poor countries, can harm human health and pollute the environment. In addition, the Environment, Forestry, and Climate Change Ministry has issued the 2016 E-Waste Management Rules. These new guidelines superseded the 2011 E-Waste (Management and Handling) Rules. In March 2021, New Delhi announced the establishment of an e-waste management park for the safe and scientific disposal of electronic devices like as appliances and e-vehicle batteries.

E-Waste in India

a) Every year, India creates around 18.5 lakh metric tonnes (MT) of electronic garbage, with Mumbai and Delhi-NCR accounting for the lion's share. By 2018, the Fig. is expected to rise to 30 lakh MT each year.
b) Other key cities that generate a significant quantity of e-waste are Bangalore, Chennai, Kolkata, Ahmadabad, Hyderabad, Pune, Surat, and Nagpur.
c) Maharashtra ranks first among the eight major e-waste producing states, followed by Tamil Nadu (2nd), Andhra Pradesh (3rd), Uttar Pradesh (4th), Delhi (5th), Gujarat (6th), Karnataka (7th), and West Bengal (8th).
d) Over half of the developed world's e-waste is sent to developing nations, mostly China, India, and Pakistan, where metals such as copper, iron, silicon, nickel, and gold are recovered during the recycling process.

e) Unlike in wealthy nations, where facilities for recycling e-waste have been particularly established, recycling in poor countries sometimes includes manual participation, exposing employees to harmful compounds included in e-waste.

Salient features:

a) Manufacturers, dealers, refurbishers, and the Producer Responsibility Organization (PRO) have been included as new stakeholders to the guidelines.
b) The regulations' application has been expanded to include EEE components, consumables, spares, and parts, in addition to the equipment mentioned in Schedule I.
c) Compact Fluorescent Lamps (CFLs) and other mercury-containing lamps are now subject to regulations.
d) For collection of e-waste by producers under Extended Producer Responsibility, a collecting mechanism-based strategy has been implemented, which includes collection centres, collection points, take-back systems, and so on (EPR).
e) PRO, e-waste exchange, e-retailer, and Deposit Refund Scheme have been proposed as additional channels for Producers to apply EPR and assure effective e-waste channelization.
f) Pan India Provision, CPCB EPR authorization has been implemented to replace state-by-state EPR authorization.
g) Collection and channelisation of e-waste in Extended Producer Responsibility - Authorisation must meet the goals set out in Schedule III of the Rules. The phase-wise collection target for e-waste, which can be either in number or weight, shall be 30 percent of the quantity of waste generation as indicated in the EPR Plan during the first two years of rule implementation, 40 percent during the third and fourth years, 50 percent during the fifth and sixth years, and 70 percent during the seventh year and beyond.
h) The Deposit Refund Scheme has been introduced as an additional economic instrument in which the producer charges an additional amount as a deposit when selling electrical and electronic equipment and returns it to the consumer along with interest when the end-of-life electrical and electronic equipment is returned.
i) The rules include an option for e-waste exchange as an independent market instrument offering assistance or independent electronic systems offering services for the sale and purchase of e-waste

generated from end-of-life electrical and electronic equipment between agencies or organisations authorised by these rules.

j) The manufacturer is now responsible for collecting e-waste created during the manufacturing of any electrical and electronic equipment and channelling it for recycling or disposal, as well as seeking authorisation from the SPCB.
k) If the dealer has been assigned the obligation of collection on behalf of the producer, he or she must collect the e-waste by supplying the customer with a box and then transfer it to the producer.
l) The sum shall be refunded to the depositor of e-waste by the dealer, retailer, or e-retailer in accordance with the producer's take back system or Deposit Refund Scheme.
m) Refurbishers must collect e-waste created during the refurbishment process and send it to an authorised dismantler or recycler via its collection centre, as well as get a one-time permit from SPCB.
n) The State Government's functions have also been incorporated into the Rules in order to safeguard the safety, health, and skill development of personnel participating in the dismantling and recycling processes.
o) The transportation of e-waste should be carried out in accordance with the manifest system, with the transporter obliged to carry a document (three copies) pre-pared by the sender outlining the information.
p) Liability for environmental or third-party damages caused by incorrect e-waste management has also been included, as has the potential for levying financial penalties for violations of the Rules.
q) Urban Local Bodies (Municipal Committee/Council/Corporation) have been tasked with collecting and directing orphan items to approved dismantlers or recyclers.

Amendment Rules, 2018

a) The extended producer responsibility (EPR) e-waste collection objectives have been changed. The phase-wise collection objectives for e-waste in weight must be 10 percent of the waste produced as mentioned in the EPR Plan in 2017-18, with a 10 percent rise per year through 2023. From 2023 onwards, the objective has been set at 70 percent of the amount of trash generated as specified in the EPR Plan.

b) For new manufacturers, separate e-waste collection objectives have been developed.
c) The government is responsible for the expense of sampling and testing under the Reduction of Hazardous Substances (RoHS) rules. If the product fails to meet RoHS requirements, the cost of the test will be paid by the manufacturers.

Plastic Waste Management Rules, 2016

Plastic waste is divided into two categories: thermoplastics and thermoset plastics. Thermoplastics account for around 80 percent of all post-consumer plastics waste created in India, with Thermoset accounting for approximately 20 percent. Polyethylene Terephthalate (PET), Low Density Poly Ethylene (LDPE), Poly Vinyl Chloride (PVC), High Density Poly Ethylene (HDPE), Polypropylene (PP), Polystyrene (PS), and other thermoplastics are recyclable plastics. Alkyd, epoxy, ester, melamine formaldehyde, phenol formaldehyde, silicon, urea formaldehyde, polyurethane, metalized, and multilayer plastics are examples of Thermoset polymers. According to reports, 15,000 tonnes of plastic garbage are created every day, of which 9,000 tonnes are collected and processed, while 6,000 tonnes are not collected. Plastic carry bags, in particular, are the most significant producers of littered garbage; every year, millions of plastic bags wind up in the environment via soil, water bodies, and water courses, and it takes an average of 1,000 years for them to degrade fully. In the absence of a dependable eco-friendly alternative capable of totally replacing plastic, the main issue is to enhance plastic waste management systems.

Salient features:

a) Increase the minimum thickness of plastic carry bags from 40 to 50 microns, and mandate a minimum thickness of 50 microns for plastic sheets as well, to enable plastic trash collection and recycling.
b) To encourage the use of plastic trash for road development as per Indian Road Congress rules, or for energy recovery, or waste to oil, or other profitable waste usage.
c) Since plastic has reached rural regions, these rules have been extended to include them. Gram Panchayat is assigned responsibility for enforcing the norms.
d) For the first time, trash producers' obligation is being established. Individual and bulk producers, such as offices, commercial

enterprises, and factories, are required to separate plastic garbage at the source, hand over segregated waste, and pay user fees in accordance with local body bylaws.

e) After public events (marriage festivities, religious gatherings, public meetings, etc.) organised in open locations, plastic items are strewn. For the first time, the organisers of such gatherings have been held accountable for the garbage created by these activities.
f) The use of plastic sheet for packaging and wrapping commodities is prohibited, unless the thickness of the plastic sheet impairs the operation of the product. A vast number of goods are packed/wrapped in plastic sheets, and these sheets are then littered. Provisions have been made to guarantee that they are collected and sent to authorised recycling facilities.
g) Producer Responsibilities Extend: Previously, EPR was left to the discretion of local governments. For the first time, producers and brand owners have been held accountable for collecting rubbish created by their products. They must engage local governments to develop a plan/system for plastic waste management within the time limit specified.
h) SPCBs will not grant/renew registration of plastic bags or multi-layered packaging unless the manufacturer submits an action plan that has been approved by the relevant State Development Department.
i) Producers must maintain track of the vendors to whom they have given raw materials for the production of carry bags, plastic sheets, and multi-layered packaging. This is to prevent certain items from being manufactured in the unorganised sector.
j) Retailers and street sellers are the primary entrance sites for plastic bags/plastic sheets/multi-layered packaging into the commodities supply chain. They have been tasked with avoiding supplying items in plastic bags/plastic sheets/multi-layered packaging that do not comply with these requirements. Otherwise, they must pay the fine.
k) Plastic carry bags will be supplied exclusively to shopkeepers/street sellers who have pre-registered with municipal bodies and paid a registration fee. The money collected as a registration fee by local governments will be utilised for trash management.
l) The CPCB has been tasked with developing thermoset plastic rules (plastic difficult to recycle). There was no particular provision for this sort of plastic in the previous Rules.
m) The production and usage of non-recyclable multi-layered plastic will be phased out during the next two years.

Amendment Rules, 2018

a) Multilayered Plastic (MLP) phasing out is now applicable to MLP that are non-recyclable, non-energy recoverable, or have no alternative application.
b) Central registration system for producer/importer/brand owner registration.
c) A nationwide registry has been mandated for producers having a presence in more than two states, whereas a state-level registration has been mandated for smaller producers/brand owners operating in one or two states.
d) Explicit pricing of carry bags has been excluded.

Amendment Rules, 2022

Classification of plastics

a) Category 1: This category will contain rigid plastic packaging.
b) Category 2: This category will contain flexible plastic packaging of a single layer or multilayer (more than one layer with different types of plastic), plastic sheets and coverings made of plastic sheet, carry bags, plastic sachets or pouches.
c) Category 3: This category will contain multi-layered plastic packaging (at least one layer of plastic and at least one layer of material other than plastic).
d) Category 4: This category includes plastic sheets or similar materials used for packing, as well as carry bags constructed of composite polymers.

Hazardous Waste Management Rules, 2016

Hazardous waste is defined as any waste that poses a risk to human health or the environment due to properties such as physical, chemical, biological, reactive, poisonous, combustible, explosive, or corrosive. It includes trash generated during the manufacturing processes of commercial items such as industries involved in petroleum refining, pharmaceutical manufacture, petroleum, paint, aluminium, and electronic devices, among others. According to CPCB data from 2015, the total hazardous waste produced in the country is 7.46 million metric tonnes per year from around 44,000 industries. For the first time, rules have been developed to differentiate between hazardous trash and other types of garbage. Other wastes include tyres, paper, metal scrap, used electronics, and so on, and are regarded as a resource for recycling and reuse. These resources augment industrial operations and lessen the country's reliance on virgin resources.

Salient features:

a) The Rules' scope has been broadened by the addition of 'Other Waste.'
b) Waste management hierarchy has been implemented in the order of priority of prevention, minimization, reuse, recycling, recovery, co-processing, and safe disposal.
c) All forms under the regulations for approval, import/export, filing of yearly returns, transportation, and so on have been drastically updated, demonstrating a severe strategy for managing such hazardous and other wastes while also simplifying procedure.
d) The essential requirement of infrastructure to protect the health and environment from the waste processing sector has been established as Standard Operating Procedures (SOPs), particular to waste type, which must be followed by stakeholders and enforced by SPCB/PCC when such licence is granted.
e) The procedure has been streamlined by combining all approvals into a single-window permission for the establishment of a hazardous waste disposal plant and the import of other wastes.
f) Co-processing has been given as a favoured alternative over disposal for the utilisation of waste as a supplemental resource or for energy recovery.
g) The licencing procedure for hazardous waste co-processing to recover energy has been simplified and is now based on emission criteria rather than trials.
h) The Rules' method for importing and exporting trash has been optimised by streamlining the document-based approach and amending the list of waste controlled for import/export.
i) The import of metal scrap, paper trash, and certain types of electrical and electronic equipment for re-use has been freed from the requirement of getting approval from the Ministry.
j) Standard Operating Procedures (SOPs) relevant to waste type have been specified as the basic need of infrastructure to protect the health and environment from the waste processing sector.
k) The following are the State Government's responsibilities for ecologically sound waste disposal of hazardous and other wastes:
1) To build-up/allocate industrial space or sheds for recycling, pre-processing, and other trash use.

2) Registration of personnel engaged in recycling, pre-processing, and other utilisation operations.
3) To organise worker groups to ease the establishment of such facilities.
4) To carry out industrial skill development activities while also ensuring worker safety and health.

l) A list of hazardous waste-generating processes has been evaluated in light of technological advancements in the sectors.
m) The list of waste constituents with concentration limits has been updated to reflect international and drinking water standards.
n) Import of the following items has been prohibited: waste edible fats and oils of animal or vegetable origin; household waste; critical care medical equipment; tyres for direct re-use; -Solid Plastic wastes including pet bottles; -Waste electrical and electronic assemblies scrap; other chemical wastes, particularly in solvent form.
o) The State Government is permitted to develop an integrated strategy for the successful execution of these rules and is required to submit an annual report to the Ministry of Environment, Forestry, and Climate Change.
p) The State Pollution Control Board (SPCB) is required to create an annual inventory of trash generated, waste recycled, recovered, used, including co-processed waste, waste re-exported, and garbage dumped and report it to the Central Pollution Control Board by September 30th of each year.

Ozone depleting substance rule

The rules are developed in accordance with the Environment (Protection) Act. These Rules provide dates for the phase-out of certain ODSs, in addition to regulating the manufacture, trading, import, and export of ODSs and products containing ODS. These Rules restrict the use of CFCs in the manufacture of different items after January 1, 2003, with the exception of metered-dose inhalers and other medical devices. Similarly, except for critical purposes, the use of halons is restricted after January 1, 2001. Other ODSs, such as carbon tetrachloride, methylchoroform, and CFC, can be used in metered-dose inhalers until January 1, 2010.

Furthermore, the usage of methyl bromide is permitted until January 1, 2015. Because HCFCs are utilised as intermediate alternatives for CFCs, they are permitted until January 1, 2040.

Wetland (Conservation and Management Rules) 2017

For effective conservation and management of wetlands throughout the country, the Wetlands (Conservation and Management) Rules 2017 supersede the Wetlands (Conservation and Management) Rules 2010.

Key features:

a) Wetland management should be decentralised. State governments have been given authority under the new guidelines to preserve and conserve wildlife at the local level. The federal government has mostly kept surveillance authority.
b) The Central Wetlands Regulatory Authority (CWRA) has been superseded by the National Wetland Committee, which serves only as an advisory body under the new laws.
c) The State or UT Wetlands Authority must compile a list of all wetlands as well as a detailed list of activities that must be controlled and authorised inside designated wetlands and their zones of impact.
d) Encroachments on wetlands, solid waste dumping, and discharge of untreated waste and effluents from industry and human settlements are also prohibited under the new guidelines.
e) It mandates that conservation and management be based on the principle of sensible usage, which will be enforced by the Wetlands Authority.

Inadequacy

We already read the Ramsar convention's definition of a wetland in this topic. The Ramsar convention was used to define the 2010 wetland guidelines. However, the 2017 guidelines, in the definition of wetland do not cover river channels, paddy fields, man-made water bodies/tanks primarily for drinking water reasons and buildings expressly created for aquaculture, salt production, leisure, and irrigation purposes. According to this revised definition (excluding the aforementioned wetlands), about 65 percent of the country's wetlands will lose their classification. Because they are not deemed wetlands, the management and protection granted to river channels and man-made wetlands will be ineffective. Wetlands are defined and included by the relevant state or territory. This may result in a lack of consistency in designating and protecting the country's wetlands. The former provision of appealing to the National Green Tribunal is no longer in effect under the 2017 Rules. There is also no timetable for eliminating the disposal of solid waste and untreated garbage into wetlands. The new laws do not take into consideration the Supreme Court's decision in Jagpal Singh vs. State of Punjab (2011) for the restoration of encroached wetlands throughout the country.

National Wetland Committee

A National Wetland Committee (NWC) will be formed for the following functions:

a) Keeping track of how regulations are being followed
b) Advising the national government on suitable policies and action plans for wetlands protection and responsible usage.
c) Recommending the Ramsar Convention designation of Wetlands of International Importance.
d) Advice on working with international agencies on wetlands concerns.
e) Making a digital inventory: All state agencies are required to compile a list of all wetlands. Based on this, a wetlands digital inventory will be produced and updated every 10 years.
f) Prohibited activities include the discharge of undesired trash from villages, towns, cities, and industry, as well as solid waste disposal into wetlands. Conversion of wetland area for non-wetland activities is prohibited, as is the erection of a permanent structure on registered wetlands.

CHAPTER 31

MAJOR ENVIRONMENTAL MOVEMENTS IN INDIA AND GREEN WARRIORS

Genesis of Global Environment Movement

There was no obvious beginning to the global environmental movement. There were no significant incidents that provoked widespread indignation, no unique leaders that inspired a widespread movement, and no abrupt shifts in human understanding. Rather, the global movement arose from a succession of autonomous responses to local concerns in many locations and times. Since the mid-nineteenth century, individuals have formed local groups in response to various concerns for a variety of reasons. Local groups gradually evolved into national movements, which merged after World War II (1939-45) to form a multi-national movement, and have since evolved into a vast complicated and diverse worldwide movement. The origins of the environmental movement, however, were found in the technological, economic, and social developments brought about by the industrial revolution in Europe and North America, which included enhanced scientific understanding, more personal mobility, and the extension of human settlement. In the early twentieth century, an organised effort was launched to share information across national lines for the protection of nature and natural resources, as well as for the preservation movement. By the end of the nineteenth century, attempts were being made to foster more international cooperation, notably in the subject of "Wildlife Protection." The primary focus of emphasis was on the conservation of birds and animals. In 1909, European environmentalists gathered in Paris for the International Congress for the Protection of Nature. After reviewing Europe's performance on nature conservation, they urged for a global response and recommended the formation of an international nature protection authority. The work of various conservationist groups, including the Society for the Preservation of the Wild Fauna of the Empire (SPWFE), the International Office for the Protection of Nature (IOPN), and the International Committee for Bird Protection (ICBO), as well as the signing of the 1900 and 1933 African conventions, as well as the Western Hemisphere convention, sowed the seeds of an interest in international cooperation.

Later, the Second World War altered people's views on internationalism. However, since the UN involvement, the conservation movement has gotten the attention it deserves. The establishment of the International Union for Conservation of Nature (IUCN), the World Wildlife Fund (WWF), and the International Union for the Protection of Nature (IUPN) are all milestones in the right direction.

The Chipko Movement

The Chipko movement, also known as the Chipko Andolan, was an Indian forest conservation mission. It established a precedent for nonviolent protest in India. It began in 1973 in Uttarakhand, then a part of Uttar Pradesh (in the foothills of the Himalayas), and went on to become a rallying point for many subsequent environmental campaigns worldwide. Though the campaign began by educating locals about afforestation, it was eventually able to include people in larger problems of ecology and environmental preservation. The Chipko movement was also instrumental in establishing a smokeless chulah and gobar gas factory in the area. Chandi Prasad Bhatt and Sundarlal Bahuguna were the men behind the effective organisation of this type of movement. The organisers of the movement now not only take after the environmental problems of the entire Garwal Himalayas, but also maintain contact with other such initiatives in the country and abroad.

Significance of the Movement

a) Gaura Devi, Suraksha Devi, Sudesha Devi, Bachni Devi, Chandi Prasad Bhatt, Virushka Devi, and others were key figures in the Satyagraha for Environment movement, which used Satyagraha tactics.
b) Sunderlal Bahuguna gave the movement a proper direction, and its success meant that the world immediately took notice of this nonviolent movement, which was to inspire many similar eco-groups in time by-helping to slow down rapid deforestation, exposing vested interests, increasing social awareness and the need to save trees, increasing ecological awareness, and demonstrating the viability of people power.
c) The majority of the movement's supporters were women. Beyond the eco-socialism tint, it is increasingly being viewed as an eco-

feminist movement. Women were not just its backbone, but also its mainstay, despite the fact that many of its leaders were males.

d) In the Western Ghats region, it was a major inspiration for the big Appiko forest-saving campaign, and it attracted support for similar environmental-protection groups.

Silent Valley Movement

The silent valley is located in Kerala's Palghat district. It is bordered by many hills in the state. The British proposed building a dam on the Kunthipuzha River in this hill range in 1929. The technical feasibility study was completed in 1958, and the project was approved by the Government of India's Planning Commission in 1973. In 1978, the idea was met with widespread opposition from all segments of the public. The initiative was started by locals and was later taken over by the Kerala Sastra Sahitya Parishad (KSSP). Many environmental organisations, including the Narmada BachaoAndolan (NBA), the Bombay Natural History Society (BNHS), and the Silent Valley Action Forum, took part in the campaign.

Appiko Movement

It is a movement inspired by the Chipko movement by the villagers of Western Ghats. During the months of September and November 1983 in the Uttar Kannada district of Karnataka, people from the Western Ghats launched the AppikoChalewali movement. The commercial harvesting of trees for wood extraction triggered forest devastation in this area. Contractors cut the region's natural woods, causing soil erosion and the drying up of perennial water supplies. Forest dwellers in Sirsi'sSaklani village were prohibited from gathering usufructs such as twigs and dried branches, as well as non-timber forest products, for the purposes of fuelwood, fodder, honey, and so on. They were denied their traditional rights to these goods. The Appiko Movement has three main objectives:

a) Ulisu (to conserve)
b) Belsu (to grow)
c) Balasu (to make rational use of)

Narmada BachaoAndolan

The Narmada River is the largest west-flowing river in the Indian peninsula, rising on the Amarkantak plateau in Madhya Pradesh's Shahdol district. It includes Madhya Pradesh, Maharashtra, Gujarat, and parts of Rajasthan. It travels 1312 kilometres to the Arabian Sea, passing through beautiful woods, hills, agricultural districts, and steep canyons. It is encircled by three mountain ranges, Satpura, Vindhya, and Maikal, and merges into the Arabian Sea on the fourth side. The villages make up 81 percent of its basin and are mostly populated by tribal people such as Bhils, Gonds, Baigas, and others whose principal employment is agriculture.

The Narmada BachaoAndolan is the strongest popular campaign opposing the construction of a massive dam on the Narmada river, which began in 1985. The Narmada Dam Project called for the construction of approximately 3000 large and minor dams along the river. More than 250,000 people were expected to be displaced by the projected Sardar Sarovar Dam and Narmada Sagar. The big fight of the Save the Narmada Movement was over the resettlement or the rehabilitation of these people.

Key points

a) Following independence, India's first Prime Minister, Jawaharlal Nehru, advocated for the construction of dams on the Narmada River, which flows through Madhya Pradesh and Gujarat, to stop surplus water from flowing into the Arabian Sea and benefit local people and national development.
b) Sardar Sarovar and Narmada Sagar were two of the major projected dams.
c) The Narmada Water Disputes Tribunal authorised the Narmada Valley Development Project, which involved the construction of 30 big dams, 135 medium dams, and 3,000 minor dams, as well as boosting the height of the Sardar Sarovar dam.
d) Medha Patkar and her colleagues visited the Narmada Dam Project site in 1985 after hearing about it. They discovered that project progress was being reviewed owing to an order from the Ministry of Environment and Forests, GOI.
e) Construction on the Sardar Sarovar Dam began in 1987, and the people who would be affected by the dam's construction were given little information other than a promise of rehabilitation.

f) The Narmada BachaoAndolan conducted a 2,000-person, five-day sit-in at Prime Minister V. P. Singh's house in New Delhi in May 1990, which prompted the Prime Minister to'reconsider' the project.
g) Approximately 6000 men and women began the Narmada Jan Vikas Sangharsh Yatra (Narmada People's Progress Struggle March) in December 1990, marching over 100 kilometres.
h) Baba Amte and the seven-member team started an indefinite hunger strike (which lasted 22 days) and a sit-in till death in January 1991.
i) The building of the Sardar Sarovar Dam resumed in 1999 and was completed in 2006. The project's height was increased from 138 to 163 metres. PM Narendra Modi inaugurated it in 2017.

Tehri Dam Conflict

The Tehri Dam is India's highest dam and one of the highest in the world. It is a rock and earth-fill embankment dam on the Bhagirathi River in Tehri, Uttarakhand, India. It serves as the main dam for THDC India Ltd. and the Tehri hydropower complex. The Tehri Dam, which was completed in 2006, has a reservoir for agriculture, municipal water supply, and the generation of 1,000 megawatts of hydroelectricity. The dam's 1,000 MW variable-speed pumped-storage plant is presently under construction, with a May 2018 commissioning date. Environmental organisations and local residents in the region have protested the Tehri Dam. From the 1980s through 2004, environmental activist Sunderlal Bahuguna spearheaded the Anti-Tehri Dam movement. The demonstration was against the eviction of town residents and the environmental consequences of the weakened ecology. The Tehri Dam is built in the highlands of Uttarakhand, India, above the Bhagirathi River, the primary tributary of the Ganges. The Tehri Hydro Development Corporation (THDC) was established in 1988 to oversee the dam.

Environmentalists and residents of Tehri and adjacent regions have been outspoken in their opposition to the project. More than 1 lakh persons have been relocated, resulting in court challenges. Concerns have also been expressed about the dam's location in the Central Himalayan Seismic Gap, a major geologic fault zone (this region was the site of a major earthquake in October 1991). Sunderlal Bahuguna is one of the project's opponents. Despite the complaints, the proposal was approved by the court.

Jungle BachaoAndolan

The Jungle BachaoAndolan began in the 1980s in Bihar's Singhbhum district (presently in Jharkhand). It was a protest against the government's plans to replace native Sal forests with commercial teak plantations. This judgement has the greatest impact on the indigenous community since it violates the rights and livelihood of Adivasis in that region. In various versions, this movement was extensively disseminated in places such as Bihar, Jharkhand, and Odisha.

Bishnoi Movement

The Rajasthani Bishnoi movement has always been a cohesive and well-articulated manifestation of ecological philosophy. The Bishnoi sect (also known as Vishnoi) was a religious order in India formed in 1485 by Guru Jambeshwar Ji Maharaj. Its origins may be traced back to the Indian state of Rajasthan. The Bishnois are an Indian religious sect dedicated to the protection of the environment, particularly animals. Guru Jambeshwar formed the group, which had 29 tenets known as the 29 principles or commandments provided by Guru Jambeshwar in 1485 AD. The subsistence economy of the Bishnoi is focused on cattle and crops.

Amrita Devi Bishnoi is the leader of the Bishnoi movement. The movement was formed in reaction to a huge forestry campaign put out by the kings of Marwar in 1485, following a seven-year drought. Amrita Devi, a Bishnoi lady, organised a protest of 363 people against the destruction of Khejri trees in order to build a palace for the Maharaja of Marwar. Amrita Devi and her companions were mercilessly slaughtered by Maharaja Abhay Singh's army, yet their sacrifice moved him to order that no further trees be chopped down. The campaign was started to safeguard Khejri as well as other species such as antelope, deer, and wildfowl. In 1730, several Bishnois were slain while attempting to preserve their trees from being chopped down by soldiers headed by Jodhpur's Abhay Singh. With its forestry initiative, the Bishnoi movement has created a magnificent environment that refills itself and supports species. While not as well-known as other environmental groups, the Bishnoi movement promotes a form of sustainability that reflects many of today's most pressing environmental challenges. It teaches everyone about the power of deliberate action.

Green Warriors of India

- In 1787, Amrita Devi sacrificed her life along with 363 others to conserve khejri trees (Prosopis cinenaria) in Rajasthan.
- Anil Agrawal is a journalist, thinker, and educator who specialises on environmental issues. Anil K. Gupta is an environmentalist. Pew Award winner in 1994.
- Ashish Kothari is a journalist, environmental activist, and educator.

- Chipko activist Chandi Prasad Bhatt campaigns against Himalayan hydroelectric projects.
- Gaura Devi inspired women to embrace trees in protest of their axing, launching the Chipko movement.
- Harivallabh Parikh – Gandhian social worker battling for the cause of forest degradation and tribal uprooting and relocation in Gujarat.
- Kishore Saint is a supporter of the "Save Aravalli Movement."
- Annaseheb Hazare, Krishna Bhaurao– In 1970, a driver in the Indian Army inspired the residents of "RaleganSidhi" in the drought-prone region of Ahmednager to build clay dams for soil and water conservation, and he performed a miracle by transforming the destitute hamlet into a thriving one.
- M. K. Prasad is an environmentalist. In the 1970s, he led the "Silent Valley Movement" in Kerala's Western Ghats to prevent the construction of a planned hydroelectric project.
- M.S. Swaminathan is an Indian conservationist and pioneer of sustainable agriculture. In 1995, he received the UNEP Sasakawa Award.
- Mahesh Chandra Mehta is India's environmental legislation champion. He has won many Supreme Court issues involving public interest. He received the Magsaysay Award in 1997.
- Megha Patekar is an environmental activist who is working for the cause of forest submergence, tribal uprooting, and relocation caused by the Narmada Sagar Dam.
- Meneka Gandhi was a staunch supporter of vegetarianism and nonviolence toward animals.
- N. D. Jayal is a thinker and environmental educator.
- Pandurang Hegde is an Appiko activist who is leading a campaign in South India to stop environmental damage.
- Priya Ratna Misra organized and persuaded the residents of "Sukhomajri" in the Sub-Himalayan Shivalik hills to build an

earthen dam, minimise overgrazing, and safeguard watershed regions.

- R. K. Pachauri is an environmental educator and activist.
- Rajendra Singh is an environmentalist. Led to the cessation of marble mining near Alwar's Sariska Tiger Reserve (Rajasthan).
- Rameshwr Das is a Gandhina activist who organises and motivates the villagers of "Seed village" in Rajasthan's Aravalli desert to control their own environment through village courts.
- Salim Ali is a well-known omithologist and conservationist.
- Sant Kumar Bishnoi is a dedicated naturalist who inspires people to protect Haryana's flora and animals.
- ShamisthaJagawat is a tribal (Bhil) lady who is spearheading the Gujarat eco-development movement.
- ShivramKaranth is an environmental activist and teacher. Winner of the Indira Gandhi Environmental Award in 1992.
- Sunderlal Bahuguna is a Chipko activist and Gandhian who is leading an anti-environmental campaign in India.
- T. N. Khoshoo is an eminent environmentalist who received the Indira Gandhi Environmental Award in 1993.
- T. Shivaji Rao is a civil engineer. To minimise ecological damage, the height of the main Sardar Sarovar dam should be reduced by 100 feet.
- V. M. Meher Homi is an environmentalist.
- Vandana Shiva is a journalist, thinker, and educator who focuses on environmental issues.
- Vasant Kanibera organises and motivates people in Andhra Pradesh to work for the regeneration and reclamation of damaged areas via traditional farming.
- Veer Bhadra Mishra is an environmental campaigner and activist who has been fighting Ganga pollution in Varanasi since 1992. He was honoured at the Rio Earth Summit in Brazil in 1992.
- Vinayak Patil is an environmental campaigner who inspired people in Maharashtra villages to build earthen dams, retain rainwater, and avoid soil erosion.
- VizayParanjpye is a thinker and environmental educator.
- Pavan Sukhdev is an Indian environmental economist who researches green economics and international finance.
- Jadav "Molai" Payeng, also known as the Forest Man of India, is an environmental activist and forestry worker from Majuli. Over

several decades, he has planted and cared for trees on a sandbar of the Brahmaputra River, transforming it into a forest reserve.

- Anuj Ramatriis the founder of Eco Freak, whose mission to share stories of Real Life Heroes who are working for the betterment of the environment.
- AbhirajRajadhyaksha is an environmental videographer, thinker and educator.
- Niyati Mavinakurve is an environmental videographer, thinker and educator.

CHAPTER 32

NATURAL HAZARDS AND DISASTER MANAGEMENT

Introduction

Natural catastrophes are causing more frequent and severe losses throughout the planet. Natural disasters have quadrupled in frequency over the last four decades, with inflation-adjusted economic losses multiplying by a ratio of nine and insurance losses multiplying by a factor of 15. Increased urbanization, settlement in and industrialization of extremely vulnerable places, susceptibility of contemporary technologies, and anthropogenic changes in the environment are the primary reasons. Estimates of future loss potentials, which for a number of credible disaster scenarios reach magnitude not seen before in history, are particularly alarming.

A natural hazard is a naturally occurring phenomenon that has the potential to harm humans and other creatures, as well as the environment. The difference between a natural hazard and a disaster may be shown by the fact that an earthquake was the hazard that triggered the 2001 Bhuj earthquake tragedy, while disasters are severe disturbances to a community's functioning that surpass the community's ability to manage using its own resources. Natural, man-made, and technical risks, as well as many elements that impact a community's exposure and vulnerability, can all contribute to disasters. Anthropogenic activities, such as land-use change, drainage, and development, can cause or alter natural hazards.

Nature of Hazards

Natural Hazards appear to happen all too often. Thousands of people are killed every year as a result of tropical cyclones, floods, earthquakes, volcanic eruptions, droughts, and other natural disasters. Settlements are destroyed or severely damaged, agricultural soils are washed away, and natural ecosystems are deteriorated. Those that survive are vulnerable to sickness, hardship, and loss, which can take years to recover from. The various types of hazards are explained below:

a) Hydro Meteorological Hazards: Hydro meteorological hazards are atmospheric, hydrological, or oceanic processes or phenomena that can result in death, injury, or other health consequences, property damage, loss of livelihoods and services, social and economic disruption, or environmental degradation.

1) Tropical cyclone: Tropical cyclones are storms that develop over a warm tropical ocean. These cyclones are distinguished by low air pressure, strong winds, and heavy rainfall. Tropical cyclones are striking the Gulf Coast of North America, northern Australia, eastern India and Bangladesh, among other places, in the Atlantic region, the Pacific Ocean, and the Indian Ocean. Cyclones strike India during the North Indian Ocean Cyclone Season, which lasts from April to November. The length of the Indian coastline is approximately 7,516 km, with 5,770 km of it vulnerable to natural disasters such as cyclones. The east coast of India is more vulnerable to cyclones than the west. Human activity interacts with dangers to cause vulnerability. Most coastal and island locations across the world are prone to tropical cyclones, with a few exceptions. Tropical cyclone vulnerability can be absolute or relative. Location determines absolute vulnerability. For example, because they are in the path of such meteorological occurrences, islands in the western Caribbean are vulnerable to hurricanes. Exposure to consequential risks results from the combination of local characteristics such as topographic relief with tropical cyclone components such as strong winds or storm surge. Flash floods, mudslides, and coastal erosion are examples of significant risks.

2) Floods: Floods are the most common sort of natural catastrophe, occurring when an excess of water submerges normally dry ground. Floods are frequently induced in coastal regions by heavy rainfall, quick snowmelt, or a storm surge from a tropical cyclone or tsunami. Flood disasters are among the most common and destructive types of catastropheson the globe. Despite some remarkable attempts and successes to avoid them, they are getting more frequent and hazardous over time. Although Science and medical technology have improved our capacity to forecast and withstand floods, population expansion, social, economic, and political dynamics have increased society's exposure and vulnerability to these disasters, making

them almost endemic. Unfortunately, flood tragedies are frequently underestimated. Severe floods are uncommon, making it difficult for communities to keep a cultural memory of the hazards that might guide future generations. Human-caused environmental changes, such as urbanization and greenhouse gas emissions that cause global warming, have insidious and misleading effects on the potential of floods.

India is located in South Asia's monsoon climatic zone. The monsoon circulation causes around 80–90 percent of the yearly rain in most regions of the nation to fall during the summer monsoon season, which lasts from June to September. During this time, cyclonic disturbances from the Bay of Bengal and the Arabian Sea cause widespread and heavy rain, resulting in serious flooding in Indian rivers. Punjab, Haryana, the majority of the Gangetic plains, including Uttar Pradesh, North Bihar, and West Bengal, the Brahmaputra valley, coastal Andhra Pradesh and Orissa, and southern Gujarat, are the most flood-prone regions in India. Kerala and Tamil Nadu are currently experiencing the full force of the floods. Two strong floods reported in India, the flood of 6 September 1970 on the Narmada River and the flood of 11 August 1979 on the Machhu River, were discovered to be world records. Floods are the worst natural catastrophes in India. India endured 235 floods between 1980 and 2017, resulting in 126,286 deaths and affecting 1.93 billion people.

3) Droughts: Drought hazard is the total of the geophysical elements that cause a lack of moisture, such as rainfall, temperature, humidity, radiation, soils, vegetation, and land cover. Drought, and the repercussions of malnutrition and famine, is one of the most serious natural disasters affecting people all over the world. According to the International Federation of Red Cross and Red Crescent Societies (IFRCRCS, 1993), based on data from the Centre for Research on the Epidemiology of Disasters (CRED), drought killed over 73,000 people, affected 58 million people, and resulted in economic losses of nearly US$ 4 billion from 1969 to 1993. Drought, one of India's most common climatic phenomena, is caused by a variety of factors. Drought in India is caused by four key factors: monsoon delay/failure, monsoon variability, a long gap in monsoon, and regional differences in monsoon persistence. From 1870 to 2018, India experienced 18

meteorological and 16 hydrological droughts. The five worst hydrological droughts happened in 1876, 1899, 1918, 1965, and 2000, whereas the five worst meteorological droughts occurred in 1876, 1899, 1918, 1965, and 2000. Interior Orissa, western Jharkhand, Chattisgarh, Madhya Pradesh, Rajasthan, Punjab, Haryana, Gujarat's Kathiawar and Kachchh, Maharashtra's Vidharbha and Marathwada, northern and eastern Karnataka, and western Andhra Pradesh are the most impacted. Droughts are becoming more linked to a mismatch between supply and demand, particularly demand for non-agricultural purposes. While proper water supply is critical for agriculture, it is also influenced by other factors such as temperature, humidity, sun radiation, and wind patterns. The different types of droughts are explained below:

i. Meteorological drought: When there is a prolonged period of insufficient rainfall with uneven distribution over time and space. Meteorological drought occurs when rainfall falls below 90 percent of the average.
ii. Agricultural drought: It is distinguished by low soil moisture, which is required to support crops, causing in crop failures.
iii. Hydrological drought: When the availability of water in various storage and reservoirs, such as aquifers, lakes, reservoirs, and so on, decreases beyond what precipitation can refill.
iv. Ecological drought: When the production of a natural ecosystem fails owing to a lack of water and as a result of ecological distress, the ecosystem suffers.

4) Extreme Temperature: The entire impact of temperature extremes has yet to be thoroughly documented and understood. Temperature impacts are frequently documented in conjunction with other meteorological occurrences and on climatological time and spatial scales. Seasonal effects are more cumulative and complicated than those caused by cold spells or heat waves. Nonetheless, the consequences are quantifiable. In order to offer relevant, intelligible, and timely information for nations and towns to prevent natural catastrophes, weather forecasting must take into account the dangers and impacts of temperature extremes. Since the beginning of the summer season in March, severe heatwaves have been observed in wide regions of India. On May 20, 2022, temperatures in several parts of Delhi and

Uttar Pradesh reached over 50 degrees Celsius, a day after Jacobabad, Pakistan, hit 51 degrees. As a climate change-induced heatwave continued to sweep broad regions of northern and central India, Delhi recorded its highest temperature since records started. The previous record for the greatest temperature in Delhi was 45.6 degrees Celsius on April 29, 1941. Heat waves are characterized by the Indian Meteorological Department (IMD) as follows: Heat waves should not be included until a station's maximum temperature reaches at least 40 degrees Celsius for plains and at least 30 degrees Celsius for hilly areas.

5) Lightning: Lightning is a random, unexpected, and capricious phenomenon. It has the following physical characteristics: an average peak current of 30 kA, with a time to peak from zero of several microseconds; temperatures of 30,000 Kelvin; voltages of 100,000,000 Volts; and the generation of an auditory shock wave (thunder). Approximately 100 lightning strikes strike the planet per second due to the 2,000 active thunderstorms throughout the world. In the United States, lightning causes more than 26,000 home fires each year, resulting in property losses of more than $5 billion (NLSI estimates). After floods, it is the second most common cause of weather-related fatalities in the United States. Lightning rods have been used to transmit stray voltages away from structures to the ground since the 1700s. According to form, height, and other parameters, lightning rods, now known as air terminals, are thought to discharge streamers upward at varied lengths and periods. Different types of lightning protection systems (LPS) can be used depending on the level of protection required. Because of the monsoon season, India had the maximum concentration of lightning in May, June, and September of 2020. Although the average rainfall in India in 2020 was 109 percent (only 1 percent less than the 110 percent average in 2019), lightning increased by 22.6 percent compared to 2019. According to government estimates, over 2,500 people die each year in India as a result of lightning strikes, compared to only 45 in the United States. According to a report by the non-profit Climate Resilient Observing Systems Promotion Council, India had over 18 million lightning strikes between April 2020 and March 2021. This is a 34 percent increase over the same time the previous year. Brazil was the country with the most lightning

strikes in 2021, with a total of 225,607,575 strikes, including both in-cloud and cloud-to-ground lightning.

b) Geological Hazards: A geologic hazard, sometimes known as a geohazard, is a dangerous geologic situation that can result in extensive property and lives loss. These dangers are caused by geological and environmental circumstances, and they can occur over long or short periods of time.

1) Earthquake: The accumulation of stress in the Earth as a result of long-term movements of its outer layers, driven by heat energy in its interior, is the underlying cause of earthquakes. When the stress builds up to a certain level, it is released by a quick release of energy from a large volume of rock, resulting in an earthquake. The earthquake is frequently accompanied by rupture, which may be recognised as a geological fault if it reaches the surface. For the largest earthquakes, the rupture can stretch up to 100 kilometres or more from the place of origin, known as the focus or hypocentre. Elastic waves propagate outward from the source, hitting the surface first at the epicentre, which lies directly above the focus. These waves can be recorded using instruments or felt as tremors. Intensity scales are used to characterise the impacts of earthquake shaking, with a maximum of XII describing the most severe damage. The magnitude of an earthquake is determined by seismograph recordings and refers to the total energy produced at the source. Plate tectonics explain a lot about how earthquakes happen all across the planet. The Earth's outer layers are split into many huge plates, each of which moves in reaction to underlying forces. Earthquakes are common in countries near plate boundaries. Most older structures have previously been shaken by earthquakes, and only the most vulnerable will have survived. Deep earthquakes are common in these places, and their impacts can be felt throughout a large area without being particularly intense in any one location. The largest earthquake in India occurred on August 15, 1950, in the India-China area, with a Richter magnitude of 8.6. 1,530 people died as tectonic plates shifted at a depth of 30 kilometres. The earthquake also triggered a tsunami, which resulted in further deaths and property damage. The year 2021 was the most seismically active since 2007, with 19 large earthquakes, three of which were over 8.0 magnitude. The earthquake design code of India splits India into four seismic zones, according to the most recent

edition of the seismic zoning map of India. Jammu and Kashmir, Gujarat, and Assam are the most earthquake-prone states in India, according to the report. These are the three states with the highest risk of earthquakes. A list of five major earthquakes that have affected India in the past is provided below in table 32.1:

Table 32.1: Major Earthquakes in India

Place	Magnitude	Date
Bihar	8.1	January 15, 1932
Gujarat	7.7	January 26, 2001
Maharashtra	6.4	September 30, 1993
Assam	8.6	August 15, 1950
Uttarkashi	6.1	October 20, 1991

2) Volcanoes: A volcano is a fracture in a planetary-mass object's crust that permits hot lava, volcanic ash, and gases to escape from a magma chamber under the surface, such as Earth. One of the most significant characteristics of volcanic eruptions to recognise is that a variety of distinct hazards can occur simultaneously or sequentially. Moderate earthquakes may do considerable damage around the vent before an eruption occurs. As magma flows closer to the surface, the earth may be deformed. Landslides can be caused by earthquakes or ground deformation. When an eruption begins, large slabs of rock can be expelled explosively from the vent, while fast-moving, hot flows of ash, rock, and gas travel laterally from the vent, and ash columns rise into the sky and drift downwind. A lava flow from the vent is also possible. Toxic gas and aerosol clouds can arise. Areas of the volcanic cone may collapse during large eruptions, resulting in rock avalanches. If they reach the sea, a tsunami might occur. Rainfall can move the ash that covers the hillslopes following an eruption, causing mudflows down drainage channels and across low-lying terrain. The Ring of Fire, also known as the Circum-Pacific Belt, is a belt of active volcanoes and frequent earthquakes that runs along the Pacific Ocean. The Ring of Fire is home to the bulk of the world's volcanoes and earthquakes. Indonesia boasts the world's largest number of active volcanoes and is one of the Pacific Ring of Fire's locations. This is a 25,000-mile-long (40,000-kilometer-

long) horseshoe-shaped area on the Pacific Ocean's western edge, where many tectonic plates collide. India has six volcanoes, including one that is still active, the Barren Islands volcano, which is also the only confirmed active volcano in South Asia.

3) Tsunami: Tsunamis are a sequence of ocean waves caused by enormous, sudden surface disturbances such as earthquakes, volcanic eruptions, landslides, slumps, and meteor strikes. Within minutes of their origin, these waves may envelop a coastal town, causing loss of life, catastrophic damage to houses and infrastructure, and significant erosion of the shoreline due to hours of repeated attacks by waves that are several minutes apart. People are swept along with other debris in tsunami-induced currents at speeds of up to 60 kilometres per hour, drowning as a consequence of various injuries such as shattered bones, lacerations, abrasions, punctures, and crushed bodily cavities. Over 120,000 people have died as a result of tsunamis in the Pacific since 1850. Tsunamis pose a severe threat to anyone living along the shore in earthquake-prone areas. Tsunamis may cause devastation throughout whole ocean basins, despite the fact that 99 percent of tsunami deaths occur near the producing site. Tsunamis that originated in Chile killed people in Hawaii and Japan just hours later. Tsunamis that inflict damage to remote shorelines do not have an earthquake indication, but the waves produce abrupt changes in sea level and a tremendous roar as they reach the shoreline. Roughly five tsunamis have occurred each year in the last 100 years, with one of these causing death and devastation. 82 tsunamis were reported in the 1990s, with eleven inflicting 4,601 fatalities and more than $1 billion in damage. Surprisingly, this decade had two of the twelve most catastrophic Pacific tsunamis since 1850. On December 26, 2004, the world's worst tsunami struck India, killing over 230,000 people in 14 nations, the majority of them were in Indonesia. The tsunami was caused by a 9.1 magnitude earthquake that struck off the west coast of Sumatra, Indonesia. This was one of the strongest Richter Scale ratings ever recorded in modern history. Thailand, India, Sri Lanka, and Sumatra have all experienced waves of up to 15 metres. The waves pounded the Andaman and Nicobar islands minutes after the earthquake. Tamil Nadu and Andhra Pradesh were the states in mainland India that were most hit. The tsunami

wreaked havoc on Chennai, one of the world's most populous cities. In India alone, about 18,000 people have died. Surprisingly, the tsunami was able to travel 5,000 kilometres to Africa and wreak casualties and property damage. Because of the global migration of people to susceptible coastal locations, tsunami casualties are likely to continue to rise.

4) Landslide: A landslide occurs when a mass of rock, debris, or ground slides down a hill. Landslides are a sort of "mass wasting," which is defined as any downward movement of soil and rock caused by gravity. Landslides can be thought of as an environmental or geomorphic result of dynamic terrain evolution processes. They reflect an ongoing effort to strike a compromise between the strength of the geological materials used to dig slopes and the gravitational forces or loads that cause mass instability. Human activity affects these natural geomorphic processes linked with landscape evolution, resulting in hazardous situations as a result of human development. The interaction between natural processes and human actions causes the possibility of disaster. As a result, the vulnerability of the items at risk will impact the risk assessment. The degree of development of the landslide morphology and the activity of the slope movement, together with the volume of the landslide, determine the landslide's power.

The Himalayas, the world's tallest mountain range, were formed by the collision of the Indian and Eurasian plates. The Indian plate's northward advance towards China generates ongoing stress on the rocks, making them friable, unstable, and prone to landslides and earthquakes. The Malpa landslide was one of India's deadliest natural disasters. Massive landslides in Uttarakhand's Pithoragarh district, then in Uttar Pradesh's Kali Valley in the Higher Kumaon region of the Himalayas, swept out the whole village of Malpa around 3 a.m. on August 18, 1998. Aside from the Himalayas, the north-eastern hill ranges, the Western Ghats, the Nilgiris, the Eastern Ghats, and the Vindhyans, which cover roughly 15 percent of the continent, landslides and avalanches are among the primary hydro-geological hazards that threaten huge sections of India.

5) Glacial: A glacier or glacier-related feature is regarded a hazard when it negatively impacts human operations, either directly or indirectly. Over 32,000 individuals were killed by glacially-related phenomena throughout the twentieth century. Hundreds

of millions of dollars have been lost, and critical infrastructure has been damaged or destroyed. Hydropower extraction is rising in the Himalayas and the Andes, putting further strain on rural development. Glacial risks, on the other hand, are most common in these high alpine places. There are thousands of glaciers in these mountains, ranging in size from less than one square kilometre to tens of square kilometres. Lakes typically occur behind moraine dams produced by the glaciers' earlier advance when the temperature changes and the glaciers recede. As a result of one or more breach processes, such as melting ice cores or seismic shaking, these dams are vulnerable to rapid and catastrophic breakdown. Many millions of tonnes of water and debris can be spilled downstream if a dam fails, sometimes at speeds of up to 80 kilometres per hour. The ensuing Glacial Lake Outburst Floods downpour tears up vegetation and rock debris, as well as defoliates trees, whose trunks become gigantic battering hammers. The combination of water, wood debris, and mud is devastating, both in terms of immediate harm and the environmental devastation left behind after the floodwave has gone. On February 7, 2021, a chunk of the Nanda Devi glacier broke off, unleashing water held beneath the ice and causing an avalanche and downpour that swiftly evolved into flash floods in Uttarakhand's Chamoli district.

c) Environmental and Technological Hazards: An environmental hazard is a form of danger caused by chemical, biological, or physical substances emitted by current or historical human activity. Technological risks are caused by technological or industrial circumstances, risky operations, infrastructural failure, or human behaviour. Chemical, biological, radiological, and nuclear (CBRN) threats are all examples of technical threats.

1) Wildfire: Wildfires affect thousands of million hectares of forest and other ventures every year. Fire has an ecologically important function in biogeochemical cycles and disturbance dynamics in some ecosystems. In some ecosystems, fires destroy forests or cause long-term site damage. Wildfires burning in harsh weather circumstances have a negative impact on economies, human health, and safety in most parts of the world, with repercussions equivalent to the severity of other natural hazards/disasters. Forest and other vegetation fires emit gaseous and particle emissions that affect the composition and function of the global atmosphere. These emissions interact

with those emitted by fossil-fuel combustion and other technical sources, which constitute the primary driver of anthropogenic climate forcing. Smoke emissions from wildfires can reduce visibility, which can lead to accidents and economic losses. Smoke from wildfires affects human health and, in certain situations, results in the loss of human life. Geologically, wildfires originated shortly after the advent of terrestrial plants roughly 419 million years ago during the Silurian time, indicated by fossil charcoal. The 1871 Peshtigo Fire was the worst wildfire in recorded history. On October 8, 1871, the whole Great Lakes area of the United States was hit by a massive inferno that extended over the U.S. states of Wisconsin, Michigan, and Illinois. However, no place was affected harder than Peshtigo. The little town in northern Wisconsin saw at least 1,500 deaths, accounting for around 85 percent of the town's population at the time. According to forest inventory statistics, 54.40 percent of India's forests are subjected to occasional fires, 7.49 percent to moderately frequent fires, and 2.40 percent to high incidence levels, while 35.71 percent of India's forests have yet to be exposed to significant fires.

2) Bhopal gas tragedy: On the night of December 2, 1984, the chemical methyl isocyanate (MIC), a colourless liquid used in the production of pesticides, exploded. MIC is safe when properly maintained. The chemical is very flammable. When MIC comes into contact with water, the compounds react, resulting in a heat reaction. The chemical that leaked from Union Carbide India Ltd's (UCIL) pesticide plant turned Bhopal into a gigantic gas chamber. It was India's first major industrial disaster. At least 30 tonnes of methyl isocyanate gas killed over 15,000 people and injured 600,000 workers. The Bhopal gas tragedy is often recognised as the worst industrial disaster in history.

d) Future Hazards: A future hazards are the hazards which are likely to caused damage to the ecosystems and the environment with the phase of time series.

1) Climate change: Future implications of global climate change might include more frequent wildfires, longer drought spells in some areas, and an increase in the frequency, length, and severity of tropical storms. Extreme weather events such as heat waves, droughts, cyclones, blizzards, and rainstorms are becoming more common as a result of global warming. Such

occurrences will become more often and more intense in the future. Climate change is to blame for certain extreme weather events.

2) El Niño and La Niña: When an El Niño or its opposite, La Niña, arises in the future, it is anticipated to have more powerful effects throughout numerous geographical regions, magnifying variations in temperature, precipitation, and wildfire danger. Severe drought and food shortages, flooding, precipitation, and temperature surges caused by El Niño are generating a variety of health issues, including disease outbreaks, hunger, heat stress, and respiratory discases.

Natural Hazard Management

The hazard mitigation plan includes methods to lessen, if not eliminate, the likelihood or frequency of occurrence and the severity of repercussions. It should be underlined that no single measure, whether done individually or collectively, can ever be successful; only a combination of measures may provide the intended effects (table 32.2). A variety of methods can be implemented to mitigate natural hazards:

Table 32.2: Measures for abatement of natural hazards

Hazards	Mitigation
Earthquakes	Protect yourself by hiding behind a solid wall or behind heavy desks or tables. Avoid any locations with glass, such as windows, mirrors, or portraits, since they may shatter. Large bookshelves and cabinets, for example, should be avoided in situations where heavy furniture might stumble over. Establishing earthquake monitoring stations (seismological stations) for frequent monitoring and rapid distribution of information to the public Creating a vulnerability map for the country, disseminating vulnerability risk information to the public, and educating them on methods and means to mitigate the negative effects of catastrophes. Modifying home types and building designs in susceptible locations, as well as prohibiting the construction of high-rise structures, major industrial operations, and huge urban centres. Finally, making earthquake-resistant designs and the use of light materials essential in large construction operations in risky locations.

Volcanism	Assessment of the risk of a volcanic eruption. Monitoring and forecasting volcanic eruptions. Volcanic danger reduction through geological engineering. Hazardous area civil evacuation. Having an evacuation plan is one of the most effective strategies to reduce the danger of a volcanic eruption. This entails ensuring that evacuation plans are in place, as well as emergency shelter and food supplies. Exclusion zones can be established based on monitoring data to guarantee that people are evacuated from regions that are likely to be affected before an eruption. Local residents should also be informed on what they can do to lessen the danger of death or harm. People are instructed what they should do if they are unable to be evacuated, such as staying indoors to avoid falling ash and rock.
Cyclones	It implies that hazard mapping may be used to anticipate the sensitive regions impacted by storms. It depicts the pattern of historical cyclones based on wind speed, impacted areas, flooding frequency, and so on. By effectively implementing land use planning, critical activities and settlements in the most susceptible locations may be avoided. A settlement on the floodplains, for example, is extremely dangerous. As a result, authorities should plan ahead of time to avoid such hazards. These structures withstand wind forces and reduce losses. The country's public infrastructure should be planned with cyclone hazard mapping in mind. Settlements in non-engineered structures should ensure that they are aware of their dwellings' resilience to wind or other potentially hazardous weather conditions. Cyclone shelter building should be undertaken at the national, state, and regional levels to assist the vulnerable people in the event of a cyclone. The shelters should be created with population density, transit and communication in mind, distance from previously damaged regions in mind, and topography in mind. Because cyclonic storms provide excessive rainfall, which causes flooding in many places, flood management should be prioritised. To prevent floods, drainage systems should be well-designed. Improving plant cover is critical for increasing water infiltration capacity. More environmentally beneficial mangroves should be planted. India accounts about 3 percent of the world's mangrove cover. Mangrove root systems aid in the

	mitigation of tsunamis, soil erosion, and other natural disasters. Saline embankments along the shore safeguard dwellings, agricultural crops, and other critical infrastructure. Levees: These structures operate as a windbreak and give refuge during floods. The amount of public awareness efforts enhances community engagement. Governments at all levels should launch initiatives to raise awareness about natural disasters and make allowances for more local engagement in the mitigation process.
Flood	Manage watershed to minimize erosion. Reforest denuded areas. Emergency flood control measures. Mangroves will operate as an ideal natural bulwark to avoid coastal flooding as a result of increasingly intense and frequent cyclones (as a result of climate change). As a result, protecting mangroves from deterioration and destruction is critical. Intelligent flood warning systems, such as IFLOWS, should be used. Embankments and dykes are examples of structural measures. Flood forecasting and community engagement in flood risk management are examples of nonstructural approaches. Building regulations must be strictly enforced, and flood-prone zones must be delineated. Restoration of wet lands and implementation of water shed management programmes should be prioritised. As a physical barrier, flood walls, embankments, and dykes should be built.
Drought	Water harvesting, safeguarding water sources from pollution, developing water sources – such as micro dams, ponds, and wells, using reserve groundwater supplies, and water rationing/allocation are just a few examples of significant mitigating methods. Restoring pastures and managing land and water resources are both good objectives.
Landslides	Hazard zones must be defined, and certain slides must be stabilised and maintained, in addition to the installation of monitoring and early warning systems at specified locations. Landslide-prone locations should be identified using hazard mapping. When dealing with landslides, it is usually best to use area-specific procedures. Plants like vetiver can be planted to minimise landslides and floods.

	Construction and other developmental operations such as roads and dams should be restricted, agriculture should be limited to valleys and places with a moderate slope, and major communities in high susceptibility zones should be controlled. Monitoring, early warning, and research and development. Terrace farming, rather as Jhumming or shifting agriculture, should be supported in the north-eastern hill states. Knowledge management and network Capacity building and training
Tsunami	Restricted growth and sustainable development measures should be part of coastal land-use planning. Construction of tsunami evacuation facilities and improvements to evacuation routes should be part of evacuation preparations. Seawalls, breakwaters, tsunami river gates, and the development of forest barriers may all aid in averting widespread devastation. To avoid harm to life and property, careful planning and the adoption of building codes should be done prior to the construction of building structures. The preservation of coastal ecosystems such as mangroves and coral reefs may aid in the reduction of tsunami-related risks. Putting in place tsunami warning and forecast systems. Population and community assets should be moved away from risk areas, and plans for post-tsunami rehabilitation should be made.
Glacial Outburst	Identifying potentially dangerous lakes Use of technology Channellingpotential floods Uniform codes for construction activity Enhancing early warning systems Siphoning Pumping Excavation of a channel

Disaster Terminology

a) Threat: The state of being exposed to a hazard, harm, injury, or damage is referred to as threat. It is also a possible danger. A deadly cobra, for example, poses a threat to the guy below who is attempting to catch it. The threat he confronts is a snake bite. This threat is erased as soon as the cobra is trapped without causing any harm. When the threat is eliminated, the danger vanishes.

b) Danger: The word Danger refers to the situation of being confronted with any type of threat that may result in major loss, harm, or death.
c) Capacity: Capacity is described as the sum of an organization's, community's, or society's capabilities, qualities, and resources available to manage and decrease catastrophe risks and enhance resilience. Infrastructure, institutions, human knowledge and skills, and collective traits such as social interactions, leadership, and management are all examples of capacity. Coping capacity is the ability of people, organisations, and systems to manage unfavourable situations, risks, or catastrophes utilising existing skills and resources. Capacity to cope necessitates ongoing knowledge, resources, and excellent management, both during normal times and during catastrophes or poor situations. Capabilities for disaster response help to reduce catastrophe risks.
d) Natural disaster: Natural disasters are frequently linked to atmospheric, geologic, seismic, volcanic, and other natural phenomena. Examples include earthquakes, tsunamis, landslides, volcanoes, avalanches, floods, hurricanes, and lightning. According to a recent UN World Bank joint book, catastrophes have killed approximately 3.3 million people in the previous 42 years, with year-to-year trends varying. Drought has killed more people than any other disaster, and poor nations have suffered the most.
e) Man-made disaster: Anthropogenic factors, including industrial, biological, chemical, nuclear, technical, mining, transportation, and other causes, are commonly connected with man-made disasters. Man-made disasters like as Hiroshima, Bhopal, Chernobyl, the Exxon Valdez, and Fukushima are hideous scars on the otherwise lovely face of mankind, and no amount of plastic surgery can hide the humiliation of these wounds. Terrorism is yet another serious threat.
f) Disaster insurance: Excessive reliance on relief and rehabilitation packages produces a system in which there are no incentives to undertake risk-reduction measures. Insurance is a potentially useful mitigation technique in disaster-prone places since it promotes excellent infrastructure and consciousness, as well as a culture of safety and prevention. Disaster insurance often operates on the idea of 'greater the risk, higher the premium,' raising awareness of susceptible places and encouraging individuals to settle in comparatively secure locations.

g) Crisis: The word crisis derives from Greek and means a moment of sudden approaching change, one way or another. The Chinese term for crisis combines the ideograms for disaster and opportunity.
h) Calamity: Calamity refers to an exceptionally severe incident that causes misery, misfortune, dreadful suffering, adversity, and tremendous loss, as well as a trail of lingering sorrow, emotional outbursts, affliction, and mourning. A calamity's anguish stems more from personal or general sadness and sorrow than from other causes.
i) Catastrophe: The term catastrophe is derived from the Greek words "kata" and "streiphen" meaning "down" and "turning over." As a result, catastrophe is another phrase for an irreversible tragedy that goes beyond disaster or calamity. Catastrophes, on the other hand, are more clearly defined in catastrophe theory as occurrences that influence systems.
j) Cataclysm: Cataclysm is a term used to describe a massive occurrence that results in upheaval and the ruin of an existing system. In contrast to the word disaster, the term cataclysm refers to wide or global events rather than particular events.
k) Hazard: The term Hazard refers to a possible source of a potentially dangerous incident that might result in a disaster. It is determined by the likelihood of an event occurring, its location, and the strike time. Natural hazards include volcanoes, earthquakes, and tsunamis; technology hazards include nuclear and chemical plant accidents; and human-caused hazards include terrorism and bombing. When hazards cascade or overlap, they become more dangerous.
l) Risk: The term "risk" refers to the outcome of a hazard, which is commonly expressed in terms of monetary loss. We need to know the amount of hazard, the elements at risk, and the accompanying susceptibility to determine the risk involved in a catastrophe. When the hazard is H and the vulnerability is V, the risk is the product of the two.
m) Preparedness: Preparedness entails being ready to deal with an impending crisis in any form 24 hours a day, seven days a week. Disasters predicted and avoided are the best demonstration of preparation. Prepared communities practise well-structured disaster management plans and work to strengthen their skills in teamwork, skill development, forecasting, early warning, and prompt evacuation.
n) Vulnerability: Vulnerability refers to a person's sensitivity to hazards or the extent to which they are exposed to them. Natural hazards such as earthquakes and volcanoes cannot be avoided, but

if we take actions to limit our susceptibility to them, we will be less affected. Individuals living in hazardous buildings, for example, will be extremely vulnerable because earthquakes are known to kill people due to building collapse. The vulnerability of buildings may be decreased from 100 percent to less than that by making them earthquake-resistant. Vulnerability quantification is a highly specialised activity that is frequently left to individual intuition and mapping by professionals. It is, however, critical for everyone of us to understand that our sensitivity to a threat might manifest itself in a variety of ways. Some examples are as follows:

1) Physical vulnerability: People who live in unsafe houses in earthquake-prone locations are at risk.
2) Economic vulnerability: People with little financial and material resources are economically weak and hence more vulnerable than the wealthy.
3) Social vulnerability: People who lack social support and connectedness are socially fragile.
4) Ecological vulnerability: Communities and individuals who are not safeguarded from ongoing environmental degradation are environmentally vulnerable.
5) Organizational vulnerability: Communities that lack organisational capacity and support are organizationally vulnerable.
6) Communicational vulnerability: Communities that lack access to information and telecommunications services have communication challenges.
7) Attitudinal vulnerability: Attitudinal vulnerability exists in communities that lack behavioural training, education, and awareness.
8) Political vulnerability: Locations that do not have political favours are politically vulnerable.

Classification of Disasters

Natural disasters and man-made disasters are the two primary kinds of disasters.

a) Natural disasters are disasters that occur as a result of natural occurrences (meteorological, geological or even biological origin). Natural catastrophes include cyclones, tsunamis, earthquakes, and volcanic eruptions that are entirely natural in origin. Landslides, floods, droughts, and fires are all examples of socio-natural disasters since their causes are both natural and man-made. Flooding can occur as a result of strong rainfall, landslides, or human waste clogging channels.

b) Man-made disasters are those that arise as a result of gross negligence. These are related to industry or energy producing facilities and include explosions, hazardous waste leaks, pollution, dam collapse, wars or civil turmoil, and so on. Many occur on a regular basis, while others occur just sporadically.

Disaster Mitigation

Disaster mitigation refers to a wide variety of efforts and measures aimed at reducing the negative impact of a disaster before it occurs. Its reach extends beyond catastrophe preparedness in that it encompasses the complete range of operations, both structural and nonstructural. Building and infrastructure retrofitting are examples of structural mitigating strategies. Disaster education, disaster policy planning, forecasting, and early warning are examples of nonstructural mitigating strategies. The breadth of catastrophe mitigation goes beyond simply protecting lives and property. It also entails minimising the detrimental impact of risks on economic development, biodiversity, and social institutions.

Disaster Management

Disaster management encompasses a wide variety of efforts and actions, including policy formation, strategic planning, administrative, financial, and legal interventions, and decision making, all with a focus on the full disaster cycle. The term is also used to describe emergency management that focuses on rescue, relief, and rehabilitation in the immediate aftermath of a disaster.

Disaster Mismanagement

When a hazard is permitted to become a disaster, and a disaster is allowed to become a bigger disaster due to a lack of preparation, lack of readiness, and delayed reaction, we get what is known as disaster mismanagement. Little minds, bad planning, horrible mismanagement, bureaucratic delays, and a lack of accountability undermine good initiatives. The subsequent blame game aggravates the situation.

Disaster Management Cycle

The concept of a disaster management cycle integrates isolated attempts by various actors, both government and nongovernment, to reduce vulnerability or mitigate disasters within the encompassing domain of disaster management, as phases occurring at different time periods in the disaster management continuum. This has permitted a planned approach to disaster management in that post-disaster recovery and pre-disaster mitigation planning are seen as intertwined activities.

Disaster management strives to limit or avoid potential losses from hazards, to provide timely and appropriate support to disaster victims, and to ensure speedy and successful recovery. The disaster management cycle depicts the continual process by which governments, corporations, and civil society plan for and mitigate the effect of catastrophes, react during and shortly after a disaster, and recover after a disaster has happened. Appropriate actions at all points in the cycle result in higher readiness, better warnings, decreased susceptibility, or catastrophe prevention during the following cycle iteration. The whole disaster management cycle involves the development of governmental policies and strategies to either change the causes of catastrophes or reduce their consequences on people, property, and infrastructure. The three key stages of activities in disaster management are given below:

a) Before a disaster: To prevent the possibility of human, material, or environmental losses as a result of risks, and to guarantee that these losses are minimised when tragedy hits.
b) During a disaster: To guarantee that victims' needs and provisions are satisfied in order to relieve and minimise suffering.
c) After a disaster: To achieve rapid and long-term recovery.

The mitigation and preparedness stages take place as disaster management improvements are implemented in advance of a catastrophic event. Developmental factors play an important part in the mitigation and preparedness of a community to face a disaster successfully. When a disaster strikes, disaster management players, particularly humanitarian groups, are drawn into the immediate response and long-term recovery stages. The four disaster management phases depicted above do not necessarily, or even usually, occur in this order or in isolation. The phases of the cycle (Fig. 32.1) frequently overlap, and the length of each phase is highly dependent on the severity of the disaster.

a) Mitigation is the process of reducing the impacts of a calamity. Building rules and zoning are two examples, as are vulnerability evaluations and public education.
b) Preparedness is the process of planning how to respond. Preparedness strategies, emergency drills/training, and warning systems are some examples.
c) Response includes efforts to reduce the risks posed by a calamity. Examples include search and rescue and emergency aid.
d) Recovery is the process of restoring normalcy to a community. Examples include temporary accommodation, grants, and medical treatment.

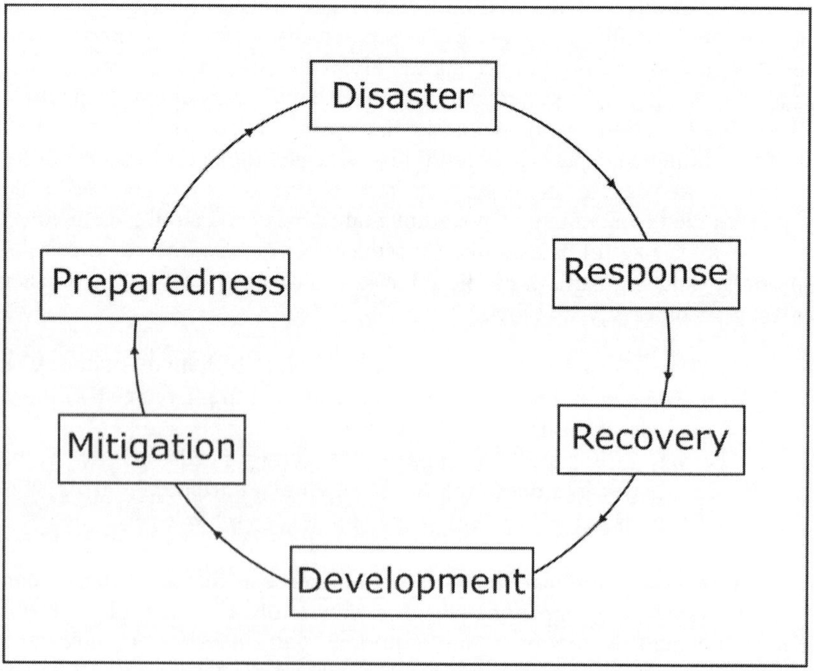

Fig. 32.1: Diagram showing various stages of disaster management cycles

Disaster Management in India

Background

In India, disaster management has progressed from an activity-based reactive structure to a proactive institutionalised framework, and from a relief-based strategy to a multi-dimensional pro-active holistic approach for risk reduction. Prior to independence, the policy was relief-oriented, and relief departments were established for catastrophes and calamities. Activities included developing relief standards and launching food-for-work programmes. Following independence, the job of disaster management remained with the Relief Commissioners in each state, who worked under the Central Relief Commissioner, with their authority confined to the delegation of relief goods and money in afflicted areas. Following the recognition of the decade of 1990 as the 'International Decade for Natural Disaster Reduction' (IDNDR) by the UN General Assembly, a permanent and institutionalised structure began in the 1990s with the establishment of a disaster management cell under the Ministry of Agriculture. Following a succession of catastrophes like as the Latur Earthquake (1993), the Malpa Landslide (1994), the Orissa Super Cyclone (1999), and the Bhuj Earthquake (2001), a high-powered Committee was formed, chaired by Mr. J.C. Pant, Secretary, Ministry of Agriculture. As a result, the disaster management division was transferred to the Ministry of Home Affairs in 2002, and a hierarchical disaster management system formed in India. For the first time, the 10th Five-Year Plan included a full chapter on Disaster Management. The Twelfth Finance Commission was also tasked with reviewing the disaster management finance arrangements.

Agencies involved in Disaster Management

a) The National Disaster Management Authority: The National Disaster Management Authority, or NDMA, is an apex disaster management authority in India, led by the Prime Minister. It is in charge of the National Disaster Response Force's oversight, guidance, and control (NDRF). The Prime Minister serves as the NDMA's ex-officio chairperson. The Vice-Chairman is the Cabinet Minister. The NDMA Secretariat, led by a Secretary, is in charge of secretarial assistance and stability. In addition, the NDMA includes eight Ministers of State as members. The Disaster Management Act, 2005, adopted by the Government of India, formed the NDMA, which was formally established in December 2006. The NDRF is a specialised reaction force that may be deployed in a catastrophe or potential disaster situation. The NDMA has overall supervision, direction, and control over this force, and the Director General of the National Disaster Response Force has command and supervision over it. The National Disaster Response Force now consists of 12 battalions, three from the BSF and CRPF and two each from the CISF, ITBP, and SSB. Each battalion comprises 18 self-contained professional search and rescue teams of 45 people each, which include engineers, technicians, electricians, dog squads, and medical/paramedics. Each Battalion currently has 1149 men. The National Disaster Management Authority (NDMA) was founded to carry out the following functions:

1) Disaster management policies are being implemented.
2) Approving a variety of plans, including national plans; plans by various ministries and departments of the Indian government, and other plans related to the national plan.
3) In accordance with the state plan, establishing rules for state government authorities.
4) Developing recommendations for various government agencies to use in incorporating disaster prevention or mitigation strategies into their development plans and initiatives.
5) Coordinating with the staff and putting in place the required policies and strategies.
6) Recommending that funds for mitigation be made available.
7) Aid and help to countries in need are dependent on the central government.
8) Developing guidelines and broad policies for the NIDM's operation (National Institute of Disaster Management).

b) National Executive Committee (NEC): The NEC is made up of high-ranking ministers from the Indian government, including the Union Home Secretary as Chairperson and Secretaries to the Government of India from Ministries/Departments such as Agriculture, Atomic Energy, Defense, Drinking Water Supply, Environment and Forests, and so on. The NEC develops the National Disaster Management Plan in accordance with the National Disaster Management Policy.
c) State Disaster Management Authority (SDMA): The SDMA is led by the appropriate state's Chief Minister. The State Executive Committee (SEC) of the State Government works with the State Disaster Management Authority (SDMA) on disaster management.
d) District Disaster Management Authority (DDMA): The DDMA is led by the District Collector, Deputy Commissioner, or District

Magistrate, depending on the situation, and is co-chaired by elected representatives of the local government. The DDMA guarantees that the recommendations established by the NDMA and SDMA are implemented by all State Government ministries at the District level, as well as municipal authorities in the District.

e) Local: Panchayati Raj Institutions (PRI), Municipalities, District and Cantonment 11 Institutional and Legal Arrangements Boards, and Town Planning Administrations are examples of local authorities that regulate and administer municipal services.

National Disaster Management Act, 2005

The Disaster Management Act of 2005 is making headlines once more. The Act was passed in the country on December 26, 2005. It went into force in the aftermath of the 2004 tsunami, when hundreds of people died. The government passed it to ensure effective disaster management, including the development of mitigation measures, capacity-building, and other associated issues. The act, which has 11 chapters and 79 sections, was signed by the President of India on December 23, 2005 and Chapter 10 of the Act deals with "Offences and Penalties."

Provisions of the Act

a) The Act intends to handle catastrophes by preparing mitigation methods, increasing capacity, and more.
b) The Agency of Home Affairs is also designated as the nodal ministry responsible for overseeing the country's entire national disaster management.
c) The Act also includes provisions for financial mechanisms, such as the establishment of emergency response funds such as the National Disaster Response Fund.

Governing bodies that were established under this Act

a) The National Disaster Management Authority (NDMA) is the primary central organisation for disaster management coordination, and it is chaired by the Prime Minister. It is in charge of developing disaster management policies, strategies, and guidelines to provide an effective and robust response in the event of a disaster.
b) The second is the National Executive Committee, or NEC, which aids the NDMA and creates the country's national disaster management plan.

c) On a state and district level, the SDMA and NDMA are in charge of developing disaster plans for their respective states and districts.
d) Then there's the NDRF, or National Disaster Response Force, which takes immediate action in the event of a disaster or calamity. For years, the NDRF has played a vital part of rescuing people from calamities.

Some Sections

a) Section 6: It empowers the National Disaster Management Agency (NDMA) to develop national disaster management strategies. It also guarantees that the strategy is carried out by the state's disaster management authority.
b) Section 10: It permits the NEC to offer countries instructions on what actions they should take.
c) Section 33: states that the District Ability has the authority to instruct any officer or department at the district level, as well as any local authority, to take steps for disaster prevention or mitigation. The officer or department in question is obligated to carry out the instruction.
d) Section 51 to 60: The Act establishes sanctions for specified infractions. Anyone who is caught hindering an officer or employee from doing their job will be sentenced to jail. The penalty may be up to a year in prison, a fine, or both.

National Policy on Disaster Management, 2009

In 2009, the NDMA issued a National Policy on Disaster Management (NPDM). It is designed with the objective "To establish a safe and disaster resilient India by building a comprehensive, proactive, multi-disaster focused and technology driven strategy via a culture of prevention, mitigation, readiness and response." Disaster management is a multidisciplinary endeavour that requires the collaboration of all stakeholders. The National Policy on Disaster Management calls for an integrated approach to disaster management, with a focus on developing strategic partnerships at all levels. It creates a welcoming environment for all and addresses the problems of all segments of society, including differently abled people, women, children, and other marginalised groups. It also seeks to increase transparency and accountability in all aspects of disaster management by involving the community, community-based organisations, Panchayati Raj Institutions (PRIs), local governments, and civil society. The main objectives of National Policy on Disaster management, 2009 are given below:

a) Using information, creativity, and education to foster a culture of prevention, readiness, and resilience at all levels.
b) Promoting mitigation strategies based on technology, traditional wisdom, and environmental sustainability.
c) Integrating disaster management into the process of development planning.
d) Creating an enabling regulatory environment and a compliance system by establishing institutional and techno-legal frameworks.
e) Ensuring an effective framework for identifying, assessing, and monitoring catastrophe risks.
f) Creating modern forecasting and early warning systems that are supported by responsive and fail-safe communication with information technology support.
g) Ensuring effective reaction and relief with a compassionate attitude to the needs of society's most vulnerable members.
h) Using rebuilding as a chance to establish disaster-resistant structures and habitat for wildlife

The National Disaster Management Plan, 2016

In 2016, the Government of India published its first National Disaster Management Plan. The Plan's Vision is to make India disaster resilient, achieve significant disaster risk reduction, and significantly reduce losses of life, livelihoods, and assets – economic, physical, social, cultural, and environmental by maximizing disaster preparedness at all levels of administration and among communities. It is closely aligned with the Sendai Framework for Disaster Risk Reduction, the Sustainable Development Goals 2015–2030, and the Paris Agreement on Climate Change at COP21. While the Sendai Framework is the first international agreement approved as part of the post-2015 development agenda, the Sustainable Development Goals also identify disaster risk reduction as an essential component of sustainable development. The Paris Agreement recognises the critical need to account for the rising frequency of extreme weather events caused by climate change.

Salient Features of the Plan

The strategy addresses all aspects of disaster management, including prevention, mitigation, response, and recovery. The strategy employed in this national plan includes the four goals enunciated in the Sendai Framework into the disaster risk reduction planning framework under the five thematic areas for actions:

a) Understanding Risk
b) Inter-Agency Coordination
c) Investing in Disaster Risk Reduction – structural measures
d) Investing in Disaster Risk Reduction – non-structural measures
e) Capacity Development

Objectives

a) To increase catastrophe risk, vulnerability, and hazard awareness.
b) To improve catastrophe risk governance at all levels, from the national to the local.
c) To invest in catastrophe risk reduction for resilience through structural, non-structural, and financial measures, as well as a comprehensive capacity building programme.
d) To strengthen catastrophe preparedness in order to respond more effectively.
e) To promote the concept of 'Building Back Better' in the areas of recovery, rebuilding, and rehabilitation.
f) To reduce catastrophe risk and loss of life, livelihood, health, and assets, including physical, economic, cultural, social, and environmental assets, by preventing disasters and ensuring a considerable reduction in disaster risk.
g) Enhance resilience and mitigate existing catastrophes while preventing new ones from arising.
h) To encourage the implementation of comprehensive and integrated structural, economic, social, legal, health, cultural, environmental, educational, political, technical, and institutional actions to decrease hazard exposure and catastrophe vulnerability.
i) To equip communities and local governments with the tools they need to manage and mitigate catastrophe risks.
j) In all aspects of catastrophe management, to improve technological and scientific skills.
k) To build capacity at all levels to respond effectively to numerous threats and to manage disasters in a community-based manner.
l) To provide clarity on the different departments' and ministries' responsibilities and tasks.
m) At all levels, establish a culture of catastrophe risk mitigation and prevention.

n) To make it possible to include catastrophe management issues into planning.

Drawbacks

a) It does not provide a clear and realistic path. It is excessively broad in identifying the tasks that the federal and state governments must undertake for disaster risk mitigation, readiness, response, recovery, reconstruction, and governance.
b) The plan does not provide a time range for completing these actions, other than to state that they must be completed in the short, medium, and long term.
c) The plan does not forecast the amount of money needed to carry out these tasks, nor does it indicate how those amounts will be raised.
d) Furthermore, there is no structure for monitoring and evaluating the strategy in the plan.

Challenges for the Future

There is a rising need to examine catastrophes from a development standpoint. Disasters may have a severe impact on communities and greatly stall development efforts. However, it may also provide a chance to engage in development activities in the aftermath of a calamity. Disasters provide communities with the opportunity to remake themselves. Four factors contribute to and benefit from the execution of sustainable development policies: disaster prevention, mitigation, readiness, and relief. These factors, as well as environmental conservation and sustainable development, are inextricably linked. The Yokohama Strategy, which emerged from the international decade for natural disaster reduction in May 1994, stresses that disaster prevention, mitigation, and readiness are superior to disaster response in terms of accomplishing vulnerability reduction goals and objectives. The Indian government has made mitigation and prevention critical components of its growth plan. The Tenth Five Year Plan highlights the importance of incorporating mitigation into the development process if growth is to be sustainable. In summary, disaster management is being included into development planning. However, there are several underlying issues throughout the entire process. In truth, societal disparities are at the root of many crises.

One of the most evident gaps in India's disaster management process has been the failure to account for natural catastrophes. Recent efforts have concentrated solely on natural catastrophes, however the current global situation necessitates actions to mitigate the effect of artificial disasters. International developments – notably civil wars and civil conflict in Eastern Europe and South America, as well as 9/11 – have pushed the subject of natural catastrophes to the forefront of disaster management. The international community has acknowledged the grave implications of Nuclear, Biological, and Chemical (NBC) warfare. This will be a significant problem for India to address in the near future.

India: Vulnerability Profile

India is one of the world's 10 most disaster-prone countries. Because of its unique geo-climatic and socio-economic characteristics, it is exposed to a wide range of natural and man-made calamities. A total of 27 of the country's 35 states and union territories are disaster-prone. According to the National Disaster Management Authority, approximately 40 million hectares of land in India are vulnerable to floods (roughly 12 percent of total land area), 68 percent of land is vulnerable to droughts, landslides, and avalanches, 58.6 percent of landmass is earthquake-prone, and tsunamis and cyclones are a regular occurrence for 5,700 km of the 7,516-km long coastal line. Such fragile conditions have pushed India to the top of the list of disaster-prone countries. The country's five main areas, including the Himalayan region, the alluvial plains, the hilly peninsula, and the coastal zone, each have their own set of issues. On the one hand, the Himalayan area is vulnerable to calamities like as earthquakes and landslides, while the plain is flooded virtually every year. Droughts and starvation harm the desert region of the nation, while cyclones and storms threaten the coastal zone. India is the 14th most susceptible country in the world owing to catastrophic weather-related occurrences, according to the Global Climate Risk Index report 2019. According to the research, catastrophes claimed the lives of 2,736 people in India in 2017, second only to Puerto Rico (2,978). Furthermore, economic damages in India as a result of such disasters were $13,789 million, the fourth largest in the world. Between 1970 and 2009, India was hit by 371 natural disasters, resulting in the deaths of 1,51,000 people and the displacement of 1.86 billion people. Floods are the most common natural catastrophe in India, accounting for 52 percent of all disasters, followed by cyclones (30 percent), landslides (10 percent), earthquakes (5 percent), and droughts (5 percent). Earthquakes of moderate to very high intensity affect 58.6

percent of the continent. Floods and river erosion threaten around 40 million hectares (12 percent) of the country's territory. Approximately 5,700 kilometres of the 7,516-kilometre-long coastline are vulnerable to cyclones and tsunamis. Droughts threaten 68 percent of the country's cultivable land. Landslides and avalanches pose a threat in the area's steep sections too.

Aside from natural factors, human-induced activities such as rising population pressure, deteriorating environmental conditions, deforestation, unscientific development, faulty agricultural practises and grazing, unplanned urbanisation, and the construction of large dams on river channels, among others, are all contributing to the country's increased disaster impact and frequency.

NDMA Guidelines for Natural Disaster Management

NDMA has issued 33 guidelines as of October 2021, some of which are listed below:

a) Earthquake management: The NDMA guidelines for increasing the efficacy of earthquake management in India are based on six seismic safety pillars:
   1) Earthquake resistant construction of new structures: All central ministries and departments, as well as state governments, will facilitate the implementation of relevant seismically safe design and construction standards for buildings and other lifeline and commercially important structures under their administrative control, such as bridges, flyovers, ports, and harbours.
   2) Selective seismic strengthening andretrofitting of existing priority structures and lifeline structures: All central ministries and state governments are expected to develop seismic strengthening programmes for priority structures through ULBs and PRIs. Nationally significant structures such as Raj Bhavans, Legislatures, Courts, essential structures such as academic institutions, public utility structures such as reservoirs and dams, and multi-story buildings with five or more storeys. State governments are responsible for identifying these buildings.
   3) Regulation and enforcement: State governments are responsible for implementing systems to guarantee that all players, including as builders, architects, engineers, and government departments, adhere to seismic safety in all design and building operations. The Home Ministry formed a national level expert

panel that suggested changes to town and country planning Acts, land use and zoning rules, DCRs, and building bye-laws that are technically rigorous and adhere to globally acknowledged principles.

4) Awareness andpreparedness: The NDMA recognises that one of the most difficult jobs in earthquake preparedness and mitigation is sensitising all stakeholders. It suggests creating earthquake safety handbooks, homeowner seismic safety guides, a structure safety audit manual, and video films for the general public. It also emphasises the importance of developing vulnerability maps for geographical regions and streamlining NGOs and Volunteer Groups.
5) Capacity development: Capacity development target groups include elected representatives and government officials, visual and print media professionals, urban planners, engineers, architects, and builders, NGOs, Community Based Organizations (CBOs), social activists, social scientists, schoolteachers, and students.
6) Emergency response: The local administration coordinates all response actions using the Incident Command System. It involves community, corporate, and specialist team participation.

b) Management of tsunamis: The various guidelines for the management of Tsunamis are given below:
   1) Tsunami risk assessment and vulnerability analysis: Based on coastal land use maps and coastal bathymetry, the NDMA suggests assessing vulnerability and risk mapping in the tsunami hazard. It suggests developing models to predict the arrival and run-up height of tsunami waves. The Indian Naval Hydrographic Department (INHD) reports to the Chief Hydrographer to the Government of India. It frequently supplies bathymetry data to approved entities for the creation of inundation maps.
   2) Tsunami preparedness: IMD intends to develop a 17-station Real Time Seismic Monitoring Network (RTSMN), and Bottom Pressure Recorders (BPRs) are utilised to monitor the propagation of Tsunami waves in the open ocean. A big issue is that unattended ocean observation platforms at sea are being damaged, either intentionally or unintentionally.
   3) Structural mitigation measures: Construction of cyclone shelters, submerged sand barriers/dykes, sand dunes with sea weeds, and coastal forest and mangrove plantation in conjunction

with specialists, construct location-specific sea barriers and coral reefs. The creation of breakwaters along the shore to give the necessary cushion, Creating a "Bio-Shield" – a limited strip of land near the shore.

4) Regulation and enforcement: Strict adherence to coastal zone restrictions (within 500 m of the high tide line with elevation of less than 10 m above mean sea level). Adoption of MHA's model techno-legal framework for assuring compliance with tsunami-safe zoning, planning, design, and construction standards and encouraging optimum land use.
5) Emergency: Because the community is the first responder, a series of public awareness campaigns may be launched throughout the coastal area through various ways. Search and rescue activities can be carried out by SHGs, NGOs, and CBOs. Inflatable motorised boats, helicopters, and search and rescue equipment are needed soon after a tsunami to seek for and rescue persons caught in waterlogged regions, on tree tops, or clinging to structures.
6) Ensuring implementation: Aggressive capacity development needed for the local people and administration for dealing with catastrophes in the aftermath of the tsunami and cyclone.

c) Flood Management: To date, the major focus of India's flood prevention effort has been on structural measures.
   1) Reservoirs, Dams, Other Water Storages: Construction of reservoirs along river channels might store additional water during floods. However, such steps have not proven successful thus far. Dams constructed to control Damodar floods were unable to do so.
   2) Embankments/flood levees/flood walls: Floodwaters can be kept from spilling the banks and spreading into surrounding regions by constructing flood prevention embankments. The construction of embankments on the Yamuna near Delhi has been successful in limiting the flood.
   3) Drainage improvement: The building of roads, canals, railway lines, and other structures typically clogs the drainage system. Flooding might be avoided if the original drainage system is rebuilt.
   4) Channel improvement: By increasing the discharge carrying capacity of a canal, it may be made to transport flood flow at levels lower than its current peak flood level. It seeks to expand the area of flow or the velocity of flow (or both) in order to

enhance carrying capacity. Selective desilting/dredging at outfalls/confluences or local reaches, on the other hand, can be used to address the problem locally.

5) Diversion: Diverting all or a portion of the flow into a natural or intentionally created channel inside or, in certain situations, outside the flood plains is a beneficial method of decreasing river water levels.
6) Catchment area treatment/afforestation: Watershed management strategies such as afforestation and soil conservation, in conjunction with structural works such as check dams, retention basins, and so on, are helpful in minimising flood peaks and regulating the abruptness of runoff.

d) Landslide and snow avalanches management: The NDMA guidelines' major goals are to institutionalise landslide hazard mitigation efforts, to educate society on the many elements of landslide hazard in the country, and to equip society to take appropriate action to decrease both the risks and costs associated with this hazard. Among the suggestions are:

1) Landslide monitoring and early warning system: It illustrates previous work, best practises, and current status in the fields of Landslide Early Warning System (LEWS), rainfall threshold-based modelling, ground-based wireless instrumentation, and real-time monitoring system for landslide prediction, earthquake-triggered landslides, monitoring mechanism of landslides, and gaps in landslide monitoring and development of early warning system. Technical recommendations for designing and implementing rainfall thresholds, Numerical Weather Prediction (NWP), Automatic Rain Gauges, Wireless Sensor Network (WSN), Micro-Electro Mechanical Sensors (MEMS), and so on have been presented for future prospects.
2) Awareness Programme: The plan specifies the need for awareness campaigns, a review of previous work and best practises, the identification of gaps, and suggestions and implementation methods. It tries to create a culture of readiness and awareness, so that individuals in society become vigilant and informed in the event of an emergency, or take some preventative actions before the disaster happens. A participatory method has been developed to ensure that all segments of the community are participating in the awareness campaign. Because the community is the first to face the tragedy before

any relief arrives, an awareness mechanism is designed to include and educate the people.

3) Capacity building: The majority of landslide-prone habitations are found in the Himalayan country side, north-eastern states, and on the Eastern and Western Ghats of the country's southern and south-western coastal states. As a result of the remoteness and roughness of the terrain, as well as the lack of infrastructure amenities, it becomes a very difficult task for the district administration and response teams, including NDRF and SDRF teams, to reach the disaster site within a short period of time. Not long ago, the need for Capacity Building and Training of Stakeholders in Landslide Risk Management was recognised in India.
4) The insight occurred following the sad occurrences of the Okhimath and Malpa landslides in August 1998.
5) Preparation of mountain zone regulation and policies: The country is prone to one or more disasters, such as landslides and related hazards such as cloud burst and flash flood, which are among the most hazardous and devastating natural hazards in terms of loss of life and property in the Himalaya and Western Ghats. The widespread property damage caused by the recent landslide and related dangers such as cloud burst and flash flood has revealed that most building designs are ill-conceived and do not adhere to established norms. Even government departments do not always adhere to design codes. This has resulted in an unsettling scenario in which a big number of dangerous building stock is added each year to an already massive number of existing unsafe structures in hostile climates, fragile environments, and tectonically active unstable mountainous terrain. There is no landuse policy in the country that can be implemented at the national, state, or municipal levels. The Himalayan cities are expanding and turning into mountains of rubbish and plastic, untreated sewage, chronic water shortages, uncontrolled urban sprawl, and even local air pollution caused by automobiles. These communities must be developed, especially with the influx of summer tourists in mind. To address these concerns, several jurisdictions have tried everything from prohibiting plastics to charging tourists. However, they require assistance and fresh perspectives on old architectural methods, local water management through lake preservation, and various sewage and trash disposal systems.

6) Stabilization and mitigation of landslides and creation of special purpose vehicle (SPV) for landslide management: Landslides affect not just humans, but also environment and the economics and development of places. As a result, there is an urgent need for a national effort to mitigate landslides and rebuild-rehabilitate damaged communities. At the moment, no Ministry or Department of the Government of India has a programme in place to control landslide risk across the country. The Geological Survey of India (GSI), which is part of the Ministry of Mines (MoM) and is the Nodal Agency for landslide management in India, has not yet implemented any mitigating measures on the ground. Their primary responsibilities have remained landslide investigations, research and development, and assistance to states on landslide control. To close this gap, this policy plan proposes to provide full technofinancial assistance to landslide-prone states, who will submit Detailed Project Reports (DPRs) to project sanctioning agencies in order to implement site-specific landslide mitigation measures. Landslides are site-specific in nature, and because the risk varies by location, the approach / technology for mitigating each landslide will change, necessitating distinct operations.

e) Drought management: The various guidelines for drought management as per NDMA are given below:

1) Institutional framework and financial arrangements: Drought is a complicated natural and man-made phenomena that need specialist management to handle various levels of complications. This calls for the development of an institutional framework to assist the coordination of inputs from various governmental organisations and institutions while preserving the functional autonomy of the institutions and organisations involved. Droughts are monitored at the national level by a variety of institutions, including the IMD, which monitors rainfall and aridity anomalies, and the Central Water Commission, which monitors reservoir storage situation (CWC). Agriculture, Irrigation, Economics, and Statistics Departments at the state level monitor crop and seasonal conditions.

2) Assessment and Early warning: Building greater institutional capacity through the improvement and application of seasonal and shorter-term forecasts, integrated monitoring and drought early warning systems, and connected information delivery

systems, developing preparedness plans at all levels of governance, adopting mitigation actions and programmes, and creating a safety net of emergency response programmes that ensure timely and targeted response will promote risk management principles.

3) Prevention, preparedness and mitigation: Pre-disaster initiatives aimed at increasing readiness and improving operational and institutional skills for reacting to a drought are referred to as prevention and preparedness. Drought prevention and preparedness include water supply augmentation and conservation (e.g., rainwater gathering techniques), irrigation facility development, effective drought management, and public awareness and education. Transport and communication linkages are essential for ensuring the delivery of food and other commodities during and immediately following a drought. Drought management needs community understanding of mitigation techniques, farmer insurance systems, agricultural contingency plans, and so on.
4) Capacity development: Human resource development, training, education, and capacity building are critical components of an efficient drought mitigation and management approach. Capacity building aims to establish a structured functional mechanism with skilled human resources. It must be interpreted broadly in order to encompass knowledge, skill, attitude, and resources in an integrated manner.
5) Relief and response: Droughts affect several regions of India every year. Drought-prone states must investigate the phenomena and devise long-term remedies, including emergency relief. The goal must be to assist them in focusing on the major concerns for policy formation and subsequent measures, such as understanding the nature of drought, altering perceptions and responses to drought, and shifting the focus from relief to mitigation of drought. The identification of susceptible places and populations, as well as their inclusion and effective land use planning, are critical concerns in drought mitigation and relief strategy. It is critical to utilise open criteria for identifying sensitive places and populations. This will allow development projects and drought assistance to be targeted to the most vulnerable and deserving people.
6) Implementation: The Government of India decided in 2002 to keep the subject of drought management with the Department

of Agriculture and Cooperation while transferring the handling of all other types of natural and man-made catastrophes to the Ministry of Home Affairs. National, state, and district comprehensive DM plans will be developed. The national strategy will focus on many areas of District Management, such as preparedness, mitigation, and reaction. These plans will explicitly specify key stakeholders' duties for each catastrophe level, as well as assessments of their own response capacity.

f) Cyclone management: In 2008, the NDMA issued its National Guidelines for Cyclone Management. The fundamental concept of these principles is that mitigation must be multi-sectoral. The following are the key points in these guidelines:
   1) Early warning systems: The government would set up a cutting-edge cyclone EWS that would include observations, projections, alerts, and customised local-scale advise for decision-makers (national/state/district level) to help them manage the effect of cyclones.
   2) Aircraft probing of cyclone (APC) facility: The government plans to launch an Aircraft Probing of Cyclones (APC) facility for India, which would use a combination of manned aircraft and high altitude Unmanned Aerial Vehicles (UAV) to cover important observational data gaps in the case of cyclones over the Bay of Bengal and the Arabian Sea. Actual flight into and around a tropical cyclone at various phases of growth and movement may give essential data for researching and comprehending the structure and movement of a cyclone, considerably lowering track and intensity forecast mistakes.
   3) National Disaster Communication Infrastructure: With the adoption of cutting-edge operational infrastructure, the government would commission National Disaster Communication Infrastructure (NDCI) at the NDMA/MHA, State Disaster Management Authorities (SDMAs) of coastal states/UTs, and District Disaster Management Authorities (DDMAs) of the 84 coastal districts vulnerable to cyclones.
   4) Expanding warning dissemination outreach: The government plans to expand warning dissemination outreach by using Direct-To-Home (DTH) transmission services in remote and rural areas (Panchayats) that cannot be reached otherwise, to introduce a weather channel and broadcast cyclone warnings from high-power coastal radio stations, as well as the use of

satellite radio services such as World Space, Ham radios, community radios, and the VHF network.

5) Establishing an integrated hazard mitigation framework that takes into account cyclones and associated storm surge, wind hazards, rainfall-runoff, river flood, and Geographical Information System (GIS) models for estimating possible areas of inundation as well as the depth of inundation (levels), potential damage to infrastructure, crops, and houses, and so on, evaluating not only the vulnerability but also the changing profile of vulnerability over time.
6) Maintaining the entire planned carrying capacity of main drains and canals, as well as feeder primary/secondary/tertiary channels, and constructing extra flood flow canals in regularly flooded regions.
7) Construction of saline embankments to avoid salty water incursion caused by cyclonic storm surge.

APPENDIX

a) Ramsar Wetland Sites in India

The Ramsar Convention is an international agreement for the conservation and sustainable use of wetlands that recognises wetlands' core biological functions as well as its economic, cultural, scientific, and recreational importance. There are 2400 Ramsar Sites worldwide, including 49 Ramsar Sites in India.

Sl.No.	Name of Site	State	Date of Declaration	Area (*sq.km*)
1	Kolleru Lake	Andhra Pradesh	19.8.2002	901
2	DeeporBeel	Assam	19.8.2002	40
3	Kabartal Wetland	Bihar	21.07.2020	26.2
4	Khijadia Wildlife Sanctuary	Gujarat	13.04.2021	5.12
5	Nalsarovar Bird Sanctuary	Gujarat	24.09.2012	120
6	Thol Lake Wildlife Sanctuary	Gujarat	05.04.2021	6.99
7	Wadhvana Wetland	Gujarat	05.04.2021	6.3
8	Bhindawas Wildlife Sanctuary	Haryana	25.05.2021	4.12
9	Sultanpur National Park	Haryana	25.05.2021	1.425
10	Chandertal Wetland	Himachal Pradesh	8.11.2005	0.49
11	Pong Dam Lake	Himachal Pradesh	19.8.2002	156.62
12	Renuka Wetland	Himachal Pradesh	8.11.2005	0.2
13	Wular Lake	Jammu & Kashmir	23.3.1990	189
14	Hokera Wetland	Jammu and Kashmir	8.11.2005	13.75
15	Surinsar-Mansar Lakes	Jammu and Kashmir	8.11.2005	3.5

16	Tsomoriri Lake	Jammu and Kashmir	19.8.2002	120
17	Asthamudi Wetland	Kerala	19.8.2002	614
18	Sasthamkotta Lake	Kerala	19.8.2002	3.73
19	VembanadKol Wetland	Kerala	19.8.2002	1512.5
20	Tso Kar Wetland Complex	Ladakh	17.11.2020	95.77
21	Bhoj Wetlands	Madhya Pradesh	19.8.2002	32.01
22	Lonar Lake	Maharashtra	22.7.2020	4.27
23	NandurMadhameshwar	Maharashtra	21.6.2019	14.37
24	Loktak Lake	Manipur	23.3.1990	266
25	Bhitarkanika Mangroves	Orissa	19.8.2002	650
26	Chilka Lake	Orissa	1.10.1981	1165
27	Beas Conservation Reserve	Punjab	26.9.2019	64.289
28	Harike Lake	Punjab	23.3.1990	41
29	Kanjli Lake	Punjab	22.1.2002	1.83
30	Keshopur-Miani Community Reserve	Punjab	26.9.2019	3.439
31	Nangal Wildlife Sanctuary	Punjab	26.9.2019	1.16
32	Ropar Lake	Punjab	22.1.2002	13.65
33	Keoladeo Ghana NP	Rajasthan	1.10.1981	28.73
34	Sambhar Lake	Rajasthan	23.3.1990	240
35	Point Calimere Wildlife and Bird Sanctuary	Tamil Nadu	19.8.2002	385
36	Rudrasagar Lake	Tripura	8.11.2005	2.4
37	Bakhira Wildlife Sanctuary	Uttar Pradesh	29.06.2021	28.94
38	Haiderpur Wetland	Uttar Pradesh	8.12.2021	69.08
39	Nawabganj Bird Sanctuary	Uttar Pradesh	19.9.2019	2.246
40	Parvati Agra Bird Sanctuary	Uttar Pradesh	2.12.2019	7.22
41	Saman Bird Sanctuary	Uttar Pradesh	2.12.2019	52.63
42	Samaspur Bird Sanctuary	Uttar Pradesh	3.10.2019	79.94
43	Sandi Bird Sanctuary	Uttar Pradesh	26.9.2019	30.85
44	SarsaiNawarJheel	Uttar Pradesh	19.9.2019	16.13
45	Sur Sarovar	Uttar Pradesh	21.8.2020	4.31

46	Upper Ganga River (Brijghat to Narora Stretch)	Uttar Pradesh	8.11.2005	265.9
47	Asan Conservation Reserve	Uttarakhand	21.7.2020	4.444
48	East Kolkata Wetlands	West Bengal	19.8.2002	125
49	Sunderbans Wetland	West Bengal	30.1.2019	4230

b) Tiger Reserves of India

As of 2022, India has 53 tiger reserves (the number has increased to 53 with the inclusion of the Guru Ghasidas National Park in Chhattisgarh).

Sl. No.	**Tiger Reserve**	**Area (*sq.km.*)**	**State / UT**
1	NagarjunsagarSrisailam	3296.31	Andhra Pradesh
2	Namdapha	2052.82	Arunachal Pradesh
3	Pakke	1198.45	
4	Kamlang	783	
5	Manas	2837.1	Assam
6	Nameri	464	
7	Kaziranga	1173.58	
8	Orang	492.46	
9	Valmiki	899.38	Bihar
10	Udanti-Sitanadi	1842.54	Chattisgarh
11	Achanakmar	914.017	
12	Indravati	2799.07	
13	**Guru Ghasidas NP &Tamor Pingla WLS**	2048	
14	Palamau	1129.93	Jharkhand
15	Bandipur	1456.3	Karnataka
16	Bhadra	1064.29	
17	Dandeli-Anshi	1097.514	
18	Nagarahole	1205.76	
19	Biligiri Ranganatha Temple	574.82	
20	Periyar	925	Kerala
21	Parambikulam	643.662	

22	Kanha	2051.791	Madhya Pradesh
23	Pench	1179.632	
24	Bandhavgarh	1598.1	
25	Panna	1598.1	
26	Satpura	2133.308	
27	Sanjay-Dubri	1674.502	
28	Melghat	2768.52	Maharashtra
29	Tadoba-Andhari	1727.591	
30	Pench (The Pench Tiger Reserve in Maharashtra includes a core habitat size of 257.3 km^2 and a buffer area of 483.96 km^2.)	741.22	
31	Sahyadri	1165.57	
32	Nawegaon-Nagzira	1894.944	
33	Bor	816.27	
34	Dampa	988	Mizoram
35	Similipal	2750	Odissa
36	Satkosia	963.87	
37	Ranthambore	1411.291	Rajasthan
38	Sariska	1213.342	
39	Mukandra Hills	759.99	
40	**Ramgarh Vishdhari**	1052	
41	Kalakad-Mundanthurai	1601.542	Tamil Nadu
42	Mudumalai	688.59	
43	Sathyamangalam	1408.4	
44	Anamalai	1479.87	
45	**Srivilliputhur-Megamalai**	**1016**	
46	Kawal	2015.44	Telangana
47	Amrabad	2611.39	
48	Dudhwa	2201.775	Uttar Pradesh
49	Pilibhit	730.2498	
50	Corbett	1288.31	Uttarakhand
51	Rajaji	1075.17	
52	Sunderbans	2584.89	West Bengal
53	Buxa	757.9038	

c) Elephant Reserves of India

Project Elephant is a government-sponsored initiative that began in February 1992. The government assists states with wild elephants in a free-ranging population in the conservation and management of elephants through the Project Elephant initiative. It guarantees that elephant corridors and habitat are protected in order for the wild elephant population to survive. Elephas maximus, the Indian elephant, may be found in the central and southern Western Ghats, north-east India, eastern India, and northern India, as well as sections of southern peninsular India. Various state governments have publicly notified 32 Elephant Reserves (ERs) covering around 58000 square kilometres as of 2022.

Sl. No	Name Of the Reserve	Area (sq.km)	State / UT
1	Mayurjharna	414	West Bengal
2	Singhbhum	4,530	Jharkhand
3	Mayurbhanj	3,214	Orissa
4	Mahanadi	1,038	
5	Sambhalpur	427	Orissa
6	Baitarni	1,755	
7	South Orissa	1,049	
8	Lemru	450	Chattisgarh
9	Badalkhol-Tamorpingla	4,216	
10	Kameng	1,892	Arunachal Pradesh
11	South Arunachal	900	
12	Sonitpur	1,420	Assam
13	DihingPatkai	937	
14	Kaziranga-Karbi	3,270	
15	Dhansiri-Lungding	2,740	
16	Chirang Ripu	2,600	
17	Intanki	202	Nagaland
18	Eastern Dooars	978	West Bengal
19	Garo Hills	3,500	Meghalaya
20	Khasi Hills	1,331	
21	Mysore	6,724	Karnataka
22	Wayanad	1,200	Kerala
23	Nilgiri	4,663	Tamil Nadu
24	Rayala	766	Andhra Pradesh
25	Nilambur	1,419	Kerala
26	Coimbatore	566	Tamil Nadu
27	Anamalai	1,457	
28	Anamudi	3,728	Kerala
29	Periyar	3,742	

30	Srivilliputtur	1,249	Tamil Nadu
31	Shivalik	5,405	Uttarakhand
32	Uttar Pradesh ER	744	Uttar Pradesh

d) Mike Sites in India

Sl. No	Name	State
1	Chirang-Ripu	Assam
2	DihingPatkai	
3	Eastern Dooars	West Bengal
4	Deomali	Arunachal Pradesh
5	Garo Hills	Meghalaya
6	Mayurbhanj	Orissa
7	Shivalik	Uttarakhand
8	Mysore	Karnataka
9	Nilgiri	Tamil Nadu
10	Wayanad	Kerala

e) List of Sacred Groves

The Sacred Groves are pockets of woodland or natural vegetation that range in size from a few trees to many acres and are generally devoted to local folk deities. The holy groves are major stores of plant and fauna variety that have been sustainably protected by local people. Approximately 14,000 sacred groves have been documented from across India, serving as reservoirs of uncommon fauna and, more commonly, unusual flora in rural and even urban environments. According to experts, the total number of sacred groves might be as high as 100,000. The holy groves in Hariyali, near Gauchar in Uttarakhand's Chamoli District, and the Deodar grove in Shipin, near Simla in Himachal Pradesh, are among the largest in India.

Sl. No	State / UT	Number of Sacred Groves	Local Names
1	Andhra Pradesh	750	Pavithravana
2	Arunachal Pradesh	58	Gumpa Forests (attached to Buddhist monestries)
3	Assam	40	Than, Madaico
4	Bihar	43	Sarnas
5	Chhattisgarh	600	Sarna, Devlas, Mandar, Budhadev
6	Goa	93	Deorai, Pann
7	Gujarat	29	Oran Mata and Sadhay Pir, Shravan Kavadia temple, Bhuj (world's only Inland Mangrove)
8	Haryana	57	Beed or Bid, Bani, Bann, Janglat, Shamlat
9	Himachal Pradesh	329	Dev Kothi, Devban, BakhuDevban
10	Jammu & Kashmir	92	Ram Mandir, Jama Masjid, Wali Rafiq Sahib Ziarat
11	Jharkhand	29	Sarana
12	Karnataka	1476	Devara Kadu
13	Kerala	1500	Kavu, Sara Kavu
14	Madhya Pradesh	170	Sayenebuda, Bajranj, Bari Mata
15	Maharashtra	1600	Devrai, Devrahati, Devgudi
16	Manipur	365	Gamkhap, Mauhak (sacred bamboo reserves)
17	Meghalaya	125	Ki Law Lyngdoh, Ki Law Kyntang, Ki Law Niam
18	Odisha	322	Jahera, Thakuramma
19	Puducherry	108	Kovil Kadu
20	Rajasthan	560	Orans, Kenkris, Jogmaya
21	Sikkim	56	Gumpa forests
22	Tamil Nadu	448	Swami shola, Koilkadu
23	Telangana	65	Jainath Sacred Grove, MaulaliDarga Sacred Grove
24	Uttarakhand	133	Deo Bhumi, Bugyal (sacred alpine meadows)
25	Uttar Pradesh	32	Bauri Samay, ChakiaMaata, Van Satti Devi
26	West Bengal	670	Garamthan, Harithan, Jahera, Sabitrithan, Santalburithan

f) Mangrove Sites in India

According to the Ministry of Environment, Forests, and Climate Change, this is a list of all 38 Mangroves locations in India in 2021. (MoEF&CC). Maharashtra has the most mangrove locations in a state, with ten, while Orissa comes in second with seven.

Sl. No.	Mangrove Areas	State / UT
1	**Sunderbans**	West Bengal
2	**Bhaitarkanika**	Orissa
3	**Mahanadi**	
4	Subarnarekha	
5	Devi-Kauda	
6	Dhamra	
7	Mangrove genetic resources centre	
8	Chilka	
9	**Coringa**	Andhra Pradesh
10	**East Godavari**	
11	**Krishna**	
12	**North Andaman**	Andaman and Nicobar Island
13	**Nicobar**	
14	**Pichavaram**	Tamil Nadu
15	**Muthupet**	
16	**Ramnad**	
17	**Pulicat**	
18	**Kaznuveli**	
19	**Vembanad**	Kerala
20	**Kannur**	
21	**Coondapur**	Karnataka
22	**Dakshin Kannada/ Hannavar**	
23	**Karwar**	
24	**Mangalore Forest Division**	
25	**Goa**	Goa
26	**Achra-Ratnagiri**	Maharashtra
27	**Dev Garh-Vijay Durg**	
28	**Veldur**	
29	**Kundalika-Revdnada**	
30	**Mumbra-Diva**	
31	**Vikroli**	

32	**Shreevardhan**	
33	**Vaitarna**	
34	**Vasai-Manori**	
35	**Malvan**	
36	**Gulf of Kutchh**	Gujarat
37	**Gulf of Khambhat**	
38	**Dumas-Ubhrat**	

g) Some Environmental Fragile Areas in India

Sl. No.	Place	State / UT	Details
1	Silent Valley	Kerala	The planned hydroelectric dam in the 1970s devastated a rich biodiversity with pristine flora and wildlife.
2	Doon Valley	Uttar Pradesh	Biodiversity destruction due to Limestone quarrying.
3	Nilgiri Hills	Tamil Nadu	Destruction due to urbanization and tourism.
4	Palni Hills	Tamil Nadu	Eco-destruction of pristine shola forest due to tourism.
5	Shivalik Hills	Himachal Pradesh	Destruction due to deforestation and mining activities.
6	Himalayas	India / Nepal	Severe Biodiversity loss and soil erosion due to deforestation.
7	Chilka Lake	Orissa	Siltation and destruction of aquatic life.
8	Damodar River	Bihar / Bengal	Polluted from coal washeries and industrial effluents.
9	Nal Sarovar	Gujarat	Due to mining activity, a "pristine wetland" is threatened
10	Western Ghats	Kerala and Karnataka	Ecosystem threatened due to development activities.
11	Manas	Assam	Ecological goldmine threatened with destruction.
12	Doimiasiat	Meghalaya	Threat of radiation hazard
13	Sundarbans	West Bengal	Eco-destruction of pristine mangrove ecosystem and loss of biodiversity.
14	Rann of Kachchh	Gujarat	Home to "Wild Ass," threatened with eco destruction.

15	Tehri (Garhwal Hills)	Uttar Pradesh	Eco-destruction and submergence of land due to Tehri dam project.
16	Thar Desert	Rajasthan	Increasing desertification due to population pressure.
17	Lake Renuka	Haryana	Overexploitation has resulted in a dying lake.

h) Environmental episodes

Sl. No	Episodes	Year	Place	Details
1	Meuse valley incident	1930	Belgium	A severe thermal inversion with heavy fog, contaminated with industrial pollutants chiefly oxides of sulphur, various inorganic acids and metallic oxides.
2	Donora smog incident	1948	Pennsylvania, USA	Heavy fog, contaminated with pollutants settled over the town.
3	Tokyo-Yokohama asthma	1946	Japan	Heavy industrialization and air pollution
4	London smog	1952	London, UK	A thick fog contaminated with smoke from industry's stacks layered over the city. The pollutants responsible for smog formation was SO_2 that resulted from extensive use of coal.
5	Poza Rica Industrial accident	1950	Mexico	Released of huge amount of H_2S gas
6	Los Angeles smog	1943	California, USA	Resulted largely from a reaction between organic compounds and nitrogen oxides activated by sunlight.

7	Bhopal MIC gas tragedy	1984	Bhopal, India	45 tonnes of Methylisocyanate (MIC) gas along with traces of phosgene leakage.
8	Oleum leakage	1985	Delhi, India	Oleum gas spilled from the industry. Oleum is a mixture of boiling SO_3(65 percent) and H_2SO_4 (35 percent).
9	Three mile island nuclear accident	1979	Pennsylvania, USA	Malfunction of the equipment and errors by operators.
10	Chernobyl nuclear accident	1986	Ukranian town of Chernobyl, the former Soviet Union	The site had four "950 MWe RBMKs" operational reactors, out of which Reactor 4 of the Chernobyle exploded.
11	Tokaimura nuclear accident	1999	Tokaimura, Japan	This was caused due to leakage of high levels of radiation from a uranium processing plant.
12	Beirut Explosion	2020	Beirut, Lebanon	The exploding ammonium nitrate came from a Russian-leased vessel that halted in Beirut in November 2013 while travelling from Georgia to Mozambique.

i) Some Environmental Safety and Permissible Limits by WHO

Sl. No.	Breathing Air Quality	
1	Sulphur dioxide	40–60 micrograms / cubic metre of air.
2	Suspended Particulate Matters (SPM)	60–90 micrograms / cubic metre of air.
	Drinking Water Quality	
1	Biological Oxygen Demand (BOD)	3 mg / litre of water
2	Dissolve Oxygen (DO)	8 mg / litre of water
3	Total Suspended Solids (TSS)	80 to 100 mg/litre of water
4	Fecal Coliform Bacteria	Zero/100 ml of water
5	Nitrates	10 mg/litre of water
6	Fluoride	0.7 to 1.2 mg/litre of water
7	pH	6.5 to 8.5
8	Total Dissolve Solids (TDS)	TDS levels less than 300 mg/liter are regarded outstanding, TDS levels between 300 and 600 mg/liter are good, 600-900 are medium, 900-1200 are bad, and TDS levels greater than 1200 mg/litre are unacceptable.
	Heavy Metals in Human Environment	
1	Lead (Pb)	a) Food- 3 mg/litre b) Drinking water- 100 mg/litre c) Air- 1 mg/cubic metre of air
2	Mercury (Hg)	a) Food- 0.05 mg/kg b) Drinking water- 1.0 mg/litre c) Air- 1.0 mg/ 10 cubic metre of air
3	Cadmium (Cd)	a) Food- 0.1 to 0.2 micrograms in every grain b) Surface water- Below 1 microgram/litre c) Rain water- Between 0.1 to 50 milligrams/litre d) Air- 1 to 2 micrograms/ cubic metre of air.

	Emission Limits for Automobiles	
1	Petrol Vehicles	Carbon Monoxide (CO) a) Two and Three wheelers-4.5 percent by volume b) Four Wheelers- 3 percent by volume
2	Diesel Vehicles	a) Smoke density-65 Hatridge Units b) Carbon Monoxide (CO)- 14 gms/kWh c) Hydrocarbons (HC)-3.5 gms/kWh d) Nitrogen Oxides (NO_x)- 18 gms/kWh

GLOSSARY

- Abyssal zone: a region of the ocean that is shallower than 6,000 m and deeper than about 2,000 m (6,600 ft) (20,000 ft). The zone is primarily identified by its remarkably consistent environmental conditions, which are mirrored in the variety of living things that call it home.
- Acid rain: Rain that is more acidic than usual due to the presence of sulphuric acid and nitric acid formed from sulphur (SO_x) and nitrogen (NO_x) oxides in the atmosphere.
- Adaptation: The acceptability of a structure, function, or entire organism in a certain environment.
- Adsorption: The adhesion of a gaseous or solution material to the surface of a solid.
- Aerosol: Haze and clouds are the most frequent atmospheric aerosols. They are gases that include suspended solid particles or droplets of liquid that may stay suspended in air due to their relatively tiny size (typically less than 1 micrometre in diameter).
- Agricultural runoff: Surface water that transports fertilisers like phosphate and nitrates from agricultural land to lakes and streams.
- Air pollution: Gaseous or aerosol material in the air that is not regarded a normal ingredient or an excess of a normal minor constituent such as sulphur dioxide, carbon dioxide, nitrogen dioxide, dust, etc., that creates issues or impairs the use of air as a resource.
- Air pollution episodes: A dramatic increase in high levels of air pollution, resulting in significant difficulties such as discomfort, disease, or even death.
- Allergens: Substances that stimulate the immune system and generate an allergic reaction may not be directly antigenic, but they may render other materials antigenic.
- Allergy: A hypersensitive condition caused by exposure to a specific allergen.
- Alpha particle: Alpha particles are made up of two protons and two neutrons that are closely bonded together. They are released from the nucleus of certain radionuclides during alpha-decay, a kind of radioactive decay.
- Alpha radiation: A kind of radiation consists of a particle containing two neutrons and two protons.

- Alveoli: The vertebrate lung's tiny, thin-walled sacs. These are the places where blood and air exchange gases.
- Amensalism: A relationship between organisms of two distinct species in which one is prevented from thriving or is killed while the other is unaffected.
- Ambient air: The air that surrounds us.
- Ambient Air Quality Standards or Environmental Quality Standards: Maximum allowed limits, maximum admissible concentrations, and maximum acceptable levels of pollutants that must not be exceeded under specific conditions in order to create pollution.
- Anaemia: Low haemoglobin levels owing to iron insufficiency or a lack of red blood cells
- Anthrax: An contagious, frequently deadly illness contracted by man by contact with infected wool or other animal products or through inhalation of airborne spores.
- Anthropogenic: Concerned with human activity.
- Anxiety: A state of uneasiness, uncertainty, and fear that is accompanied by physiological changes (sweating, tremor, etc.).
- Asbestos: A soft greyish-white material that does not burn and is used in construction, clothing, and other applications as a fire retardant and insulator.
- Ash: Mineral content of a product left behind after full combustion.
- Asthma: Shortness of breath, wheezing, and bronchial muscular systems are all symptoms of this uncomfortable condition.
- Atmosphere: The gaseous envelope that envelops a planet. The earth's atmosphere is made up of 79.1 percent nitrogen, 20.9 percent oxygen, and 0.03 percent carbon dioxide, as well as traces of noble gases (argon, krypton, xenon, neon, and helium), water vapour, traces of ammonia, organic materials, ozone, different salts, and suspended solid particles.
- Bathyal: Referring to the section of ocean between the continental shelf and the abyssal zone.
- Beta particles: Radioactive decay produces high-energy electrons.
- Beta radiation: Radiation made up of electrons emitted from the nuclei of numerous fissionable atoms.
- Bioaccumulation: Cellular selective absorption and concentration of molecules.
- Biochemical Oxygen Demand/Biological Oxygen Demand (BOD): It is a common test for determining the amount of dissolved oxygen required by organisms for aerobic breakdown of organic materials in

water. It is expressed in terms of the quantity of oxygen used by a unit volume of water during biological oxidation at a certain temperature and time. The BOD value may be used to assess waste strength and the degree of water pollution; the more organic matter a sample contains, the more oxygen its microbes need.

- Biodegradation: The oxidative degradation of manufactured or natural chemical compounds by microorganism activity (mainly aerobic bacteria).
- Bio-depletion: The word mainly alludes to the ongoing mass extinction of species.
- Biodiversity: The number of distinct organism species found in a given location.
- Biological monitoring: The direct measurement of changes in a habitat's biological state based on the quantity and distribution of individuals or species before and after a modification.
- Biological pest control: Use of naturally occurring substances or organisms to minimise pest organism populations.
- Biomagnification: Increase in the concentration of certain stable substances (for example, heavy metals or fat-soluble herbicides) in higher trophic levels of a food chain or web.
- Biotechnology: The use of industrial-scale biochemical processes, most notably recombinant DNA technology, to manufacture medications or bulk consumables for humans or livestock, often through waste recycling.
- Black frost: Clear ice that lacks the usual whiteness of hoar frost, snow, and so forth. Sometimes seen at night on damp roads; alludes to super-cooled rain that collects on rigging, frequently with deadly results in polar waters.
- Black smoke: Smoke generated by carbon particles expelled during hydrocarbon cracking and then rapidly cooled.
- Cadmium (Cd): Soft silvery-white metal discovered in association with zinc ores. Used in the production of fusible alloys, electroplating, and the fabrication of control rods in nuclear reactors. Cadmium contamination has been blamed for the Japanese sickness 'itaiitai.'
- Carbon absorption: Carbon particles are used to treat compounds by attaching the chemicals to the carbon particles.
- Carbon dioxide: Minor component of air, accounting for around 0.32 percent of the atmosphere. Carbon dioxide is necessary for living systems; it is produced by respiration and, to a lesser extent, volcanic activity, and it is removed from the atmosphere through photosynthesis.

- Carbon monoxide: Colourless gas occurs in trace amounts in the natural environment and created by incomplete combustion, most notably in automobiles and cigarettes. It can create a stable molecule, carboxyhaemoglobin, with blood haemoglobin. In tiny dosages, it is completely safe.
- Carbon sink: Carbon-accumulation sites, such as large forests (organic compounds) or ocean sediments (calcium carbonate).
- Carbon source: Carbon source that returns to the carbon cycle; cellular respiration and combustion.
- Commensalism: A relationship between two organisms where one gains and the other has neither gain nor loss.
- Condensation: When the saturation concentration is surpassed, water molecules aggregate from vapour to liquid or solid.
- Condensation nuclei: Tiny particles that float in the air and facilitate the condensation process.
- Coral reef: An underwater environment known as a coral reef is characterised by corals that construct reefs. Coral polyp colonies are bound together by calcium carbonate to build reefs. Stony corals, whose polyps gather together, make up the majority of coral reefs.
- Corrosion: The process by which air and water react to turn iron and other metals into oxides and carbonates.
- Cosmic rays: Complex system of radiation from space that reaches the earth.
- Cost of pollution: It refers to socioeconomic losses brought on by deteriorating human health, decreased crop output, safeguarding of fine instruments, preservation of cleanliness in food and beverage manufacturing, soiled dwellings and national landmarks, and pollution prevention.
- Crop rotation: The process of planting various crops one after another to assist sustain production; cover crops that fix nitrogen are frequently utilised in succession with other crops.
- Decomposer: An organism that consumes the remains, waste, or body parts of other species in order to decompose and reuse the nutrients they contain.
- Degradable: Able to be dissolved by natural processes (usually by decomposer organisms).
- Detergent: A synthetic organic cleaning agent used to remove dirt and grease from surface.
- Dew: When the dew point is reached, generally in the evening when surfaces cool more quickly than the surrounding air, water vapour

condenses on solid objects. Hoar frost will occur via sublimation if surfaces cool down below the freezing point.

- Dew point: The temperature at which condensation happens for a specific amount of atmospheric water vapour.
- Ecological balance: A natural community's components are in equilibrium (ecological balance) if their relative numbers remain relatively constant, providing a stable ecosystem.
- Ecology: The scientific study of the interactions between living organisms and their surroundings. It is concerned with individual species' life cycles, distribution, and behaviour, as well as the structure and operation of natural systems at the population, community, and ecosystem levels.
- Ecotone: A region that serves as a separation or transition between two ecosystems is known as an ecotone.
- Effluent: A fluid released by a source in general.
- Electromagnetic radiation: Radiation that travels in waves and possesses both an electric and magnetic field. Both natural and artificial sources contribute to it. From low energy to high energy, electromagnetic radiation can have different intensities. It consists of x-rays, gamma rays, infrared light, visible light, ultraviolet light, radio waves, and microwaves.
- Emission standards: Regulations that limit the quantity of air pollutants that can be emitted by certain point sources.
- Endangered species: A species that is on the verge of extinction over all or a large portion of its range.
- Environmental costs: When a resource is utilised, it causes environmental damage.
- Environmental ethics: The quest for moral ideals and ethical principles in human interactions with nature.
- Environmental impact assessment: EIA refers to the assessment of the impacts that are anticipated to result from a substantial project (or other activity) that has a significant impact on the natural or/and artificial environment. It is a systematic and integrative procedure created in the United States as a result of the Environmental Policy Act of 1970 for analysing potential implications prior to deciding whether or not to approve a proposal.
- Environmental quality standards: Pollutant maximum limits or concentrations allowed in the environment.
- Environmental resources: Everything an organism need may be obtained from its surroundings.

- Environmental science: The methodical, scientific investigation of our environment and our participation in it.
- Erosion: The loss of soil due to erosion caused by moving water or wind.
- Eutrophic: Lakes that are very prolific in terms of organic matter formation and are well nourished.
- Eutrophication: Enrichment of a water body with organic material including nitrate and phosphate, resulting in increased development of aquatic plants and, in some cases, algal bloom.
- Evaporation: The process of converting a liquid to a vapour (gas phase).
- Evapotranspiration: Water vapour loss from a given surface over time, both directly and through plant leaves.
- Extinction: The demise of a species; the elimination of all individuals in a given kind; can be a natural process that occurs when environmental conditions change.
- Filters: A porous mesh of cotton fabric, spun glass fibres, or asbestos-cellulose that enables air or liquid to flow through while keeping solid particles at rest.
- Filtrate: Substance that has gone through a filter.
- Fission (nuclear): A nuclear process that occurs when a heavy atomic nucleus divides into two lighter nuclei, releasing neutrons and energy.
- Flora: All of the plants found in a certain area.
- Flotation: A method of separating suspended particles of grease or oil from a solution using natural processes or by injecting tiny air bubbles that bring the pollutants to the surface and remove them.
- Fly ash: Finely divided ash particles carried over in exhaust gases produced by the burning of coal, oil, or waste material.
- Fog: Visible moisture in the atmosphere reducing horizontal sight to less than 1000 metres. Fog is created by the cooling of relatively warm, moist air as it comes into contact with a colder land or sea surface, lowering the temperature of the air in direct contact with the surface to below the dew point.
- Fossil fuels: From the organic waste and dead carcasses of once-living biological species, geological forces produce petroleum, natural gas, and coal.
- Garbage: Household wastes, primarily made up of leftover food and vegetables, animal waste, and any other useless item.
- Geological time scale: The "calendar" for historical events on Earth is the geologic time scale. It divides all of time into designated units of

abstract time that are referred to as eons, eras, periods, epochs, and ages in decreasing order of duration.

- Glacier: When snow falls on higher terrain, it presses together under pressure, creating an ice flow in a high mountain valley.
- Greenhouse effect: The atmosphere's gases, which capture infrared light from the planet and keep it from escaping into space, are what warm the globe.
- Greenpeace: International environmental group that advocates against whaling and nuclear weapons testing in the atmosphere.
- Ground water: Water that is present below the surface and above a layer of impermeable material, occupying pores and crevices in rock and soil. The water table sets the higher limit for groundwater.
- Half-life: The time required for a material to decay or vanish after half of it has done so in an exponential, regular manner.
- Hazardous waste: Any useless, unwanted, or discarded material that could endanger the health of people or the environment. This includes materials that are ignitable, corrosive, explosive, or highly reactive. It also includes materials that are ignitable, mutagenic, carcinogenic, or teratogenic to humans or other life forms.
- Hazardous-waste dump: A location where hazardous material is disposed of without consideration for potential environmental or health problems, such as a dump, landfill, or surface impoundment.
- Hertz (Hz): A measure of sound frequency.
- Hot brine: A brackish underground water body that may be exploited as a source of geothermal energy since its temperature is significantly greater than that determined by the typical geothermal gradient.
- Humification: Humus is created by the microbial degradation of organic materials in soil.
- Hydrocarbons: A group of organic molecules with carbon and hydrogen atoms that are emitted from burning fuel or are leftovers from incomplete combustion and serve as a major source of air pollution.
- Hydrogen sulphide: A headache-inducing gas that is toxic and has an unpleasant scent when inhaled in small doses. It can result in mortality in high quantity.
- Hypoxia: A decrease in a tissue's oxygen levels below normal levels despite a sufficient blood supply to the tissue.
- Ice cap: A layer of ice may be 2500m to 3000m thick over a contour, such as a mountain peak, the planet's poles, etc.
- Incineration: Method of burning solid waste for disposal.

- Industrial revolution: Mechanization of production causes a move from small-scale factory production to large-scale factory production, resulting in social and economic changes; this is a historical time when technology supplanted human labour.
- Industrial waste: Waste produced during the manufacturing of particular items, whether it be liquid, solid, or gaseous.
- Inversion: A temperature inversion in the atmosphere is a situation when the temperature rises with height above the ground rather than declining.
- Ionizing radiation: Electromagnetic radiation with a high energy or energetic subatomic particles produced as a result of nuclear decay.
- Isotopes: Different isotopes of the same element that may be recognised by minute integral changes in atomic mass and, frequently, by minuscule variations in their physical and chemical characteristics.
- ISRO: The national space agency of India is the Indian Space Research Organisation, with its main office in Bangalore. The Chairman of ISRO also serves as the Executive of the Department of Space, which is directly under the direct control of the Prime Minister of India.
- Knock-knee syndrome: A condition of the leg or thigh in which the ankles are spaced apart more than usual and the knees are unusually near together.
- Landfill: A technique for getting rid of solid wastes that entails burying the garbage in locations that have been especially made for it.
- Landsat: The Landsat programme has been collecting satellite images of Earth for the longest. A combined NASA and USGS effort is involved.
- Leaching: Percolating water is used to remove soluble components from soil or sludge deposits. Porous strata are also used to dispose of liquids. the downward migration of soil water, which causes minerals from the A horizon to the B horizon to shift.
- Liquefied petroleum gas: Natural gas-derived mixture of propane and butane that has been compressed into a liquid.
- Mangroves: These tropical and subtropical intertidal areas are home to salt-tolerant forest ecosystems.
- Maximum permissible level: A term that is frequently used to describe to the highest level of contamination allowed from any source, but particularly from radioactive compounds.
- Mesosphere: The portion of the atmosphere that is between the stratosphere and thermosphere, or between 40 and 80 kilometres above the earth's surface.

- Methane: Often referred to as marsh gas, this flammable, odourless gas makes up the majority of regular gas. It is used in some sewage treatment plants for heating and electricity. It occurs naturally when organic waste decomposes.
- Minamata: As a result of being exposed to the fish and shellfish captured in the bay in 1959, a disease of the central nervous system brought on by mercury poisoning has been dubbed "Minamata" after a bay and town in Japan. The contamination was caused by the discharge of effluents from the Chisso plant, which produced vinyl chloride and acetaldehyde. Dimethyl mercury, which was consumed by marine species, accumulated as a result of the pollution. 43 people passed away between 1953 and 1960, while several more were rendered ill.
- Mist: Visibility between 1 and 2 km is decreased by liquid (water) droplets floating in the atmosphere with a diameter less than 10 micron.
- Monitoring: To assess a substance's level over time either quantitatively or subjectively.
- Monitoring programme: A programme created to assess a substance's level quantitatively or quantitatively over time. These programmes frequently offer data on how contaminants are distributed across time or place so that practical solutions may be devised to reduce their negative impacts.
- Municipal waste: All garbage generated by a community's citizens.
- Mutualism: An association where both the organisms are mutually benefited.
- NASA: The National Aeronautics and Space Administration is a US government agency in charge of satellite launches and administration as well as space exploration.
- Natural gas: Fossil fuel that burns as a gas and contains between 50 and 90 percent methane as well as smaller amounts of other organic gases like propane and butane.
- Natural pollutant: A naturally occurring substance that, when present in excess, might be considered a contaminant to the environment (e.g., volcanic dust, sea salt particles, ozone formed photochemically or by lightning, products of forest fires).
- Nausea: An unpleasant sensation with a tendency to vomit.
- Noise pollution: Unwanted sound energy is introduced into the environment at a level that might harm human activities.
- Nondegradable: Unaltered by natural processes (or broken down only over a time span of hundreds of years).

- Non-governmental organization: A non-profit organisation with formal standing.
- Nutrient: Any substance consumed by living things that fosters development; commonly speaking, this refers to the nitrogen and phosphorus in water, but it can also refer to other necessary and trace elements.
- Oil pollution: Pollution brought on by marine oil spills.
- Oligotrophic: Refers to lakes that are deficient in plant nutrients and are not very productive in terms of the formation of organic material.
- Ozone: A harmful air pollutant made up of three oxygen atoms and being very reactive. However, ozone produces an ultraviolet-absorbing screen in the stratosphere that shields humans from carcinogenic radiation.
- Ozone hole: When the temperature approaches 0 degrees Celsius in the Antarctic spring, it is a sign of the ozone layer's thinning. This causes nitrogen oxides to freeze, which causes a build-up of chlorine monoxide as a result. Members of the British Antarctic survey team made the discovery of the hole.
- Ozone layer: The higher stratospheric layer where sunlight causes some oxygen atoms to change into ozone atoms A large amount of UV light is absorbed by ozone, keeping it from harming the earth. As a result, it effectively blocks harmful UV rays.
- Paralysis: Loss or reduction in motor function caused by damage to the brain or muscular system.
- Particulate material: Dust, ash, soot, lint, smoke, pollens, spores, algal cells, and other suspended materials are examples of atmospheric aerosols; the term was originally solely used to describe solid particles but is now used to describe liquid droplets.
- Pathogen: An organism that causes illness in its host, which is characterised by changes to one or more metabolic processes as a result of the organism's existence.
- Pedology: The study of the origins, makeup, distribution, and use potential of soil resources is known as pedology.
- Percolation. The movement of water through soil layers within pores.
- Photochemical smog: A yellowish-brown haze caused by the reaction between hydrocarbons, nitrogen oxides, and sunlight.
- Photodegradable plastics: Plastics that degrade when exposed to light of a certain wavelength or to the sun.
- Phytoplankton: Floating or freely suspended plants.

- Point sources: Specific areas where pollution is discharged in significant concentrations, such as industries, power plants, sewage treatment facilities, underground coal mines, and oil wells.
- Pollutant: A pollutant is, strictly speaking, anything that alters air, water, or any other resource in such a manner as to make it less usable.
- Pollution costs: The expenses paid for health care and decreased use of public resources as a result of pollution, as well as any private or governmental investments made to mitigate harm caused by pollution after it has already happened.
- Pollution: Changes in the physical, chemical, or biological characteristics of the air, the water, or the soil that might have a negative impact on people's health, their ability to survive, or their daily activities.
- Pollution tax: A fee levied against a polluter based on the type and volume of pollutants they release into the environment.
- Porosity: The amount of void space in a rock is measured by its porosity. It is determined by dividing the volume of the pores or voids by the overall volume.
- Primary treatment: The majority of the floating and settleable particles in waste water are removed throughout the waste water treatment process, which lowers the concentration of suspended solids. It is done by employing tanks for heavy materials to settle in and screens to trap floating items.
- Productivity: Productivity is the measure of how quickly an ecosystem produces biomass. It is also known as the energy that plants store during photosynthesis. The two categories of productivity are Primary Productivity and Secondary Productivity.
- Radiation: Electromagnetic waves, such as light waves, or particle rays emitted from a source, such as alpha, beta, and gamma rays.
- Radioactive: An unstable isotope that spontaneously decays and discharges energy or subatomic particles.
- Radioactive decay: A transformation of the radioactive isotopes' nuclei that progressively transforms into another isotope or other element while emitting high-energy electromagnetic radiation and/or subatomic particles.
- Radioactive half-life: Period required for the radioactive material to spontaneously decay by half.
- Radioactive waste: Radioactive waste that is created at labs, hospitals, and other facilities that use radioactive materials, but mostly as a by-product of nuclear power generation.
- Radioactivity. The emission of radiation by atomic nuclei.

- Radio nuclides: Isotopes that exhibit radioactive decay.
- Radon: Certain types of rock release radioactive gas, which can build-up in particularly tightly-sealed structures.
- Remote Sensing: By measuring the reflected and emitted radiation of an area from a distance, remote sensing is the process of identifying and keeping track of the physical features of a location (typically from satellite or aircraft).
- Risk assessment: The use of data and hypotheses to calculate the likelihood of harm to human health or the environment that might be brought on by exposure to particular pollutants, hazardous substances, or management choices.
- Risk: Probability that anything bad will happen as a result of being exposed to a risk.
- Risk management: A method of decision-making that considers factors including risk assessment, technical viability, economic implications, social issues, and legal constraints.
- Ruminant animals: Animals that chew cud, such as cattle, sheep, goats, and buffalo, have many chambers in their stomachs where microbes help to break down cellulose.
- Salinity: A measurement of the total quantity of salts that have dissolved in a particular amount of water; also known as the salt solution's level of concentration.
- Satellite: Any object that circles another object, such as the Earth orbiting the sun, is referred to as a satellite. There are a large number of satellites in use. They serve a variety of functions, including the Global Positioning System, amateur radio, television transmission, and weather forecasting.
- Secondary treatment: Bacterial breakdown of leftover dissolved organic components and suspended particulates from basic sewage treatment. To do this, sewage and bacteria are combined in trickling filters or during the activated sludge process.
- Sedimentation: The process of setting under the action of gravity or centrifugation to remove an insoluble solid from a liquid in which it is suspended.
- Sensor: A sensor is a device that detects input of any kind from the physical world and reacts to it. Light, heat, motion, moisture, pressure, or any of the many other environmental phenomena might be the specific input.
- Sewage: The materials in sewers that transport a community's watery wastes.

- Sewage sludge: The volume of solid trash that is eliminated during wastewater treatment.
- Sewage treatment: The process of treating sewage to make it more environmentally friendly.
- Silicosis: A lung condition brought on by exposure to silica dust
- Sludge: Components that settle out of wastewater at a sewage treatment facility that are a semisolid combination of organic and inorganic materials.
- Smog: H.A. Des Voeux, the founder and first president of the British National Smoke Abatement Society, coined the term in 1905 to describe the mixture of smoke and natural fog, which in metropolitan areas may have unpleasant and even disastrous effects. Nowadays, any type of urban air pollution or photochemical pollution products are frequently referred to as "smog."
- Smoke: A tiny solid or liquid aerosol that is typically less than 1 millimetre in diameter that is created when a fuel burns partially. The burning of coal is mostly to blame for air pollution.
- Solid waste: Materials produced by human activity that are useless, undesired, or wasted. Animal raising and slaughter, as well as the processing of animal products and orchard and field crops, all result in the production of agricultural waste. Stores, offices, and other businesses that don't actually produce anything produce commercial garbage. Manufacturing and industrial operations produce industrial waste. Municipal garbage includes both business and domestic waste. Domestic solid waste is waste from households.
- Solid waste management: Control of solid waste creation, storage, collection, transport, processing, recycling, recovery, and disposal that is intentional and systematic.
- Stratosphere: The portion of the atmosphere between the tropopause and about 50 km above the surface of the planet where temperatures are steady or very slightly increase with altitude and where there is little water vapour but plenty of ozone.
- Stress: Animals can be stressed out by physical, chemical, or emotional causes; plants can also be physiologically stressed by unfavourable climatic circumstances.
- Sunburn: Unfiltered UV rays cause skin damage with erythema, discomfort, and even blistering after prolonged exposure to sunshine.
- Sustainable development: Arise in human welfare that enables us to satisfy our wants in the here and now without endangering the capacity of future generations to satisfy their own needs.

- Tax incentives for environment protection: Incentives offered by the government to encourage the purchase of pollution control equipment in industrial facilities, such as tax credits, tax exemptions, or expedited depreciation.
- Thermal inversion: The circumstances in which warm air in a valley serves as a lid over the valley by being sandwiched between two levels of cold air.
- Thermal pollution: Pollution from heated wastewater discharged at temperatures that may be harmful to aquatic life.
- Thermal stratification: The presence of a series of clearly distinct layers, each with a different temperature, lying at varying depths, with the bottom layer being the coldest, within a body of water. Under the ice, inverse stratification takes place.
- Thermoplastics: Soft plastics may be melted and reshaped to create usable goods. They are made of single-chain, unlinked polymers like polyester, vinyl chloride, polyvinyl chloride, polyethylene, and polypropylene.
- Thermosphere: The zone of hot, diluted gases above the mesosphere that reaches out to around 1,600 km from the earth's surface. The upper atmosphere where temperature rises with height.
- Fourth World: Not meant to be disparaging, but less developed nations that do not have industrialised, capitalist, or (in the Second World) centrally planned communist economies. In actuality, the phrase "Third World" was used to refer to those nations whose economies were less developed than those of the other nations.
- Threshold levels: The bare minimum of something needed to have discernible effects.
- Tillage: The mechanical alteration of soil for crop production known as tillage has been shown to dramatically change soil properties including soil temperature, infiltration, and evapotranspiration processes.
- Toxic waste: Toxic substances that, when released into the environment, kill or seriously hurt both humans and animals.
- Toxicity: A characteristic of some materials that, in large enough quantities, can endanger the environment or human health.
- Toxins: Poisonous substances, frequently hazardous even at low quantities, that interact with certain biological components to kill cells or change growth or development in unfavourable ways.
- Transpiration: Water loss from plant surfaces, particularly through stomata.
- Trauma: Either accidental or violent injury.

- Tropopause: The boundary separating the stratosphere from the troposphere.
- Troposphere: The lowest point in the atmosphere, when temperature and pressure both drop with height, is defined by the tropopause above and the land or sea surface below.
- Tumour: A swelling or pathological enlargement characterised by unchecked and increasing cell proliferation.
- Turbidity: A liquid's cloudiness brought on by the presence of suspended, finely split particles.
- Turbine: A mechanical device with fan-like bodies fixed on a shaft that spins when water, air, or both flow through the blades in order to produce mechanical energy that may be utilised to create electricity.
- Waste: Any human-produced solid, liquid, or gaseous emission that cannot be used by the system or organism that produced it and for which a disposal strategy has to be developed.
- Waste sludge: Transferred to sludge conditioning for final disposal, biological sludge.
- Waste treatment: To remove dissolved and suspended particles from wastewater, physical, chemical, and biological techniques are used.
- Waste water: Water from municipal or industrial sources that contains suspended and dissolved solids.
- Waterborne diseases: Diseases that are frequently spread by polluted water supplies include cholera, typhoid fever, dysentery, gastroenteritis, and hepatitis.
- Water logging: Overwatering results in soil saturation, saturating all air spaces and killing plant roots by depriving them of oxygen.
- Water table: The water table is an underground boundary separating the soil's surface from the region where groundwater seeps into rock crevices and voids between sediments.
- Weathering: The mechanical and chemical degradation of rocks brought on by contact with water, air, or organic material.
- Wetland: Wetlands are places where water covers the soil or is present at or close to the soil's surface all year long or intermittently throughout the year, including during the growing season.
- Wind erosion: The wind's effect in removing material from the ground or structures.
- Zoogeography: The field that studies how animals are distributed geographically.
- Zooplankton: Tiny creatures that live in watery ecosystems and eat phytoplankton.